The Practice of Consumer Exposure Assessment

Gerhard Heinemeyer • Matti Jantunen •
Pertti Hakkinen
Editors

The Practice of Consumer Exposure Assessment

Springer

Editors
Gerhard Heinemeyer
Department of Exposure
German Federal Institute for Risk Assessment
Berlin, Germany

Matti Jantunen
Department of Environmental Health
National Institute for Health and Welfare (THL)
Kuopio, Finland

Pertti Hakkinen
National Library of Medicine and the National Institutes of Health
Bethesda, MD, USA

ISBN 978-3-319-96147-7 ISBN 978-3-319-96148-4 (eBook)
https://doi.org/10.1007/978-3-319-96148-4

This Springer imprint is published by the registered company Springer Nature Switzerland AG.
The registered company address is: Gewerbestrasse 11, 6330 Cham, Switzerland

Disclaimer

The legal background in several chapters of this book is described for information purposes to the best of the knowledge of the editors and authors. It can, however, not be considered legally binding. In case of any doubts the competent authorities, such as the European Commission services, should be consulted.

Preface

This textbook is the first one that compiles the current knowledge in exposure science, with the focus on risk assessment. A number of experts working for many years in this field brought their knowledge together for what is needed for exposure estimations in their practical and daily work. In risk assessment, the results of the exposure assessment are compared with the results of the hazard assessment. Toxicologic evaluation covers the identification of toxic properties of substances and the toxic doses; exposure assessment estimates the amount of the substance (dose) that can be incorporated. While the evaluation of toxic effects is a very old and well-established scientific task, scientifically based exposure evaluation is very young. Due to the fact that many scientific disciplines are involved in exposure assessment, it might be logical that many opinions exist about how exposure assessment should be carried out due to the objective of the respective evaluations. The different experimental instruments are used to identify the concentrations of substances in numerous consumer products and in other items. Also, the habits of the exposed persons and the characteristics of the populations and individuals involved have influence on the results of exposure assessments. This might also be a reason why the scientific instruments to estimate and measure exposure are so different, in accordance with the current development of the particular scientific discipline, e.g., analytical laboratory work, statistics, public opinion evaluation, pharmacokinetics and pharmacodynamics, chemistry, physics, and medicine. Because exposure science brings together all these different disciplines, the understanding of the experts representing them is sometimes difficult and controversial. For the uninformed observer, exposure science might occur as a house which is still under construction. However, it has a roof and rooms, and the tenants are already living in it.

The intention of this book is to compile the current knowledge to show how the lack of information needed for exposure assessment, differences, and controversies might be overcome in the practical work.

The editors

Berlin, Germany — Gerhard Heinemeyer
Kuopio, Finland — Matti Jantunen
Bethesda, MD — Pertti Hakkinen

Acknowledgements

The editors and contributors thank gratefully Dr. Stefan Fabiansson for critical reading and revision.

The assistance of Katrin Pfeiffer (MRI) in preparing the manuscript of Sect. 4.6 is gratefully acknowledged by the author.

Contents

Editors and Contributors

Editors

Gerhard Heinemeyer Department of Exposure, German Federal Institute for Risk Assessment (Retired), Berlin, Germany

Matti Jantunen Department of Environmental Health, National Institute for Health and Welfare (THL) (Retired), Kuopio, Finland

Pertti Hakkinen National Library of Medicine and the National Institutes of Health, Bethesda, MD, USA

Contributors

Klaus Abraham Department of Food Safety, German Federal Institute for Risk Assessment, Berlin, Germany

Andreas Ahrens European Chemicals Agency (ECHA), Helsinki, Finland

Davide Arcella European Food Safety Authority (EFSA), Parma, Italy

Jos Bessems VITO Health, Mol, Belgium

Annette Bitsch Department of Chemical Safety and Toxicology, Fraunhofer Institute for Toxicology and Experimental Medicine (ITEM), Hannover, Germany

Annegret Blume Department of Exposure, German Federal Institute for Risk Assessment, Berlin, Germany

Katrin Blume Department of Exposure, German Federal Institute for Risk Assessment, Berlin, Germany

Polly E. Boon Department Food Safety, Institute for Public Health and the Environment (RIVM), Bilthoven, The Netherlands

Gianfranco Brambilla Department Food Safety, Nutrition and Veterinary Public Health, National Institute of Health (ISS), Rome, Italy

Alison Connolly ISES-Europe Board, Freiburg, Germany

Francesco Cubadda Department of Food Safety, Nutrition and Veterinary Public Health, National Institute of Health (ISS), Rome, Italy

Yuri Bruinen de Bruin European Commission Knowledge Management Service, European Commission Joint Research Centre, Ispra, Italy
ISES-Europe, Freiburg, Germany

Christiaan Delmaar Institute for Public Health and the Environment (RIVM), Centre for Safety of Substances and Products (VSP), Bilthoven, The Netherlands

Birgit Dusemund Department Food Safety, German Federal Institute for Risk Assessment, Berlin, Germany

Stefan Fabiansson European Food Safety Authority (EFSA), Dietary and Chemical Monitoring (Retired), Parma, Italy

Peter Fantke ISES-Europe Board, Freiburg, Germany
Technical University of Denmark, Lyngby, Denmark

Mary Gilsenan Evidence Management Unit, European Food Safety Authority (EFSA), Parma, Italy

Matthias Greiner Department of Exposure, German Federal Institute for Risk Assessment, Berlin, Germany

Ursula Gundert-Remy Humboldt-University, Charité, Berlin, Germany

Rainer Gürtler Department of Food Safety, German Federal Institute for Risk Assessment, Berlin, Germany

Stefan Hahn Department of Chemical Safety and Toxicology, Fraunhofer Institute for Toxicology and Experimental Medicine (ITEM), Hannover, Germany

Pertti Hakkinen National Library of Medicine and the National Institutes of Health, Bethesda, MD, USA

Astrid Heiland Department of Exposure, German Federal Institute for Risk Assessment, Berlin, Germany

Gerhard Heinemeyer Department of Exposure, German Federal Institute for Risk Assessment (Retired), Berlin, Germany

Matti Jantunen Department of Environmental Health, National Institute for Health and Welfare (THL) (Retired), Kuopio, Finland

Stefanie Klenow Department of Exposure, German Federal Institute for Risk Assessment, Berlin, Germany

Sven Knüppel Department of Epidemiology, German Institute of Human Nutrition Potsdam-Rehbrücke (DIfE), Nuthetal, Germany

Wolfgang Koch Department of Chemical Safety and Toxicology, Fraunhofer Institute for Toxicology and Experimental Medicine (ITEM), Hannover, Germany

Majlinda Lahaniatis International Affairs, German Federal Institute for Risk Assessment, Berlin, Germany

Oliver Lindtner Department of Exposure, German Federal Institute for Risk Assessment, Berlin, Germany

Friederike Neisel Department of Exposure, German Federal Institute for Risk Assessment, Berlin, Germany

Birgit Niemann Department of Food Safety, German Federal Institute for Risk Assessment, Berlin, Germany

Demosthenes Papameletiou European Commission, Joint Research Centre (JRC), Ispra, Italy

Ralph Pirow Department of Chemicals and Products Safety, German Federal Institute for Risk Assessment, Berlin, Germany

Artur Radomyski European Commission, Joint Research Centre (JRC), Ispra, Italy

Vittorio Reina European Commission, Joint Research Centre (JRC), Ispra, Italy

Thomas Rüdiger Department of Exposure, German Federal Institute for Risk Assessment, Berlin, Germany

Klaus Schneider Forschungs- und Beratungsinstitut Gefahrstoffe GmbH (FoBiG), Freiburg, Germany

Michael Schümann Hamburg, Germany

Urs Schlüter ISES-Europe Board, Freiburg, Germany
BAUA, Dortmund, Germany

Christian Sieke Department of Pesticide Safety, German Federal Institute for Risk Assessment, Berlin, Germany

Yasmin Sommer Department of Exposure, German Federal Institute for Risk Assessment, Berlin, Germany

Andrea Straßburg Max Rubner-Institut (MRI), Federal Research Institut of Nutrition and Food, Karlsruhe, Germany

Wouter ter Burg Centre for Safety of Substances and Products (VSP), Institute for Public Health and the Environment (RIVM), Bilthoven, The Netherlands

Anne Theobald European Food Safety Authority (EFSA), Parma, Italy

Natalie von Goetz ISES-Europe Board, Freiburg, Germany
ETH Zurich, Zurich, Switzerland

Michal Wiecko Department of Exposure, German Federal Institute for Risk Assessment, Berlin, Germany

Alexandre Zenié European Commission, Joint Research Centre (JRC), Ispra, Italy

Abbreviations

24hDR	24-hour dietary recall
2NPPD	2-nitro-p-phenylenediamine
3-MCPD	3-*M*ono*c*hlor*p*ropane-1,2-*d*iol esters
A.I.S.E.	Association Internationale de la Savonnerie, de la Détergence et des Produits d'Entretien
Acropolis	Cumulated and aggregated exposure to pesticides
ADI	Acceptable Daily Intake
AEGL	Acute Exposure Guideline Level
AFC	Panel on Food Additives, Flavourings, Processing Aids and Materials in Contact with Food (EFSA)
ALARA	As Low As Reasonably Achievable
ANS	Panel on Food Additives and Nutrient Sources (EFSA)
ANSES	National Authority for Food Safety, Nutrition and Environment (France)
AR	Average Requirement
ARfD	Acute Reference Dose
AUH	Ausschuss für Umwelthygiene (Germany)
AVV DatA	Allgemeine Verwaltungsvorschrift über den Austausch von Daten im Bereich der Lebensmittelsicherheit und des Verbraucherschutzes
BAuA	Bundesanstalt für Arbeitsschutz und Arbeitsmedizin (Germany)
BBN	BetaBinomial-Normal (model)
BfR	Bundesinstitut für Risikobewertung (Federal Institute for Risk Assessment, Germany)
BH	Body Height
BLS	Bundeslebensmittelschlüssel (German Nutrient Database)
BMD	Benchmark Dose
BMDL	Benchmark Dose Lower Confidence Limit
BMEL	Bundesministerium für Ernährung und Landwirtschaft (Ministry for Nutrition and Agriculture, Germany)

BMJ	Bundesministerium für Justiz und Verbraucherschutz (Ministry for Justice and Consumer Protection, Germany)
BMU	Bundesministerium für Umwelt, Naturschutz und nukleare Sicherheit Ministry for Environment, Nature Conservation and nuclear Safety, Germany)
BNN	Beta-binomial Normal (BBN))
BoA	Board of Appeal (REACH)
BS	Body Surface
BVL	Federal Office for Consumer and Food Safety (Germany)
BW	Body Weight
CAC	Codex Alimentarius Commission
CARES	Cumulative and Aggregate Risk Evaluation System
CCFA	Codex Committee on Food Additives
CDC	Centers for Disease Control and Prevention (USA)
CEF	Panel on Food Contact Materials, Enzymes, Flavourings and Processing Aids (EFSA)
CF	Conversion factor
CFCD	Comprehensive Food Consumption Database
CHAFEA	Consumers, Health, Agriculture and Food Executive Agency
CHESAR	Chemical Assessment and Reporting Tool (REACH)
CI	Confidence Interval
CLP	Classification, Labelling and Packaging
CMR	Carcinogenic, Mutagenic, Reprotoxic (substances)
CONCAWE	Conservation of Clean Air and Water in Europe
CONTAM	Panel on Contaminants in the Food Chain (EFSA)
CoRAP	Community Rolling Action Plan (REACH)
COT	Committee on Toxicity of Chemicals in Food, Consumer Products and the Environment
CSA	Chemical Safety Assessment (REACH)
CSA	Category System for Agents (see also TKS)
CSPA	Consumer Specialty Products Association
CSR	Chemical Safety Report (REACH)
DEEM-FCID	Dietary Exposure Evaluation Model - Food Commodity Intake Database
DEGS	German Health Examination Survey (Deutscher Ernährungs u. Gesundheits-Survey)
DEHP	di-Ethyl-Hexyl-Phthalate
DG	Directorate-General of the European Commission
DISHES	Diet Interview Software for Health Examination Studies
DIY	Do it yourself
DNEL	Derived No-Effect Level
DRV	Dietary Reference Values
EAR	Estimated Average Requirement
ECDC	European Centre for Disease Prevention and Control

ECHA	European Chemicals Agency
ED	Exposure Determinants
EDI	Estimated Daily Intakes
EEA	European Environmental Agency
EEC	European Economic Community
EFCOSUM	European Food Consumption Survey Method
EFSA	European Food Safety Authority
EIS-ChemRisks	European Information System on chemical risks
EMA	European Medicines Agency
eo-FCS	Exposure-oriented Food Consumption Survey
EPA	Environmental Protection Agency (USA)
EPIC	European Prospective Investigation into Cancer and Nutrition
ERNA	European Responsible Nutrition Alliance
ESD	Emission Scenario Document (OECD)
ESD	Exposure Scenario Determinant (REACH)
EsKiMo	Ernährungssurvey als KIGGS Modul (Nutrition Study as KiGGS Module)
ETAD	The Ecological and Toxicological Association of Dyes (and Organic Pigments Manufacturers)
EU	European Union
EuPCS	European Product Categorisation System
EVA	Ethyl Vinyl Acetate
FACET	Food flavourings, food additives and food contact materials exposure tool
FAIM	Food Additives Intake Model
FAO	Food and Agriculture Organization
FCS	Food Consumption Study (Survey)
FDA	Food and Drug Administration (USA)
FERA	Food and Environmental Research Agency (UK)
FFQ	Food Frequency Questionnaire
FIAP	Food Improvement Agents Package
FNB/IOM	Food and Nutrition Board of the Institute of Medicine (USA)
FOPH	Federal Office of Public Health (Switzerland)
FSA	Food Standards Agency (UK)
GEADE	Global Estimate of Acute Dietary Exposure
GECDE	Global Estimate of Chronic Dietary Exposure
GEMS	Global Environment Monitoring System
GerES	German Environmental Survey for Children
GMO	Genetically Modified Organisms
GPSD	General Product Safety Directive
GSFA	General Standard for Food Additives
HBGV	Health-Based Guidance Value
HBM	Human Biomonitoring
HCPA	Household & Commercial Products Association

HD	House dust
HERA	Human and Environmental Risk Assessment
HR	Highest Residue (pesticides)
IARC	International Agency for Research on Cancer
ICP	Import Control Plan for food of animal origin
ICRP	International Commission on Radiological Protection
IEDI	International Estimated Daily Intake
IESTI	International Estimated Short-Term Intake
IF	Intake Fraction
ILO	International Labour Organization
ILSI	International Life Sciences Institute
INTERA	Integrated Exposure for Risk Assessment (in Indoor Environments)
IOM	Institute of Medicine (USA)
IPCC	Intergovernmental Panel on Climate Change
IPCS	International Programme on Chemical Safety
IRAG	Improvement of the Risk Assessment Guidelines under the General Product Safety Directive 2001/95/EC
ISO	International Organization for Standardization
ISUF	Iowa State University Foods (model)
IUCLID	International Uniform Chemical Information Database
JECFA	Joint Expert Committee on Food Additives (FAO/WHO)
JMPR	Joint FAO/WHO Meeting on Pesticide Residues (FAO/WHO)
JRC	Joint Research Centre of the European Commission
JUST	Justice and Consumers
KEMI	Kemikalieinspektionen (Swedish Chemicals Agency)
KIESEL	Kinder-Ernährungsstudie zur Erfassung des Lebensmittelverzehrs
KiGGS	German Health Interview and Examination Survey for Children and Adolescents
KUS	Kinder Umwelt Survey (children's environmental survey)
LANUV	Landesamt für Natur, Umwelt- und Verbraucherschutz Nordrhein-Westfalen
LCIA	Life cycle impact assessment models
LNN	LogisticNormal Normal
LOAEL	Lowest-observed-adverse-effect level
LOD	Limit of Detection
LOQ	Limit of Quantification
LP	Large Portion (size)
MAK	Maximale Arbeitsplatz Konzentration (Commission for the Investigation of Health Hazards of Chemical Compounds in the Work Area)
MANCP	Multi-Annual National Control Plan
MCRA	Monte Carlo Risk Assessment
MINTEL	Market Intelligence (Agency)

MLE	Maximum Likelihood Estimation
MMAD	Mass Median Aerodynamic Diameter
MOS	Margin of Exposure
MPL	Maximum permitted level (food additives)
MPRL	Minimum Performance Required Limit
MRI	Max Rubner Institute (Germany)
MRL	Maximum Residue Level (or Limit)
MSC	Member State Committee (REACH)
MSM	Multiple Source Method
NCI	National Cancer Institute (USA)
NDNS	National Diet and Nutrition Survey
NEDI	National Estimated Daily Intake
NEL	No Effect Level
NEMONIT	Nationales Ernährungsmonitoring (German National Nutrition Monitoring)
NGO	Non-Governmental Organisation
NHANES	National Health and Nutrition Examination Survey (USA)
NICNAS	National Industrial Chemicals Notification and Assessment Scheme (Australia)
NOAEL	No-Observed-Adverse-Effect Level
NOEL	No-Observed-Effect Level
no-FCS	Nutrition-oriented Food Consumption Survey
NRC	National Research Council (USA)
NRCP	National Residue Control Plan
NVS	Nationale Verzehrsstudie (National Nutrition Survey, Germany)
OECD	Organisation for Economic Co-operation and Development
OHT	OECD Harmonised Template
OIE	World Organisation for Animal Health
OIM	Observed Individual Mean (model)
PACEM	Probabilistic Aggregate Consumer Exposure Model
PAH	Polycyclic Aromatic Hydrocarbons
PBDE	Polybrominated diphenyl ethers
PBPK	Physiologically based pharmacokinetic (modelling)
PCB	Polychlorinated Biphenyl
PCS	Poison Control Centre
PHE	Public Health England
PKPD	Pharmacokinetics/Pharmacodynamics
POD	Point of Departure
PPP	Plant Protection Products
PPR	Panel on Plant Protection Products and their Residues (EFSA)
PROMISE	Probabilistic Methodology for Improving Solvent Exposure Assessment
PVC	Polyvinyl chloride
QA/QC	Quality Assurance/Quality Control
RAC	Risk Assessment Committee (REACH)

RAC	Raw Agricultural Commodity (pesticide regulation)
RAPEX	Community Rapid Information System for the rapid exchange of information system between the Member States and the Commission
RASFF	Rapid Alert System for Food and Feed
RCR	Risk Characterisation Ratio
RCT	Randomized Controlled Trial
RD	Residue definition
RDA	Recommended Dietary Allowance
REACH	Registration, Evaluation, Authorisation and Restriction of Chemicals
RIP	REACH Implementation Project
RIVM	Rijksinstituut voor Volksgezondheid en Milieu (The Netherlands)
RMM	Risk Management Measures
RP(A)	Reference Point (for Action)
RPA	Reference Point of Action
SANCO	Directorate Health and Consumer Protection (substituted by JUST for consumers' related policy)
SCC	Scientific Committee on Cosmetics
SCCNFP	Scientific Committee on Cosmetic Products and Non-food products
SCCS	Scientific Committee on Consumer Safety
SCEDs	Specific Consumer Exposure Determinants
SCENIHR	Scientific Committee on Emerging and Newly Identified Health Risks
SCF	Scientific Committee on Foods
SCOOP	Scientific Co-operation (on Questions Relating to Food)
SEAC	Socio-Economic Analysis Committee (REACH)
SED	Systemic Exposure Dose
SEEM	Systematic Empirical Evaluation of Models
SEv	Substance Evaluation (REACH)
SHEDS	Stochastic Human Exposure and Dose Simulation (model)
SIEF	Substance Information Exchange Forum (REACH)
SPADE	Statistical Program to Assess Dietary Exposure
SRMA	Systematic reviews and meta-analyses (of epidemiologic studies)
STMR	Supervised Trial Median Residue
SVHC	Substance of Very High Concern (REACH)
SVOC	Semivolatile Organic Compound
TDI	Tolerable Daily Intake
TDI	Toxikologischer Dokumentations und Informationsverbund (Germany)
TDS	Total Diet Study
TDTK	Toxicodynamics/Toxicokinetics
TEQ	Toxicological equivalent

TG	Test Guideline (OECD)
TKS	TDI-Kategorie-System (Germany, see also CSA))
TMDI	Theoretical Maximum Daily Intake
TNO	Toegepast Natuurwetenschappelijk Onderzoek (the Netherlands)
TTC	Threshold of Toxicological Concern
UA	Uncertainty Analysis
UBA	Umweltbundesamt (Federal Environmental Office, Germany)
UFI	Unique Formula Identifier
UNECE	United Nations Economic Commission for Europe
U_{RAC}	Unit weight (weight of food commodity of concern; pesticide regulation)
US	United States
USDA	United States Department of Agriculture
VR	Ventilation rate
VWA	Voedsel en Warenauthoriteit - bureau Risicobeoordeling
WHO	World Health Organization
WTO	World Trade Organization
WUR	Wageningen University & Research Centre

List of Figures

List of Tables

Chapter 1
General Introduction

Pertti Hakkinen, Gerhard Heinemeyer, and Matti Jantunen

No substance is without poison. Only the dose limits the extent of the poisonous action. This famous quote from the medieval physician Paracelsus is the basis of toxicology as used in risk assessment. Risk assessment, in the context of regulation, consists of several pillars, including hazard identification, hazard assessment and exposure assessment, followed by risk characterisation and risk communication. The development of toxicology as a scientific discipline has been dominated for a long time by the assessment of the hazards, which is identification of the toxicological characteristics expressed as toxicological endpoints (see Sect. 2.2.2) of a substance. Hazard assessment characterises the toxicological properties and the corresponding toxic doses of a chemical agent, which, however, cannot be modified. This is partly "genetically" determined and limited. The only element that controls toxicity is the current dose by which a person is exposed. Therefore, exposure assessment, which is dose evaluation, has become an important pillar of risk assessment. Further, the results of the exposure assessment play a primary role in the risk communication occurring when the risk assessment results are discussed between the different stakeholders. Many handbooks and guidance documents have been published about exposure assessment, but there is no publication covering the complexity and the different aspects of exposure assessment and their interdependencies.

P. Hakkinen
National Library of Medicine and the National Institutes of Health, Bethesda, MD, USA
e-mail: berthakkinen@gmail.com

G. Heinemeyer (✉)
Department of Exposure, German Federal Institute for Risk Assessment (Retired), Berlin, Germany
e-mail: gerhardheinemeyer@o2mail.de

M. Jantunen
Department of Environmental Health, National Institute for Health and Welfare (THL) (Retired), Kuopio, Finland
e-mail: matti.jantunen@janding.fi

G. Heinemeyer et al. (eds.), *The Practice of Consumer Exposure Assessment*,
https://doi.org/10.1007/978-3-319-96148-4_1

The title of this textbook is “The Practice of Consumer Exposure Assessment”. This terminology is used according to EU chemicals regulation which differentiates three types of risks, namely, of (1) the environment, (2) the people at work (workers, occupational exposure) and (3) people in their private surroundings (consumers). The term “consumer” therefore means the general population when not at work. This includes driving a car, being at leisure at home, doing hobbies, eating foods, and using products intended for private use (consumer products), e.g. cleaning products, textiles, hobby articles, toys and others which are not used at the workplace. In this context consumer exposure includes activities in kindergartens and schools for children’s exposure and other private facilities outside the home. Human exposure via the environment is also included, if it occurs in the private surroundings. For example, many food exposures are indeed environmental exposures, because of the occurrence of environmental contaminants in food as the last element of the food chain.

1.1 Aims and Objectives of This Textbook

This book covers all aspects of exposure assessment under a common roof. To understand what exposure assessment means, the general concepts of risk assessment must be explained, followed by the approaches used in exposure assessment and its major elements.

Exposure assessment should include all the different situations, sources, and pathways by which substances may be incorporated into the human body. Therefore, exposure assessment has to describe these differences and consider them in the scenario characterisation. There are also different statistical methods that can be applied in the presentation of exposure data and exposure results. The major approach to estimate the extent of exposure is exposure modelling, which requires adequate model variables (exposure parameters). The exposure parameters in scenario models depend on each other. Although total consumer exposure is seldomly measured, the parameters used in exposure models include those that are the subjects of measurements, either by chemical or physical analysis, as well as epidemiological and public opinion research studies.

This book is directed to anyone interested in consumer exposure assessment, either from a regulatory institute, a university or other educational institution, a chemical(s) or consumer product manufacturing company or a member of the public. The objective is to give a general introduction into exposure assessment and to point out the major implications of exposure assessment for risk assessment. It should be understood as a textbook, giving guidance for undertaking exposure analyses and giving help for interpretation of the data and results and information about the background and methodology of exposure analysis. In this context, one objective is to foster the understanding of common issues and opportunities of exposure assessment.

The authors in this book see exposure assessment playing a primary role in scientifically based risk assessment. Along with hazard identification and hazard assessment, exposure assessment is a critical pillar in the risk assessment process. Exposure assessment delivers the actual information on the extent of the dose people are in contact with. Therefore, exposure is embedded in the risk assessment, the basics of which should be explained first:

- The basic and common understanding of consumer exposure assessment, which is scenario description, model building and characterisation of the model variables (exposure parameters and - data)
- Special issues of consumer exposure assessment, e.g. food exposure and consumer product exposure
- The availability and generation of information used for exposure assessment
- Regulatory issues in consumer assessment, and national and international activities in relation to the regulation and stakeholder interests

Exposure assessment involves many different scientific disciplines, ranging from medicine/toxicology/pharmacology, nutrition, chemistry and physics, mathematics and statistics, as well as epidemiology, public opinion research and psychology. Laboratory analytics and consumer questionnaire studies are important instruments to measure exposure issues and gather information. Also, exposure assessment is part of risk assessment regarding a great number of regulations. Therefore, the communication between stakeholders of these different fields is crucial. Instruments for exposure assessment are still not well harmonised, with terminology as a central issue and opportunity. The authors of this book therefore emphasise the importance and use of a consistent terminology. Guidelines to use harmonised terminology have been developed within the frame of the WHO/IPCS "Harmonization Programme". All three editors have worked in different exposure projects on an international level (EU and WHO) that have developed guidance documents for documentation of exposure data, harmonisation with regard to terminology and exposure modelling and on uncertainty analysis. Although a quasi "Babylonian terminology confusion" is still present in exposure assessment, it is one of the key issues to apply the principles of these guidance documents in the entire fields of exposure science. This book encourages the use of harmonised terminology and methodology.

Chapter 2
Major Principles and Concepts of Risk Assessment

Gianfranco Brambilla, Matthias Greiner, Ursula Gundert-Remy, Gerhard Heinemeyer, Friederike Neisel, and Wouter ter Burg

2.1 Introduction

Ursula Gundert-Remy

Risk has two components: a known threat, which in the context of toxicology is called a hazard, and an exposure situation. In toxicology, risk assessment requires quantification of the two components: (1) the hazard which is the possible health impairment, given by the inherent toxicological and physicochemical properties of a substance. This is called hazard identification, and combined with dose-response characterisation it becomes, hazard assessment, and (2) the exposure which has to be

G. Brambilla
Department of Food Safety, Nutrition and Veterinary Public Health, National Institute of Health (ISS), Rome, Italy
e-mail: gianfranco.brambilla@iss.it

M. Greiner · F. Neisel
Department of Exposure, German Federal Institute for Risk Assessment, Berlin, Germany
e-mail: matthias.greiner@bfr.bund.de; Friederike.Neisel@bfr.bund.de

U. Gundert-Remy (✉)
Institute of Clinical Pharmacology and Toxicology, Universitätsmedizin Berlin, corporate member of Freie Universität Berlin, Humboldt-Universität zu Berlin, and Berlin Institute of Health, Berlin, Germany

G. Heinemeyer
Department of Exposure, German Federal Institute for Risk Assessment (Retired), Berlin, Germany
e-mail: gerhardheinemeyer@o2mail.de

W. ter Burg
Centre for Safety of Substances and Products (VSP), Institute for Public Health and the Environment (RIVM), Bilthoven, The Netherlands
e-mail: Wouter.ter.Burg@rivm.nl

G. Heinemeyer et al. (eds.), *The Practice of Consumer Exposure Assessment*, https://doi.org/10.1007/978-3-319-96148-4_2

assessed and quantified. The ratio of hazard assessment and exposure assessment is the probability that under a given exposure situation, a specified health impairment will materialise.

Risk assessment is aimed at an objective, traceable and transparent result. Even if the data are evaluated in an objective way, implicit assumptions are made and the outcome is subject to uncertainties. In the risk assessment document, these assumptions and uncertainties have to be presented, and their influence on the assessment's outcome has to be discussed.

Risk management is based on the outcome of the risk assessment. In addition, risk management considers also other factors: for example, the costs which would result from risk management measures, a comparison between the alternatives, and the preferences of the public and its concerns and choices (socioeconomic evaluation). In most of the regulatory decision schemes, risk assessment is separated from risk management. Whereas persons specialised in scientific matters perform the assessment, management is performed by persons specialised in political decision-making.

Several steps have to be performed for risk assessment. In the first step, the hazard identification includes, the inherent toxicological (and physicochemical) properties of the substance are determined based on the results of toxicological testing in silico, in vitro and in animals, experimental data in humans and additional human data from epidemiological studies if available (Bauch et al. 2012). The second step, hazard assessment, is the analysis of the data to evaluate the relationship between dose and the probability or the incidence of the effect for a given type of adverse response (dose-response assessment). This step is performed with the aim to identify the [highest] dose that does not exhibit an adverse effect (no-observed-adverse-effect level, or NOAEL).

The third step is the quantification of the exposure. It aims to determine the amount of a chemical to which the population and subgroups of it will be exposed.

The results of the dose-response assessment and the exposure assessment are then combined to develop an estimate of risk (risk characterisation), expressed as the probability of health impairment under the existing exposure including an analysis of the uncertainties.

2.2 Approaches for Risk Assessment

The aim of the risk assessment is to make a scientific evaluation on the consequences for the human health caused by a given exposure scenario.

Two approaches for risk assessment are used in regulatory circumstances. The first approach is used to set health-based guidance values, and the second approach assesses the distance between the exposure and the dose without an appreciable effect (margin of safety, MOS). In both cases, the first step is hazard identification and hazard assessment which is the analysis of the dose-response relationship.

2.2.1 Hazard Identification

Hazard identification relies mainly on experimental and epidemiological studies of a substance. To identify genotoxic properties, in vitro and in silico methods are accepted tools for hazard identification. In a critical evaluation of the results in comparison to a control group, the toxicological profile of a substance can be described. In epidemiological studies, the number of endpoints which can be assessed is limited, whereas in animal experimental studies with toxicological endpoints, a broad range of effects may be investigated. The standard for animal studies in all regulatory frameworks is the performance of studies according to OECD testing protocols. Details of the OECD testing protocols concerning type and endpoints of the studies are given below.

Most regulations stipulate a preference to available human data for risk evaluations. Data from epidemiological studies have been used in a small number of food health risk assessments. For example, the update of limit levels for lead has been assessed by EFSA exclusively on the basis on some epidemiologic studies that show correlation to IQ measurements in children with blood lead levels (EFSA 2010b). Also, EFSA experts have performed cadmium assessment by taking measurements of beta-2-microglobulin performing a meta-analysis of several studies in humans (EFSA 2009a). A limited number of larger epidemiological investigations have been performed, e.g. after chemical disasters. The very intensive scientific discussion about dioxins started after the Seveso accident. Health hazards are also related to perfluorinated hydrocarbons. Some case-control studies have been started to evaluate sperm count and quality in Scandinavia (Nordström Joenssen et al. 2009), birth weight and other birth indicators (Fei et al. 2008), and after continuous release of perfluorooctanoic acid (PFOA) into the drinking water by the producer, a DuPont plant in West Virginia[1] (US). A big epidemiological study including up to 60,000 persons was started for PFOA (C8-Panel 2017).

It is obvious but should nevertheless be noted here that the absence of a toxicological effect can never be proven. The sensitivity to detect adverse effects of a substance clearly depends on the kind of the tests and the array of parameters investigated. The catalogue of toxicological studies that is available for a substance may differ depending on the area of regulation. For example, in the European chemicals regulation, REACH, the number and extent of required studies depend on the production volume per year; for food additives, a tiered approach is taken, and the pesticides regulation stipulates that the whole array of studies is performed (Directive 91/414/EEC). Table 2.1 informs about the requirements according to the European REACH regulation (Regulation (EC) No 1907/2006; Table 2.1). According to this regulation, for example, information is lacking on adverse effects by prenatal exposure for substances with a production/import volume of less than 100 tons per year.

[1]http://ohiovalleyresource.org/2017/02/13/dupont-offers-670m-settlement-teflon-chemical-contamination-water/

Table 2.1 Testing requirements according to regulation (EC) No 1907/20061

Production or imported volume per year	Required tests (without waiving arguments)					
1 tons and more	Acute toxicity testing programme[a]					
10 tons and more	Acute toxicity testing programme[a]	Subacute toxicity testing (28 days)	TOXICOkinetic assessment	In vitro cytogenicity study in mammalian cells or in vitro micronucleus study In vitro gene mutation study in mammalian cells, if a negative results in former testing	Reprotoxicity screening test	
100 tons and more[b]	Acute toxicity testing programme[a]				Prenatal developmental toxicity study/two-generation reproductive toxicity study	Subchronic toxicity testing (90 days)
1000 tons and more[c]	Acute toxicity testing programme[a]				Developmental toxicity study	Carcinogenicity testing

[a]Includes irritation/corrosion skin; eye irritation tests, sensitisation tests, acute toxicity testing (single dose), in vitro mutagenicity test in bacteria
[b, c]Shorter term repeated dose testing not necessary if longer term studies are provided covering the endpoints

Type of Studies and Endpoints for Hazard Identification

The types, duration and endpoints of studies have evolved over the years to establish the toxicological profile of a substance.

In the regulatory context, study protocols are used which are standardised within the framework of OECD (1981–2016). The protocols are published in the series of OECD Guidelines for the Testing of Chemicals.[2] The test guidelines are updated regularly, and the series no. 4 describes studies for the characterisation of health effects. These guidelines are an internationally recognised tool for assessing the inherent toxicity of substances. The OECD publication contains about 150 of the most relevant internationally agreed testing methods (100 for human health endpoints), used by industry and research institutions mainly in the context of regulatory safety testing. These protocols are used with the aim to identify and describe the inherent toxicity of chemicals. The guidelines are accepted by governments as being suited for this aim. Results of testing according to the OECD guidelines can be used for submitting dossiers for notification of chemicals, pesticides and chemical products and also for the risk assessment of substances in the area of food safety.

The following studies may be part of a dossier for addressing human health endpoints.

Acute Toxicity Testing

A package of tests is used to characterise properties of the substance in question for the lowest production volume in the REACH regulation. This package contains acute toxicity testing, testing for dermal and ocular irritation and corrosivity, skin sensitisation and one genotoxicity test in bacteria.

Genotoxicity

In the OECD test methods inventory, 26 test methods are described which could be used to evaluate substances with respect to their genotoxic properties. The number and the kind of tests required depend on the legislation, and in REACH depending on the production volume. In general, the test methods address adverse effects on the genetic material whereby the alteration can have direct or indirect effects on the DNA. The strategy of testing requires that in vitro tests are performed first. In the absence of effects, no further testing is required. If a positive result was found, in vivo tests will follow.

[2] http://www.oecd.org/chemicalsafety/testing/oecdguidelinesforthetestingofchemicals.htm

Subacute/Subchronic Toxicity Testing/Chronic Toxicity Testing

These tests are aimed at assessing the systemic toxicity of a substance and its potential to elicit in a dose-dependent manner lesions of organs and tissues as well as functions of the body. The duration of the exposure varies from 28 days (subacute), over 90 days (subchronic) to 12 months (chronic) study. The preferred animal species tested is the rat. The route of administration is oral (by gavage, in food, in drinking water). However, also inhalation and dermal exposure are addressed by the relevant guidelines of OECD.

Carcinogenicity Studies

The aim of this type of study is to observe test animals for a major portion of their life span for the development of neoplastic lesions during or after exposure to various doses of a substance.

Fertility Studies

Several types of studies are described in the OECD guideline inventory. The spectrum covers screening tests for reproduction/developmental toxicity, one- and two-generation reproduction toxicity tests and combinations of repeated dose toxicity testing assessing reproduction and developmental toxicity.

Developmental Toxicity

The main aim is to investigate effects on the foetus and the dam. Any effects on the foetus can be considered as being caused by a single or several doses, but this conclusion is usually not justified by the design of standard studies. Given the relatively short dosing period, any effects reported could be relevant to the derivation of an acute reference dose (ARfD), especially if developmental toxicity is the most critical effect.

Neurotoxicity

These single or multiple dose studies are aimed to identify critical neurotoxic effects for substances with a specific mode of action (e.g. organophosphates).

Developmental Neurotoxicity Study

The aim of the study is to obtain information on neurodevelopmental effects of repeated intrauterine and early postnatal exposure to a substance.

Immunotoxicity

Extensive immunotoxicity testing is not a regular requirement of chemical substances, and it is restricted to special investigations. However, several parameters covering immune pathology are included in the subacute repeated dose test that gives some indication on effects on the immune system.

Use of the Results of Hazard Identification

The results of hazard identification are used for classification, labelling and packaging (CLP) according to the global harmonised system (GHS), which is an internationally agreed system. In this system, the toxicological properties as resulting from the testing, described above, are classified using globally agreed rules. Classification has downstream consequences for labelling, e.g. standardised indication of hazard classes and categories to be used in the form of a pictogram on the *label*, and for packaging, e.g. the request to use specific packaging material.

All tests under the heading acute toxicity testing are used only for this purpose. This is due to the fact that the effects are local as with the effects of skin and eye irritation/corrosivity and also the sensitisation. Testing for genotoxicity results in a "yes" or "no" answer, but at present, only in rare instances, a dose-response relation can be derived from the data. In acute toxicity testing, data are obtained which can be analysed with respect to their dose-response relationship. However, the endpoints measured in this type of testing are limited, and single administration does not allow for the prediction of effects after repeated administration.

In the tests in which the substance is given by a repeated dose administration, a broad spectrum of endpoints is measured. This spectrum embraces pathological changes of organs, tissues and haematological and laboratory parameters. In addition, in special studies, functional parameters are observed, and influences on fertility and the (neuro-) development are investigated. From the data obtained, a dose-response relationship can be derived, if an effect is observed. Hence, these data can be used for CLP and also for hazard assessment.

2.2.2 *Hazard Assessment*

The WHO (2004) defines hazard assessment in the following way:

> A process designed to determine the possible adverse effects of an agent or situation to which an organism, system, or (sub)population could be exposed. The process includes hazard identification and hazard characterisation. The process focuses on the hazard, in contrast to risk assessment, where exposure assessment is a distinct additional step.

In this process the information and data obtained in the studies undertaken to identify the hazard after repeated dose administration are evaluated to describe the dose-response relationship. The aim is to estimate a dose level which can be used as the reference point (RP) (nomenclature of EFSA 2012b) or point of departure (POD) for deriving a health-based guidance value or for performing a margin of safety analysis. This dose without appreciable adverse health effects in the test animals critically depends on the experimental conditions such as number of animals tested, dose range and dose interval, and the range of endpoints tested (e.g. histopathology, clinical chemistry and functional tests; see above).

The level of the dose without appreciable adverse health effects may differ, for example, between a subacute and a subchronic/chronic study due to the fact that the number of animals and also the number of endpoints investigated are lower in subacute studies compared to subchronic/chronic studies. In addition, adverse effects may only be developing after a long-time exposure. One explanation may be the kinetics. Substances (and/or their metabolic reaction products) with long half-life will accumulate and reach the maximum level only after prolonged exposure. Also, toxicodynamic effects may occur only after continued long-term exposure, e.g. effects on the testes or thyroid as secondary effects with a primary effect on the liver.

Subchronic and chronic toxicity studies are the most commonly used types of "standard" study for setting RP/POD for health-based guidance values. Subchronic and chronic toxicity studies could be used for setting the RP/POD for acute reference doses (ARfDs), if it can be justified that there is no difference between acute, subchronic and chronic effects. Otherwise acute studies have to be used for this purpose.

Box 2.1 Example 1: Derivation of an Acute Reference Dose (ARfD) for Imazalil (The EFSA Journal (2007) 460, 1–15)

The NOAEL for maternal toxicity was 10 mg/kg of BW per day based on reduced body weight gain and food consumption at the next higher dose of 20 mg/kg of BW per day. The NOAEL for foetal toxicity was 5 mg/kg BW per day based on the increased number of resorptions at the next higher dose of 10 mg/kg BW per day.

ARfD 1: 0.05 mg/kg BW based on foetal toxicity (NOAEL 5 mg/kg BW) using a safety factor of 100 for pregnant and nursing women. The panel could not exclude the possibility that the increased resorptions were due to a single exposure.

It was considered that this ARfD would be protective of the potential effects observed during gestation and lactation. As these effects were observed only following in utero exposure, this ARfD was not considered applicable to other groups of the population such as prepubertal children and adult males.

(continued)

Box 2.1 (continued)

ARfD 2: 0.1 mg/kg of BW based on adult (maternal) toxicity (NOAEL 10 mg/kg of BW) using a safety factor of 100 for the general population excluding pregnant and nursing women. It is not known whether the sensitivity of the rabbit is due to species differences (NOAEL lower than after acute exposure of rats) or because the animals were pregnant (similar NOAELs were in pregnant rats and mice). In the absence of information to the contrary, it is assumed that the NOAEL for adult (maternal) toxicity is applicable to the general population, excluding pregnant and nursing women. It is to be noted that an ARfD based on maternal toxicity evaluated in a developmental study following repeated administration is very likely to be conservative.

Examples for the derivation of an ARfD and an ADI are given at the end of this chapter. It was discussed whether the RP/POD to derive health-based threshold values should be a no-observed-effect level (NOEL) or a no-observed-adverse-effect level (NOAEL). The no-effect level (NEL) is an effect level irrespective if the effect has to be considered as an adverse or an adaptive or a compensatory effect or just a biochemical change. For example, an adaptive effect is the induction of the endoplasmic reticulum in the liver that will lead to an enhanced metabolic capacity and thus a quantitative and qualitative change of the toxicity of the parent compound and its possible metabolites. An example for compensatory changes is the increased inhalation rate in metabolic acidosis. The definition of adverse effect of the International Programme on Chemical Safety (WHO 2004) is:

> Change in the morphology, physiology, growth, development, reproduction, or life span of an organism, system, or (sub)population that results in an impairment of functional capacity, an impairment of the capacity to compensate for additional stress, or an increase in susceptibility to other influences.

Box 2.2 ADI for Sorbic Acid (E200) (EFSA Journal 2015; 13(6); 4144)

Short-term and subchronic toxicity studies performed in rats and mice did not show any adverse effects at the concentrations tested (up to 9200 mg/kg of BW per day in rats).

A two-generation reproduction toxicity study was performed in CD/Crl:CD rats in accordance with the Organisation for Economic Co-operation and Development (OECD) guideline 416 and Good Laboratory Practice (GLP). Sorbic acid was administered by gavage at a dose of 0, 300, 1000 or 3000 mg/kg of BW per day. Several adverse effects were described in pups, such as decrease in man litter body weight, milestones of physical development in F1 pups, a delay in functional development in F1 pups and a decrease in anogenital distance in F2 pups in the mid- and high-dose groups. Considering the aforementioned observations, the panel concluded that, by gavage, the

(continued)

Box 2.2 (continued)
NOAEL for developmental toxicity is 300 mg/kg BW/day. Furthermore, the panel noted that, by gavage, the NOAEL for parental toxicity is 1000 mg/kg BW/day based on effects on body weighs of the parental male animals.

In a developmental toxicity study performed in rabbits in accordance with the OECD guideline 414 and GLP at doses of 0, 300, 1000 or 3000 mg sorbic acid/kg of BW per day by gavage from day 6 to 29 of gestation, maternal and foetal toxicity was observed in the mid- and high-dose groups. The panel considered that the maternal NOAEL was 300 mg sorbic acid/kg of BW per day and the NOAEL for the foetuses was also 300 mg/kg of BW per day.

The panel considered that the NOAEL of 300 mg sorbic acid/kg BW/day from the two-generation reproductive toxicity study in rats can be used to allocate a temporary group ADI for sorbic acid. By applying an uncertainty factor of 100, the panel established an ADI expressed as 3 mg sorbic acid/kg of BW per day.

In the era of genomics, changes can be observed which cannot be classified to be adverse. A recent scientific report of EFSA (2014) discusses the future perspectives for the application of the OMICs technologies as valuable tools to measure changes which are indicative for a mode of action or an adverse outcome pathway at the level of DNA/RNA (transcriptomics), proteins (proteomics) and the whole metabolome (metabolomics). Currently, the results of OMICS testing provide additional information on the potential mechanism of action; they may be used in the future to detect subtle changes, which will at higher doses lead to probable adverse effects. However, more work has still to be done to connect changes at the OMICS level quantitatively (and not only qualitatively) to the adverse effect.

Derivation of a Reference Point or Point of Departure

The most preferred procedure to derive the dose without appreciable adverse health effects as the reference point (RP) or point of departure (POD) is from modelling the dose-response data in the so-called benchmark dose approach (BMD).

The BMD is a dose level, estimated from the fitted dose-response curve, at which a pre-specified effect size, the benchmark response (BMR), is observed. By convention and by tradition, a 5% or 10% increase or decrease in response compared with the background response is set as the BMR. The BMD's lower confidence bound (BMDL) is used as the RP/POD. Currently available BMD software from US-EPA and from RIVM (BMDS, US-EPA benchmark software; PROAST) can be used for fitting purposes and for calculating the BMDL (EFSA 2009b, 2017a). This technique has the advantage that all experimental data can be utilised for the modelling and it also gives information on the uncertainty in the data.

An alternative procedure is to select the dose for which no adverse effect could be observed (no-observed-adverse-effect level, NOAEL). The dose without an (adverse) effect is the dose in an experimental study for which no effect has been observed compared to effects in the group with the control treatment. The limitation of this procedure is obvious. We must determine the dose with an adverse effect in order to be able to define the neighbouring lower dose as the dose without an effect. Hence, the number of animals and the spacing of doses are crucial in this respect. It is well known that not seeing an effect is not proof for the absence of an effect. In animal studies, the number of animals can be limited (reduced) due to animal protection requirements. In contrast, statistical power considerations would require a higher number of animals with the aim to increase the power of the study. Regulatory studies are performed according to the OECD guidelines for testing chemicals. These guidelines are a compromise between the two principles, and most of the studies have the limitation of a low statistical power. It is to be noted that using NOAEL as the RP/POD does not utilise all available information on the dose-response relationship.

Derivation of a Health-Based Guidance Value

When the RP/POD is derived from animal data, adjustment has to be made to the human situation. In this adjustment step, the interspecies difference between animal and human is taken into consideration. Although we use the word "difference" to characterise the situation, the factor by which this is done is actually a set of factors. This is because the underlying processes are kinetic and dynamic processes that by their nature are multiplicative processes, e.g. area under the concentration time profile of the substance and the derived clearance. By convention, the factor to express the underlying uncertainty accounting for the interspecies difference is set as 10 as a default value. The factor is divided into two subfactors from which one subfactor accounts for the difference in toxicokinetics, in particular the clearance, and the second one for the difference in the response of the effect, i.e. the difference in toxicodynamics (Renwick 1999a, b). As the clearance of a substance is correlated to the surface of the body, the relationship of the body surfaces is used to account for the difference in clearance between species (Boxenbaum 1982). In the guidance document of ECHA (2012), the factor is given as species-dependent according to the body surface of the animals. Hence, a factor of 4 is used for the toxicokinetic difference if the animal employed in the toxicological studies was the rat. Whereas other authorities (e.g. EFSA 2012a) always use a factor of 4, irrespective of the animal species, ECHA proposes to use different factors according to allometric (surface area) scaling (ECHA 2012, 2014), e.g. a factor of 7 is used for mice. The remaining subfactor of 2.5 is said to account for the difference in internal concentration that is necessary to elicit the same size of the effect, the toxicodynamic difference. When using the toxicodynamic subfactor of 2.5, the basic assumption is that humans are 2.5-fold more sensitive than animals.

A second factor of 10 accounts for the variability within the human population (intraspecies variability) because the health-based guidance level should be applicable for the population as a whole ("general population"). The default value of 10 is also divided into subfactors accounting for toxicokinetic variability and for toxicodynamic variability. Based on analysis of data, it is proposed to use the value of 3.2 for both subfactors (Renwick and Lazarus 1998).

In cases where appropriate data are available, the default factors and their subfactors can be superseded by data-derived factors/subfactors. The WHO document on chemical-specific adjustment factors describes what information can be used to replace the defaultfactor/subfactors by data-derived factors, e.g. kinetic data in animals and/or in humans (WHO 2005). The chemical-specific factors may, dependent on the chemical, be lower or higher than 100 if data are available.

The health-based guidance value is then derived by dividing the POD/RP by the adjustment factors (in cases where no specific information is available, the composite factor is the default factor of 100), accounting for the interspecies difference and also for the variability in the human population (EFSA 2012b; ECHA 2012). The result is the human (exposure) dose that is considered safe for a daily lifetime oral exposure by food additive (ADI, accepted daily intake) or food contaminant (TDI, tolerated daily intake), daily lifetime exposure by dermal contact or daily lifetime exposure via air (ECHA 2012). The resulting health-based guidance value is named the derived no-effect level (DNEL) in the context of REACH (ECHA 2012).

For the derivation of health-based guidance value at the working place (OEL, occupational exposure limit), special considerations apply, and the intraspecies default assessment factor is reduced to 5 (ECHA 2012), the composite factor being 50.

If human data are available, the POD/RP can be derived from those data; in this case, the use of an interspecies extrapolation factor is not appropriate, and only the intraspecies variability has to be taken into consideration. A default factor of 10 is used if no further information on the variability within the human population is available.

In the last step in risk assessment, the health-based guidance value is compared with the exposure for the chemical in question. An exposure below the health-based guidance value is considered safe. If the exposure is higher than the safe dose, measures have to be considered to reduce the exposure to a level that will not result in adverse health effects in the exposed population.

2.2.3 *Margin of Safety*

The second approach for the risk assessment, called the margin of safety (MOS) approach, is the comparison of the exposure with the dose of the substance without an observable effect (BMDL, NOAEL). This approach is also called MOE (margin of exposure). The terms can be used synonymously.

The MOS is the distance to be assessed between the exposure estimate and the BMDL/NOAEL. It can be also expressed as a ratio. Based on the considerations already discussed in the paragraph on health-based guidance values, a ratio (MOS) of 100 is in most of the cases considered to be sufficient by convention (WHO 1994; EFSA 2010a). Depending on the nature of the effect, the slope of the dose-response relationship and other factors which may have an influence on relevant health endpoints, a larger MOS can be required, e.g. if reprotoxicity studies are lacking, a MOS of 300 (WHO 1994; EFSA 2010a) is sometimes considered as being required.

The MOS approach can also be used if the exposure exceeds the health-based guidance value, e.g. ADI, to inform the safety assessment on the probability and severity of the health impairment that the exceedance may cause.

Alternatives

In cases in which toxicological data are insufficient to derive a NOAEL or BMDL, a risk assessment cannot be performed. However, there is an approach that can be applied without knowing the full toxicological profile.

The threshold of toxicological concern approach (TTC, Kroes et al. 2000, 2005) takes advantage of the results of in vivo animal studies of several hundreds of chemical substances. The empirical distribution of the NOELs from the studies is used to derive an exposure value at which the probability of the occurrence of an adverse effect is very low. TTC is the level of exposure that is at the low end of the distribution of known NOAELs/100 (SF/UF). Generally a distribution threshold of 5% is selected and hence the probability that the NOAEL of a substance with unknown toxicological properties is below 5%.

When applying the TTC approach, the exact assessment of the exposure becomes a crucial prerequisite (EFSA 2019). If the exposure is lower than the threshold dose (expressed as a dose in mg/kg of BW/day), the probability that the substance will cause an adverse effect on human health is low. This approach is appropriate in cases where a chemical contamination is detected, the chemical has been identified, but the toxicological data are insufficient for risk assessment, and the exact exposure has been estimated with high confidence.

2.2.4 *Exposure Assessment*

The assessment of exposure is of the same importance as the hazard identification and hazard assessment. The following text provides a brief overview. The principles of the exposure assessments are identical irrespective of the population under consideration. However, there are differences in the specifics of the exposure assessment of workers (occupational exposure) compared with the exposure assessment of the general population and the exposure assessment of food consumption (consumer exposure). Consumers are exposed by the environment (air, soil, water), from consumer products, or by food and drinking water. Inhalation and dermal contact have to be considered as routes of exposure when assessing the occupational exposure, whereas the consumer can also be exposed via oral intake.

In particular, it should be noted that at the place of work, also technical measures can be utilised to reduce the exposure, e.g. protective clothes, which cannot be required for the general population. At the work place, exposures are typically measured by personal sampling that reduces the uncertainty in the assessment.

In the exposure assessment, the first step is the identification of the appropriate exposure scenario: exposure pathway, route of entry, time pattern and duration.

In the second step, the exposure scenario is to be parametrised. For food intake, databases have been collated, and for most of the food items, the content of food additives/contaminants are known, allowing estimation of exposures. For other exposures, e.g. to consumer products via dermal route, the exposure scenarios and parameters are far less known. For, e.g. certain types of cosmetics, special rules have been developed to estimate the exposure assuming standardised scenarios (e.g. shower gel, toothpaste).

In order to take a cautious approach, so-called "worst-case" assumptions are made that bear the potential of grossly overestimating the exposure. However, if no concern is raised by the exposure estimate based on "worst-case" assumptions, a refined exposure assessment is not required. It should be noted that worst-case assessments (for further implications of applying "worst-case" approaches, see Sect. 3.2.1) for products that are alternatives to each other should be compared very cautiously, because their respective levels of overestimation may differ by orders of magnitude.

2.2.5 Risk Characterisation

As pointed out in Fig. 2.1, risk characterisation combines the hazardand the exposure assessment. Risk characterisation compares the quantitative results of the hazard assessment, from animal studies or derived from human studies, with the quantitative exposure estimate. When using an NOAEL or NOEL, the assessment has to consider the interspecies differences and the intraspecies variability. By convention, it is assumed that no health concern is given if the exposure is less than 1/100 of the NOAEL or NOEL. When using health-based guidance values (HBGV, ADI, TDI, DNEL), the exposure should be lower than the HBGV. An exposure exceeding the HBGV indicates a risk which must be clarified and removed, either by a higher tier assessment or by introduction of new/additional exposure reducing measures.

In hazard assessment, various endpoints are addressed in particular with respect to the duration of exposure. In the exposure evaluation, different scenarios are considered. Hence, in the risk characterisation, the decision which hazard data should be compared to which exposure estimate is a special issue of its own.

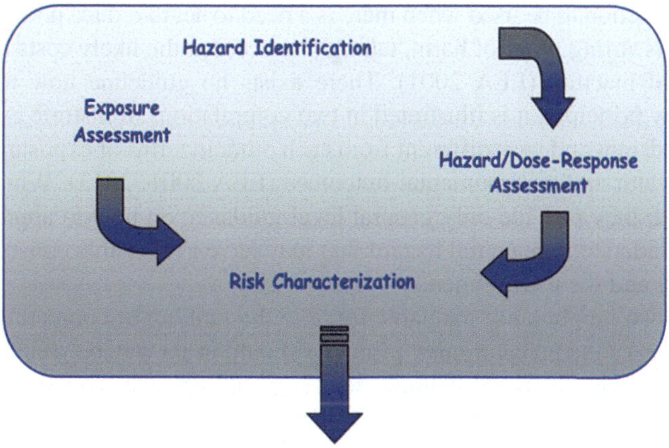

Fig. 2.1 Steps in the process of risk assessment

Particular rules are pointed out in the respective chapters in the special part of this book.

In general, acute hazards should only be compared with short-term exposure and chronic hazards to continuous long-term exposure. Still another question is which endpoint is most relevant for irregular or inconstant exposures. Risk characterisation should consider the characteristics of the populations that are in the focus of the assessment, e.g. whether there is a special endpoint relevant for a subpopulation with a special exposure situation (general population in comparison to a subpopulation characterised by different strata). This is described in more detail in the exposure scenario characterisation (see Sect. 3.2.1).

The principle of risk characterisation cannot be applied to endpoints for which it is currently assumed that no threshold exists, i.e. genotoxicity and genotoxic carcinogenicity. In these cases, it is assumed that every exposure is of concern.

2.2.6 *Risk Management*

In practice, risk management equals exposure management. Vaccination is one of the few contrary examples where, instead, the dose response is being managed.

Risk assessment is a scientific approach that does not include risk management. However, in some regulatory contexts (e.g. REACH), risk assessment also includes the elaboration of risk management options with a discussion of their pros and cons, including costs. To elaborate on the consequences for the society, however, is not in the remit of risk assessment.

In cases where the toxicological profile is not fully known but indications exist for potentially severe, widespread and irreversible risks, the risk management may apply an approach called "precautionary principle". Its main element is a general rule of public policy action to be used when there is a need to act to reduce potential hazards before there is strong proof of harm, taking into account the likely costs and benefits of action and inaction (EEA 2001). There exists no guideline how to apply the precautionary principle; it is illustrated in two compilations of historic examples, all based on hindsight and very different from each other in terms of exposure levels and scenarios, health and environmental outcomes (EEA 2001, 2013). While these are well reported, they provide only general level guidance on how to approach a new and poorly understood potential hazard that may have irreversible consequences for public health and the environment.

There is also no guidance available for, e.g. the ranking of competing risks, and the application of the precautionary principle should in no way be seen as a shortcut for risk management without adequate dose-response and risk assessment. It should also be kept in mind that a precautionary action based on an early warning may itself carry health risks. One such example is the rapid mass vaccination that resulted from the swine flu epidemic (EEA 2013). While the mortality risk of swine flu turned out to be much smaller than the first alarm indicated, this vaccination turned out to exhibit a totally unexpected increased risk for narcolepsy in several countries (Johansen et al. 2016).

In some countries, e.g. Germany, risk communication, i.e. to communicate the results of risk assessment, is an area of activity officially assigned to risk assessment institutions. In other regulatory circumstances and countries, risk communication is the responsibility of the risk managers only.

2.3 Risk Communication

Gerhard Heinemeyer

Risk communication is a crucial process that goes along with risk assessment and risk management and ensures that each party, risk assessors, risk managers and stakeholders gain access to relevant information and that the whole process is transparent. The basic principles for risk communication have been introduced by Covello (1992). Although the aim of the communication activities and the audience may differ, there may be a focus on risk management, the dialogue between partners of risk communication or an emphasis on risk monitoring (US Department of Homeland Security 2012).

The homepage of the Federal Institute for Risk Assessment (BfR 2016) states

> It (risk communication) is defined as a continuous and interactive process and characterised by a participatory dialogue with various stakeholders.

The basis for this dialogue is current information to be communicated. BfR (2016) mentions three principles to raise confidence between communication partners:

Transparency
Reliability
Greatest possible openness

In the view if the CDC, risk communication is a primary approach for providing information to support people for making decisions in case of crises and emergency situations (CDC 2016).

In a document regarding Zika virus infections, the WHO (2016) defines risk communication as:

> the real-time exchange of information, advice and opinions between experts, community leaders, or officials and the people who are at risk. The ultimate purpose is to enable everyone at risk to take informed decisions to act to protect themselves and others from infection and mitigate the effects of Zika virus and its potential complications.

This example clearly shows that risk communication is primarily communication of exposure.

The level of scientific details that can be communicated about the hazard depends on the competence of the target group. From the public health point of view, communication about exposure is more straightforward and relevant. It typically conveys explicit recommendations about risk mitigation and prevention measures to be considered by managers and business operators and "take-home messages" for the lay audience.

Risk communication can therefore be understood as a measure to react under acute situations (e.g. outbreaks of infections) or as a permanent dialogue between players and stakeholders. In any case, exposure is in the centre of this dialogue.

The characterisation of risk communication should include three questions:

1. What is the purpose of risk communication?
2. Who communicates?
3. What is communicated?

Figure 2.2 illustrates the audience that is involved in risk communication. First of all, the purpose of risk communication is to inform risk managers about possible risks from contacts with agents with the aim to reduce them by risk management measure recommendations. It should be mentioned that risk management equals exposure control.

Secondly, risk communication takes place between the risk managers and the affected groups of the audiences, (1) the players which are affected directly by the risk management measures, namely, the producers of products, and (2) the part of the audience which is indirectly affected by the risk management measures, i.e. the population which consumes the products.

The players of the audience between which risk communication takes place can be separated into three parts: (1) the risk assessors and risk managers, represented by government and subordinated agencies (regulators), between which a functional separation exists to distinguish political and scientific aspects in the communication; (2) the group which is affected by regulations, represented by the producers and occupational users of substances and products (industry, handcraft) and in last

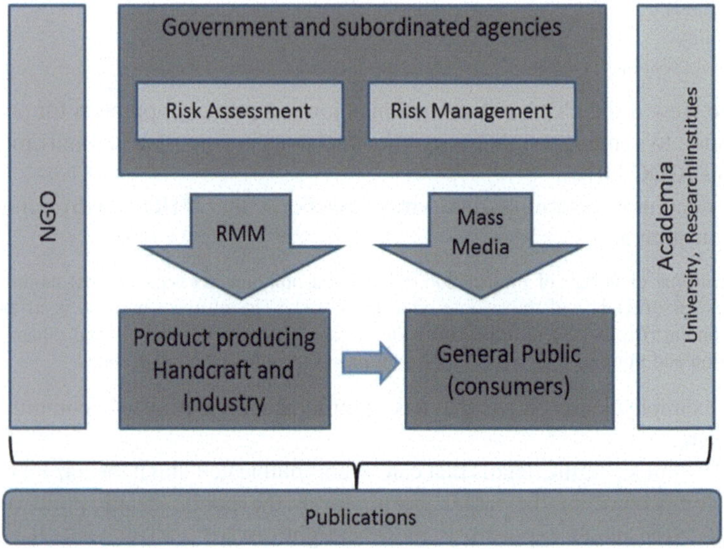

Fig. 2.2 The scenario of risk communication

consequence the general public, represented as the consumers of products; a (3) third group of players that comments the assessment and informs the public. This group consists of non-governmental organisations, academia and the media.

The issue of the risk communication consists of the result of the risk assessment, i.e. the quantitative result of the risk characterisation and the risk reduction options. The latter are normally recommendations for reduction or avoidance of exposure. The scientific basis for risk assessment is also supported by publications by every player in scientific journals and generally available publications.

For example, in Germany risk management is currently (Jan 2019) the responsibility of certain federal ministries (BMEL, BMJ and BMU). The primary subordinated agency for food and product safety is the Federal Office for Food and Consumer Safety (BVL). Risk assessment is performed at the governmental level by the independent working Federal Institute for Risk Assessment (BfR), which performs the hazard assessment and the consumer exposure assessment. The BfR mostly works on demand of the BVL and the responsible ministries by preparing risk assessment reports, which are communicated to the general public by issuing press releases and other scientific publications.

The competent authorities for environmental and occupational safety are the Umweltbundesamt (UBA) and the Bundesanstalt für Arbeitsschutz (BAuA), respectively.

In practice, much of the risk assessment and management actions are conducted by industries themselves, with regulatory supervision. For example, chemical industry performs risk assessments for chemical substances and notifies them to the ECHA for Registration, Evaluation, Authorisation and restriction of chemicals, known as the European Chemicals Regulation (REACH). This approach covers a risk assessment with the inclusion of risk management measures, according to guidance documents worked out by consortia of governmental and industry officials and NGOs. In the meantime, the technical guidance documents which have been developed in this programme (RIP) have been amended by ECHA, in co-operation with the other players (Sects. 5.6 and 7.2). The impact of risk communication between industry and consumers by giving use advice has been focused by Bruinen de Bruin et al. (2007), as discussed in Sect. 5.6.3.

Food safety assessment is performed by the European Food Safety Agency (EFSA) which prepares scientific opinions with the help of a number of scientific panels on demand of the European Commission for risk management (Sect. 7.1).

Risk assessment results are communicated to the interested public by means of press releases and publications via the Internet or social media.

The scientific community is involved in this process by own projects, some in co-operation with industry and with governmental institutions.

Lastly, the result of the risk communication process is the basis for implementation of regulations and directives on different levels for chemicals, consumer products, toys, food, pesticides and others. They will be characterised in the respective chapters.

Risk assessment is also performed by stipulation of law by producers of, e.g. chemical products. This regards in particular to the chemical safety and pesticides/biocides regulations and is characterised in the respective Sects. 5.6 and 4.21.

In addition, on the international level, risk assessment activities are also performed by the WHO and its various international activities (e.g. International Programme on Chemical Safety), the International Labour Organization (ILO) and the Food and Agriculture Organization of the United Nations (FAO).

2.4 Impact of Exposure on Risk Management

Gerhard Heinemeyer

Risk management is not the primary focus of this book; however, exposure assessment delivers options for risk management. Therefore, risk management measures and the responsibilities should be shortly characterised.

Risk management actions are needed when the result of the risk assessment indicates a "concern". This means that the risk characterisation has revealed an exposure which cannot be judged as safe:

- The margin of exposure (MOE) falls below an accepted value.
- Limit value is exceeded according to the exposure assessment.
- The source of exposure to the substance is not approved for consumer products, or risk management measures for an approved source are not existing or effective.

- Internal exposure assessment indicates health hazards, for example, by exceedance of health-based risk values, e.g. HBM limit values or health impairment based on epidemiologic findings.
- The exposure assessment delivers an exposure level that should be lowered and proposes risk management options, i.e. measures to be enacted to eliminate or reduce the exposure and thus the risk to an acceptable level.

The last one might apply for sources of exposure that contribute to an extent which is not acceptable and which can be avoided or reduced, as well as for pathways that contribute to unacceptable excess to the exposure.

Risk management aims to control the possible risks from exposures to agents from any source. When risk is defined as the ratio of the dose caused by the exposure and the dose that causes hazard, a ratio of <1 means that the exposure causes less than the toxic dose, and the ratio of >1 indicates risk which should be acted upon.

The task of developing risk management measures (RMM) is to make rules that prevent any exceedance of the ratio of 1. This implies a number of measures that can be taken which are basically independent from the particular regulation, although they are expressively mentioned in the regulation. Some risk reducing measures are in very close relation to the regulation, and others are not.

Table 2.2 lists some important risk management measures which play central roles and are stipulated by the manufacturers according to existing rules and regulations. First of all, classification and labelling are a very important RMM which is not based on exposure considerations but solely on hazard assessment.

A differentiation should be made between communicated and non-communicated risk management measures. The latter are measures that are controlled by the manufacturers of consumer products.

Positive and negative lists are based both on hazard and exposure assessments. Some substances may be prohibited solely due to their hazardous character. Such measures are typically stipulated on substances that are classified as CMR substances or allergens. This is due to many regulations, e.g. REACH, cosmetics, products safety regulation and toys regulation. For details see Chap. 5. Restrictions for hazardous substances can be enacted for particular exposure situations, e.g. if a certain concentration will not be exceeded, or for particular types of products, where the substance is needed as the active part.

Product composition limitations can lead to reductions of exposure. If the exposure estimation reveals the exceedance of limits (e.g. TDI), then the concentrations may be reduced to levels which lead to meet the limit levels. This approach is broadly applied to setting limit levels in foods. Starting at the TDI, the contribution of the different foods, based on food consumption data, is back calculated to reveal the highest levels in all respective food groups that will not lead to an exceedance of the TDI.

The product composition may lead to reduction of risks, e.g. by setting a particular pH value in the product by a special composition or by assembling a particular physical characteristic form.

Table 2.2 Selection of non-communicated risk management measures (RMM) to control risks from chemical

RMM	Remarks	Regulatory applications
Classification and labelling	Independent from exposure issues, only regarded to substance characteristics	Chemicals regulations Products safety CLP regulation
Positive lists	List of ingredients which are allowed in products, often not justified by exposure	Cosmetics Product safety REACH
Authorisation	To authorise the use of the substance for certain products and uses	REACH
Negative lists	List of ingredients which are not allowed in products, often not justified by exposure	Cosmetics Product safety REACH Pesticides
Concentration limits in products	Definition of maximum levels in products incl. food Based on keeping the estimated aggregated exposure from all sources within the acceptable intake levels	Food regulation Cosmetics regulations Toys regulation Pesticides regulation
Design of product composition	A special composition, e.g. pH	
Package design requirements	Child proof sealings Blister package Colouring	To prolong the time a child needs to open a package or a sealing
Package size limits	Products exceeding a certain size not be marketed	Lamp oils are only allowed in packages of <1 l for private customers
Special release/dosing systems	Dispensers for, e.g. liquids (disinfectants, cleaners)	The dispenser will only release a defined and small volume of the liquid

Exposure reduction or avoidance is the objective of design of packaging, which is aimed for prolongation and/or complicating its opening (e.g. child proof sealings and packaging) or for limiting the release of the content from the package.

Special dosing systems by use of dispensers may reduce the exposure because only small portions of the products are released at a time. Blister packages, frequently used for pharmaceuticals but also for consumer products, e.g. cleaners for dishwashing machines, are only releasing one particular tablet and thus reduce exposure.

Restriction of the package size is used as a management reduction measure that is also aimed to reduce the amounts of liquids to be, e.g. swallowed.

In addition to non-communicated RMMs, the manufacturers of products provide use instructions to the consumers. This captures a broad range of instructions such as amount of product to be taken for use, the recommendation to open windows for increasing the room ventilation rate, the instruction not to use the product in closed environments due to inhalation risks from sprays or wearing protecting material which is common in occupational scenarios. These include protective clothing, facemasks, gloves or special spectacles.

Risk management measures can be considered in exposure models. The reduction of concentrations in models is a central issue, as well as the limitation of the amounts to be used. The control of the latter, however, should be differentiated. For example, reducing the package size may be more successful than giving recommendations to the user who must not convincingly follow the resp. instructions. It was therefore proposed that official risk management should consider only the non-communicated risk management measures. Communicated risk management measures should not be considered due to the possible non-compliance of the users.

Governmental bodies perform risk management. Different levels of regulation are possible, e.g. the law, respective regulatory ordinances and technical guidance documents. Positive and negative lists are discussed and decided by expert panels on the EU and on national levels, depending whether the legislation is on the national or international level.

Risk assessment and risk management are separated from each other in several national and supranational bodies due to legislation. The responsible authorities in the EU for food safety have been compiled, e.g. in the BfR food safety almanac (BfR 2014). Competent authorities responsible for chemicals regulation can be found, e.g. at https://echa.europa.eu/about-us/partners-and-networks/member-states-and-competent-authorities.

2.5 The Precautionary Principle

Gianfranco Brambilla

Within the structured risk analysis process (evaluation, management, communication), there are cases where scientific evidence is insufficient, inconclusive or uncertain and a preliminary scientific evaluation indicates that that there are reasonable grounds for concern that the potentially dangerous effects on the environment, human, animal or plant health may be inconsistent with the high level of protection chosen by risk managers to protect the community (EU 2000).

In such cases, risk managers employ the precautionary principle defined as

> a general principle of Community law requiring the competent authorities to take appropriate measures to prevent specific potential risks to public health, safety and the environment, by giving precedence to the requirements related to the protection of those interests over economic interests (Bourguignon 2015).

Decision-makers need to be aware of the degree of uncertainty attached to the results of the evaluation of the available scientific information. Judging what is an "acceptable" level of risk for society is an eminently political responsibility. Decision-makers faced with an unacceptable risk, scientific uncertainty and public concerns have a duty to find answers. Therefore, all these factors have to be taken into consideration when a risk management option based on the precautionary principle is taken (EU 2000) (Table 2.3).

Also within the World Trade Organization Sanitary and Phytosanitary Agreement (WTO 1994), the Article 5.7, although it does not explicitly uses the term of precautionary principle, reads:

> In cases where relevant scientific evidence is insufficient, a Member may provisionally adopt sanitary or phytosanitary measures on the basis of available scientific information, including that from the relevant international organisations as well as from sanitary and phytosanitary measures applied by other Members. In such circumstances, Members shall seek to obtain the additional information necessary for a more objective assessment of risk and review the sanitary or phytosanitary measure accordingly within a reasonable period of time.

The precautionary principle should not be confused with the risk assessment factors leading to conservative estimates that scientists usually apply in their evaluation of scientific data.

Rather, the precautionary principle may be a consequence of one or more of the following factors affecting the production and the evaluation of scientific evidences: insufficient peer review process of published scientific data, insufficient transparency about data and how they have been obtained (i.e. study design, methodology, data handling and processing, uncertainties).

Critical situations relevant for the precautionary principle are those determined by single or highly controversial observations, i.e. those reporting lowest no-observed-effect level in a study, regardless of study quality and the reproducibility, when such studies are considered to derive points of departure for risk assessment (Dekant 2016). To this purpose, the European Food Safety Authority is proposing the benchmark dose level approach that considers all the range of the dose-response curve and not only the NOAEL and LOAEL doses, to derive on probabilistic basis, the point of departure for risk evaluation (EFSA 2017a). Another critical point relies on the robustness of the data considered for evaluation: a recent survey reports more than 70% of researchers failed to reproduce results of other scientists experiments and more than half failed to reproduce their own observations (Baker 2016).

A more comprehensive evaluation of all the available scientific evidences about the toxicity of a chemical should be carried out under the weight of evidence (WoE) approach (Lutter et al. 2015). However, detailed guidance regarding study design, applied methodology and toxicological relevance and consistence of results and effects assessment in WoE needs to be developed and should be supported by an adequate experience in such a toxicological assessment (Rhomberg 2015).

The adoption of measures based on the precautionary principle should be inter alia proportional to the chosen level of protection, non-discriminatory in their application, consistent with similar measures already taken, based on an examination of the potential benefits and costs of action or lack of action (including, where appropriate and feasible, an economic cost/benefit analysis), subject to review - in the light of new scientific data - and last but not least capable of assigning responsibility for producing the scientific evidence necessary for a more comprehensive risk assessment (EU Commission 2000).

Table 2.3 Minimum, median and maximal interpretation of the precautionary principle

Probability that the risk of concern is real (%)	IPCC scale	Scientific usage	Example of definition	Interpretation of the precautionary approach		
				1/minimal	2/median	3/maximal
99	"Virtually certain"	Rigorously proven	Rio declarationUNFCCC			
90–99	"Extremely likely"	Substantially proven				
80–90	"Likely"	Very probable				
67–80		Probable	Swedish environmental code French environmental charter			
50–67	"About as likely as not"	"More likely than not"				
33–50		Evidence points in this direction, but not fully proven				
20–33	"Unlikely"	Evidence is beginning to accumulate in this direction	Ministerial Declaration of the third international conference on the Protection of the North Sea			
10–20		Plausible, supported by some evidence				
1–10	"Extremely unlikely"	Attractive				

From Bourguignon (2015) on the basis of C. Weiss (2006), M. Haritz (2011), Intergovernmental Panel on Climate Changes (IPCC) synthesis report (2008)

Examples of the case studies involving the consideration of the precautionary principle can be found in the USA–EU dispute on the use of hormones in food producing animals, in the EU withdrawal of the insecticide imidacloprid for professional application on crops considered attractive to bees (a long list including oilseed rape and maize) and on cereals apart from seed treatment of winter cereals, in the ban of feed additives with antimicrobial activity, in the use of GMO and nanotechnologies (Bourguignon 2015).

The last consideration about the precautionary principle is related to the risk communication: the transparency in taking a decision based on the precautionary principle should avoid to consider it as a product of "pseudo-science" under political pressure or the result of a "zero-risk" approach. Rather, it provides a basis for action when science is unable to give a clear answer and underlines that determining what is an acceptable level of risk for the EU is a political responsibility.

2.6 Risk Assessment for Intermittent or Continuous Exposures

Wouter ter Burg and Friederike Neisel

2.6.1 Introduction to Risk Assessment of Long-Term Exposure

The risk assessment paradigm includes as a first step the hazard identification followed by the exposure and effect assessments. Finally, in the risk characterisation, the exposure and effect assessments are compared to assess the probability of an adverse outcome to occur following a given exposure to a chemical substance (van Leeuwen and Vermeire 2007). Traditionally in risk assessment, this comparison is made for the long-term duration of exposure. A long-term or chronic-based health-based guidance value (HBGV) is typically derived in the effect assessment, even though the underlying toxicity studies are performed with subchronic durations (Belkebir et al. 2011).

In real life, however, the exposure to a substance is seldom continuous and at a constant level. Both in actual exposure situations and in controlled animal studies, the exposure or dose administration is not continuous either. As a result, the extrapolation steps within the exposure or effect assessment and ultimately in the risk assessment are prone to uncertainty leading to possible over- or underestimates of the true risk. Under conditions where taking conservative assumptions consider uncertainty and risks are sufficiently controlled, one does not need to take any further actions. In cases where risks cannot be excluded, a refinement is deemed necessary. Refinement steps require more in-depth knowledge on the toxicity and exposure of a substance to support the decisions taken for refinement (Geraets et al. 2016; Haber et al. 2016).

In regulatory frameworks such as the REACH legislation, the Biocides and Plant Protection Products regulations, relatively extensive toxicological profiles of the substance are requested, whereas the provided exposure related information, and quality thereof, is relatively poor. The exposure information is either very focused on specific intended uses (Biocides and PPP) or described in a high abstraction level to cover as many processes or product uses as possible (REACH). In food regulatory frameworks, food consumption surveys are performed at a national level and combined in a European or national database (see also Sect. 4.6). The food consumption data generally includes information on food intake data from a large population for several days, from which the short-term and long-term consumption is deduced. The latter, long-term dietary exposure is the result of the average of eaters and possibly including non-eaters over a lifetime, but does not account for seasonal products or certain preferences for a food product group (Zarn et al. 2010, 2011). Where the level of detail of the exposure information may differ per framework, overall the exposure assessment requires time extrapolation steps to acquire a long-term exposure estimate as exposure information seldom describes a long-term exposure.

In this chapter, the elements of time extrapolation on the exposure and toxicity part and how they may impact the risk assessment are described. Several options for refinement are presented also indicating when it is best not to apply time extrapolations, but to seek alternative approaches.

2.6.2 Long-Term, Intermittent and Fluctuating Exposures

Long-term exposure can have more than one explanation and is also often referred to as continuous exposure over lifetime. Lifetime exposure describes the exposure from birth (sometimes even prenatal exposure) to death. Occasionally, it also means a major of a lifetime such as a working life (40 years) or adult life. Long-term exposure is often described as an exposure that at least spans 1 year and is typical for several years of exposure, e.g. adolescence. Long-term exposure does not necessarily mean that the exposure is continuous or even occurs daily; however the exposure should at least have a high frequency. For example, long-term worker exposure typically exists of working 5 days per week for 8 h per day and is assumed as such in regulatory frameworks. In the long term, one may consider a high frequent exposure of at least weekly up to several times per day, though there is no fixed definition. In the latest ECHA guidance update of Chapter R.15, a cut-off of 15 days in a year is considered as long-term exposure. It may be discussed whether exposure during childhood may be considered a long-term exposure or not. In general, across regulatory frameworks, long-term exposure spans at least 1 year, which is typical for the years of exposure including a certain life stage. Hence, exposure of a child, e.g. from 1 to 4 years, may also be considered as long-term exposure if this definition is followed and is generally compared directly to an acceptable daily intake (ADI) or tolerable daily intake (TDI) in the food safety assessment which is in fact referring to a lifetime exposure (VWA 2008).

Where exactly the lines are drawn between long-term exposure and short-term exposure is unclear. Exposure every other day generally falls under long-term exposure, but it is not continuous. Exposure with a more or less on-off exposure profile, albeit against a background exposure, is referred to as intermittent exposure. Intermittent exposure includes various types of exposure profiles, showing regular or irregular frequencies of exposure that may occur on consecutive days for a limited period and periods without exposure. Seasonal products, such as sunscreen or seasonal foods, can also be considered as products that are applied or consumed intermittently, despite the fact that they are used daily for a certain period. This may even extend to increased exposure during child years, as a result of high food intake relative to body weight, and thus is sometimes also referred to as intermittent exposure on the scale of a lifetime. Other examples of intermittent use of a consumer product are home decoration paint, which is used for a couple of days and perhaps only twice per year, or products used at regular intervals such as hair dyes, which due to hair growth need to be applied every 4–6 weeks. It must be considered that the toxicokinetic profile of substances also determines whether or not an exposure is long term or short term, as discussed in Sect. 3.5, and whether it should be compared to chronic-based HBGVs values, such as ADIs, or not.

Fluctuating exposure is used to describe exposures that due to increased use or consumption show elevated exposure levels, compared to background levels. The exposure is continuous and thereby discriminates itself from intermittent exposures, though the different terms are quite often interchanged. The brief elevated exposure levels could be considered as peak exposures in case the increase is much higher than background levels or limit values.

In contrast to the human exposure, in animal experiments the dose administration is daily (not always including weekends) for the preset duration. The labels of acute, subacute, subchronic and chronic given to the toxicity studies are really based on life expectancy of the test species. This seldom matches the real-life exposures of humans showing intermittent and fluctuating exposures.

2.6.3 *Time Extrapolation*

Assessing long-term exposures by time extrapolation has an impact on the risk assessment and is subject to uncertainty. Especially in case datasets on both exposure and toxicity are limited, uncertainty is high. The way that time extrapolation is dealt with in many regulatory frameworks is, at first, to simply ignore that differences over time may exist. In other words, the exposure estimate generated for an event or a day exposure is assumed to occur daily and thus set equal to a long-term/chronic exposure. In case the frequency is higher than once per day, the event is multiplied by this factor; however, the frequency is ignored if it is less than once a day. Ignoring that an exposure may only last for a part of the day and possibly infrequent, the assumption is for practically all substances worst case. In cases where risks are sufficiently controlled, this is fine. On the other hand, it may lead to unnecessary actions, such as too strict conditions of use, or it can lead to a conservative conclusion that risks are not controlled. Disregarding the fact that perhaps other elements in the exposure assessment or hazard assessment can be refined as well, the focus in this chapter is on refinement by applying time extrapolation in the risk assessment.

In general there is quite some confidence in the hazard assessment and derivation of the chronic-based HBGV. Even when there is no chronic toxicity study available, empirically derived extrapolation factors have been established to be used instead. Several studies have done research to the critical effect levels, either in NOAELs or LOAELs (no- or lowest-observed-adverse-effect level), and/or the benchmark dose (BMD), providing ratios between fixed durations in toxicity studies, e.g. to be able to extrapolate between subacute and chronic durations (Kramer et al. 1996; Pieters et al. 1998; Bokkers and Slob 2005; Batke et al. 2011). Although it should be noted that uncertainty underlying such factors is acknowledged both by regulators (ECHA 2012) and scientists (Pohl et al. 2010; Batke et al. 2011). The extrapolation factors currently applied by default under REACH (Pohl et al. 2010; Batke et al. 2011) have been appointed as 2 for subchronic to chronic, 6 (six) for subacute to chronic and 3 (three) for subacute to subchronic duration.[3] Chronic usually refers to a 1.5–2-year study, subchronic to a 90-day study, and subacute usually to a 28-day study (all for rodents).

The refinement that is often pursued is on the other side of the spectrum. The amortising (averaging out) of the exposure over the time span one is interested in, in this case the long-term exposure, simply divides the exposure estimate by the duration. By example, the estimated daily exposure is multiplied by the frequency of use and divided by the number of days in 1 year. The dilution of the exposure can reach very high factors when the frequency of use on a yearly basis is very low, i.e. a factor of 1/365 is applied to the exposure estimate if the exposure occurs once per year. Note that this (scientifically not justified) approach is commonly applied in some regulatory risk assessments without any further considerations of the actual exposure nor the toxicity of a substance. In the risk assessment paradigm, where exposure is compared to a considered safe level, amortising the exposure is possibly leading to an underestimation of the risk. It should be realised that extrapolating the hazard PoD from short to longer duration by a linear fashion is a conservative case approach, covering for uncertainty in the extrapolation, whereas if it is applied to the exposure estimate, the opposite is achieved.

Inseparable from today's risk assessment where time extrapolation is performed is Haber's rule (Haber 1924; Witschi 1999), which has been the golden standard for many years and the basis for linear extrapolation (Belkebir et al. 2011). The principle

[3]It should be noted that the terms "(sub)acute" and "(sub)chronic" are terms which are not favoured by the authors of this book to characterise the duration of exposure.

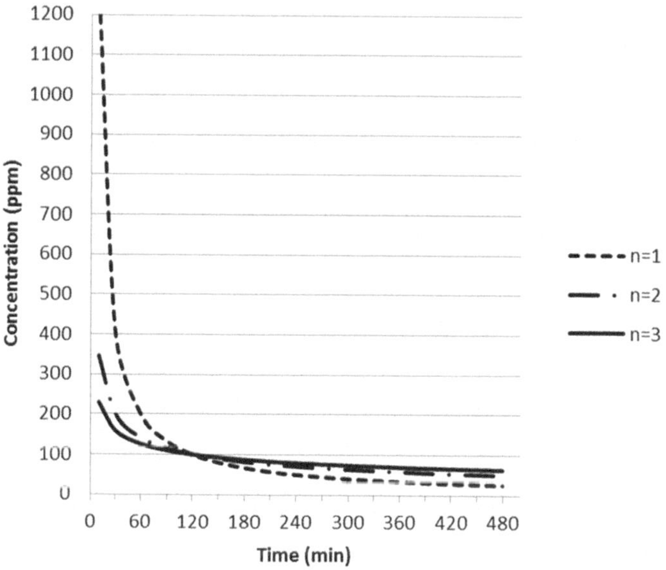

Fig. 2.3 Modified Haber's law, $C^n \times t = k$, shown with three different values for n, starting from the x, y coordinates of 120 min, 100 ppm. Time is presented in minutes, whereas the same would apply if it were days, weeks or months

of Haber's rule is that any combination of concentration or dose with time leading to a constant will result in the same magnitude of effect for a specific effect; $C \times t = k$. Haber established his theory in the field of acute inhalation toxicity and the lethal effects of war gases on animal species. This principle has been extrapolated to the other fields in toxicity and to the fields of exposure assessment and risk assessment. Moreover, where Haber's rule was established for inhalation toxicity, it is also applied to other routes of administration.

To date, awareness exists that Haber's rule is much more complex and underlying elements in relationship between exposure (concentration, dose), time and effect play a role (ten Berge et al. 1986, Belkebir et al. 2011). Haber's rule has been modified as $C^n \times t = k$ or $C \times t^n = k$, indicating that either concentration or time can have a larger influence on the constant effect if the n-value is not (close to) 1. This was already shown in Haber's own field of acute inhalation toxicity where ten Berge et al. (1986) showed that the n-value generally varied between 0.5 and 3.5. As a consequence, when applying time extrapolation to derive acute exposure guideline levels (AEGL), the AEGL committee decided to apply an n-value of 3 to shorter duration and an n-value of 1 to longer duration, if no substance specific n-value could be derived (AEGL 2001). Figure 2.3 provides a graph showing the extrapolation of a constant from the PoD of 100 ppm at 120 min. to shorter and longer durations using different n-values. Moreover, for slight (or first occurring) irritation effects on the respiratory tract, generally no time extrapolation is applied at all, but concentrations are flatlined across the 10-min to 8-h time span. Similar approaches have been followed on the longer time scales as well for irritation but also for other types of substances where the concentration is the determinant factor rather than time.

Elements that play a role in the relationship between exposure (or concentration), duration (time) and effect are the way the body deals with the substance and how the substance may affect the body, i.e. the toxicokinetics and toxicodynamics (Renwick 1999a). Therefore, refinement of the risk assessment should only be done with knowledge on the exposure profile and toxicokinetics and toxicodynamics or stick to the default approach.

2.6.4 Internal Exposure: Role of Toxicokinetics and Mode of Action

The exposure profile may be short term regarding the way food is consumed or products are used, but the exposure of a substance can still lead to a relevant long-term exposure internally to the mother compound, its metabolites or formed adducts. Under the basic assumption in risk assessment that a test species is a good model for humans, such elements should be covered in the animal toxicity testing. However, (1) differences between species on toxicokinetics and mode of action (toxicodynamics) exist, which need to be accounted for; (2) the way of and routes of administration can be very different from normal human exposure, e.g. in short-term oral toxicity studies, exposure to a chemical might be achieved via gavage resulting in daily peak-levels, while, in contrast, drinking water or dietary studies result in more gradually increased levels or different exposure routes leading to differences in the systemic availability; and (3) animal toxicity studies are terminated leaving out the possibility to check for postexposure effects. It is acknowledged that some studies include observation periods or recovery periods, but these periods can be too short (Geraets et al. 2016). Intermittent exposures are very uncommon in toxicity testing, though recently Sweeney et al. (2014) did consider it in their acute inhalation toxicity testing of hydrogen cyanide (Sweeney et al. 2014, 2015).

One of the basic but critical elements in toxicokinetics is the metabolism and half-life, indicating the elimination rate of a substance and the formation of metabolites. It provides insight in the duration of effective doses within the animal and human. In case of high elimination rates (thus low half-life values), each dose or exposure on the next day is a new exposure as accumulation of the substance is not expected. This may not hold true for the metabolites, which can become very important if the metabolite is the toxic agent, rather than the mother compound, nor if long lasting adducts are formed. Lipophilic substances will tend to concentrate in adipose tissue and depending on the exposure profile may accumulate independent of the kind of exposure routes. In the absence of information on the elimination half-life, the parameter log Kow, as measure for the lipophilicity of a substance, might be used. In general, substances with a high log Kow have long biological half-lives.

Substances with a log Kow value of around 4 or higher could result in a built up of that chemical within the body (ECHA 2017).

A short-term internal exposure might be obtained in case of low elimination half-life in combination with enough space between two consecutive exposure periods. In case consumers only paint once or twice each year for 1 or 2 consecutive days, this intermittent exposure is less likely to lead to an internal long-term exposure, but given the terminal elimination half-life ($T_{1/2}$), it could be seen as short term (Geraets et al. 2016). As a rule of thumb, the majority of a substance is removed from the body after 5 times the half-life (see Sect. 3.5). An exposure after 5 times $T_{1/2}$ will not lead to accumulation of the substance (Renwick 1999a). Care should be taken though whether or not effects resulting from exposure are long-lasting, irreversible or accumulate. Therefore, knowing the mode of action is very important.

Toxicity studies are designed to evaluate potential toxicological effects, and usually the highest dose levels are chosen with the aim to induce toxicity, as prescribed by several OECD testing guidelines (OECD 1991–2016; Halle 1998). The set of parameters evaluated may give information on the type of effect and mode of action. Evaluation of the toxicodynamics gives crucial information that can be used to scrutinise possible risks. Information as to whether the effect would be concentration- or dose-dependent (which are also subject to toxicokinetics as doses/concentrations may overload normal detoxification processes) might provide valuable input for the evaluation of less than lifetime exposures (Geraets et al. 2016; Haber et al. 2016). The total dose (or AUC) is the area under the blood/plasma concentration time (C_t) curve on an internal level. It is also referred to as the toxic load. It determines the toxic effect, and the concentration (i.e. blood/plasma-concentration on an internal level) of a substance is most relevant. In some cases (e.g. methylene chloride), the toxic effects of a chemical may be driven both by the maximum concentration (C_{max}) and the AUC (Bos et al. 2006). Whether the chemical acts via a dose-dependent or concentration-dependent mechanism is important for the application of time scaling under Haber's law. Substances exerting effects that depend on the exposure concentration do not obey to Haber's law. For these chemicals, "flat lining" the concentration over time is applied, $C = k$. In case where the systemic dose is more important, Haber's rule (adapted or no) can apply. Sweeney showed that it even applied under intermittent exposure profiles (Sweeney et al. 2014, 2015). Hence, it may be concluded that the toxic load model predicts well systemic toxicity related lethality even when exposure was not continuous.

Knowing how one recovers (if not an irreversible effect) or knowing that effects can be significantly delayed is important to consider whether or not the repeated dosing and effect in animals or in human situations may in fact be accumulating or not. For example, the kidney compensates for loss of renal function due to acute exposure to a chemical and the compensatory mechanisms remain operative for a considerable period of time; these alterations can be maladaptive in the long term (Solecki et al. 2005). A note of caution as given by example: the adverse effects of organophosphates (i.e. neurotoxicity) are primarily the result of irreversible binding to the enzyme acetylcholinesterase. Due to this irreversibility, the decline in plasma organophosphate levels is steeper than the recovery and increase in acetylcholinesterase activity. This example shows the necessity of information on the kinetics and mode of action.

2.6.5 Matching Exposure and Limit Values in Risk Assessment

In most if not all regulatory frameworks, a chronic health-based guidance value needs to be derived, and under certain conditions, acute or short-term health-based guidance values are set. On the other hand, the exposure assessment generally focuses on exposure per events, per day exposures and extrapolated long-term exposure estimates. The limit value based on animal toxicity studies and the exposure assessments never perfectly match regarding the durations they supposed to cover; moreover, as a starting point generally the less than lifetime exposure is compared to a chronic-based HBGV. In those cases where risks are concluded, one must consider to refine the risk. In principle, the human exposure assessment should describe real-life exposure. Therefore, refinement of exposure would focus on the assessment if the internal exposure profile resulting from a short-term exposure profile is indicative for a long-term exposure or not. The elements to consider are the external exposure profile describing the frequency of exposure, maximum duration without exposure and elements such as the $T_{1/2}$, which are indicators for whether or not the exposure can accumulate and how long the substances will remain biologically available in the human body. If the elements indicate that the internal exposure is indeed also short term, then alternative PoD for HBGVs could be used that are more appropriate in the risk assessment (Geraets et al. 2016). It should be noted, however, that one must always consider the critical effects of the substance of interest. Certain toxic effects occur when a subject is exposed in a critical time window, e.g. certain developmental toxic effects. Consequently, the HBGV for chronic and acute durations are the same. In that case, refinement is not possible, and care should be taken for possible accumulation of the exposure.

Theoretically, (modified) Haber's rule could be applied to a chronic-based HBGV to derive a HBGV for a desired duration matching the exposure, under the condition that the critical effect does not change over time. However, animal toxicity testing for a single substance will generally not provide the means to derive the substance-specific equation, hence remaining a theoretical solution.

2.7 Epidemiology and Exposure

Matthias Greiner

Exposure assessment provides the basis for risk characterisation through direct comparison of the estimated exposure with the appropriate health-based guidance values. If information on the dose-response relationship is available, the results of exposure assessment can be used as input for the quantification of adverse health effects in the exposed population. In both cases, the aim is to characterise a risk for a defined population resulting from a defined exposure scenario. This approach follows the US National Academy of Sciences (NAS) principles for governmental risk assessment (National Research Council, NRC 1983), adopted by the CAC (1999). Since risk is a function of exposure, the latter is always included when referring to risk assessment. The quantification of population health risks in relation to exposure is the domain of epidemiology. Epidemiologic studies, in particular those that involve human biomarkers for exposure and/or adverse health effects can provide valuable information about realistic levels, sources or health effects of expose in humans under realistic conditions. Systematic reviews and meta-analyses of epidemiologic studies (referred to as SRMA from now on) have been facilitated by appropriate reporting standards for epidemiologic studies (Vandenbroucke et al. 2014). Reporting standards do also exist for SRMA studies (Moher et al. 2015). Epidemiologic studies may therefore contribute enormously to high-quality risk assessment. In the following section, the epidemiologic approach will be briefly described, and its possible application in the context of chemical safety will be illustrated using examples (Sect. 2.7.1). Differences and similarities between the epidemiologic approach and risk assessment will be elaborated with focus on the principles of causation and the role of statistical inference (Sect. 2.7.2). The concept of evidence will be outlined since it plays a central role in both areas (Sect. 2.7.3). Further comparative aspects that may support good practices in risk assessment will be outlined (Sect. 2.7.4). The aim of this section is to raise awareness that the epidemiologic approach is neither contradictory nor competing with the risk assessment approach and that useful insights can be obtained by looking at exposure assessment from an epidemiologic perspective.

2.7.1 Principles and Application Examples of Epidemiologic Studies in the Context of Chemical Safety

A distinction is to be made between experimental and observational epidemiologic studies. Randomised controlled trials (RCTs) follow an experimental design, whereas cross-sectional studies, case-control studies and cohort studies are observational.

Experimental Epidemiologic Studies

Experimental epidemiologic studies are often used to investigate therapeutic interventions and require by design that treatment be randomly allocated to the patients or study subjects ideally with blinding of participants and investigator. In order to study the effect of exposure under controlled conditions, experimentally defined levels of exposure, in analogy to a treatment factor, can be administered to the study participants. Ethical and practical considerations often preclude RCTs in exposure science due to anticipated health risks for participants and difficulties to effectively blind the exposure event. Nevertheless, RCTs are common in nutrition research and may also generate relevant evidence for dietary exposure assessment. For example, Gibson et al. (2010) have reviewed the evidence for carcinogenesis associated with food constituents such as dietary fat, meat, fibre, antioxidant nutrients and calcium provided by RCTs and cohort studies. One important result was the high degree of discrepancies between findings from experimental and observational studies. Although hierarchies of evidence have been disputed (see, e.g Borgerson 2009), RCTs have been ranked higher than cohort studies under some conditions (Howick et al. 2011). Specific recommendations for the use of RCTs have been elaborated in nutrition research (Hebert et al. 2016). The RCT may in fact be a favourable design for hazard identification and hazard characterisation, i.e. to establish a dose-response relationship. Typical limitations beyond those mentioned above include high costs and difficulties to recruit volunteers, which in turn may result in low sample sizes and selection bias. However, the biggest limitation of the RCT design for exposure assessment arises from its major design feature. The controlled exposure provides no information for quantifying the prevalence and degree of exposure in the population and little information for characterising the effect of exposure as observed under realistic conditions.

Observational Epidemiologic Studies

Cross-sectional studies, cohort studies and case-control studies are the main observational epidemiologic study designs for investigating the association between exposure and adverse health outcomes. These study designs have in common that the exposure to a risk factor and associated health effects are recorded on individual level and reflect some important conditions of occurrence as prevailing in the study population (see below for details). The fact that the event of exposure is not experimentally controlled is an important distinction from experimental epidemiologic studies. Excellent textbooks are available that describe the properties of epidemiologic study designs, their pros and cons and the measures they provide (e.g. Rothman et al. 2008a). These properties hold true irrespectively of the actual case definitions for the "risk factors" and the "outcomes". It is noted that neither terminology nor study methodology implies a causal relationship between risk factor and outcome. Using these generic terms, some distinctions should be made as to

Table 2.4 Examples for various definitions of risk factor and outcomes in relation to a given research question[a]

Research question	Risk factor(s)[a]	Outcome	Example
Does the exposure to the agent constitute a health risk (hazard identification)?	Exposure to an agent	Adverse health effect or biomarker for the health effect	Systematic review of neurodevelopmental and/or neurobehavioural effects in children due to PBDE and PFC exposure (Roth and Wilks 2014)
Does the behaviour constitute a health risk (hazard identification of the agent only if exposure from such sources is well established)? To what extend can the occurrences of adverse health outcomes in the population be attributed to various recorded behaviours?	Consumption of food, use of product or other behaviours that can be linked with exposure to the agent	Adverse health effect or biomarker for the health effect	Cohort study for exploration of relationships between diet and chronic diseases in women in the UK (Cade et al. 2015) Cohort study for estimation of the population attributable fraction of adherence to Mediterranean diet in the UK (Tong et al. 2016)
Is the exposure level of the agent correlated with the probability of the health effect (hazard identification or hazard characterisation)?	Quantitative level of exposure to an agent (external exposure) or biomarker for exposure (internal exposure)	Adverse health effect or biomarker for the health effect	Systematic review and meta-analysis: association between maternal caffeine intake and risk of pregnancy loss (Chen et al. 2016)
To what extend do the behaviours contribute to the internal exposure to the agent?	Consumption of food, use of product or other behaviours that can be linked with exposure	Quantitative level of biomarker for exposure to an agent (internal exposure)	Cohort study of blood levels of arsenic, cadmium, lead, total mercury and selenium in relation to sociodemographic factors, diet and living habits in Finland (Abass et al. 2017)

[a]Besides the relevant risk factors for the given research question, other factors are considered simultaneously to account for confounding, effect modification or interaction

characterise possible purposes of epidemiological studies in exposure science (Table 2.4). The examples in this chapter are provided for illustrative purposes and not selected based on any assessment of the study quality. The choice of a study design depends on the given research question, in particular on the prevalence and the nature of the exposure and the adverse health outcome. Study design aspects have also been discussed in nutritional epidemiology (Margetts and Nelson 1997), occupational health (Checkoway et al. 2007) and environmental epidemiology (Morgenstern and Thomas 1993). Some basic properties of the study types and basic measures of occurrence and association are sketched below. It is emphasised that multivariable statistical modelling is necessary to derive valid estimates of epidemiological measures that account for confounding and other effects. This typically leads to the use of generalised linear models such as multivariable logistic regression. For more details on the statistical methods, the reader is referred to the cited application examples and epidemiologic textbooks. If designed, conducted, analysed, interpreted and reported according to established standards (Vandenbroucke et al. 2014), epidemiologic studies are able to provide valuable information that can be integrated with other sources of information using a weight of evidence approach (Adami et al. 2011).

A comprehensive account of the contribution of epidemiology to risk assessment has been given by van den Brandt et al. (2002). It is commonly understood that the observational epidemiologic studies may inform the hazard identification step. Epidemiologic studies have been used extensively to investigate chemical hazards associated with occupational or environmental exposures. Dietary risk factors for health effects have been studied in nutrition research. Epidemiologic studies including those that use biomarkers for exposure can provide evidence for hazard characterisation, whereby substances with short physiologic half-lives are a challenge (LaKind et al. 2014). Biomonitoring data are ideal markers for aggregate exposure provided that the single (time) point concentration of the marker in the bio sample well captures multiple exposure events of the substance under evaluation via different pathways (Sect. 3.5). A more refined exposure evaluation is typically required in the area of consumer safety. For this application, food consumption or use of products has to be evaluated on individual level in order to be able to attribute adverse health effects to the specific exposure. The challenges of sampling and collection of the required information in epidemiologic studies on consumer safety have been studied by Papadopoulou et al. (2016).

Cross-Sectional Studies

Cross-sectional studies use a representative sample from the study population defined in terms of demographic features (e.g. gender and age groups), time (sampling time point or period) and space (study region). The study participants are not selected for their exposure status or health outcomes. Therefore, representativeness can be presupposed with regard to occurrence of exposure (pathways, prevalence, exposure levels) and adverse health outcome (prevalence, severity levels) as well as for other factors (e.g. other risk factors, comorbidity) unless selection biases are in operation. For example, Zota et al. (2016) used a cross-sectional design to study the association between biomarkers for bisphenol A and phthalates and fast-food intake in the US population. Prevalence of exposure and disease as well as epidemiologic measures of association can be estimated from cross-tabulated dichotomous study results. Prevalence odds ratio (Greenland et al. 2008) and prevalence ratio, estimated using Poisson regression and log-binomial regression models (Barros and Hirakata 2003), are common measures of association used in cross-sectional studies. A typical difficulty in a cross-sectional study is to assess the chronologic order of exposure and disease events, which limits the support of a causal interpretation. Biases may occur due to the fact that cases with long duration of illness are expected to be overrepresented in a cross-sectional sample and vice versa (Rothman et al. 2008b). Cross-sectional studies are relatively inexpensive and fast and thus effective if exposure and disease are not rare in the population and representativeness of the study population is warranted.

Cohort Studies

The participants of cohort studies are recruited from the study population, which can be the general population or specific subgroups depending on the research question. At the time of their enrolment into the study, the participants are free from the diseases under consideration. For example, Cade et al. (2015) recruited apparently healthy women into vegetarian, fish-eating and meat-eating cohorts to study the association between dietary preferences and the occurrence of cancer and other chronic diseases in women. Cohort studies are prospective by design and thus allow recording of cases that become incident during the observation period. In contrast to cross-sectional studies, which can only evaluate the co-occurrence of indicators for exposures and disease at one given time point, the exposure status can be established in a cohort study before the recorded disease outcome occurs in study participants. This allows the direct estimation of a relative risk, i.e. the ratio of the number of new cases per person-months at risk in exposed and nonexposed persons, respectively.

Case-Control Studies

An efficient design to study rare diseases and common exposures is the case control study. Cases are selected from the target population using appropriate case definition and selection criteria. For each case, one or more controls are selected from the same population with matching on demographic or socioeconomic factors. Case-control studies have been conceptualised as a more efficient version of a corresponding cohort study in the sense that cases in a case-control study should be the same people who would have been considered in a cohort study of the same population (Rothman et al. 2008b). Case-control studies are not able to provide a direct estimate of a relative risk. However, the odds ratio for exposure estimated in a case-control study is a good approximation of the relative risk if the prevalence of the disease condition is low (Hogue et al. 1983). For example, Den Hond et al. (2015) estimated an odds ratio for exposure to endocrine disrupting chemicals in men with (cases) and without (control) reduced sperm quality recruited through four fertility clinics.

2.7.2 *Differences and Similarities Between the Epidemiologic Approach and Scenario Modelling*

Causation: Goal of Inference Versus Model Requirement

Epidemiologic studies can be used for both explorative and inferential purposes in the sense of generating and testing hypotheses, respectively, about associations between exposure to a risk factor and putative health effect. Depending on the type of study, epidemiological studies may provide various measures of association between exposure to a risk factor and a disease outcome such as odds ratio or relative risk (see above). Using a set of criteria (e.g. Bradford-Hill) and under consideration of study design and quality, such associations can be interpreted in terms of causality. Parascandola and Weed (2001) have reviewed concepts of causation in the epidemiologic literature and proposed to differentiate conditions referred to as (a) "production" and that play essential parts in producing the occurrence of disease; (b) "necessary cause", without which the effect cannot occur; (c) "sufficient-component cause" that guarantees that its effect will occur when the cause is present and which is made up of a number of components, no one of which is sufficient on its own but which taken together make up a sufficient cause; (d) a "probabilistic cause", which is neither required to be necessary nor sufficient but increases the probability of its effect; and (e) a "counterfactual" cause, which is neither required to be necessary nor sufficient but makes a difference in the outcome (or the probability of the outcome) when it is present, compared with when it is absent. Risk assessment, on the other hand, requires prior evidence for causation and generally does not contribute any evidence for causation. Information and uncertainties about the relationship between cause and effect may be expressed in the hazard characterisation part, for example, when characterising a dose-response function. An important point to make here is that a p-value for testing an exposure effect should by no means be misinterpreted as a quantitative expression of likelihood for a risk factor being causal. Epidemiologic studies may be "overpowered" in the sense that very small effects are statistically significant although the size of effect has no biological significance. In fact, p-values and hypothesis testing can be omitted from the further discussion as explained in the following. Table 2.5 shows possible explanations for

Table 2.5 Possible explanations for causal relationship not being reflected by significant statistical association and vice versa

Causal relationship	Statistical association	Possible explanation
Yes	No	Sample size too small (underpowered) Misclassification of exposure or health effect Confounding
No	Yes	Sample size too large (overpowered) Statistical significance misinterpreted as biological significance Confounding

causal relationship not being reflected by significant statistical association and vice versa.

Statistical Inference: Changing the Perspective May Be Fruitful

Risk assessment may benefit from both the research results and methodological foundations of epidemiology. Conversely, a risk assessment with a clear risk question may clarify the fitness-for-purpose requirements of epidemiologic research output that is to be used as input into a risk assessment. In modern epidemiology, the emphasis is on estimation and on characterising the estimate along with its statistical uncertainty (Greenland et al. 2016). Estimation rather than hypothesis testing provides a strong conceptual link between the areas of epidemiology and risk assessment. To illustrate this and to understand the different perspectives on uncertainty, it is useful to consider population parameters that are typical input quantities in risk models such as prevalence (e.g. the proportion of the population that consume a certain food stuff) or concentration of a chemical hazard in the foodstuff. Prevalence and concentration are estimated in observational studies, for instance, in a cross-sectional consumption survey and in a food monitoring programme, respectively. Internal and external validity are key concepts in epidemiology (Rothman 2008) and address two important questions: Does the estimate actually reflect the quantity of interest in the study population or does it provide a biased reflection of the hypothetical true quantity due to sampling biases, measurement biases or inadequate statistical analysis (internal validity)? Can the estimate, provided that it is internally valid, be generalised to some other population which is different from the original study population (external validity)? The judgement of these questions is possible using information derived from the meta-data of the study results, covariate and documentation of the analysis. A few key aspects without claiming completeness are elaborated below to illustrate how concepts of epidemiology can inform the risk assessment. A direct link between the two areas occurs when using population parameters as input in risk models, whereby "problematic" properties of estimates such as bias and inaccuracy are inherited by the risk assessment and feature there in the form of model flaws or uncertainties:

- Sampling design issues: Various biases may be in operation, such as inappropriate sampling frame, inappropriate study design, selection bias, referral bias, spectrum of disease bias or intentional bias due to risk-based sampling. Ignoring the presence of such biases may invalidate the results of risk assessment. Potential biases can be addressed in the form of an uncertainty analysis (EFSA 2016; BfR 2015).
- Small sample size: Lack of precision in parameter estimates and large uncertainty intervals are the results of small sample sizes in observational studies. The impact on risk assessment is a wide uncertainty range of the outcome due to parameter uncertainty. High percentiles or extreme values are of particular interest in risk assessment and cannot be estimated reliably based on small samples. For

example, estimation of a 95th percentile (P95) of the individual daily intake of rice based on data from 100 consumers depends on a single observation.
- Large sample size: High precision of the parameter estimates and a small statistical uncertainty interval are the results of large sample sizes. The impact is a small uncertainty range of the outcome of the risk assessment due to small parameter uncertainty. This may lead to wrong conclusions about association parameters when confusing statistical significance and biological relevance. In explorative observational studies, it may be overlooked that a large sample may be available by convenience and still be compromised through a flawed sampling design or lack of external validity.
- Measurement bias or information bias: The quantity estimated in the observational study may be biased due to imperfect measurement or diagnostics. Consider, for example, the limit of detection, limit of quantification, analytical errors, diagnostic sensitivity and specificity. The impact is a biased outcome of risk assessment. Ignoring the presence of such biases and lack of adjustment for such biases may invalidate the results of risk assessment. Potential biases can be addressed in the form of an uncertainty analysis.
- Lack of external validity: Depends on differences between the study population and the target population. The impact is that the outcome of the risk assessment may not apply to the target population depending on the risk question and relevance of the difference

2.7.3 Epidemiology, Risk Assessment and Evidence

Evidence is information available for the purpose of answering a specific research question or management question that requires scientific support. The term evidence-based implies that some formal methodology has been used for the collection and appraisal of information. Evidence-based methods may result in new knowledge by extracting information from multiple primary sources, e.g. published studies, which is not apparent in any individual primary source. Systematic reviews and meta-analyses of epidemiologic studies are therefore considered original research. Weight of evidence assessment involves some form of integrating information from multiple sources that may draw from different types of studies and different scientific areas such as molecular biology, toxicology and epidemiology. An overview of and guidance for applying weight of evidence in various areas of risk assessment have been elaborated recently (EFSA 2017b).

Systematic reviews and meta-analyses (SRMA) of epidemiologic studies and risk assessments are conducted for the purpose of answering a clearly framed question, often within a limited time. These features are typical in application areas for evidence-based approaches. Original research papers in epidemiology provide new evidence and, supported by additional evidence, may lead to new knowledge. SRMA for combining evidence from individual primary studies may generate new knowledge as well. However, perhaps even more importantly, the evidence-based approaches can be seen as best practice for decision support in terms of their transparency, accountability and scientific quality. This holds true for risk assessment as well, which generally does not provide new evidence but may lead to a better understanding. In this regard and similar to SRMA of epidemiologic studies, risk assessment may generate new knowledge. Risk assessment can be seen as one example for a weight of evidence approach of using complementary lines of evidence (EFSA 2017b). Conceptualising risk assessment and epidemiological studies from the viewpoint of weight of evidence allows a better insight into the quality requirements for and the intersection between the two methodologies. Reliability, relevance and consistency are key criteria in weight of evidence assessment and have important connotations in epidemiologic research and risk assessment. The terms have been defined as the extent to which information is correct (reliability), as contribution the information, given it was reliable, could make to answer a specified question (relevance) and as the extent to which information from different sources, given that it was reliably and relevant, are compatible (consistency) (EFSA 2017a).

Reliability

Adherence to guidelines for the conduct and reporting of epidemiologic studies (Vandenbroucke et al. 2014) allows that information can be extracted from original research papers and used for SRMA or in risk assessment. This also supports an appropriate judgement of "quality aspects" such as lack of independence/conflict of interest, sampling biases, selection bias, measurement bias, information bias, lack of precision, lack of repeatability and reproducibility. Reliability issues may be addressed by exclusion of the study/information, statistical adjustment in some cases and through uncertainty analysis in risk assessment.

Relevance

If observational studies are reported according to pertinent reporting standards, the relevance of the information can be judged with regard to the specified question of the risk assessment. Thus, the relevance depends on what has been studied. Relevance does not depend on actual results or outcomes of the study and in general does not even depend on how a result was obtained, which is an aspect of reliability. In epidemiologic studies, it is important to consider whether the case definitions and measurements provide relevant information for the given research question. In risk assessment, the choice of a scenario, the elements of the scenario and data that inform model parameters should be discussed under consideration of relevance.

Consistency

This criterion relates to the actual outcome of a study and should be assessed independently from reliability and relevance. Inconsistent results should be assessed using exactly similar criteria as the "consistent" results. Information that is not compatible with other relevant and reliable information may receive less weight in the integration step. Great care should be taken because generally speaking, exclusion or weighing should in principle be independent from the study results. In some situations, the study results may be seen as an indication of compromised reliability or relevance cannot be identified such that the inconsistency of the outcome is the only indication that such limitations actually apply. This situation may be avoided through better adherence to reporting standards in the first place and require a case-by-case expert judgement. Accounting for inconsistency should be supported by statistical methodology that in principle treats all information equally. Subjectivity in rejecting unusual or undesired outcomes should be avoided.

2.7.4 Further Comparative Aspects that May Support Good Practices in Risk Assessment

Ecologic Fallacy

Epidemiologic studies that use aggregate measures (e.g. means or proportions of individual observations, environmental or global measures that apply to a group of persons) are referred to as "ecologic studies" and are prone to important methodological limitations known as "ecologic fallacy" (Hal Morgenstern 2008). One well-known example relates to the association between the introduction of aspartame as artificial sweetener and increased reported incidences of brain tumours (reviewed by Weihrauch and Diehl 2004). Parallel trends over time in disease rate and putative risk factor are not sufficient to indicate causality as described above. The important point to recognise is that individual-level data are required for biological inference (Hal Morgenstern 2008).

Risk assessment, although not used for causal inference, has similar features as an ecologic study. Consider, for example, a food monitoring programme and a consumption survey as typical data sources in exposure assessment for estimating the occurrence of a hazardous agent in food and the frequency and amount of food intake, respectively. This is aggregated data in the sense that it has not been observed at the same individual level. The specific variant of the ecologic fallacy in exposure assessment arises from the fact that consumption is informed by individual-level data, while external exposure is estimated via occurrence data that is not individual level. A correlation between food intake and concentration of the hazard may be plausible. For example, consumers who strictly adhere to a specific diet may differ from an average consumer in terms of both intake and sourcing of certain food categories, whereby concentration profiles might be affected by the source of the food. Consequently, the exposure of high-level consumers might be underestimated. Duplicate diet studies are suitable to estimate not only the intake quantities but also the individually consumed concentration of the chemical under consideration. Both components, occurrence and intake, are observed at the individual level (see, e.g. Papadopoulou et al. 2016). Such studies are expensive and therefore very rare. Likewise, exposure assessment for multiple pathways (dietary, environmental, occupational) is rarely supported by appropriate studies. Another variant of the ecologic fallacy in risk assessment is due to the use of separate data sources for estimation of exposure and hazard characterisation. Epidemiologic studies that use biomarkers for the health effect and combine an extensive questionnaire for exposure information and a duplicate diet approach would be ideal to provide the required information and thereby overcome the ecologic fallacy in risk assessment.

Survey Sampling and Analysis

A wide range of approaches and statistical methods are available that support the investigator in making inferences that are valid for the specific sampling method. Basic principles, properties of simple random sampling, systematic sampling, stratification and cluster sampling along with appropriate statistical methodology are described by Levy and Lemeshow (2011). It should be noted that the statistical concepts of populations are not at all restricted to human populations and apply as well to food, products and hazardous agents. Estimates of population parameters are input quantities in risk assessment models. Therefore, it is beneficial for the risk assessor, to be aware of the appropriate methodology for survey sampling and analysis.

Bias and Uncertainty

Systematic errors (biases) of epidemiologic studies are of paramount interest in epidemiologic research. A good overview has been provided by Greenland and Lash (2008), who also make the point that systematic errors typically outweigh the random errors. The authors also describe the link to uncertainty analysis. Typical biases in epidemiologic studies include the bias through unmeasured confounders, selection bias, measurement bias and misclassification bias. For example, in the ideal observational study, accurate and precise measurements of exposure and disease outcome would be available for each individual participant. In practice, this information is often derived from questionnaires, interview or other surrogate sources of information. Misclassification occurs if the dichotomous (yes/no) classification of exposure and disease is subject to diagnostic error. The latter can be expressed in terms of the diagnostic accuracy. In both areas, epidemiology and risk assessment, similar strategies are applicable to deal with biases: identification, adjustment if possible, sensitivity analysis and transparent description. In risk assessment, this would be considered as part of uncertainty analysis.

References

Abass K, Koiranen M, Mazej D, Tratnik JS, Horvat M, Hakkola J, Järvelin MR, Rautio A (2017) Arsenic, cadmium, lead and mercury levels in blood of Finnish adults and their relation to diet, lifestyle habits and sociodemographic variables. Environ Sci Pollut Res 24(2):1347–1362. https://doi.org/10.1007/s11356-016-7824-5. Accessed 27 Oct 2019

Adami HO, Berry SCL, Breckenridge CB, Smith LL, Swenberg JA, Trichopoulos D, Weiss NS, Pastoor TP (2011) Toxicology and epidemiology: improving the science with a framework for combining toxicological and epidemiological evidence to establish causal inference. Toxicol Sci 122(2):223–234. https://doi.org/10.1093/toxsci/kfr113. Accessed 27 Oct 2019

AEGL (2001) Standing operating procedures for developing acute exposure guideline levels for hazardous chemicals. National Academy of Sciences., Washington, DC

Baker M (2016) 1,500 scientists lift the lid on reproducibility. Nature 533:452–454

Barros AJ, Hirakata VN (2003) Alternatives for logistic regression in cross-sectional studies: an empirical comparison of models that directly estimate the prevalence ratio. BMC Med Res Methodol 3(1):21. https://doi.org/10.1186/1471-2288-3-21. Accessed 27 Oct 2019

Batke M, Escher S, Hoffmann-Doerr S, Melber C, Messinger H, Mangelsdorf I (2011) Evaluation of time extrapolation factors based on the database RepDose. Toxicol Lett 205(2):122–129

Bauch C, Kolle SN, Ramirez T, Eltze T, Fabian E, Mehling A, Teubner W, van Ravenzwaay B, Landsiedel R (2012) Putting the parts together: combining in vitro methods to test for skin sensitizing potentials. Regul Toxicol Pharmacol 63(3):489–504

Belkebir E, Rousselle C, Duboudin C, Bodin L, Bonvallot N (2011) Habers rule duration adjustments should not be used systematically for risk assessment in public health decision-making. Toxicol Lett 204(2–3):148–155

BfR (2014) 3rd updated and revised edition of the EU Food Safety Almanac. http://www.bfr.bund.de/en/search.html?search%5Bquery%5D=food+safety+almanac. Accessed 27 Oct 2019

BfR (2015) Guidelines on uncertainty analysis in exposure assessments. Recommendation of the committee for exposure assessment and exposure standardisation of the Federal Institute for Risk Assessment (BfR). Available at: http://www.bfr.bund.de/cm/350/guidelines-on-uncertainty-analysis-in-exposure-assessments.pdf. Accessed 27 Oct 2019

BfR (2016) Risk communication. http://www.bfr.bund.de/en/risk_communication-1834.html. Accessed 28 Oct 2019

Bokkers BG, Slob W (2005) A comparison of ratio distributions based on the NOAEL and the benchmark approach for subchronic-to-chronic extrapolation. Toxicol Sci 85(2):1033–1040

Borgerson K (2009) Valuing evidence bias and the evidence hierarchy of evidence-based medicine. Perspect Biol Med 52(2):218–233

Bos PM, Zeilmaker MJ, van Eijkeren JCH (2006) Application of physiologically based pharmacokinetic modeling in setting acute exposure guideline levels for methylene chloride. Toxicol Sci 91(2):576–585

Bourguignon D (2015) The precautionary principle: definitions, applications and governance. European Parliamentary Research Service, European Union, December 2015—PE 573.876. isbn:978-92-823-8480-0; https://doi.org/10.2861/821468. Accessed 27 Oct 2019

Boxenbaum H (1982) Interspecies scaling, allometry, physiological time, and the ground plan of pharmacokinetics. J Pharmacokinet Biopharm 10:201–227

Bruinen de Bruin Y, Hakkinen PB, Lahaniatis M, Papameletiou D, Del Pozo C, Reina V, Van Engelen J, Heinemeyer G, Viso AC, Rodriguez C, Jantunen M (2007) Risk management measures for chemicals in consumer products: documentation, assessment, and communication across the supply chain. J Expo Sci Environ Epidemiol 17(Suppl 1):S55–S66

C8 Panel (2017) The C8 panel website. http://www.c8sciencepanel.org/. Accessed 27 Oct 2019

CAC (1999) Principles and guidelines for the conduct of microbiological risk assessment (CAC/GL 30-1999, adopted 1999, amendments 2012, 2014)

Cade JE, Burley VJ, Alwan NA, Hutchinson J, Hancock N, Morris MA, Threapleton DE, Greenwood DC (2015) Cohort profile: the UK women's cohort study (UKWCS). Int J Epidemiol 5:567–587. https://doi.org/10.1093/ije/dyv173. Accessed 27 Oct 2019

CDC (2016) Gateway to health communication & social marketing practice. http://www.cdc.gov/healthcommunication/Risks/index.html

Checkoway H, Pearce N, Kriebel D (2007) Selecting appropriate study designs to address specific research questions in occupational epidemiology. Occup Environ Med 64(9):633–638

Chen L-W, Wu Y, Neelakantan N, Chong MF-F, Pan A, van Dam RM (2016) Maternal caffeine intake during pregnancy and risk of pregnancy loss: a categorical and dose–response meta-analysis of prospective studies. Public Health Nutr 19(7):1233–1244. https://doi.org/10.1017/S1368980015002463. Accessed 27 Oct 2019

Covello VT (1992) Risk communication: an emerging area of health communication research. In: Deetz SA (ed) Communication yearbook 15. Sage, Newbury Park, CA, pp 359–373

Dekant W (2016) Toxicology and the reproducibility crisis: Scientific publishing, hazard assessment and risk characterisation. Toxicol Lett 263:76–77

Den Hond E, Tournaye H, De Sutter P, Ombelet W, Baeyens W, Covaci A, Cox B, Nawrot TS, Van Larebeke N, D'Hooghe T (2015) Human exposure to endocrine disrupting chemicals and fertility: a case-control study in male subfertility patients. Environ Int 84:154–160. https://doi.org/10.1016/j.envint.2015.07.017. Accessed 27 Oct 2019

ECHA (2012) Guidance on information requirements and chemical safety assessment Chapter R.8: characterisation of dose [concentration]-response for human health https://echa.europa.eu/documents/10162/13632/information_requirements_r8_en.pdf. Accessed 27 Oct 2019

ECHA (2014) Guidance on information requirements and chemical safety assessment Chapter E. https://echa.europa.eu/documents/10162/13632/information_requirements_part_e_en.pdf. Accessed 27 Oct 2019

ECHA (2017) Guidance on the biocidal products regulation volume III: human health, part b: risk assessment. Version 1.1, April 2015. https://echa.europa.eu/documents/10162/23036412/biocides_guidance_human_health_ra_iii_part_bc_en.pdf/30d53d7d-9723-7db4-357a-ca68739f5094. Accessed 27 Oct 2019

EEA (2001) Late lessons from early warnings: the precautionary principle 1896–2000. Environmental issue report No 22, EEA (European Environmental Agency) Copenhagen, 211 pp. https://www.eea.europa.eu/publications/environmental_issue_report_2001_22. Accessed 8 Mar 2018

EEA (2013) Late lessons from early warnings: science, precaution, innovation. Summary. EEA (European Environmental Agency) Report No 1/2013, EEA (European Environmental Agency) Copenhagen, 44 pp. https://www.eea.europa.eu/publications/late-lessons-2. Accessed 8 Mar 2018

EFSA (2007) Opinion of the Scientific Panel on Plant protection products and their Residues on a request from the commission on the Acute Reference Dose (ARfD) for imazalil. (Question N° EFSA-Q-2006-202) adopted on 7 March 2007. https://efsa.onlinelibrary.wiley.com/doi/10.2903/j.efsa.2007.460. Accessed 27 Oct 2019

EFSA (2009a) Scientific opinion of the panel on contaminants in the food chain on a request from the European Commission on cadmium in food. EFSA J 980:1–139. http://www.efsa.europa.eu/en/efsajournal/pub/980. Accessed 27 Oct 2019

EFSA (2009b) Use of the benchmark dose approach in risk assessment. Guidance of the Scientific Committee. EFSA J 1150:1–72. http://onlinelibrary.wiley.com/doi/10.2903/j.efsa.2009.1150/abstract. Accessed 8 Mar 2018

EFSA (2010a) Statement on the safety evaluation of smoke flavourings primary products: interpretation of the margin of safety EFSA. EFSA J 8(1):1325. http://onlinelibrary.wiley.com/doi/10.2903/j.efsa.2009.1325/full. Accessed 27 Oct 2019

EFSA (2010b) Scientific opinion on lead in food. EFSA J 8:1570, 1147 pp. https://doi.org/10.2903/j.efsa.2010.1570. http://onlinelibrary.wiley.com/doi/10.2903/j.efsa.2010.1570/full. Accessed 27 Oct 2019

EFSA (2012a) Scientific committee guidance on selected default values to be used by the EFSA scientific committee, scientific panels and units in the absence of actual measured data. EFSA J: 10(3):2579. http://onlinelibrary.wiley.com/doi/10.2903/j.efsa.2012.2579/abstract. Accessed 27 Oct 2019

EFSA (2012b) Scientific opinion on exploring options for providing advice about possible human health risks based on the concept of Threshold of Toxicological Concern (TTC). EFSA J;10 (7):2750 http://onlinelibrary.wiley.com/doi/10.2903/j.efsa.2012.2750/abstract. Accessed 8 Mar 2018

EFSA (2014) Modern methodologies and tools for human hazard assessment of chemicals. EFSA J 12(4):3638. http://onlinelibrary.wiley.com/doi/10.2903/j.efsa.2014.3638/full. Accessed 8 Mar 2018

EFSA (2015) Scientific opinion on the re-evaluation of sorbic acid (E 200), potassium sorbate (E 202) and calcium sorbate (E 203) as food additives. EFSA J 13(6): 4144. http://onlinelibrary.wiley.com/doi/10.2903/j.efsa.2015.4144/full. Accessed 8 Mar 2018

EFSA (2016) Guidance on uncertainty in scientific assessment. Revised Draft. https://www.efsa.europa.eu/sites/default/files/160321DraftGDUncertaintyInScientificAssessment.pdf. Accessed 8 Mar 2018

EFSA (2017a) Update: guidance on the use of the benchmark dose approach in risk assessment. EFSA J;15(1):4658, 41 pp. http://onlinelibrary.wiley.com/doi/10.2903/j.efsa.2017.4658/full. Accessed 8 Mar 2018

EFSA (2017b) Guidance on the use of the weight of evidence approach in scientific assessments. EFSA J;15(8): 4971, 69 pp. https://www.efsa.europa.eu/en/efsajournal/pub/4971. Accessed 8 Mar 2018

EFSA (2019) Guidance on the use of the threshold of toxicological concern approach in food safety assessment. EFSA J 17:4. https://www.efsa.europa.eu/en/consultations/call/181112

EU (2000) Communication from the Commission about the precautionary principle Brussels, 02 February 2000 COM. http://eur-lex.europa.eu/legal-content/EN/TXT/HTML/?uri=CELEX:52000DC0001&from=EN. Accessed 8 Mar 2018

Fei C, McLaughlin JK, Lipworth L, Olsen J (2008) Prenatal exposure to perfluorooctanoate (PFOA) and perfluorooctanesulfonate (PFOS) and maternally reported developmental milestones in infancy. Environ Health Perspect 116(10):1391–1395. https://doi.org/10.1289/ehp.11277

Geraets L, Nijkamp MM, ter Burg W (2016) Critical elements for human health risk assessment of less than lifetime exposures. Regul Toxicol Pharmacol 81:362–371

Gibson TM, Ferrucci LM, Tangrea JA, Schatzkin A (2010) Epidemiological and clinical studies of nutrition. Semin Oncol 37(3):282–296

Greenland S, Lash TL (2008) Bias analysis. In: Rothman KJ, Greenland S, Lash TL (eds) Modern Epidemiology, 3rd edn. Lippincott, Williams & Wilkins, Philadelphia, PA, pp 345–380

Greenland S, Rothman KJ, Lash T (2008) Measures of effect and measures of association. In: Rothman KJ, Greenland S, Lash TL (eds) Modern epidemiology, 3rd edn. Lippincott, Williams & Wilkins, Philadelphia, PA, pp 51–70

Greenland S, Senn SJ, Rothman KJ, Carlin JB, Poole C, Goodman SN, Altman DG (2016) Statistical tests, P values, confidence intervals, and power: a guide to misinterpretations. Eur J Epidemiol 31(4):337–350. https://doi.org/10.1007/s10654-016-0149-3. Accessed 27 Oct 2019

Haber F (1924) Zur Geschichte des Gaskrieges. In: Fünf Vorträge aus den Jahren 1920–1923. Verlag von Julius Sprenger, Berlin, pp 23–29

Haber LT, Sandhu R, Li-Muller A, Mohapatra A, Petrovic S, Meek ME (2016) Framework for human health risk assessment of non-cancer effects resulting from short-duration and intermittent exposures to chemicals. J Appl Toxicol 36:1077–1089

Halle W (1998) Toxicity tests in cell cultures for the purpose of predicting acute toxicity (LD50) and reducing the number of animal experiments. Schr Forschungszent Juelich, Lebenswiss/Life Sci 1:1–92

Haritz M (2011) An inconvenient deliberation: the precautionary principles contribution to the uncertainties surrounding climate change liability. In: Climate change law, policy, and practice series, vol 5. Wolters Kluwer Law & Business, Alphen aan den Rijn, pp 391–437

Hébert JR, Frongillo EA, Adams SA, Turner-McGrievy GM, Hurley TG, Miller DR, Ockene IS (2016) Perspective: randomized controlled trials are not a panacea for diet-related research. Adv Nutr 7(3):423–432

Hogue CJ, Gaylor DW, Schulz KF (1983) Estimators of relative risk for case-control studies. Am J Epidemiol 118(3):396–407. Retrieved from http://www.ncbi.nlm.nih.gov/pubmed/6613982

Howick J, Chalmers I, Glasziou P, Greenhalgh T, Heneghan C, Liberati A, Moschetti I, Phillips B, Thornton H, Goddard Hodgkinson M (2011) Oxford Centre for evidence-based medicine 2011 levels of evidence. Retrieved January 31, 2017, from http://www.cebm.net/index.aspx?o=5653

IPCC (2008) Climate change 2007: synthesis report. Core Writing Team, Pachauri RK, Reisinger A (eds) Contribution of working groups I, II and III to the fourth assessment report of the intergovernmental panel on climate change. IPCC, Geneva, 104 pp. isbn:92-9169-122-4

Johansen K, Brasseur D, MacDonald N, Nohynek H, Vandeputte J, Wood D, Neels P (2016) Scientific Committee of International Alliance for Biological Standardisation. Where are we in our understanding of the association between narcolepsy and one of the 2009 adjuvanted influenza A (H1N1) vaccines? Biologicals 44(4):276–280

Kramer HJ, van den Ham WA, Slob W, Pieters MN (1996) Conversion factors estimating indicative chronic no-observed-adverse-effect levels from short-term toxicity data. Regul Toxicol Pharmacol 23(3):249–255

Kroes R, Galli C, Munro I, Schilter B, Tran L, Walker R, Würtzen G (2000) Threshold of toxicological concern for chemical substances present in the diet: a practical tool for assessing the need for toxicity testing. Food Chem Toxicol 38(2–3):255–312

Kroes R, Kleiner J, Renwick A (2005) The threshold of toxicological concern concept in risk assessment. Toxicol Sci 86(2):226–230

LaKind JS, Sobus JR, Goodman M, Barr DB, Fürst P, Albertini RJ, Weisel CP (2014) A proposal for assessing study quality: biomonitoring, environmental epidemiology, and short-lived chemicals (BEES-C) instrument. Environ Int 73:195–207. https://doi.org/10.1016/j.envint.2014.07.011

Levy PS, Lemeshow S (2011) Sampling of populations: methods and applications, 4th edn. Wiley, Hoboken, NJ. https://doi.org/10.1002/9780470374597

Lutter R, Abbott L, Becker R, Borgert C, Bradley A, Charnley G, Dudley S, Felsot A, Golden N, Gray G, Juberg D, Mitchell M, Rachman N, Rhomberg L, Solomon K, Sundlof S, Willett K (2015) Improving weight of evidence approaches to chemical evaluations. Risk Anal 35:186–192

Margetts BM, Nelson M (1997) Design Concepts in nutritional epidemiology. Oxford University Press, New York. Retrieved from https://books.google.de/books?hl=de&lr=&id=ep3FAgAAQBAJ&oi=fnd&pg=PR7&dq=epidemiologic+study+designs+nutrition&ots=QUbB5zzEc3&sig=DD-sUExilaSKqmgEYeEwaAPtIao#v=onepage&q=epidemiologic study designs nutrition&f=false

Moher D, Shamseer L, Clarke M, Ghersi D, Liberati A, Petticrew M, Stewart LA (2015) Preferred reporting items for systematic review and meta-analysis protocols (PRISMA-P) 2015 statement. Syst Rev 4(1):1. https://doi.org/10.1186/2046-4053-4-1

Morgenstern H (2008) Ecologic studies. In: Rothman KJ, Greenland S, Lash T (eds) Modern epidemiology, 3rd edn. Lippincott, Williams & Wilkins, Philadelphia, PA, pp 509–531

Morgenstern H, Thomas D (1993) Principles of study design in environmental epidemiology. Environ Health Perspect 101(Suppl 4):23–38

Nordström Joenssen U, Bossi R, Leffers H, Astrup Jensen A, Skakkebæk NE, Jørgensen N (2009) Do perfluoroalkyl compounds impair human semen quality? Environ Health Perspect 117 (6):923–927. https://doi.org/10.1289/ehp.0800517

NRC (1983) Risk assessment in the federal government: managing the process, environment: science and policy for sustainable development. National Academy Press, Washington, DC. https://doi.org/10.1080/00139157.1983.9931232

OECD (1981a) Guideline for testing of chemicals. Repeated dose dermal toxicity: 90-day study. TG 411. http://www.oecd-ilibrary.org/environment/test-no-411-subchronic-dermal-toxicity-90-day-study_9789264070769-en. Accessed 27 Oct 2019

OECD (1981b) Guideline for testing of chemicals. Repeated dose dermal toxicity: 21/28-day study. TG 410. http://www.oecd-ilibrary.org/content/book/9789264070745-en. Accessed 27 Oct 2019

OECD (1991–2016) OECD guidelines for the testing of chemicals, section 4: health effects

OECD (1998) Guideline for testing of chemicals. Repeated dose 90-day oral toxicity study in rodents. TG 408. http://www.oecd-ilibrary.org/content/book/9789264070707-en. Accessed 27 Oct 2019

OECD (2001) Guideline for testing of chemicals. Prenatal developmental toxicity study. TG 414. http://www.oecd-ilibrary.org/content/book/9789264070820-en. Accessed 27 Oct 2019

OECD (2008) Guideline for testing of chemicals. Repeated dose 28-day oral toxicity study in rodents. 407. http://www.oecd-ilibrary.org/content/book/9789264070684-en. Accessed 27 Oct 2019

OECD (2009a) Guideline for testing of chemicals. Carcinogenicity studies. TG 451. http://www.oecd-ilibrary.org/content/book/9789264071186-en. Accessed 27 Oct 2019

OECD (2009b) Guideline for testing of chemicals. Combined chronic toxicity\carcinogenicity studies. TG 453. http://www.oecd-ilibrary.org/content/book/9789264071223-en. Accessed 27 Oct 2019

OECD (2009c) Guideline for testing of chemicals. Subacute inhalation toxicity: 28-day study. TG 412. http://www.oecd-ilibrary.org/content/book/9789264070783-en. Accessed 27 Oct 2019

OECD (2009d) Guideline for testing of chemicals. Subchronic inhalation toxicity: 90-day study. TG 413. http://www.oecd-ilibrary.org/content/book/9789264070806-en. Accessed 27 Oct 2019

Papadopoulou E, Padilla-Sanchez JA, Collins CD, Cousins IT, Covaci A, de Wit CA, Leonards PEG, Voorspoels S, Thomsen C, Harrad S, Haug LS (2016) Sampling strategy for estimating human exposure pathways to consumer chemicals. Emerg Contam 2(1):26–36. https://doi.org/10.1016/j.emcon.2015.12.002

Parascandola M, Weed DL (2001) Causation in epidemiology. J Epidemiol Community Health 55:905–912. Available at: https://www.ncbi.nlm.nih.gov/pmc/articles/PMC1731812/pdf/v055p00905.pdf. Accessed 27 Oct 2019

Pieters MN, Kramer HJ, Slob W (1998) Evaluation of the uncertainty factor for subchronic-to-chronic extrapolation: statistical analysis of toxicity data. Regul Toxicol Pharmacol 27 (2):108–111

Pohl HR, Chou CH, Ruiz P, Holler JS (2010) Chemical risk assessment and uncertainty associated with extrapolation across exposure duration. Regul Toxicol Pharmacol 57(1):18–23

Renwick AG (1999a) Subdivision of uncertainty factors to allow for toxicokinetics and toxicodynamics. Hum Ecol Risk Assess Int J 5(5):1035–1050

Renwick AG (1999b) Duration of intake above the ADI/TDI in relation to toxicodynamics and toxicokinetics. Regul Toxicol Pharmacol 30(2 Pt 2):S69–S78

Renwick AG, Lazarus NR (1998) Human variability and noncancer risk assessment—an analysis of the default uncertainty factor. Regul Toxicol Pharmacol 27:3–20

Rhomberg L (2015) Hypothesis-based weight of evidence: an approach to assessing causation and its application to regulatory toxicology. Risk Anal 35:1114–1124

Roth N, Wilks MF (2014) Neurodevelopmental and neurobehavioural effects of polybrominated and perfluorinated chemicals: a systematic review of the epidemiological literature using a quality assessment scheme. Toxicol Lett 230(2):271–281. https://doi.org/10.1016/j.toxlet.2014.02.015

Rothman KJ (2008) Validity in Epidemiologic Studies. In: Rothman KJ, Greenland S, Lash TL (eds) Modern epidemiology, 3rd edn. Lippincott, Williams & Wilkins, Philadelphia, PA, pp 128–147

Rothman KJ, Greenland S, Lash TL (2008a) Modern epidemiology, 3rd edn. Lippincott, Williams & Wilkins, Philadelphia, PA

Rothman KJ, Greenland S, Lash TL (2008b) Case-control studies. In: Rothman KJ, Greenland S, Lash TL (eds) Modern epidemiology, 3rd edn. Lippincott, Williams & Wilkins, Philadelphia, PA, pp 111–127

Solecki R, Davies L, Dellarco V, Dewhurst I, van Raaij M, Tritscher A (2005) Guidance on setting of acute reference dose (ARfD) for pesticides. Food Chem Toxicol 43:1569–1593

Sweeney LM, Sommerville DR, Channel SR, Sharits BC, Gargas NM (2014) Impact of non-constant concentration exposure on lethality of inhaled hydrogen cyanide. Toxicol Sci 138(1):205–216

Sweeney LM, Sommerville DR, Channel SR (2015) Evaluating the validity and applicable domain of the toxic load model: impact of concentration vs. time profile on inhalation lethality of hydrogen cyanide. Regul Toxicol Pharmacol 71(3):571–584

ten Berge WF, Zwart A, Appelman LM (1986) Concentration-time response relationship of irritant and systemically acting vapours and gases. J Hazard Mater 13:301–309

Tong TYN, Wareham NJ, Khaw K-T, Imamura F, Forouhi NG (2016) Prospective association of the Mediterranean diet with cardiovascular disease incidence and mortality and its population impact in a non-Mediterranean population: the EPIC-Norfolk study. BMC Med 14(1):135. https://doi.org/10.1186/s12916-016-0677-4

U.S. Department of Homeland Security (2012) START. Understanding risk communication best practices and theory. START releases two risk communication guides for emergency managers and communicators. http://www.start.umd.edu/news/understanding-risk-communication-best-practices-and-theory. Accessed 27 Oct 2019

van den Brandt P, Voorrips L, Hertz-Picciotto I, Shuker D, Boeing H, Speijers G, Guittard C, Kleiner J, Knowles M, Wolk A, Goldbohm A (2002) The contribution of epidemiology. Food Chem Toxicol 40(2):387–424. https://doi.org/10.1016/S0278-6915(01)00114-4

van Leeuwen C, Vermeire TE (2007) Risk assessment of chemicals: an introduction. Springer, Dordrecht

Vandenbroucke JP, von Elm E, Altman DG, Gøtzsche PC, Mulrow CD, Pocock SJ, Poole C, Schlesselman JJ, Egger M (2014) Strengthening the reporting of observational studies in epidemiology (STROBE): explanation and elaboration. Int J Surg 2(12):1500–1524. https://doi.org/10.1016/j.ijsu.2014.07.014

VWA (2008) Overschrijding van de gezondheidskundige advieswaarde voor chemische stoffen in de voeding van kinderen. Een afwegingskader voor het vaststellen van gezondheidsrisico's voor kinderen van een half tot twaalf jaar. Oktober 2008. In dutch. http://docplayer.nl/18418089-Kinderen-en-chemische-stoffen-in-de-voeding.html. Accessed 27 Oct 2019

Weihrauch MR, Diehl V (2004) Artificial sweeteners—do they bear a carcinogenic risk? Ann Oncol 15(10):1460–1465. https://doi.org/10.1093/annonc/mdh256

Weiss C (2006) Precaution: the willingness to accept costs to avert uncertain danger. In: Coping with uncertainty. Lecture notes in economics and mathematical systems, vol 581. Springer, Berlin, pp 315–330

WHO (1994) Assessing human health risks of chemicals: derivation of guidance values for health-based exposure limits. International Programme on Chemical Safety (IPCS). Environ Health Criteria 170

WHO (2004) IPCS risk assessment terminology. Part. 1, IPCS/OECD key generic terms used in chemical hazard/risk assessment and Part. 2, IPCS glossary of key exposure assessment terminology. IPCS harmonization project document no. 1. WHO, Geneva. http://www.who.int/ipcs/methods/harmonization/areas/terminology/en/. Accessed 27 Oct 2019

WHO (2005) Chemical-specific adjustment factors for interspecies differences and human variability guidance document for use of data in dose/concentration-response assessment (IPCS harmonisation project document; no. 2). Geneva

WHO (2016) Risk communication in the context of Zika virus. Interim guidance, WHO/ZIKV/RCCE/16.1. 1 Mar 2016. http://www.who.int/emergencies/zika-virus

Witschi H (1999) Notes on the history of Haber's law. Toxicol Sci 50:164–168

WTO (1994) The WTO agreement on the application of sanitary and phytosanitary measures (SPS agreement). Available at: https://www.wto.org/english/tratop_e/sps_e/sps_e.htm

Zarn JA, Hänggi E, Kuchen A, Schlatter JR (2010) The significance of the subchronic toxicity in the dietary risk assessment of pesticides. Regul Toxicol Pharmacol 58(1):72–78

Zarn JA, Engeli B, Schlatter JR (2011) Study parameters influencing NOAEL and LOAEL in toxicity feeding studies for pesticides: exposure duration versus dose decrement, dose spacing, group size and chemical class. Regul Toxicol Pharmacol 61(2):243–250

Zota AR, Phillips CA, Mitro SD (2016) Recent fast food consumption and bisphenol A and phthalates exposures among the U.S. population in NHANES, 2003–2010. Environ Health Perspect 124(10):1521–1528. https://doi.org/10.1289/ehp.1510803

Chapter 3
General Aspects of Exposure Evaluation

Christiaan Delmaar, Gerhard Heinemeyer, Matti Jantunen, Klaus Schneider, and Michael Schümann

This part is directed to the general instruments that are needed for an appropriate consumer exposure assessment. It is based on the recommendations of international organisations for risk assessment. The International Programme on Chemical Safety (IPCS) of the WHO (2004) has proposed the basic requirements for exposure assessment. Paustenbach (2000) already described a coherent approach to exposure estimation. Every exposure assessment should consider the three main elements of exposure assessment, (1) scenario, (2) model and (3) parameter characterisation, in combination with uncertainty analysis. Also, the tiered approach should be considered, where needed and possible. Further, combined exposures should follow the concept of toxicological equivalents (TEQ).

In the following chapters, the main requirements for the characterisation of exposure scenarios, models and parameters will be described, in combination with

The original version of this chapter was revised. A correction to this chapter can be found at https://doi.org/10.1007/978-3-319-96148-4_8

C. Delmaar
Centre for Safety of Substances and Products (VSP), Institute for Public Health and the Environment (RIVM), Bilthoven, The Netherlands
e-mail: christiaan.delmaar@rivm.nl

G. Heinemeyer (✉)
Department of Exposure, German Federal Institute for Risk Assessment (Retired), Berlin, Germany
e-mail: gerhardheinemeyer@o2mail.de

M. Jantunen
Department of Environmental Health, National Institute for Health and Welfare (THL) (Retired), Kuopio, Finland
e-mail: matti.jantunen@janding.fi

K. Schneider
Forschungs- und Beratungsinstitut Gefahrstoffe GmbH (FoBiG), Freiburg, Germany
e-mail: klaus.schneider@fobig.de

M. Schümann
Hamburg, Germany
e-mail: michael.schuemann@hamburg.de

G. Heinemeyer et al. (eds.), *The Practice of Consumer Exposure Assessment*,
https://doi.org/10.1007/978-3-319-96148-4_3

accompanying measures to ascertain an adequate exposure evaluation. As explained in the previous chapter, exposure evaluation is the instrument that identifies the source of exposure, including the release and transport, and the quantitative estimation of the incorporated amount. In the context of risk assessment, exposure assessment has at least two aims, product safety assessment and health risk assessment. Product safety assessment has the objective to evaluate whether a product is safe under certain use conditions. Health risk assessment is focused on the safety of substances in the view of exposed populations.

3.1 Introduction

The purpose of exposure assessment is to estimate quantitatively the amount of a chemical to which an individual person or a population is in contact. This quantitative estimation reveals a dose that is incorporated in the human body and which is needed to perform a risk characterisation.

Exposure assessment follows an approach that is based on three interdependent pillars:

- The characterisation of an exposure scenario
- Approaches to estimate the numerical values of concentrations and amounts of the substance in contact with consumers, which is mostly done by applying exposure models but also by measurements
- The numerical characterisation of the parameters (variables) needed to be used in exposure models

Exposure assessment should always focus these three steps although in some cases they may be combined and/or mixed up. However, the performer of an exposure assessment should always identify which parts of the assessment are regarded to it. Every assessment starts with the characterisation of an exposure scenario. This view of exposure science is published by the National Research Council (NRC) of the National Academy of Sciences (Lioy and Smith 2013). The approach of exposure assessment starts from outside (environment) to inside (the human organism). In this concept, the point of departure in the exposure (with an agent) is the source. After release, the substance reaches the human body via the environment by external exposure, followed by internal exposure and reaching the target site of exposure. This concept can be taken as a general rule in the field of exposure assessment and is in line with the definition of exposure as pointed out by WHO:

> Concentration or amount of a particular agent that reaches a target organism, system, or (sub) population in a specific frequency for a defined duration. (IPCS document on Risk Assessment Terminology WHO 2004)

The exposure scenario is a textual description of the circumstances that characterise the contact of a substance with a target. According to WHO, this can be a system (e.g. the environment) or a population. It has been proposed to support the exposure characterisation by establishing a conceptual model (WHO 2005) summarising all sources, paths and exposed populations. In this context, individuals or groups of people at their private level characterise the consumer. This covers every type of contact of persons, e.g. by eating food and playing and at leisure and hobby and cleaning and maintenance of homes. Contacts in kindergartens and at school are also included.

An exposure model represents the quantitative numerical estimate of the scenario by means of a mathematical algorithm. Modelling exposure needs an empirical and mechanistic basis for establishing exposure data and exposure parameters (WHO 2005). Total exposure can also be measured, e.g. by analysis of the concentration of certain substances in room air for inhalation exposure. However, measurements are often not representing the primary approach for consumer exposure evaluation.

Exposure parameters are numerical values representing the variables of the exposure model. Exposure parameters can be very different regarding to its empirical basis and scientific justification. On the one hand, conservative defaults estimated by expert judgement are used, and on the other, data resulting from sophisticated statistical analyses and analytical measurements may be available.

Taking these three elements of exposure evaluation, further concepts that have been accepted as valuable accompanying approaches in the assessment include:

- Exposure and/or exposure parameter measurements
- Quality assessment
- Uncertainty analysis

Although scenario, modelling and exposure parameters are representing the main pillars of the assessment, the role of measuring exposure and/or exposure parameters should be given attention, too. In the first issue of the technical guidance document for risk assessment of new and existing chemicals, it was stated that:

> actual exposure measurements, provided they are reliable and representative for the scenario under scrutiny, are preferred to estimates of exposure derived from either analogous data or from the use of exposure models. (EU 2003)

The direct measurement of concentrations in environmental media (indoor and outdoor air, soil) is a frequently used approach to control the indirect exposure of populations (see next paragraph). However, only few measurements of direct consumer exposures exist and they do not often meet the above-mentioned criteria.

On the other hand, using measured exposure model parameters is of great and increasing importance. The term "measurement" covers the physical analysis of concentrations of substances in different media as well as questionnaire studies to evaluate the behaviour and habits in a population. The most appropriate data that are available are data from assessments of food consumption-related exposures, due to the existing food consumption surveys and concentrations in food from official surveillance and market control programmes. The Danish EPA (2001–2016) performed a systematic programme of measurements of substances in a wide variety of consumer products. On the other hand, studies of migration of substances from consumer products deliver valuable data for exposure assessment (Danish EPA 2016). For many exposure situations, however, adequate data do simply not exist and must be replaced by estimates and assumptions.

In the past 10 years, international organisations have produced guidelines for uncertainty analysis (BfR 2015; ECHA 2012; EFSA 2006, 2018a; WHO 2008). Therefore, uncertainty analysis has become an increasing importance to make exposure assessments transparent (EFSA 2009b). In addition, first attempts for uncertainty analysis in hazard assessment have been made (WHO 2014).

The contact of an individual or a population with a substance may occur directly or indirectly. Direct exposure can be defined as contact with a substance that is released from a product or a matrix during active use, e.g. the dermal exposure during use of a cosmetic product, or the inhalation of a volatile solvent during painting. Direct exposure is regarded to the active use of the product by the exposed individual(s). Indirect exposure occurs after substances have been released from emitting sources, often via the environment. Indirect exposure occurs also if, e.g. a substance is entering the food chain. Food exposure represents the most important source for indirect exposure. Bystanders are also exposed indirectly. Smokers are exposed directly to cigarette smoke, while non-smokers are exposed indirectly. Differentiation of direct and indirect exposure does not have a primary impact on exposure but, from practical considerations, for exposure scenario formulation. Indirect exposure is often lower that direct exposure, for example a bystander watching another person who is painting. The direct exposure to a volatile compound occurs within a short distance to the place of action, while during indirect the distance of the bystander may be greater which can be considered in an exposure scenario by choosing, e.g. a bigger room volume.

Exposure can occur as aggregated and cumulative exposure (Meek et al. 2011). Aggregated exposure means contact with the same agent from different sources. Consumer exposure normally is aggregated exposure, due to the occurrence of the same agent as ingredient and contaminants of many and heterogeneous materials.

In contrast, cumulative exposure is characterised by contact to different agents having the same or similar toxic properties (Meek et al. 2011). The most important and well-known example for cumulative exposure is the contact with different dioxin congeners or organophosphate pesticides. A working group at WHO/IPCS developed a framework for cumulative exposure in connection with cumulative assessment of hazard evaluation and with regard to the concept of toxicological equivalents (Ahlborg et al. 1994).

A tiered approach should be generally applied in regulatory exposure assessments (Fig. 3.1) with the aim to avoid unnecessary evaluations. The tiered approach is characterised by a step-by-step procedure. It can be applied for every exposure assessment and for scenario characterisation, exposure modelling and the selection of exposure parameters as well, dependent on each other, as an integrated and iterative process in exposure evaluation (see also Sects. 3.2.10 and 3.4). This means that every iteration step must include consideration of scenario and model and parameters hand in hand.

Furthermore, exposure assessment often uses data having no empirical background. It is often based on expert judgement, or it uses data that have been created for other purposes than exposure assessment. In addition, the exposure scenarios and the respective models may be incomplete. Many concepts and data used for exposure

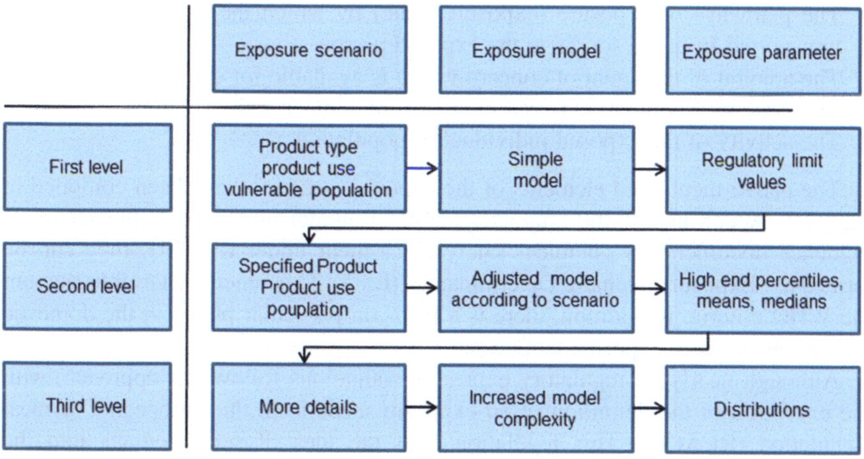

Fig. 3.1 Tiered approach for scenario characterisation, model building and parameter choice

assessment are of low quality, due to a lack of knowledge and thus uncertainty. From this point of view, tiering in exposure evaluation is a makeshift approach that needs a systematic evaluation of quality and uncertainty.

Exposure evaluation is embedded in the process of regulatory risk assessment. A "conflict of interest" may exist between the objections of the regulatory assessment and the opinion of the producer or by requirements of scientific evaluations. Rules in the regulations and the respective guidance documents and instructions often result from conventions that are based on compromises between different groups of stakeholders. On the other hand, a sound empirical and scientific basis is needed which does not necessarily match those agreements.

3.2 Exposure Scenario

Gerhard Heinemeyer

According to IPCS guidelines on terminology (WHO 2004), an exposure scenario can be defined as:

> A set of conditions or assumptions about sources, exposure pathways, amounts or concentrations of agent(s) involved, and exposed organism, system, or (sub)population (i.e, numbers, characteristics, habits) used to aid in the evaluation and quantification of exposure(s) in a given situation

In accordance with this definition, the most important information that should be addressed in an exposure scenario characterisation is:

- The source (product) of the agent to which an individual will be exposed

- The pathways of exposure (exposure route) by which the agent of concern is transported from the source to the exposed person
- The amount of the agent of concern which is available for contact
- The system or population (target) which is in contact with the agent
- The activity of the exposed individual or population

The above-mentioned elements of the exposure scenario have been compiled in Fig. 3.2. According to the characterisation of model elements in the technical guidance document for chemical safety assessment under REACH, they can be termed as "Exposure Scenario Determinants" (ESD). This scheme takes into account the WHO criteria. In addition, there is a focus on particular places in the domestic environment where exposure can occur.

Although nearly all regulatory exposure evaluations follow this approach, with the exception of the definition of an exposure scenario in the European Chemical Regulation (REACH). This regulation adds the idea of safe products into the exposure scenario concept. The manufacturers of consumer products are requested to guarantee that their products are safe. By means of consumer exposure, this approach means that exposure must not exceed certain levels. If these levels are exceeded, effective risk management measures (Bruinen de Bruin et al. 2007) must be implemented that secure the users of the products. This concept is considered by adding risk management measures (RMM) to the ESDs in Sect. 5.6.3.

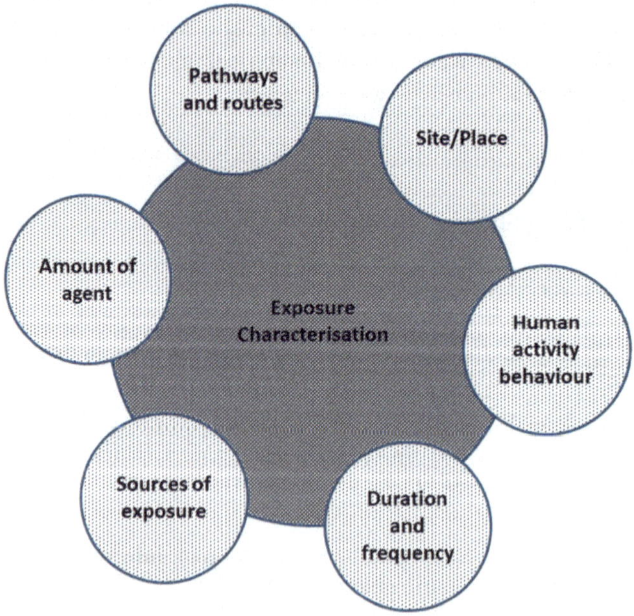

Fig. 3.2 Building elements of the exposure scenario for consumers

3.2.1 Characterising Exposure Scenarios

Exposure can be characterised as internal or external exposure representing alternative concepts of approaching an exposure estimate that nevertheless complement each other.

External consumer exposure evaluation is based on a mechanistic concept that describes the origin (source) and the release of the substance, the pathway of transport from the source to the exposed individuals and the place where exposure happens (WHO 2005). Normally, the estimate is related to bodyweight as an expression of systemic availability with individual results adjusted to represent the whole population.

Internal exposure is following the general concept of measuring the amounts of substances in the body under consideration of systemic absorption, distribution, metabolism and elimination. This can be done by analyses of the substances and their metabolites in body fluids and other compartments (blood, urine, faeces, body fat), known as human biomonitoring (HBM) and evaluated by applying pharmacokinetic methods. These methods are described in detail in Sect. 3.5.

External exposure estimates will deliver data on sources and pathways of exposure, while internal exposure indicates the amounts of substances that are actually present in the body irrespective of source, which might be unknown. Therefore, parallel estimations of both pathways are of great importance for scenario and model evaluation and validation. For risk managers, the information from mechanistic external models in combination with internal exposure and sensitivity analyses delivers valuable information about which risk-reducing strategies are most effective in lowering risk.

The steps in external exposure assessment are characterised by Heinemeyer (2004). Briefly, they are as follows:

1. The use scenario identifies handling practices for the composite product that is the source of the hazardous substance and describes quantitatively the amount that can be released from the product.
2. The disposition scenario characterises the release, distribution and decrease of the substance in the immediate environment of consumers.
3. The behaviour scenario informs about the habits of the exposed individuals of the population, their eating behaviour, hobbies, household maintenance, sports and other relevant activities.

Conservatism Versus Central Tendencies in Exposure Scenarios

Exposure scenarios can be categorised as conservative or with central tendency, with any variation in between. The understanding of exposure assessment related to estimates with conservative or central tendency may be related to the meanings of "worst case" and "average case" characterisations of the exposure scenario but without a statistical background. This is basically due to the deterministic approach or the assessment involving single numerical determinants for exposure quantification. The terms "realistic" and "worst case" may vary due to many reasons and are hardly comparable. They include some kind of an "a priori" assessment and should therefore be avoided in exposure assessments.

A conservative scenario describes a situation based on high-end exposure parameter assumptions. Conservative scenarios are often characterised by their simplicity, but this is often not convincingly portrayed as a realistic case (see example below). A realistic scenario should be based on the real "world", covering a detailed characterisation of the living conditions, e.g. by mentioning the type of rooms in which the exposure occurs, and by choosing data about room sizes which are based on data which are representative for real house and room sizes. This scenario will also include special situations that are characterised by higher exposures. This could involve the brand loyal consumer who always uses high amounts of a special product containing high concentrations of the harmful substance leading to high exposure.

The term "average case" includes conditions characterised by mean, median values or other averages for the exposure parameters that may be used synonymously with the term central tendency.

In contrast to this rather difficult characterisation of deterministic calculations, probabilistic exposure assessment may be more realistic because it includes every step of realistic deterministic exposure estimation as a distribution of possible outcomes and cover the whole range of exposure. Probabilistic estimations reveal determinants that are based on empirical data and a characterisation of their distributions using standard statistical rules.

A simple exposure scenario may be characterised as follows:

Simple Exposure Scenario A

An individual person paints the walls of a room having a volume of 25 m^3. He uses an amount of 20 kg of paint containing 2% of substance X that can be released into the indoor room air. It is assumed that the entire substance X will be released at once. There is no air ventilation. The calculation of the exposure by inhalation is done by dividing 400 g (= 2% of 20 kg) by 25 m^3 resulting in a concentration of 16 g/m^3 that is the highest possible concentration of the substance in the surrounding air. No ventilation, no vapour pressure is considered. A breathing volume of 20 m^2 per day and a body weight of 80 kg may be assumed to estimate the systemic intake by inhalation.

In this scenario, the difference between the conservative case and average case conditions can be expressed by the choice of model variables, which is room volume of 25 m^3 (e.g. $3.0 \times 3.0 \times 2.8$ m) vs. a more normal 54 m^3 (which is $4.0 \times 5.0 \times 2.8$ m) for a living room. This means that the step from conservatism to reality is going from a rare situation to a generally applicable one. Assuming no room air ventilation is a conservative case assumption, because modern houses all may have real low ventilation.

Simple Exposure Scenario B

It is assumed that the daily consumption of food is 1 kg. This food contains a concentration of a substance Y of 1 g/kg. The exposure (oral intake) of the substance will result in 1 g per day.

This scenario is indeed conservative, due to its orientation to highest possible intake rates. The step to a realistic scenario would include the differentiation into food groups, as well as their consumed amounts and the concentrations of the substance in the respective food.

Because scenario A is not a realistic one, an iterative procedure can be initiated to narrow reality (tiered approach). Iteration means a step-by-step adjustment of the scenario, for example, by introducing ventilation and time-dependent evaporation of the substance. For scenario B more realistic parameters can be introduced by differentiating the food items and their respective consumption amounts and substance concentrations. The exposure model to estimate the exposure can then be adjusted accordingly. This procedure can be very nicely performed by use of the ConsExpo tool (Delmaar et al. 2005), which offers an assessment by iterating the scenario with subsequent adjustment of the model algorithm (shown in Sect. 3.3.4, Fig. 3.6). On the other hand, model parameters can be varied without iteration of the scenario and the models, e.g. by exchanging the single numerical value by a distribution. It remains to discuss, however, whether it makes sense to use a well-characterised distribution in a low-tier simple model should the conservative scenario provide sufficient protection. A more detailed and deeper presentation of these techniques is presented in the special chapters.

3.2.2 Site/Place of Exposure

To derive an estimate of exposure that is as realistic as possible, it is appropriate to have as much information for characterisation of the exposure situation as possible. Normally, consumer exposure takes place at home or around home and some exposures are typically related to certain places. It is important to observe that rooms in a household are used differently and the duration of stay differs.

Consideration of room differences in houses and flats is an issue for the distribution of volatile substances in room air. The concentrations in the room of active work will be higher because of dispersion than in the other rooms, but with time a steady state will be established which can be modelled by tools such as MCCEM/CEM (US-EPA 1991) and the more up-to-date SHEDS model (Zartarian et al. 2008). Both models are characterised in Sect. 5.10.

Depending on the scenario and the objective of the evaluation, differences in rooms can also be a reason for estimating exposure, e.g. from house dust which can be accumulated differently in the rooms of a house/flat due to its use and the activities of their inhabitants.

Therefore, it might be valuable to consider the different rooms/places where exposure occurs.

In the process of developing and using exposure tools and models, US-EPA very early developed an approach using multi-chamber scenarios and respective models such as the MCCEM (US-EPA 1991) model and to differentiate macro- and micro-environments. Even today this concept is often used in the scientific evaluation and characterisation of exposure scenarios and models to reflect the higher tiers of exposure assessment. Exposure tools taking multi-chamber exposure into consideration are at the moment officially not available from European institutions or agencies.

3.2.3 Activity Pattern Data: Human Activity and Behaviour

The quality and quantity of exposure is considerably influenced by the behaviour and the habits of the people. This holds true for the amounts of products used (see next paragraph) and the frequency of use. In contrast to workers exposure, this descriptor is mostly related to the individual and therefore a very important determinant of consumer exposure. Whether or not an exposure is short or long-term is primarily determined by the frequency of use (see also Sect. 3.2.4). Particular behaviours are also important elements for exposures, e.g. the mouthing behaviour of children. Frequent eating of high amounts of certain foods may place a small group of consumers at particular risk in comparison to the general population. The scenario characterisation should also include the "foreseeable misuse" which means that a product is used for another purpose than intended by the manufacturer, for example, a dishwashing product is used for hair washing.

3.2.4 Duration and Frequency of Exposure

The exposure duration and frequency (time activity) are mainly determined by the use/consumption of products. Both parameters are important parameters to decide whether an exposure is short- or long-term. In short-term exposures, the duration of contact is the main determinant, e.g. for inhalation exposure, while the frequency of use or consumption determines whether the exposure is short-time or long-time. Details will be discussed in the later chapters.

3.2.5 Sources of Exposure

Characterising the sources of exposure is part of the use scenario as suggested by Heinemeyer (2004). Figure 3.3 illustrates a schematic model of consumer exposure sources and paths. The sources of exposure are, e.g. food, floor coverings (textiles, PVC flooring, parquetry), building materials, fittings (wallpaper, furniture), household, hobby and leisure products as well as cosmetics. Contact can occur directly (by touching, mouthing) and indirectly via air and house dust.

Taking the exposure scheme published by the National Research Council (NRC) of the National Academy of Sciences (Lioy and Smith 2013), the primary source of exposure can be an environmental medium that releases substances to the environment. This view is basically substance oriented. In fact, most chemicals have their origin in the wider environment. To characterise a typical exposure chain, elements like mercury, lead, cadmium and others are released from the earth crust, dispersed, partly with help of humans, to the larger environment, to plants and animals and thus to food, which is a secondary source of exposure for individuals and populations. Other chemicals are formed directly by plants, e.g. mycotoxins, plant glycosides and pyrethrum, and in animals, e.g. hormones and animal poisons. A third kind of substances are man-made, mostly from mineral oil, e.g. hydrocarbons in the pure form and as chlorinated and fluorinated compounds.

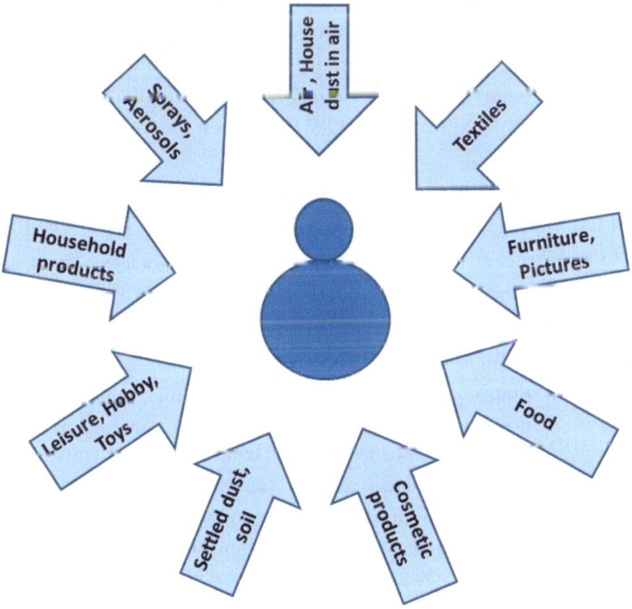

Fig. 3.3 Schematic characterisation of exposure sources and possible pathways from the source to the exposed individual or population

Consumer exposure is product oriented. Consumers are very seldom exposed to pure substances. The characteristic consumer exposure situation is focused on the use or consumption of a product, which is the central issue of exposure evaluation and the point of departure to describe a consumer exposure scenario. Consumer exposure is a small part of the general concept: it happens - via product use - with all substances mentioned in the chapter above, all of which may be dispersed also via the environment.

Due to the fact that sources of exposure are mostly products with which consumers handle or which they consume, describing their use can support the characterisation of products. Any products can be characterised by classification systems that systematically categorise them by the use of a thesaurus and further indices and descriptors. Classification systems are a key instrument for characterising products and have been developed in detail for food characterisation and in less detail (with gaps) for other products. The use of classification systems is described in Sects. 4.4 and 5.2.1 and several regulatory inventories (Heinemeyer and Hahn 2005). Knowledge of brand names is often an advantage because it allows more precise information about the composition.

Food represents a major source of chemical contaminants, but other sources may be important as well. These may be represented by the huge number of products of daily life, e.g. household, hobby and leisure products, toys, clothing but also fittings and building products. Some examples are listed in Table 3.1 (furniture, textiles, wallpaper, etc.).

During the last 15 years, contaminants formed during processing of food (process contaminants) like heating (cooking, broiling) have come into focus in risk assessment. Acrylamide and 3-MCPD represent two of the most prominent agents of this group of chemicals. Heinemeyer et al. (2012, 2013) calculated exposure in the German population to the softener di-ethyl-hexyl phthalate (DEHP). It took into consideration all possible sources of DEHP, which is the most frequently used softener worldwide.

In the WHO guidance document for exposure modelling (WHO 2005), a conceptual model to explain all the sources has been proposed. The conceptual model is displayed in Fig. 3.4.

Table 3.1 Sources of substances for consumer exposure estimation

Source of exposure	Examples
Food	Milk products, meat, fish, fruits, vegetables, grain products
	If certain substances are in a process (e.g. cooking), other substances can be formed (so-called process contaminants)
Cosmetic products	Cleaning products, makeup, skincare
Building materials	Emissions from concrete, floor and wall covering, wooden materials
Hobby, leisure	Paints, cleaning products, glue, building and modelling products, car maintenance, gardening materials, rubber and plastic articles, camping products
Furniture, indoor equipment	Textiles, leather, emissions from wooden and plastic materials
Toys	Wooden and plastic materials, often in combination with textiles

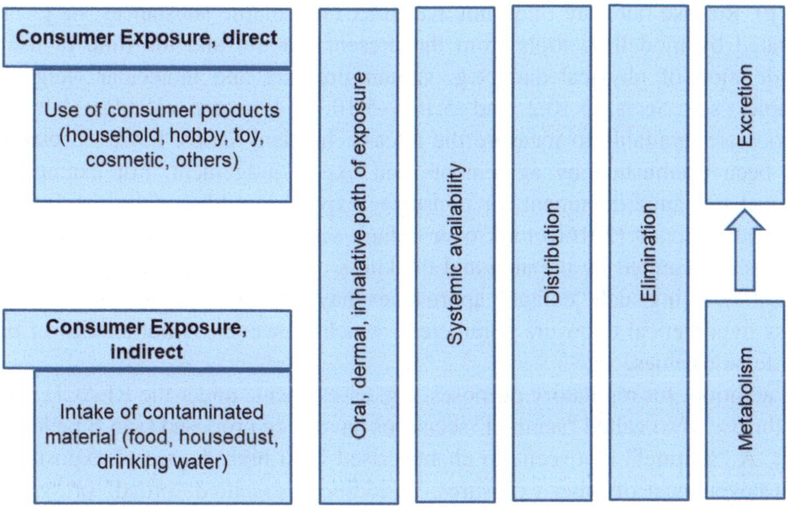

Fig. 3.4 Conceptual model of exposure

The scenarios that can be derived from this conceptual model are, for example, "playing with toys", "wearing textiles", "storing food" and for indirect exposure "eating food" contaminated with DEHP and oral intake of house dust. This means that an aggregated model to compile the different sources of exposure for model estimation must be calculated by assessing the exposures related to the different sources. Establishing a model can start with a simple and aggregated presentation of the model.

3.2.6 Amount of Agent

As a general rule, the concentration of a substance in an item must be multiplied with the mass of that item to reveal the amount of a substance that may lead to exposure.

The knowledge about the amount of the substance is therefore key information for quantification of exposure. The respective numerical value needed for exposure estimation can be taken from information about the concentration of the substance as an ingredient of a product. Product databases are very useful sources of information for that purpose (Heinemeyer and Hahn 2005). To estimate how much of a substance is accessible, it would be appropriate to measure the amount that is released within a certain time. If the substance is highly volatile, it can be assumed that the entire amount can evaporate during and after application. For example, solvents in paint can be evaporated just during and shortly after application. The weight fraction of the substance in that product (e.g. 10%), together with the total amount of the product used (e.g. 1 kg), reveals the amount of substance accessible to exposure

(100 g). Release rates are often not available. For volatile substances, they can be estimated by modelling tools from the present amount and the time of use, by consideration of physical data, e.g. vapour pressure and molecular weight (for examples see Sects. 5.10.2 and 5.10.5–5.10.7). For non-volatiles, laboratory methods are available to measure the release. In many cases, however, estimates have been established by agreements and expert judgement. For example, the technical guidance document for consumer exposure published by the European Chemicals Agency (2016b) took over default values for a fictive contact volume, which is represented by the area and thickness of 0.1 mm from earlier documents (EU 2003). Using such "default" approaches may motivate assessors to create more or less hypothetical exposure parameters, which, however, is not an aim of using such default values.

In addition, for regulatory purposes, e.g. assessments under the REACH regulation, the use of so-called "sentinel" scenarios have been proposed (van Engelen et al. 2007). A "sentinel" is a scenario characterised by a high degree of exposure that might cover most of other exposures. In addition, so-called "broad" or "generic" scenarios are more or less aggregated scenarios that cover an unknown number of specific, but not further detailed, scenarios to characterise a very conservative situation. Subsequently, the mineral oil industry (CONCAWE 2015) proposed to use so-called Specific Consumer Exposure Determinants (SCEDs), which basically compile a number of scenarios for oil-based products and substance amounts having a particular focus on REACH exposure assessments. This approach is suitable to make broad exposure scenarios more specific which looks like iteration in a tiered approach. Sometimes, the total amount of substances in groups of products (e.g. textiles) is taken as assumptions for conservative scenarios. This approach, of course, is not leading to a realistic exposure estimate, but represents only the first step of a tiered approach.

In food exposure, the situation is quite more comfortable, due to a number of food contaminant databases and food consumption studies offering data for exposure analysis (see also Sect. 4.6).

3.2.7 Pathways and Routes of Exposure

According to WHO (2005), the pathway of exposure is defined as:

> The course an agent takes from the source to the target

The exposure route is defined as:

> The way in which an agent enters a target after contact (e.g., by ingestion, inhalation, or dermal absorption).

Figure 3.5 shows the relationship between the issues that determine the "journey" of an agent after being released from the source as described in Sect. 3.2.5 to the target. In between we have the pathway of exposure, which includes like a journey the agents' departure at the source by release at a particular mass, the transport via

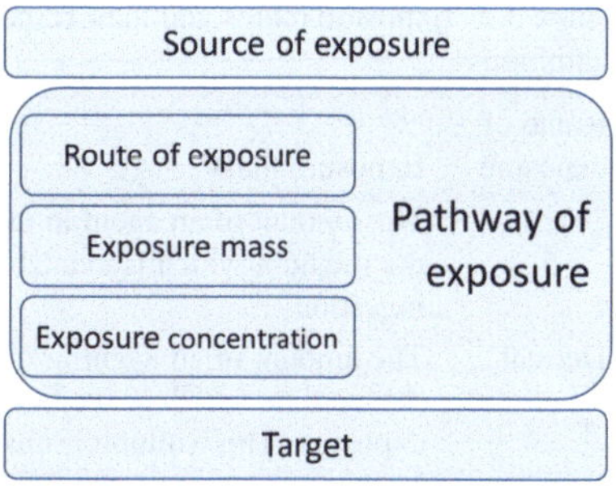

Fig. 3.5 The transfer of agents for exposure

the pathway of exposure and arrival at the entrance (route of exposure) to the human body, then passing the absorption barrier and entering the systemic circulation and the physiological binding sites in the cells which is an issue of the assessment of internal exposure (Sect. 3.5). The pathway of exposure is characterised by direct contact (touching a product) or by indirect contact, e.g. by release to the air with subsequent inhalation or via soil or dust. The route of exposure characterises the entrance to the human body, either by the mouth and skin or via the lungs. The amount of the agent is characterised by its exposure mass and the concentrations in contact media by the exposure concentration.

The most prominent routes by and the respective masses and concentrations of consumer exposure are shown in more detail in Table 3.2.

Other possible routes such as injections or infusions are not consumer exposures, but medical use or abuse. They are not included in the consumer exposure assessment.

The pathways of exposure are mostly determined by how the products are used. At first, tasks like sitting on a couch, wearing clothing, painting a wall, gluing, playing with toys, etc. are representing uses which can be characterised by the close contact with the product and the frequency and length of time of use. By introducing the concentration of the substance of interest into the scenario, the level of exposure to the substance can be estimated (Table 3.3).

Secondly, the way the substance will be transported from the source to the target should be explored. If it has a high vapour pressure, the substance will be evaporated and transported in its gaseous form via air, depending on the temperature. This scenario can be described by the release, the distribution and the subsequent disappearance of the substance from the environment (Heinemeyer 2004). Low vapour pressure is indicative of low volatility, but the substance can still be released and adsorbed to particles. Consumer relevant particles include house dust, which consists of a mixture of a number of different small-sized items, below 200–500 μm. A description of the qualitative composition of house dust can be found in Sect. 5.8. Particular pathways through inhalation comprise fine dust particulates carrying non-volatile substances, as a part of aerosols and sprays (Sect. 5.9). Such scenarios are given specific focus in later chapters of this book. Soil can also be a carrier of hazardous substances, particularly in the outdoor situation.

Table 3.2 Exposure routes and their respective exposure masses and concentrations due to WHO definitions

Route of exposure	Exposure mass	Exposure concentration
Oral	The amount of an agent in the volume of the medium which is available for oral ingestion	The concentration of the agent in the volume of the medium which is available for oral ingestion
Dermal	The amount of an agent in the volume of the medium which is available for dermal exposure. This volume is characterised by the surface area and a fictive thickness of the layer on skin	The concentration of the agent in the volume of the medium which is available for dermal absorption
Inhalation	The amount of an agent in the surrounding medium (typically the room volume)	The concentration of the agent in the surrounding medium (= room volume)

Table 3.3 Primary exposure pathways of some product groups

Product category	Exposure route	Type of exposure
Substances in food	Oral exposure	Indirect exposure (from the environment) Direct exposure (process contaminants)
Substances in products intended for dermal use	Dermal exposure	Direct exposure
Volatile substances from consumer products	Inhalation exposure via air	Direct exposure (user)
		Indirect exposure (bystander)
Non-volatile substances from consumer products	Oral, dermal and inhalation exposure from house dust	Indirect exposure, substances are released from primary sources, while house dust is a vehicle

Consumer exposure often occurs through direct contact with a product, e.g. by using cosmetic products, eating food, wearing cloth, playing with toys, etc., as well as washing and cleaning products.

3.2.8 Aggregated and Cumulated Exposure Scenarios

Exposure can occur at an aggregated or disaggregated product level. Possible cumulative exposure (Meek et al. 2011) should also be taken into account when constructing an exposure scenario. At the aggregated level, exposure should cover the entire sources, if possible, of one particular substance. This can be done by putting all of the sources together in one broad and generic scenario without further specification of detailed (sub) scenarios (van Engelen et al. 2007) with the outcome focused at an high exposure estimate. This approach follows a conservative concept and can be used for low-tier assessments (see next chapter). Realistic estimates, however, will involve individual sources and the respective (sub)scenario. According to the iteration level, aggregation of exposure can be calculated at several different levels of detail. In a study on DEHP exposure, Heinemeyer et al. (2012) showed that the level of aggregation of food groups considered in the exposure calculation had a significant impact on the resulting estimation (see also Fig. 4.3 in Sect. 4.4.3). The more detailed the scenarios are, the more realistic is the estimate, provided that sound data are used in the assessment. Consumer exposure normally is aggregated exposure, due to the occurrence of the same substance as ingredient or contaminant in many heterogeneous products where occurrence and consumption can only be matched at a group level. However, different paths of in the chain of distribution must be kept separate. Flame retardants, for example, may be released from those materials that are made nonflammable, such as building materials, furniture and textiles. However, the chemicals may also leak into the general environment and can thus enter the food chain. This way the food chain represents an important pathway for the substances to be included as a separate scenario in the exposure calculation. Exposure is easier to calculate when food is defined as the only source, although this is only relevant for process contaminants and other naturally occurring substances in food.

Cumulative exposure is relevant when there is contact to different substances that have the same or similar toxic properties exerting their effect jointly. The most important and well-known example of cumulative exposure is contact with different dioxin congeners of varying toxic potential. A working group at WHO (Meek et al. 2011) developed a framework for cumulative assessment of exposure applying a concept of toxicological equivalents (TEQ; Ahlborg et al. 1994). This approach has been used to enable a common assessment of exposure to a range of dioxins and dioxin-like PCBs of varying toxic potencies. It can be basically applied to other substances having similar action but different potencies, e.g. organophosphate pesticides or polybrominated diphenyl ethers (PBDE; Meek et al. 2011). Because the TEQ concept is based on toxicological methodology, it is not extensively explained in this book, but can be taken from other publications, e.g. in van den Berg et al. (2006). This concept has also been enlarged for different substances having unknown potencies but similar exposure.

3.2.9 Background Exposure Levels

Background exposure can be characterised as the secondary (indirect) contact with an agent from other sources than the actively used ones. Typical background exposures are those that occur from migration from sources, e.g. home textiles, furniture, building materials or others. From release to contact, there is a transport of the agent, e.g. via air or house dust. Background exposure is a special type of aggregated exposure from several sources. It accounts for the total exposure and should be considered in the characterisation of the exposure scenario. Background exposures can be formulated in separate scenario descriptions if the exposure person is a bystander or might be included in an aggregated exposure scenario that includes all sources and pathways of exposure for active users.

3.2.10 Iteration of Exposure Scenarios (Tiered Approach)

Use of a tiered approach is one of the key approaches in exposure assessment. The objective is to ensure the safety of a large part of a population by means of a step-by-step assessment starting with a conservative scenario. Another purpose is to avoid unnecessary evaluations should the conservative scenario prove to provide sufficient protection. Performing iterations of different exposure levels might increase costs of acquiring sufficiently detailed data. Therefore, a tiered assessment always starts with a scenario that uses rough data characterising conservatism. This means that the scenario involves compiling aggregated sources of exposure that enables the assessor to develop an overview of the entire exposure, but without differentiation of individual sources and paths. The so-called sentinel scenarios mentioned above (van Engelen et al. 2007) and other broad scenarios are characterised by products that contain unlikely high and/or aggregated amounts. Some particular scenarios that are used in calculation tools in the context of the REACH regulation are characterised by such sentinels (ECETOC 2005). The model in this assessment will be rather simple and, in accordance with the scenario, at a first-tier level. The defaults used in the assessment are intended to result in an overestimation of the real exposure level. To a certain extent, such conservative assumptions seem unrealistically high and have high uncertainty, but the conservatism will provide a margin for safe use. Iteration of low to higher tiers always requires justification. This is due to the effort needed to collect data with a clear focus on exposure evaluation and use of modelling concepts with an empirical background. Such data derived from detailed studies and statistical analyses will be less uncertain. Often it might be possible to derive distributions of the data that deliver, e.g. means, medians and 95th or higher percentiles. If the risk characterisation does not reveal a clear distance between exposure levels and, e.g. the limit value (compare Sect. 2.2.5), then a further detailed evaluation should follow as listed below. Tiered assessments involve using iterative steps of the information with characteristics as shown in Table 3.4.

It should be considered that the iteration steps for scenario, model and parameters must not follow a strict and systematic approach taking fixed relations of scenario, model and parameters. For example, a simple model can be fed with either numerical (deterministic) data or distributions (probabilistic). Default values can also be used in a complex model, and parameters having different qualities and statistical backgrounds may be mixed up in a model, too. It might happen that in a food consumption study, different sources of data exist, e.g. from different references that can complement each other.

This procedure is thought of as an internal working approach done step by step. It should not appear in regulatory official documents; however, the level of iteration should be made transparent and understandable, in relation to the problem. It must be mentioned that iteration always considers a combination of scenario, model and parameters.

The example below demonstrates iteration steps for the simple scenarios A and B shown earlier. Iteration includes the consideration of air ventilation, vapour pressure

Table 3.4 Iteration steps in the tiered approach of exposure evaluation

Level of tier	Characteristics of the tiered level and iteration steps
Tier 1	Broad scenario, information about product use very generic, no further specification of population Use of simple model according to the scenario Use of default values for worst case estimations to prepare a conservative estimate
Tier 2	Specified scenario and extended model, introduction of detailed information related to product use and time. Specification of population involved Use of conservative estimates derived from empirical data and including central tendencies of exposure Make difference between short-term (infrequent) and long-term (frequent) use
Tier 3	As tier 2, but further specified scenario, focus on specific product use and populations, consideration of substance characteristics, use of deterministic exposure factors to estimate ranges of exposure
Tier 4	As tier 3, but with use of probabilistic exposure parameters

(EU 2009) and some adjustments of use, such as paint used for a certain area and the duration of work. The iteration steps mentioned in the example below may be done step by step, varying their order.

For the food scenario, the different foods eaten are presented in a more detailed manner, according to results from a food consumption survey, as well as concentration data from a market surveillance programme.

Iterated Exposure Scenario A

An individual person paints the walls of a room having a volume of 25 m^3. The amount of 20 kg of paint is used containing 2% of substance X, which can be released into the indoor room air. It is assumed that the substance will be released over a certain time from the paint matrix into the room. A first iteration step considers room air ventilation and evaporation of the substance over the time of use; a second step of evaporation considers additionally the vapour pressure and molecular mass.

The air ventilation rate is 0.5 h^{-1}, the vapour pressure is 1 Pa, the area of painting is 40 m^2 with a corresponding amount of 10 kg of paint, and the duration of painting is 8 h. The substance will evaporate over several days. Model calculation of this scenario will yield a much lower indoor air concentration.[1]

[1]The calculation steps described here can be calculated by use of the ConsExpo tool.

Iterated Exposure Scenario B

Food consumption is characterised in detail for milk products, meat, fish, different types of vegetables and fruits, as well as for liquid foods, according to a current food consumption survey. Concentrations of the substance are taken from market surveillance study. Differences are considered for children, adolescents and adults.

Tiering the exposure assessment should be made in accordance with the toxic properties of the substance, e.g. if the hazard assessment has revealed acute or/and chronic toxicity. In practice, it might happen that first of all, a broad number of products containing the substance of concern can be identified. In this case, it makes sense to aggregate all the products to produce an overall estimate. A second step may include only products where higher concentrations and higher frequency rates of contact are identified. Issues of internal exposure (e.g. due to half-lives or clearance) should be added, e.g. "what is the time a substance is present in the human body?"; "is there an accumulation even at low frequencies of contact?" This issue is particularly important if chronic toxicity occurs. This example shows that tiering is an approach involving iterated scenarios by means of a continuous process. The levels (tiers) in this process are providing frames in which the iterations can be performed.

3.2.11 Short-Term Versus Long-Term Exposure

Exposure can occur over a wide range of occasions and time periods. Short-term exposure is often defined as acute exposure and long-term exposure as chronic. This terminology is taken from hazard identification to distinguish between acute and chronic effects. However, the exposure assessor should avoid this terminology because it incorrectly combines acute effects with short-term (acute) exposure, and chronic effects with long-term (chronic) exposure. Short-term contact of substances that remain over long times in the body may certainly lead to long-time exposure (which is chronic), whereas frequent exposures over longer times with substances having very short-time presence in the body might not necessarily lead to chronic risk. The impact of short and long-term exposure is discussed extensively in Sect. 2.6.

3.2.12 Human Populations

Table 3.5 shows descriptors that can be aligned to individual persons or to populations. Exposure assessment of distinct individual persons is needed if somebody is in contact with a substance or is suffering from symptoms that might be caused from a contact with hazardous substances. In this case, the individual must be described by her/his personal characteristics such as body weight, body height or body surface.

First of all, the entire range of people (the general population) in a country or region should be characterised. For sound exposure assessments, the populations and (sub)populations in a survey should be evaluated at a representative level (see also Sects. 3.7 and 3.8). The consideration of characterising the main descriptors of representativeness in a population (e.g. age, gender, season and region) requires a high number of participants, e.g. in a food consumption survey. Due to compromises, many surveys are not representative in this understanding, because not all (sub)populations from an entire general population can be considered.

Stratification of the general population due to gender is required as a general rule to characterise it adequately. Nearly every risk assessment has to differentiate the results for males and females, due to the characteristic differences in their body weights to which any exposure assessment will be related.

Age is very important to distinguish populations due to their physical and mental development and behaviour. Normally, the whole population can be separated year by year of age, while for very small children month by month. It is also appropriate

Table 3.5 Characteristics of populations for statistical stratification

Population particularity	Descriptor
General population	The entire people in a country or nation or any other geo graphically characterised region
Consuming population	Part of the population consuming a product (e.g. eating food)
Gender	Female vs. male
Pregnancy and/or breastfeeding	Pregnant and/or nursing women
Age	Newborns, babies, small children, infants, school children, adolescents, adults
Body weight, body height, body surface area	In relation to age
Region	Geographical, national, international
Ethnic	Native population, migration background, religious particularities
Education	School, university
Behaviour	Food consumption, housing, car driving, etc.
Mouthing behaviour	Primarily children but also all other age groups
Hobbies and other habits	Sports, leisure
Populations otherwise characterised	Populations characterised by their particular vulnerability and sensitivity

to differentiate the population according to their development: babies, toddlers, infants, small children, school children, adolescents and adults. Differentiation by age is very important particularly if substances may act on the development, as well as in cases where age-dependent behaviour must be considered, e.g. playing, house dust intake or eating habits.

Regional differences may characterise a population, for example, the European population may be stratified according to the regional (north-south-east-west) and national affiliation, or to populations with a specific ethnic background.

Habits may also be a reason to differentiate populations. For example, it is believed that children may be exposed to house dust due to their frequent mouthing behaviour. Vegetarians and vegans also represent populations for whom the total dietary exposure might be different compared to the general population.

From the background of regulatory exposure assessment, it is of primary interest to identify the population that might be at the highest risk. Normally, these are people who are using products or consuming particular foods at a very high level. A high degree of exposure can therefore be characterised by a very frequent use or consumption of high amounts of products. It is essential to use sufficient statistical methodology to characterise these extremes of use. The frequent and extensive use of a product, e.g. glues in a hobby, may represent such an extreme scenario. Probabilistic estimations using an appropriate database can deliver percentiles at the 95th and higher levels. For example, food consumption survey data can reveal this information as well as food concentration data. It should be taken in mind that food consumption surveys have normally a representative statistical basis, while food concentration data often do not (compare Sects. 4.6 and 4.8). Data characterising the use of consumer products are nearly always missing. Therefore, risk assessors have to work with conservative defaults, as described above. In this context, it should also be discussed what vulnerable and sensitive populations are under consideration. In both cases, toxic effects may occur at lower doses than in the general population, due to higher physiological sensitivity (e.g. children) or because of pathological existing illness (e.g. liver disease).

The evaluation of highly exposed populations should always be done in relation to the "normal" (typical) exposed population. What is "normal"? In representative surveys having a sound statistical basis, the normal population may be represented by the central tendency of the data, characterised by means and medians and the variability indicated by standard deviations and confidence intervals. It should be remembered that central tendencies are sometimes based on expert judgement or on mean data from insufficient sources not fulfilling the basic statistical criteria. Considerable differences of median and mean indicate that the population sample is not homogeneous. Such data lacking a sufficient empirical basis can be described only under consideration of uncertainties, including a description of the ranges and confidence limits. The statistical presentation and description of data used for exposure analysis, including the quality and uncertainty discussion, is presented in later chapters.

3.2.13 *Documentation of Exposure Scenarios*

Exposure scenarios should be documented in a clear and transparent manner. All information and data should be considered that explain the entire situation by which exposure takes place. Table 3.6 shows a template based on a proposal for documentation of exposure scenarios that was developed as part of the project EIS-ChemRisks of the Institute for Health and Consumer Protection at the Joint Research Centre of the European Commission in Ispra, Italy (Papameletiou et al. 2005, 2006; Schwela et al. 2003). In addition, lists and thesauri of the parameters to fill out the lists have been developed. Unfortunately, this project due to lack of financial support could not be continued, and the database is not publicly available. The example for the basic requirements of documentation of an exposure scenario

Table 3.6 Documentation of exposure scenario according to the proposal in the EIS-ChemRisks database

Documentation of exposure scenario General information	Remarks
Purpose of the evaluation	Characterisation of the reasons and the objective and aim of the assessment Reference to possible regulation
Stakeholder/rapporteur	Name of the legal entity, body or institution preparing the document
Issue at stake	Characterisation of question or problem
Review level	No review, internal, external review, peer review
Specific information	Any information not specified otherwise
Source, product	Name of product or product category
Agent, chemical	Name and identification of substance under evaluation
Target/subject	Consumer, German population, vegetarians
Macroenvironment	Houses, playing ground, kindergarten, school, etc.
Microenvironment	The rooms in a home
Geographic applicability	Harz region
Seasonal applicability	Not specified
Exposure activity	Eating food
Chemical release	Substance in food
Exposure pathway	Oral intake of entire food
Exposure event	Daily intake of the food item
Exposure route	Oral intake
Chemical-specific data	E.g. molecular mass, vapour pressure, saturation pressure
Non-chemical-specific data	Duration of exposure, manufacture of product, storage, body weight of exposed person
Exposure assessment calculations	Due to accepted model estimation
Exposure results	The presentation of the main exposure results and quantitative estimates to be taken for risk characterisation

given here is made according to the database EIS-ChemRisks. This project is presented in Sect. 7.3.1.

In addition, it is appropriate to prepare a textual characterisation of the exposure scenario (see examples above).

The statistical presentation of exposure parameters and the exposure estimation result should reflect the objective of the assessment. If, for example, the exposure of an individual person is evaluated in terms of the tiered approach, the iterations should include the variability of the exposure parameters, but with regard to the individual behaviour. If a population is in the focus of the assessment, the central tendency and a conservative value should be estimated to characterise the "general" population and the so-called high-level consumers. High consumption may be reflected by consuming or using high amounts of products and high frequencies as well or both.

3.2.14 Special Aspects of Exposure Scenario Characterisation

Special aspects of characterisation of exposure scenarios are typically due to regulatory prescriptions. In some regulations, very distinct requirements are formulated regarding to the exposure scenario (pesticide regulation, cosmetics). Basically the rules follow the scenario concept characterised in this chapter and their models as described below.

In the REACH regulation, the exposure scenario characterisation is due to particular regulatory rules. It is discussed in Sect. 5.6.3 in detail. In general, the REACH exposure scenario describes how the producer of chemical products and substances will control the safety of their products. This includes as an important element the consideration of risk management measures (RMM). Risk management measures (Bruinen de Bruin et al. 2007) are launched as important descriptors besides the exposure scenario conditions (ESC) that are discussed in this chapter. Some ESC become of particular regard and others less as pointed out in Sect. 5.6.3.

3.3 Exposure Models and Measurements

Matti Jantunen and Klaus Schneider

3.3.1 Definition of Exposure Model

In 2009 US-EPA adopted the National Academy of Sciences' (NAS) definition of a model as

> a simplification of reality that is constructed to gain insights into select attributes of a particular physical, biological, economic, or social system. (NRC 2007)

According to WHO an exposure model is

> a conceptual or mathematical representation of the exposure process. (WHO 2004)

that often includes estimates of intake (e.g. amount of chemical inhaled or ingested) and the amount of a chemical that is absorbed into the systemic circulation (e.g. amount of chemical that penetrates the gut wall, skin or the lining of the lung). An exposure model is, thus, a mathematical representation of exposure scenario delivering a quantitative estimate.

Fate and transport models are used to estimate the transport and transformation of agents in the environment and yield concentrations in different environmental media. The outputs of these models represent potential exposures to receptors, although these estimates often are used as proxies or surrogates for actual exposures or serve as inputs to other exposure models.

Integrated fate/transport exposure models are used to predict both concentrations in the contact media and actual exposures, i.e. by incorporating exposure factors, such as time activity of the target individuals or populations.

Human exposure models incorporate information on environmental concentrations and exposure factors and yield predictions of exposures based on actual or assumed contact between a receptor and the contaminants in the environmental media.

Intake fraction models are used to estimate the relative proportion of an emitted contaminant that is (eventually) taken in via inhalation, oral or dermal absorption by a defined target individual or population (usually the entire exposed population). Intake fractions are dimensionless.

Dose estimation models predict internal doses at target tissues or organisms that result from exposure to an agent. In the case of reverse dosimetry, dose estimation models reconstruct exposure levels that are consistent with measurements inside an organism or biological material. Note that all exposures do not necessarily lead to a dose.

Life cycle impact assessment (LCIA) models consider that humans are exposed to chemicals in consumer products through product use (near field) and environmental emissions (far field) over the product life cycle.

In practice software tools and applications such as ConsExpo are often also called exposure models. This is of course incorrect because models are mathematical formulas and not tools.

We recommend two supporting background documents. First, IPCS has published a general report on the principles of characterising and applying human exposure models (WHO 2005). Second, EC/JRC has published a comprehensive guidance for the harmonisation of consumer exposure modelling (Kephalopoulos et al. 2007).

Table 3.7 WHO model categories [taken from WHO (2005)]

	Mechanistic	Empirical
Deterministic	Mathematical constructs of physical/chemical processes that predict fixed outputs for a fixed set of inputs	Statistical models based on measured input and output values (e.g. regression models that relate air concentrations and blood levels of a chemical or ambient pollutant concentrations with personal exposures)
Probabilistic	Mathematical constructs of physical/chemical processes that predict the range and probability density distribution of an exposure model outcome (e.g. predicted distribution of personal exposures within a study population)	Regression-based models, where model variables and coefficients are represented by probability distributions, representing variability and/or uncertainty in the model inputs and parameters

3.3.2 *Concepts of Modelling Exposure*

An exposure model is a computational framework designed to reflect real-world human exposure scenarios and processes. On the one hand, exposure models can be divided into mechanistic and empirical and on the other hand into deterministic and stochastic, also called probabilistic (see Table 3.7). Mechanistic models mimic specific exposure scenarios, i.e. the physical and chemical processes that transmit contaminants from points of release to the target individuals or populations and the behavioural patterns of these targets. Empirical models are based on the statistical associations of the measured exposures within the target populations to the selected input parameters. Deterministic models, mechanistic or empirical, produce one output value for one set of input data, such as product use, time-activity and other exposure factors, source strengths, dispersion parameters, concentrations in macro- and microenvironmental media, etc. Probabilistic models apply statistical uncertainties and/or variations of the input variables, expressed as probability distributions, to produce probability density distributions for the output values. They are characterised by their treatment of the variability and/or uncertainty of the input variables, not by the model structures or algorithms.

An obvious yet, when using complex computerised models, sometimes forgotten fact is that no mechanistic model predicts the exposure impact of a variable that is not included in the model.

3.3.3 *Empirical Models*

Principles of Empirical Models

An empirical model is based on a data set containing measured values of exposure and respective (expected) exposure determining factors. The algorithm of an empirical model is not established on physical and chemical principles. Instead it is a statistical best fit (typically multiple regression) between the exposure and its determinants for that data set. Therefore, empirical models are, in principle, descriptive for a particular set of conditions and not predictive in a more general sense. The regression coefficient (β) and accuracy (p-value) of the relationship between the exposure and its determinants in an empirical model may be indicative for other similar exposure scenarios, but they are strictly valid only for the original data set. These facts being acknowledged, an empirical model can, however, be cautiously applied for a similar exposure scenario in a new time, population or location.

A significant advantage of an empirical model is that (for the setting it is drawn from) it describes reality, including the influences of the expected and, indirectly, also some ignored determinants of exposure.

Credibility and Applicability

Although empirical models are not based on physics or chemistry, the credibility of an empirical model depends, in addition to the p-value, on how logical or comprehensible its structure and predictions are in comparison to the user's understanding of the physical world. An empirical model that is based on a small and poor database or badly selected input variables may, regardless of its initial β-coefficient or p-value, end up being structurally illogical and generate counterintuitive input-output relationships.

If you cannot make sense of the structure and behaviour of an empirical model, you should not use it.

The *applicability* of the predictions of an empirical model depends on its β-coefficient. A low β-coefficient means either that the model input determinants do not capture those that determine the exposure variability or that the database the model was derived from does not contain enough input and output data variability. An empirical model should be applied for new input data sets only when these have been collected identically with the original data, and their values do not lie markedly outside of the original data ranges.

Examples

Empirical consumer exposure models are not frequently found in the literature. Empirical models are applied to some extent in investigating the relationship between ambient air contaminant concentrations and various influencing parameters (e.g. Jerrett et al. 2005). The lack of empirical models in consumer exposure modelling is probably due to the fact that because empirical models describe the exposure-determinant relations in specific data sets, their potential to be generalised is restricted. Yet, an empirical consumer exposure model, based on, e.g. insecticide exposure data from HBM and respective questionnaire collected determinant (insecticide application, household characteristics, time-activity data, etc.) from one community and target population, could be applied for exposure modelling for other similar communities and target populations for which identical questionnaire have been applied. The model could also be applied in a follow-up study on the original target population.

Mohorovic (2003) investigated the relationship between methaemoglobin levels in women and ambient air pollution in a region close to a coal power plant. He found a good correlation between individual methaemoglobin levels and sulfur dioxide air concentrations. If such an empirically found correlation would be used for predicting sulfur dioxide concentrations, this could be considered an empirical exposure model.

Egsmose et al. (2016) established a correlation between PCB 28 (polychlorinated biphenyl congener 28) plasma concentrations in mothers and school children and the construction period in which their respective homes and/or schools were built. Critical construction periods can then be used to identify sources of potentially higher PCB exposure.

Another prime example of empirical consumer exposure model development and application is the Finnish drinking water mutagenicity exposure model (Vartainen and Liimatainen 1988). 97 drinking water samples representing different seasons and 39 waterworks were tested for TA 100 mutagenicity. Routine water quality control parameters were collected simultaneously, and a best fit equation to explain the measured mutagenicity by the most influential water quality parameters (TOC, Cl_2 and NH_3) was:

$$f = A\left(1 - e^{-kC_R}\right) + A\left(1 - e^{-kC_D}\right) \tag{3.1}$$

A and k are constants, and R refers to raw water and D to drinking water.

For this empirical model, the correlation coefficient R between the modelled and measured mutagenicity was 0.880 and the β-coefficient 0.984. The model was subsequently used to estimate the drinking water mutagenicity for 56 Finnish municipalities in an epidemiological study that established the association of gastrointestinal and urinary cancers to drinking water mutagenicity (Koivusalo et al. 1994).

3.3.4 Mechanistic Models

Principles of Mechanistic Models

A mechanistic model (also called physical model) is driven by chemistry and physics. It is based on a given exposure scenario and understanding of the relation between the parameters and processes, which define on the one hand the fate of the substance being used or released into the environment and on the other hand the behaviour of the target individual or population within that scenario. The algorithm of a mechanistic model is a mathematical description of the relationship between the values of input parameters and the resulting exposure. As each input parameter (e.g. use of a substance, release of contaminant into indoor air, concentration in food, time activity of the target individual) has a specific physical meaning, it can be set to reflect different starting conditions. A mechanistic model is therefore, in principle, predictive, i.e. it can be applied to predict (and compare) exposures in situations that are controlled by similar phenomena but for which no empirical data exist, e.g. for alternative future exposure scenarios. In practice even the mechanistic models are not truly universal, but developed for specific exposure agents, sources, pathways and routes of entry.

Models can be route-specific (as the food exposure model above) or integrate various sources and exposure pathways, thus providing aggregated exposure estimates for these sources and routes addressed.

Tiered Approaches to Exposure Modelling

A mechanistic model is based on a mechanistic understanding of the relationship between the source and how the substances transfer via different pathways to the exposure target. As this understanding can be imperfect and omit relevant parameters, it needs to be validated against empirical data (monitored exposure data) for specific settings of input parameters.

If the predictive accuracy of the model is not sufficient or the applicability domain is not wide enough for a specific purpose, it can undergo refinement following a tiered approach.

In Sect. 3.2.1 "Exposure Scenario", a "Simple exposure scenario B" was given as an example for taking up contaminants from food.

The algorithm for this oral food model is:

- Tier 1: Exposure to substance A from food is

$$\text{Exposure (A)} = \text{Concentration(A)} * \text{Consumption} \tag{3.2}$$

where:

Exposure (A) is the intake of A via food (in mg).
Concentration (A) is the highest concentration of A in any food item (in mg/kg food).
Consumption is the average total amount of food consumed by the adult population (in kg food).

If A characterises an aggregated scenario, this simple model will create a substantial overestimation of exposure to A from food. This assessment can be refined in a tier 2 assessment, if data on consumption and occurrence in various food items are available.

- Tier 2: Exposure to substance A from food is

$$\text{Exposure (A)} = \sum \text{Concentration}(\text{A}_\text{i}) * \text{Consumption(i)} \tag{3.3}$$

where:

Exposure (A) is the intake of A via food (in mg).
Concentration (A) is the concentration of A in food item i (in mg/kg food).
Consumption (i) is the average amount of food item i consumed by the adult population (in kg food).

Tier 3: Further refinements could include differentiation between age groups, seasonal variation, etc. Also, as mentioned above, probabilistic modelling could consider the variation of both concentration and consumption data.

The intake models could also be combined with PBPK models to predict dynamic short-term body burdens and long-term accumulation in critical organs, etc. In general, refinements improve accuracy and information value of the output (exposure prediction) at the cost of elevated requirements for the input data and mechanistic understanding. Refinement can be repeated until the model satisfies the model application requirements, but care has to be taken.

Refining a model tends to increase its complexity, narrow its applicability and increase its need for input data. Therefore, a mechanistic model should be kept as simple as possible without compromising its applicability.

Examples

In order to deepen the understanding of principles of mechanistic models, two examples are given on models published in the literature and how they were established.

In an attempt to describe the transfer of various organic substances from consumer products (polymer-coated textiles, PVC floorings, paper) to cotton wipes (as a surrogate for the receiving skin), Clausen et al. (2016) performed experiments to measure the transfer of various substances from article surfaces to dry wipes and wipes soaked with artificial sweat or methanol. From these experimental data, the diffusion coefficient D was calculated for all substance-matrix combinations. D was used to establish two mechanistic models. Both are based on Fick's second law of diffusion, but the second allows for adherence of substance to the surface film and, consequently, reduced absorbance by the receptor surface. Predictions from both models were compared with mass transfer from articles to wipes under varying conditions.

Shen et al. (2014) established a skin absorption model for fragrance materials. The model uses absorption ratios in percent-applied dose. Substances are assigned to three groups with different absorption ratios by estimating substance-specific physicochemical properties [octanol/water partition coefficient (Kow), water solubility (S) and permeability coefficient (Kp)] with QSAR models. The established skin absorption model was validated using a set of existing data.

In the following second example, the estimation of room air concentrations from painting is illustrated by four iterations using deterministic parameters. They follow the tiered approach that is indicated based on the example given in the scenario in Sect. 3.2.10.

The resulting exposure for scenario A is 16 g/m^3, over the whole time, due to a fixed rate of zero ventilation.

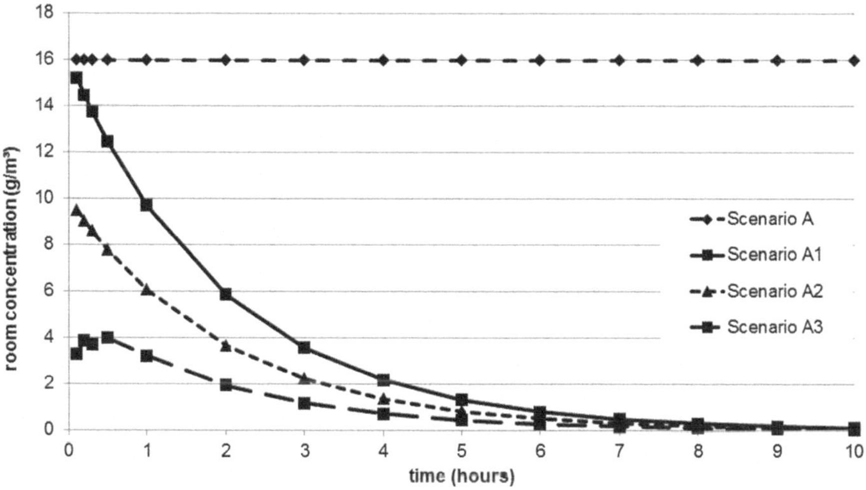

Fig. 3.6 Iterations of model estimation of room concentrations

The first iteration step from scenario A to scenario A1, as pointed out in Sect. 3.2.1, adds a nominal room ventilation rate of 0.5 h^{-1} into the model. A second step, scenario A2, could incorporate real ventilation rate and room volume, which change the model output from a conservative to more realistic one. Scenarios A, A1 and A2 are similar to the "source and ventilation" model in the ConsExpo tool and can be calculated by, e.g. MS Excel.

As a next step in the tiered approach, the evaporation of the substance that depends on the vapour pressure, molecular weight and room temperature may be added. Such calculations can be made by the ConsExpo "evaporation" model (Van Veen 1999).

Figure 3.6 demonstrates the change in the exposure estimate from the lowest and most conservative to higher-tier models, i.e. scenarios A–A3. The decreasing exposure estimates result from increasing realism of the model and additional input parameters as one goes from low- to higher-tier models.

Progression from low- to higher-tier models should result in exposure estimates which approach a realistic value. To add a further step in the assessment, the model parameters may be expanded to distributions leading to ranges of exposure. This will be pointed out in later chapters.

Probabilistic Models

For many purposes the relevant exposure value is the average exposure of a population, expressed as single value (point estimate). More often, however, the genuine interest is the variability and uncertainty (Sect. 3.8) of an exposure estimate, the full distribution of the exposure levels within a population or, e.g. its 95th percentile. Probabilistic modelling techniques respond to such needs. In rare, simple cases, probabilistic modelling may be possible analytically. Normally, however, computerised Monte Carlo or Latin Hypercube software packages are used. They sample the model input values randomly from databases or mathematically defined statistical distributions, calculate the exposure and repeat it a large number of times to produce a probability density distribution for the modelled exposure estimate. Average, median, 95th percentile and other values of interest can then be extracted from the distribution.

The same technique can be used to quantify the uncertainty of the exposure, and variability and uncertainty analyses can be combined in a two-way probabilistic modelling technique.

The simplest Monte Carlo modelling tools assume that all input parameters are statistically independent of each other. When this is not the case, i.e. quite often, the assumption would lead to more or less erroneous risk distribution estimates. Time-activity allocations serve as good examples. More time spent on one activity may increase (positive correlation) or decrease (negative correlation) the time spent on another activity. An example of the former could be the time spent outdoors and gardening, and, of the latter, the time spent in the kitchen and in the living room. More advanced probabilistic modelling tools incorporate intercorrelations between the variables as inputs. Note, however, that when the input data are drawn from complete individual-level data sets, such as time-activity diaries, the interdependencies of the times spent in different microenvironments and activities are built in and application of the respective intercorrelation coefficients would lead to double accounting.

The general application probabilistic modelling tools include @RISK and Crystal Ball.[2] There are also several published and available tools specifically for probabilistic exposure and consumer exposure modelling, such as PACEM (the probabilistic aggregate consumer exposure model) (Dudzina et al. 2015), PACEM-KD (Extended, *k*inetic-*d*ermal, version of the original PACEM) (Garcia-Hidalgo et al. 2018), PROMISE (Probabilistic Methodology for Improving Solvent Exposure Assessment; American Chemistry Council) (Silken Inc. 2003), LifeLine Version 5 (Aggregate and Cumulative Exposure/Risk Assessment Software[3]), APES/TRIM (US-EPA 2016a), CARES,[4] ConsExpo 4.0 (Delmaar et al. 2005), Calendex (Young et al. 2012) and SHEDS (Zartarian et al. 2008). In fact, most of the recent higher-tier exposure models incorporate probabilistic exposure (and risk) modelling capability. The German XProb project (UBA 2014) created a whole framework for probabilistic exposure (and risk) modelling based on simultaneously developed exposure factor databases.

---

[2]See, e.g. http://fpfn24.wdfiles.com/local%2D%2Dfiles/suplementos-de-clase/@RISK5.0_v_CrystalBall7.3.pdf

[3]See, e.g. http://fpfn24.wdfiles.com/local%2D%2Dfiles/suplementos-de-clase/@RISK5.0_v_CrystalBall7.3.pdf

[4]www.caresng.org

3.3.5 Life Cycle Impact Assessment Models

Life cycle impact assessment (LCIA) models are often based on intake fraction (iF) modelling techniques.

Exposure to chemicals in consumer products is not limited to near-field exposure, i.e. exposure as product user and consumer. Humans can be exposed also via environmental emissions over the whole product life cycle including far-field exposure from industrial production and manufacturing to waste disposal. Such comprehensive exposure pathways are often complex; chemicals can transfer directly from products to humans during use or exchange between various indoor and outdoor compartments until subfractions reach humans. A LCIA model can evaluate complex exposure pathways along a product life cycle with a mass balance-based assessment framework that structures multimedia chemical transfers in a matrix of inter-compartmental transfer fractions. Cumulative multimedia transfer fractions are then quantified by matrix inversion, and exposure pathway-specific product intake fractions (iF) are defined as chemical mass taken in by humans per unit mass of chemical in a product. LCIA models have typically low spatial resolution (Fantke et al. 2016, 2018).

Finally, prioritising the potential risk posed to human health by chemicals requires tools that can estimate exposures from limited information, i.e. in the absence of extensive databases for high-tier exposure modelling. Ring et al. (2019) created such a tool, a meta model, SEEM (systematic empirical evaluation of models), based on 2 near-field and 2 far-field pathways, 6 near-field and 7 far-field data inventories and exposure models, and tested it against CDC NHANES human biomarker data for 114 chemicals. SEEM was then applied for median iF's with credible (95%) intervals of relevant exposure pathways for 479,926 chemicals as population median intake rate (mg/kg bw/day) per kg of chemical released (or produced as a surrogate of release). For 1880 chemicals, the confidence intervals exceeded 100 μg/kg of BW per day; for 474,527 chemicals, it stayed below 1 μg/kg of BW per day; and for the remaining 3474, it was between 1 and 100 μg/kg of BW per day. Once the release or production of a chemical is known or estimated, application of the SEEM and the already-generated database can provide a fast first-order estimate of population exposure.

3.3.6 Model Selection Criteria

Ryan (1991) listed the following criteria affecting the selection of a model and stated that they are often in direct conflict with each other:

Mathematical and computational simplicity depend on the complexity of the modelled phenomena. The relevance of computational simplicity, however, decreases as the used software packages improve in power and usability. Of importance is that the user should be able to understand the model. Yet, the models should be no more complex than necessary for the purpose, i.e. what has been characterised in the scenario description.

Interpretability requires that the model system reflects our understanding of the structure and workings of the target system. It should be transparent—i.e. its assumptions and limitations should be easily seen. For consistency the model should include all relevant aspects of the target system. Such a model can be used to answer a whole range of different questions.

Accuracy of the model prediction depends on the model's construct validity but also on the validity of the model scenario, default values and the often internal (in the modelling toolbox) databases, from which many of the model input values are often drawn.

Our final criterion, *accessibility of the input variables and data*, is being increasingly met by reference materials such as the US-EPA's *Exposure Factors Handbook* (US-EPA 2011), the US-EPA's *Child-Specific Exposure Factors Handbook* (US-EPA 2008) and the US-EPA's Consolidated Human Activity Database (CHAD) (US-EPA 2016b).

At a national level, for a long time, the only compilation of available exposure factors had been done in Germany (AUH 1995). In 2003, the French database CIBLEX was published (Beaugelin-Seiller et al. 2004). Neither of these national data sources contains many distributional data to support probabilistic modelling.

The European Exposure Factors (JRC 2015) (http://expofacts.jrc.ec.europa.eu/) database (since 2004) is a new approach for exposure factor data availability (Jantunen 2006). It is a continuously updated Internet-based interactive free access database with data from 30 European countries and links to dozens of original data sources (Vuori et al. 2006). More recently results of the German Environmental Surveys from 1986 to 2016 and the XProb project (Mekel et al. 2007) of the German Ministry of the Environment were compiled into the updated German RefXP (exposure factor database for probabilistic exposure modelling) database (UBA 2014).

3.3.7 Exposure Measurements vs. Modelling, Is There a Contradiction?

Ideally the exposure assessment incorporates all available sources of information while considering their relevance and quality.

We are not aware of population representative monitoring surveys which target consumer exposures. Population-based biomonitoring surveys, on the other hand, such as the German Environmental Surveys (I to V) since 1986, the CDC NHANES (CDC 2016) surveys in the USA since 2000, the Flemish HBM Programme since 2002 and the French ENNS-based programme since 2010 (Dereumeaux et al. 2016) have collected data on growing numbers of chemicals (currently approx. 200) in human blood and urine. Except for some occupationally exposed individuals, most of these chemicals can be assumed to originate from consumer exposures, and certainly they include the impacts of consumer exposures. Consequently, such biomonitoring data

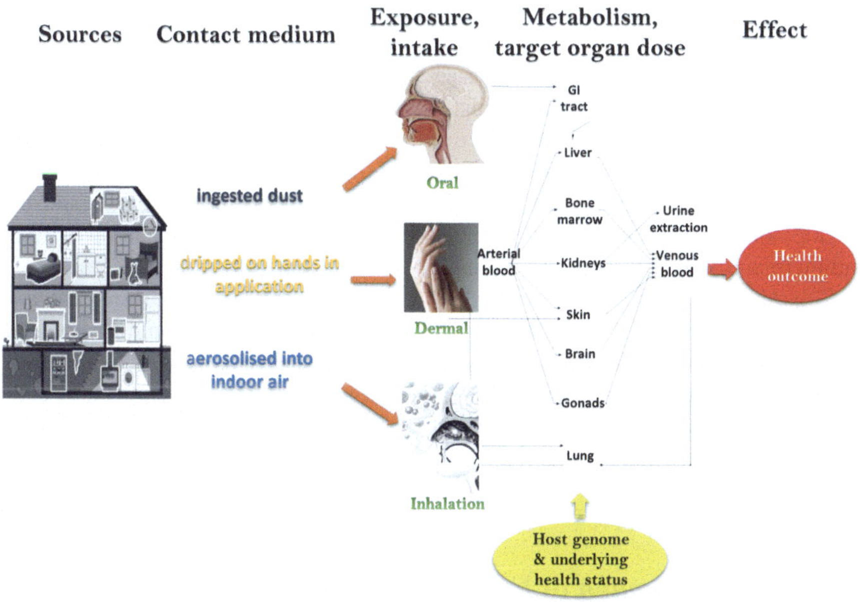

Fig. 3.7 The INTERA full-chain model from sources via exposure to health outcomes

provide an upper-bound estimate of the population-level exposure to the chemicals that have been surveyed.

Most measurements on human exposure to specific chemicals from specific consumer products originate by necessity from small-scale laboratory or controlled field experiments that have been performed for permission or compliance purposes.

Many more experimental uptake and target organ dose measurements have traditionally been required and performed with experimental animal studies. This practice, however, has recently been changed in Europe via, e.g. the EU directives on the protection of animals used for scientific purposes (EU 2010), the Regulation on cosmetic products, REACH (EU 2006) and CLP (EU 2008), all of which require, or strongly encourage, the replacement of animal testing. To adequately predict the potential risk, especially for normal life low-dose exposure, new prediction models are needed and are based on in vitro testing in combination with computational modelling. For the validation of alternative methods to animal testing, EU established, in 2011, EURL ECVAM, the European Union Reference Laboratory for Alternatives to Animal Testing (EU 2011). The Laboratory is actively involved in the search for test methods that replace, reduce or refine (the "Three Rs") the use of laboratory animals in the test process. While these computational models (typically PBPK models) focus on the chains of events from external exposure to tissue concentrations, upstream models, such as described in Sects. 3.2.3–3.2.5, are used to estimate the external exposures, which then can be linked to the PBPK models. Together they form so-called full-chain models. Figure 3.7 presents the conceptual full-chain model developed in the European INTERA project (Asikainen et al. 2011).

Because most consumer exposure assessments are needed for regulatory permission purposes, to ensure that the use of a particular product or substance is safe, the assessments are necessarily prospective and done via modelling. Depending on the modelled safety margins and the importance of the product for its manufacturer and end user, assessments from tier 1 to tier 3 (see Sect. 3.3.4) are used. Actual exposure and biomonitoring measurements are much more expensive and demanding to make and therefore used only occasionally and then for model validation.

Consequently, there is no contradiction between exposure measurement and modelling, i.e. models are the tools for everyday risk assessment and compliance verification, and measurements are essential for model development and validation.

Modelling is the primary approach to assess consumer exposure, and Huang et al. (2017) present a comprehensive review of models for near-field exposure pathways of chemicals in consumer products.

3.3.8 *Documentation of Exposure Models*

As exposure models may be complex and difficult to understand, comprehensive and transparent documentation and communication are key for their correct use. Ten requirements for model documentation were formulated within the IPCS. These requirements are summarised in Table 3.8 (for more details see WHO 2005).

Recently, Ciffroy et al. (2016) proposed a standard documentation protocol (SDP) for exposure models, following the outcome of a CEN (European Committee for Standardisation) workshop. The CEN workshop agreement resulting from this activity details minimum requirements for the amount and type of information that should be documented for exposure models and provides a structure for communicating the information.

Table 3.8 Documentation requirements for exposure models according to WHO (2005)

General topic	Requirement
General model description	1. General description of the purpose of the model and its components
	2. Individual- or population-level analysis (level of aggregation)
	3. Modelled time resolution
	4. Applicability to diverse exposure scenarios
Model inputs	5. Description of data inputs
Model processes	6. Modelling tool methodology
	7. Model code and platform
	8. Model performance and evaluation summaries
Model outputs	9. Description of model outputs
	10. Model sensitivity and uncertainty analysis

3.4 Exposure Parameters

Gerhard Heinemeyer

Every exposure model must be fed with qualitative, semi-quantitative or quantitative values to prepare an estimate of exposure.

A number of different terms exist to delineated the values that are used as input data in exposure models. The term "exposure parameter" is used in the WHO guidance documents on exposure modelling (WHO 2005) and on uncertainty analysis (WHO 2008) as a general term for model input data. The title of the *Exposure Factors Handbook* edited by the US-EPA (2011) implies that the term "exposure factor" is characterising any information about exposure parameters. The handbook itself uses the term "exposure data" in a way to describe the information needed for exposure assessment. This source also informs broadly about data that cover most of the information related to physiological and behavioural parameters commonly needed for exposure assessment.

A different meaning of the term "exposure data" was used by Papameletiou et al. (2005) in the EIS-ChemRisks project (Schwela et al. 2003; Papameletiou et al. 2006). It was proposed to use the term "exposure data" for specific product information, such as weight fractions, packaging information, ingredient concentrations, etc. The term "exposure factors" should be used to describe behaviour and anthropometrics. For practical reasons, this differentiation seems appropriate and is used also in this work (see Sect. 7.4.1).

Therefore, in connection with exposure models, the variables used in a model should be delineated as exposure parameters, as done in the WHO guidance documents on exposure models and the guidance on uncertainty evaluation. Thus, the term exposure factor means any information that relates to non-product-specific information, and exposure data mean the product-specific information.

In general, parameters used for exposure analysis should fulfil the scientific and statistical criteria for high quality (see Sect. 3.7).

3.4.1 Exposure Factors

Due to the above-mentioned definition, the term "exposure factor" is attributed to any parameter in an exposure scenario/model that describes populations and the pattern of behaviour of people using products or eating food, physiological data characterising the exposed population and the environmental conditions under which the exposure takes place.

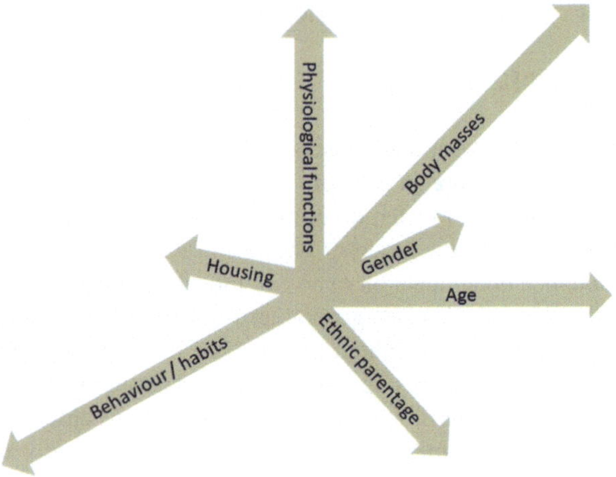

Fig. 3.8 Determinants that characterise populations and their living conditions for exposure assessments

Populations

Populations can be assigned at first as the general population which covers every person within a target area and which can be very heterogeneous. A target population is any defined subsection of the general population, e.g. adolescent males.

Figure 3.8 is expressing the multidimensional character by which all descriptors are dependent on each other (see also Sect. 3.3.4), which must be considered in scenario characterisation and in the model building.

In the following the descriptors of populations are listed that characterise persons that are exposed to substances of concern. It should be noted that all of these variables are dependent amongst each other.

Age

Nearly every physiological parameter used in exposure assessments is age-dependent. Therefore, age is the most important determinant of individuals and populations. Consequently it is important to separate individuals and populations into age classes. There are numerous systems to differentiate populations and individuals by their age. A broad differentiation into age classes may be made by separating populations into newborns, babies, toddlers, infants, schoolchildren, adolescents, adults and elderly persons (Table 3.9). Due to inaccuracy of these categories, the numerical characterisation for ages should be preferred.

Gender

Any populations and individuals should be generally characterised by gender due to respective body weight and height, as well as to gender dependent metabolic activity.

Table 3.9 Age categories to be used in exposure assessments (numerical values can be taken, e.g. from the XProb database) (UBA 2014)

Age group	Range of age
Newborn	<12th week after birth
Babies	>12th week to ≤12th month of life
Toddlers	Children during the phase of learning to walk
Small children	>12th months to ≤5th year
Children	>5 to ≤12 years
Adolescents	>12 to <18 years
Adults	>18 to ≤70 years, can be separated into 10-year groups
Elderly	>70 years

Body Masses [Body Weight (BW), Body Height (BH), Body Surface (BS) and Body Mass Index (BMI)]

Exposure estimates are normally related to body weight. Body weight should be documented for every individual person with age and gender. Actual body weights for populations can be taken from numerous statistics presenting body weights for different ages and populations. Data that are based on representative statistical investigations, under consideration of at least age and gender should be preferred. Representative data on body weights are, for example, available from the particular national food consumption studies, performed for children, adolescents and adults. Otherwise, default data should be used only if no other data are available.

Body weight should be measured by using a standard quality scale, equipped with a digital display. The person should wear light dress (underwear).

In general, food consumption surveys provide actual and precise information about body weights in a population. The most current data on bodyweight of the German population are published in the report of the second German food consumption survey (MRI 2008). This report focuses on the ages between 14 and 80 years. The mean values of body weight are presented for the different age classes (Fig. 3.9). An increase of body weights can be observed from the adolescent age up to the age of 30–40. In the elderly a slight decrease is obvious. The mean body weight for all age groups is 83 kg for men and 67 kg for women, and the 95th percentiles are 110 kg and 97 kg, respectively.

In many of the existing risk assessments, a standard default value of 60 kg has been used for body weight. Related to the actual body weights, use of this body weight overestimates the exposure.

Body height is needed to estimate the body surface. Several formulas were established to estimate the body surface (BS) from body weight (BW) and body height (BH). The first proposal was published by DuBois and DuBois (1916).

$$\mathrm{BS} = 0.007184 * \mathrm{BH(cm)}^{0.725} * \mathrm{BW(kg)}^{0.425} \tag{3.4}$$

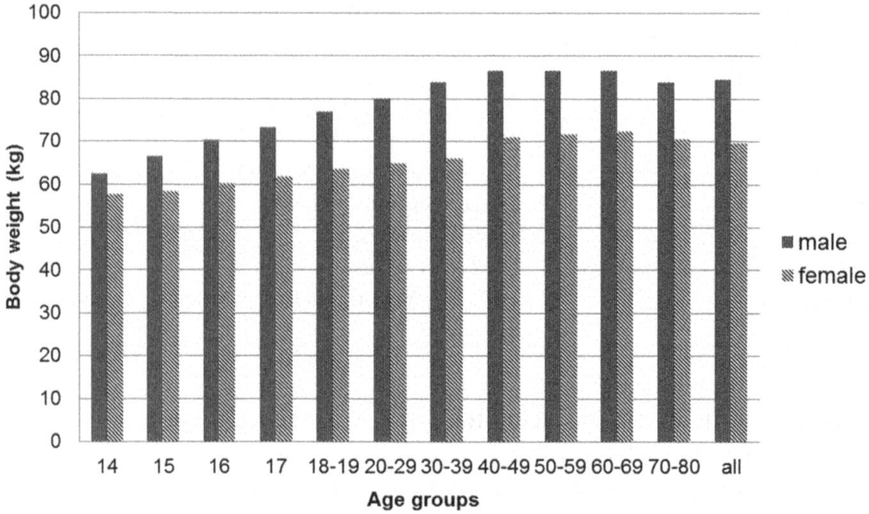

Fig. 3.9 Mean body weights in the German population (14–80 years). Numbers were taken from the report of the second German food consumption survey (MRI 2008)

Body surface correlates more linearly to basic physiologic body function (e.g. basic metabolic rate, liver size, extracellular fluid), than body weight. Therefore, the body surface area is an appropriate parameter to compare age-dependent doses. Although doses are normally related to body weight, body surface is particularly used, e.g. for estimation of drug doses in paediatrics and for pharmaceuticals having very small therapeutic ranges, e.g. antineoplastic substances. Relating doses to surface area is still not established in regulatory toxicology and risk assessment, although a better comparison of age-dependent doses would be achieved.

Figure 3.10 shows the relation of body surface to body weight over the whole span of ages showing that at younger age, the surface is much greater as compared to the weight.

For estimating dermal exposure doses, areas of different body surfaces (e.g. the entire body, trunk, arms, hands, legs, feet, head) are needed. Numerical data can be taken, e.g. from the US-EPA's *Exposure Factors Handbook* or the German AUH report (1995) and its update XProb report (Mekel et al. 2007), the latter being proposed for setting standards.

The body mass index (BMI = body weight/body height^2) provides information about the share of fat to the total body weight. It is needed for estimations of internal doses of lipophilic substances that are primarily distributed in the body fat.

Physiologic Functions

Excretion of substances in urine is often used as an exposure measure [internal exposure, human biomonitoring (HBM)]. Many examples exist at national levels of

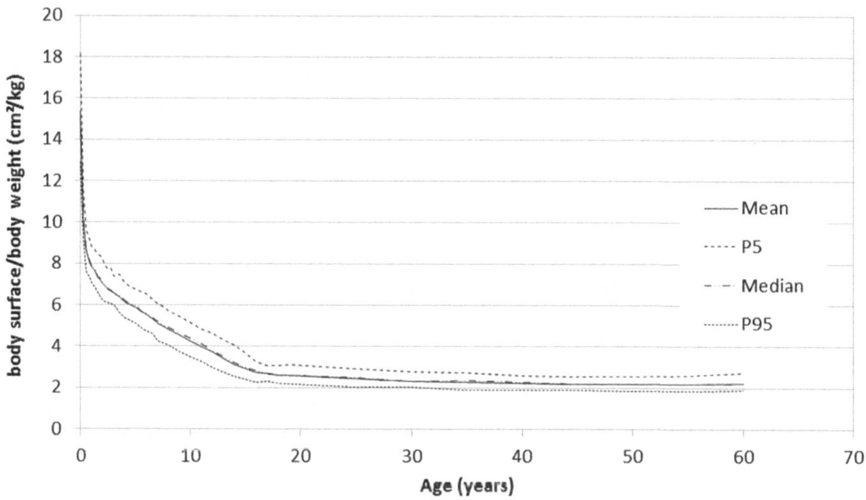

Fig. 3.10 Quotient of body surface area and body weight over age (data from AUH report, estimated by the DuBois formula)

measuring substances in urine. It should be noted that the primary intention of HBM studies is not exposure assessment. It is the observation whether or not substances of concern are present in populations, which is a qualitative approach, and to derive intervention levels (UBA-commission). Human biomonitoring data are discussed in Sect. 3.5.2.

The uptake of substances via inhalation depends on the respiration volume. The higher the respiration volume is over time, the higher the pulmonary diffusion is of substances from the inner lung volume to pulmonary blood vessels. Respiration volumes are dependent on age and on the degree of physical effort. The mean respiration volume of a resting adult person is about 10–13 m^3 per day, but the short-term inhalation rate during heavy work can exceed the average rate and can be increased during work up to seven times. For a differentiated estimation, the changes in daily exercise should be considered.

Blood volume and extracellular volume are needed in toxicokinetic modelling (see also Sect. 3.5.1), as well as metabolic capacity and organ function of circulation, liver and kidneys, including genetic particularities (e.g. poor metabolisers).

Particular groups of the population with certain illnesses that can influence the toxicokinetics of substances must be considered.

Ethnical Background

In multicultural societies, the population should be differentiated not only by regional issues (e.g. people living in the southern or northern part of a country) but also according to ethnicity. Such population differences are indicative for issues like eating habits or consumption of regionally characteristic food like “Weißwurst” or "potato balls". Fish is often consumed at higher amounts in coastal compared to inland regions. Subpopulations in many countries are increasingly characterised by ethnicity, e.g. the Turkish population living in Germany, or Pakistani in the UK, or Algerians in France, or Hispanics in the USA. Behaviours and habits of these subpopulations should be differentiated in exposure scenarios, as well as in studies to generate exposure data, e.g. food consumption surveys.

3.4.2 Activity Pattern Data: Behavioural Data

Most consumer products are not used by everyone, and some are used only by a relatively small subpopulation. Therefore all consumer exposure assessments should first identify the actual users and focus the data collection and exposure modelling to the users only. Obviously, the average values for a population consisting of 10% of users and 90% of non-users are, obviously, meaningless at best and misleading at worst. This includes information about behaviour of individuals and populations that influences the contact conditions to substances and thus exposure. For evaluation of food consumption, harmonised approaches are available and generally applied. For other consumer habits, however, only a few studies exist that inform about consumer behaviour. Some BfR questionnaire studies focused on the use of household cleaners and of shoe sprays, as well as the occurrence of cinnamon as an ingredient in cookies and consumption of energy drinks (Abraham et al. 2011; BfR 2012; Heinemeyer et al. 2006; Lindtner et al. 2014).

Particular questionnaire approaches are currently tested in a pilot study to evaluate consumer behaviour in general (Schneider et al. 2019; see also Sect. 5.2.2).

Amounts Used

First of all, the amount of an item that is used is the most important information to estimate exposure. These items may be any kind of consumer product that is used by consumers and which should be identified and characterised in the consumer exposure scenario (see Sect. 3.2.1). Together with the information about concentrations of a substance in that item, the exposure of an individual or a population can be estimated. Information about amounts of products used can be taken, e.g. from use instructions. This information may be rather general, and it is unclear how far consumers follow the recommendations or even limit the application of the products for the intended uses. Consumers may follow the principle "a lot helps a lot" and thus may use considerably higher amounts or on the contrary use less to save cost. Exposure can therefore be under- or overestimated when based on use instructions. Questionnaire studies performed by BfR (Heinemeyer et al. 2006) showed that users may exceed the amounts of products considerably if they do not expect greater risks (household cleaners), but for products which are expected to be dangerous (shoe sprays), people will follow the basic tenets of the manufacturers' recommendations more closely.

The use of food survey data for exposure assessment of household pesticides was also suggested by Lecomte and Auburtin (2006) (see also Sect. 4.21).

It can be concluded that questionnaire surveys that ask for special uses may deliver adequate data to improve consumer exposure data. The most advanced experiences of using questionnaires are available for food consumption data. Respective surveys have been performed in numerous countries, e.g. France (Dubuisson et al. 2010), mostly following the principles of statistics and epidemiology and thus complying with most of the quality criteria outlined in Sect. 3.7. On the European level, EFSA has built up a food consumption database, which comprises most of the available data from the member states of the European Community (EFSA 2015c) This database is based on the food classification system FoodEx2 (EFSA 2015b). EFSA is working to develop a universal database for information related to food exposure [EFSA Data Warehouse, (EFSA 2015a)]. This project is still under development (see Sect. 7.1).

Exposure Frequency: Frequency of Use

Exposure frequency and use frequency should be discussed separately. Exposure frequency is determined by the frequency of contact with an agent, directly or indirectly. The use frequency covers the time interval between one and the following use(s) for an active user. In many cases, use frequency and exposure frequency are similar. Exposure can occur continuously or discontinuously (see Sect. 2.6). Continuous exposure means that the intervals between exposure events are short and consistent. The intervals in discontinuous exposures are long with rarely occurring events. Frequencies of use should be expressed as events per time unit, i.e. per day, week, month or year, or even per longer periods.

For the risk assessment, the assessor must decide which exposure is continuous and which not to characterise the exposure as short-term or long-term (see Sect. 2.6). This differentiation is crucial because it must be decided whether the exposure estimate should be related to acute or to chronic toxic effects. It has been generally accustomed to delineated exposures as acute and/or chronic which is not adequate in view of the editors of this book.

Due to the above-mentioned examples of frequencies, the everyday consumption of food will lead to a continuous exposure of a certain contaminant in that food and leads to long-term exposure. Particular foods are eaten only at special events, e.g. Christmas. The paint example in Sect. 3.2.1 shows short-term exposure for an occasional wall or furniture painter, but it can be long-term in case of painting small figures as a hobby.

In the risk characterisation, short-term exposure is linked to acute and long-term exposure to chronic toxic effects. This means that an exposure event occurring rarely should be defined as short-term.

In some cases only data on chronic toxicity are available. However, long-term exposure scenarios might be missing, and only products or habits that characterise short-term exposure exists. In this case a risk characterisation regarding to chronic effects is not possible. However, it has been proposed (ECETOC 2015) to extrapolate short-term use to a more extended use to allow for a tentative risk characterisation. In practice, (single) short-term use is spread into several repeated occasions over a defined time interval. However, due to the discussion in Sect. 2.6, there will be no chronic risk if only acute exposure exists, except when the substance of concern or its impact (e.g. mutation) has a long persistence in the human body.

Duration of Use: Exposure Duration

The duration of use may be characteristic for using a special chemical product. For workers, the working time is limited by working hours. Under consumer conditions, the duration of a single exposure event is often longer.

In consumer exposure, duration of use may also be influenced by frequent repetition of single events, e.g. eating a particular food repeatedly, or by smoking one cigarette after the other. It is the purpose of the scenario characterisation to point out the specificity of every scenario with adequate data.

The best way to evaluate durations and frequencies of use is to ask respondents to fill in a protocol. However, systematic surveys and projects following the daily work activities of respondents are rare. There is a need to initiate such studies at a national level by government institutions or universities. The BfR (Schneider et al. 2019) has performed a pilot study to evaluate the practical performance of such a study and to elaborate the extensiveness and costs of a national German survey (see also Sect. 5.2.2).

Duration of (External) Contact

Exposure occurs not only for the active users but also for bystanders and passive users (smokers vs. non-smokers). These individuals are inside the room(s) where the use takes place, Thus they will be exposed to substances that are released and should be included in the exposure assessment.

House Dust Intake

An overview on house dust as an important part of consumer exposure is given in Sect. 5.8. House dust intake is the most frequent pathway for secondary indoor exposures. House dust is the carrier for almost all non-volatile substances that are released from articles and other chemical products, either as household products, toys, pesticide products or equipment (furniture, electronic devices) and other sources (see Sect. 5.4). The extent of exposure depends on the amount of the substance in house dust and the amount and frequency of house dust intake. The intake of house dust is not really known, but defaults have been extrapolated from soil intake (Calabrese et al. 1989, 1997). House dust intake rates have been developed by different national agencies, often based on the same data.

Housing/Environmental Conditions

The environmental conditions cover information about surroundings and conditions of the circumstances under which exposure takes place and on which the exposed person has only partial direct personal influence. They include information about characteristics of houses and flats where people live and thus inform about the place of exposure and regional and seasonal particularities.

Housing conditions are not uniform across the population. The social and economic conditions in populations are indicative of living characteristics. People having low economic means will live predominantly in flats, while those having higher economic means have houses or, at least larger flats. Therefore, the type of residences where people are living may considerably influence exposure by variation of the characteristics of those places, e.g. by variation in ventilation rates.

Place of Exposure

Consumer exposure can occur at various places. Therefore, numerical data used for exposure should relate to the particular situation. Consumer exposure does not include the workplace, but covers staying at home, in school and kindergarten, public buildings (hospital, shops, offices) in cars and other transport and also outdoors. Places of exposure directly govern the choice of data, e.g. room volumes, but also particular habits such as frequency and duration of stay.

Size of the Place of Exposure

The amount of a volatile substance applied or released in a room divided by the room size equals the concentration of a substance that can potentially be inhaled. Room size for various parts of a residence (a bedroom as an example) and for a particular type of residence (a flat as one example) at the national level can be taken from Chapter 19 (Building Characteristics) of the US-EPA's *Exposure Factors Handbook*. The Dutch RIVM has published a fact sheet that provides data on general conditions for exposure assessment for the Netherlands. Room sizes are partly ruled by standard norms and in some building regulations, e.g. the Berliner Bauverordnung (Land Berlin 2005) specifying that the height of rooms in new built houses shall not fall below 2.50 m. This value can be used for estimations of room volumes, together with data about room areas, e.g. from statistical offices, such as the "Statistisches Bundesamt" or, for the European level, Eurostat. However, the presentation of the parameter "room size" explains the problem of gathering adequate data for exposure assessment. For outside conditions, hypothetical room sizes can be assumed, together with realistically high rates of air exchange.

Ventilation (Air Exchange) Rate

The decline of concentrations of volatile substances is primarily limited by ventilation rate expressing the air exchange of a room. This is in particular of importance because in exposure models (see Sect. 3.3) ventilation rates are in the exponent of the model equation.

Figure 3.11 shows the results of calculations made by means of the "constant rate" model, implemented in the ConsExpo 4.1 tool (Delmaar et al. 2005). In this calculation, all parameters in the model equation were maintained unchanged, but the ventilation rate is varied between 0.2 and 4.0 m^3 per hour. There is only little variation in the peak concentrations from changing the ventilation rates (VR) between 0.2 and 1.0 m^3/h. After longer times, e.g. 250 min, the concentration at a VR of 0.2 is ~ 270 $\mu g/m^3$, and at 1.0 it is ~ 20 $\mu g/m^3$. These differences have consequences for short-term and long-term exposures and should be considered when choices are made for the VR. Ventilation rates higher than 2.5 h^{-1} are only of theoretical value without open windows and doors.

Type of Room

The use of products may be related to specific rooms in a house, with varying room volumes. For example, particular product types are used in respective rooms, e.g. bathroom or toilet. Other products may be used under particular conditions, e.g. shoe sprays are recommended for use in bigger rooms with adequate air ventilation. Very conservative estimations for these products may assume small rooms having low or no ventilation. Furthermore, the frequency of use of products may vary between rooms. Playing with toys may be regarded primarily in children's rooms.

In addition, some tools estimating consumer exposure are differentiating the number and sizes of rooms in flats or houses (e.g. US-EPA CEM, MCCEM, paint emission model, Sect. 5.10).

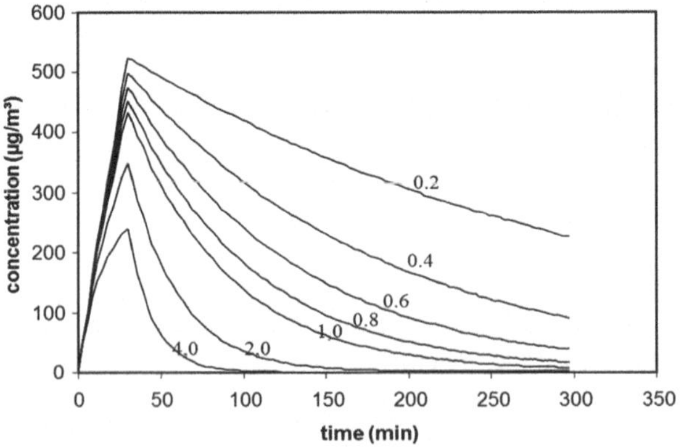

Fig. 3.11 Influence of air ventilation rates on room concentrations of a volatile compound, calculated by the ConsExpo tool

3.4.3 Region of Exposure and Other Areas

Regional differences may be relevant to specific food consumption habits, e.g. people in regions near to the seaside may eat more fish, which, in consequence, may lead to higher exposures from, e.g. mercury. Food consumption studies are often performed at the national levels that might mask regional differences without a particular focus.

3.4.4 Seasonal Influences on Exposure

Air ventilation rates may be lower in winter than in summer. However, they may also be changed due to habits of opening and closing windows and doors. Adequate data, however, are lacking for such differences.

3.4.5 Exposure Data

Information that is directly related to a particular type of product has been called product-specific. Exposure data include:

- Data about sources (occurrence data)
- Data about product types
- Data about concentrations in and released from products

These data cover information about the substances in any source that contain and release them.

Consumer Products

The information about consumer products covers data about any mixtures/preparations and articles and includes information on the names, the list of ingredients and/or composition and specific data about concentrations of the substances and their release rates. The term "chemical products" which is also sometimes used covers consumer and workers products. This book addresses only the consumer products. For details see Chap. 5. Since 2002, the Danish EPA (2001–2016) has performed surveys of substances in consumer products. These data offer a large number of measurements of substances in a great variety of consumer products and thus give hints of occurrence (Table 3.10). There are worldwide big differences in the use of products, e.g. cleaning products in private homes (Nielsen 2016), which should be considered when estimating consumer exposure.

For mixtures (also named preparations), product information can be taken from partly publicly available product databases (Heinemeyer and Hahn 2005). For more information see Chap. 5.

Food

From the understanding that the sources of substances leading to exposure are delineated as "products", food is also a product. Data from studies that analyse the concentration of substances in food can therefore be assigned as "product data". These studies include national surveillance studies as well as total diet studies (see also Sect. 4.8.4). The latter is directly aimed at generating data for food exposure assessment. Food surveillance data are also used in exposure assessment. The purpose of generating these data is, however, market control and is in many cases risk-oriented. Exposure from consuming food is given a broad focus in Chap. 4.

Other

Other important sources from which substances can be released are house dust, soil and surface water. Systematic measurements of concentrations of substances in house dust are available from survey studies, e.g. German Environmental Survey (Becker et al. 2004; Gaitens et al. 2009; Health Canada 2015), and also reported in a large number of scientific publications.

Physical and chemical characteristics, e.g. molecular mass, density, boiling point, flash point and vapour pressure, are needed for assessment of exposure by inhalation.

Products should be characterised by classification systems that systematically categorise products by the use of a thesaurus and further indices and descriptors. Classification systems are developed in detail for food characterisation, but they are needed also for other products. The use and impact of classification systems are described in Sects. 4.4, 5.2.1 and 6.6.2 and several regulatory inventories (Heinemeyer and Hahn 2005). Classification systems have been developed to characterise food. Knowledge of brand names is an advantage because it allows more precise information about the composition if available.

3.4.6 Data from Human Biomonitoring Studies

The objectives and aims of human biomonitoring are explained in detail in Sect. 3.5. Here only a short reference is made to urinary excretion data from national surveys generating concentration data of a large number of substances in human urine and blood. These concentration data can be used to estimate internal exposure by means of pharmacokinetic models. Due to the kinetic properties of the substance, exposure data should be generated from urine samples or blood samples that are collected over at least 24 h to reflect a daily balance. However, data from HBM surveys are normally produced from single samples collected from people of any age. This approach is much easier than 24 h collection, but is associated with a number of uncertainties. For example, a spot urine sample only represents a short frame of a broader scenario. Also, due to the nature of the kinetics of a high number of different substances, the concentration value does not represent quantitative exposure. Even at short half-lives, a 6:00 morning urine sample does not represent the amount of substance in the body, because, in the meantime after the contact on the day before, the substance or its metabolites might have already left the body to a large extent.

3.4.7 Sources of Exposure Factors and Data

It should be stated that most data used for exposure assessment are not primarily generated for exposure assessment but taken from sources having very different purposes. As pointed out later, different statistical types of exposure data exist. In addition, the aim and objective of the assessment may also govern the choice of the source of information and the respective numerical value. Some information may be taken from regulatory sources. For example, a default for room height of new buildings may be taken from an administrative directive that regulates planning and construction of buildings. Other data may be general official statistics, e.g. for body weights, heights and areas. Limit values are used to regulate the concentrations of substances, e.g. in cosmetic products, toys and food. These data can be used for conservative estimations on a low-tiered assessment level. A relatively adequate good database is available in food exposure. In many countries concentration data of substances in food are available, and there are food consumption studies (FCS), which, however, do not have the primary aim of exposure assessment. The concentrations are mostly monitored with a risk-oriented objective, i.e. market control and control of limit value exceedances. In some countries, however, data on chemicals in food are measured with a representative and exposure-oriented background, e.g. the German "Lebensmittelmonitoring" BVL (2002–2016).[5] Primary data from these measurements are available for the authorities and, in compiled form, to the public[6] (Table 3.10; Sect. 4.8).

To close greater gaps of knowledge in food exposure assessment, "total diet studies (TDS)" have been developed having a specific methodology and strategy. All approaches for food risk assessment are extensively described in Chap. 4. Questionnaire studies are also increasingly used to study the behaviour of people regarding the use of consumer products and articles and thus increase the knowledge in this area of exposure assessment.

In Table 3.10, a comprehensive list of studies and projects is given from which data for exposure assessment can be taken. Due to the huge number of publications,

---

[5]BVL (2016). http://www.bvl.bund.de/DE/01_Lebensmittel/lm_node.html

[6]BVL (2012). http://www.bvl.bund.de/SharedDocs/Downloads/01_Lebensmittel/01_lm_mon_dokumente/02_Monitoring_Tabellen/lm_monitoring_tabelle_2012.html

Table 3.10 List of references for information about exposure data, handbooks and databases being publicly available

Name of data source	Issue and nature of data	References, remarks
General information		
XProb and database RefXP	Project report about availability and reliability of exposure data for the use in German authorities	Consortium of project partners
AUH-Report	Collection and justification of standard values for exposure assessment in Germany	Ausschuss für Umwelthygiene, Arbeitsgemeinschaft der leitenden Medizinalbeamtinnen und <Beamten der Länder. Editor: Behörde für Arbeit, Gesundheit und Soziales, Hamburg
EC-JRC: ExpoFacts[a]	Open-access Internet database of European exposure factors and links to European data sources	EC-JRC The European Commission's science and knowledge service http://expofacts.jrc.ec.europa.eu/
US-EPA's *Exposure Factors Handbook*: 2011 Edition	Comprehensive collection about exposure factor data for the general population	US-EPA (Environmental agency) https://cfpub.epa.gov/ncea/risk/recordisplay.cfm?deid=236252&CFID=90162579&CFTOKEN=99932770
US-EPA's *Child-Specific Exposure Factors Handbook*: 2008 Edition	Comprehensive collection about exposure factor data for children	US-EPA (Environmental agency) https://cfpub.epa.gov/ncea/risk/recordisplay.cfm?deid=199243
RIVM fact sheets	Collection of data for exposure assessments General and product information Support of the ConsExpo internal database	www.rivm.nl
Chemical products		
E.g. the Swiss product register	Product database, information about substances and their concentrations in chemical products	Use of product databases (no more active)
The Nordic product register	Product database, information about substances and their concentrations in chemical products	The Nordic council of ministers, cooperation of Denmark, Sweden, Norway and Finland, www.norden.org
SPIN	Product database of Substances in Preparations In Nordic Countries	http://spin2000.net/

(continued)

Table 3.10 (continued)

Name of data source	Issue and nature of data	References, remarks
The Danish EPA	Danish surveys on chemicals in consumer products	Measurements of 146 substances and substance groups in consumer products
Food exposure		
EFSA comprehensive database	Compilation of food consumption surveys of EU member states	http://www.efsa.europa.eu/en/food-consumption/comprehensive-database
German "Lebensmittelmonitoring"	Systematic evaluation of substances in food	www.bvl.bund.de
VELS Studie BfR entwickelt neues Verzehrsmodell für Kinder	Nutrition survey to evaluate food consumption in children	Information Nr. 016/2005 des BfR vom 2. Mai 2005 http://www.bfr.bund.de/cm/343/bfr_entwickelt_neues_verzehrsmodell_fuer_kinder.pdf
National food consumption survey II	Results of the German food consumption survey	https://www.mri.bund.de/de/institute/ernaehrungsverhalten/forschungsprojekte/nvsii/?sword_list%5B0%5D=nvs also: Heuer et al. (2015)
• Official report		
• Scientific use file		
EsKiMo Das Ernährungsmodul des Kinder- und Jugendgesundheitssurveys (KiGGS)	Food consumption survey in children and adolescents	Ernährung-Wissenschaft und Praxis, July 2007, Volume 1, Issue 5, pp 225–229 https://link.springer.com/article/10.1007/s12082-007-0048-9
French food consumption survey	Données de consommations et habitudes alimentaires de l'étude INCA 2	ANSES alimentation, environne,emt, travail https://www.data.gouv.fr/fr/data sets/donnees-de-consommations-et-habitudes-alimentaires-de-letude-inca-2-3/ Dubuisson et al. (2010)
French national survey on nutrition and health, ENNS (2006)	Étude nationale nutrition santé ENNS (2006)	INVS (Institut de veille sanitaire) http://invs.santepubliquefrance.fr/publications/2007/nutrition_enns/RAPP_INST_ENNS_Web.pdf
NHANES study	National Health and Nutrition Examination Survey Repetition for every 2 years Questionnaires, data sets and related documentation	https://www.cdc.gov/nchs/nhanes/nhanes_questionnaires.htm

(continued)

Table 3.10 (continued)

Name of data source	Issue and nature of data	References, remarks
Human biomonitoring studies and surveys		
German Environmental Survey	HBM data for the German population, contains also data on substances in house dust	Survey has been repeated for several times www.uba.de
Fourth National Report on Human Exposure to Environmental Chemicals 2009	HBM data for the US non-institutional civilian population	Centers for Disease Control and Prevention/Department of Health and Human Services https://www.cdc.gov/exposurereport/
Canadian HBM monitoring	Human Biomonitoring of Environmental Chemicals	http://www.hc-sc.gc.ca/ewh-semt/contaminants/human-humaine/index-eng.php
Diverse directives and regulations		
Cosmetics regulation	Entered into force 2009; replacing Council Directive 76/768/EEC of 27 July 1976 (EU 2009)	http://eur-lex.europa.eu/LexUriServ/LexUriServ.do?uri=OJ:L:2009:342:0059:0209:en:PDF
European Regulation on Registration, Evaluation, Authorisation and Restriction of Chemicals	It entered into force in 2007, replacing the former legislative framework for chemicals in the EU	http://ec.europa.eu/
IUCLID	International Uniform Chemical Information Database to record, store, submit and exchange data on chemical substances in the format of the OECD Harmonised Template	https://iuclid6.echa.europa.eu/

[a]ExpoFacts contains links to European databases that contain exposure factor data

this table can only list some key reports and publications. The reader is requested to prepare her/his own list with respect to the current question and task.

Data generation is one of the challenges to improve exposure assessment and thus risk assessment.

3.4.8 Statistical Characterisation of Exposure Parameters

For use in exposure models, the data and factors presented in the chapters above must be expressed as numerical variables. It was also mentioned that these numerical values can be given as deterministic and probabilistic values. A deterministic variable is a single numerical value of any origin. Deterministic values may be defaults from expert judgement or empirically based data.

Deterministic Parameters

Using deterministic variables in an exposure model produces a single numerical exposure estimate usually representing a single case, average, typical or high-end exposure.

Probabilistic Parameters

Using summary statistics (distributions) in an exposure model produces a distribution of the exposure estimate.

The use of deterministic and probabilistic data should be understood as a tiered approach to use exposure parameters in an exposure model. In fact, the use of determinants of an exposure model characterises the shift from a deterministic to a probabilistic model. While doing this, all determinants can be shifted from a single value (deterministic) to the distribution (probabilistic) variable. Figure 3.12 explains this step-by-step approach: It starts with the use of a default value. In many cases, only very few data are available.

Example

In the EFSA opinion on the marine biotoxins, the EFSA CONTAM (EFSA 2008a) panel refers to 485 g of mussels mentioned in a FAO/WHO report on marine biotoxins (FAO/IOC/WHO 2004) as a possible 95th percentile of consumption. This value is typical for expert judgement. It has never been measured or evaluated in a food consumption survey. However, it has been used as a conservative default value that is high enough to secure the population against acute poisoning from mussel toxins.

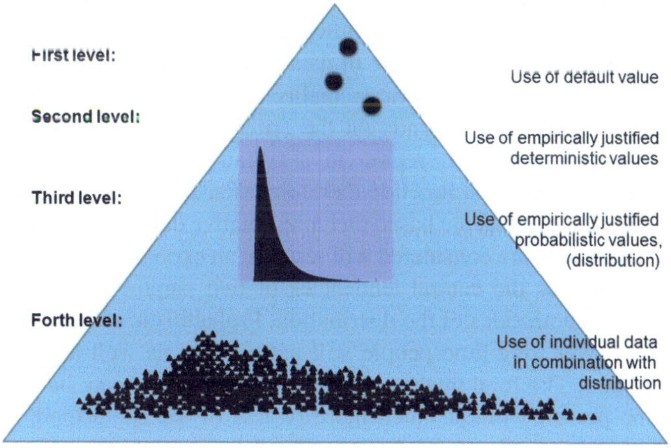

Fig. 3.12 Tiered approach for use of determinants of exposure estimation

In the BfR (2015) guidance document for uncertainty analysis, a default is defined as follows:

> Default assumptions are used in exposure models as substitutes for parameters if the assessor has no or an insufficient empirical basis for a direct estimate. This may include surrogate values (theoretical derivations from auxiliary variables - e.g. the body surface calculated from body weight and height), extrapolations (e.g. the transfer of findings from other populations to the target population) or expert opinions. The analysis of uncertainty should cover not only parameter uncertainty but also model uncertainty with regard to transfer, extrapolation or expert judgement. (BfR 2015)

In the next step of the tiered approach, deterministic statistically justified empirical values are used for exposure assessment. These values represent means, median or percentiles calculated from numerical data that are characteristic for a certain parameter of an exposure model. The source of these data may be heterogeneous: First of all, many studies have generated data that are not intended for exposure analysis. Therefore, the quality of these data is varying. In the report on DEHP exposure, Heinemeyer et al. (2012, 2013) evaluated data from different publications to arrive at a practically sufficient set of measured values as a basis for calculations. However, a major degree of representativeness was lacking. This clearly shows that sufficient data for exposure evaluations are often lacking. The exposure assessor must take what is available. This situation marks the major reason for the often-used data of low quality and high uncertainty. Exposure assessment always should therefore include an assessment of the data quality and the related uncertainty. The requirements for quality assessment are addressed in Sect. 3.7 and for uncertainty analysis in Sect. 3.8.

The third step of using parameters in exposure assessment is to use the entire available statistical data. If possible they can be expressed by distributions. If, for instance, a small set of data is available, it might be more appropriate to take all these data into consideration, e.g. by characterising a uniform or triangular distribution. Both are describing a window in which the data may vary. On the other hand, if appropriate data are available, distributions should be simulated according to predefined distribution functions, as shown in Fig. 3.13, e.g. by means of the computer tool @RISK. If the sampling of data in the source study is representative, then distributions will narrow towards reality of the respective parameter in a population or country. This depends on the criteria for which representativeness has been defined, e.g. gender, region in a country, season, ethnicity, kind of agriculture, etc. A distribution function therefore reflects the entire situation. Calculations with distributions that are based on representative data and where correlations between the parameters are considered will result in an exposure estimate that is near to reality. It combines the central tendencies of two parameters but also that of extreme values, on both sides of the distribution. Probabilistic estimations are based on the fact that over long-time people will eat average as well as low or highly contaminated food. The food intake also will be near to the mean, but from time to time, the same person will eat much more or less of the same food. The use of distributions therefore reflects the best approach to characterise exposure over a prolonged period of time.

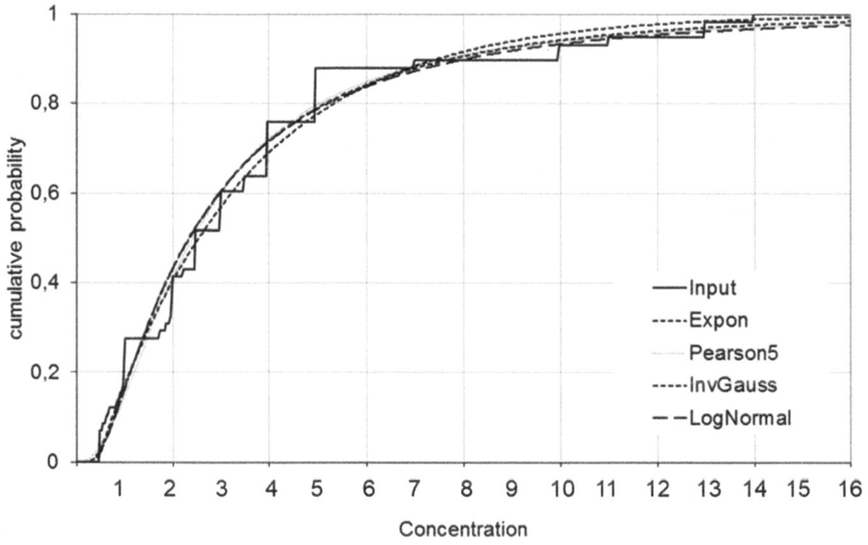

Fig. 3.13 Example for a fit of a data set of concentrations in food. Fit of the distribution functions to the input data by means of the best fit tool of @RISK

Distributions functions include two basics facts. First of all, the variability in the actual sample is characterised by the natural differences in a data set. For example, the body weight is almost solely characterised by its variability, due to representative and sufficient measurements. During recent years, the importance of including uncertainty into risk assessment has been increasingly discussed. Uncertainty is also included in the frame of a distribution. For example, a bigger difference of mean and median would indicate that a distribution is not normally distributed, with an indication of uncertainty. Uncertainty analysis is the matter of Sect. 3.8. The ranges of a distribution include both the variability of a data set and the non- or incomplete knowledge (uncertainty)

Modern concepts of risk assessment therefore are using probabilistic methodol ogy to describe the variability and the uncertainty. Methods to separate variability and uncertainty have been developed (see Sect. 3.8).

Using individual data (e.g. consumption data) in combination with distributions (e.g. concentrations) produces distributions of exposure for every individual. In summary the individual distributions can be aggregated to distributions of the statistical descriptors.

Distributions of exposure parameters can be obtained by fitting the data with a number of distribution functions. In Fig. 3.13, four distribution functions have been fitted against a set of concentrations. The results have a close overlap. All distributions are well fitted and can be used as a variable in the probabilistic exposure model. A number of powerful software tools are available. These tools are also able to calculate the exposure estimate by, e.g. Monte Carlo simulations. Respective tools are also integrated in dedicated software tools for exposure assessment (Sect. 5.10).

Basic statistical knowledge is a prerequisite to understand the processes used in the probabilistic calculations and simulations. Further details can be found in Sect. 3.9.

Any distribution of an exposure parameter in a model is representing the variability and uncertainty. Uncertainty may be expressed explicitly by the choice of a distribution. Very simple distributions may be used with an objective to express the uncertainty, e.g. by choosing triangular distributions to characterise a range in which the concentrations may vary and one value having high probability to represent a suspected central tendency. Taking this into consideration, it is possible to express even low-quality data in form of a distribution, which however is showing uncertainty rather than variability.

Data should be expressed in form of their distribution functions as presented by the tools with outputs in many forms, as functions, figures and numbers that allow any kind of subsequent calculations.

Sample Size

Sieke et al. (2008) made an assessment of appropriate sample size for measurements of pesticide residue in food that can be adopted for other measurements of substances in exposure media. To ensure appropriate calculations, the number of measurements should not fall below 188 to cover the 95th percentile of a sample. The authors estimated a sample size of 94 to characterise a mean value with a ratio of 1:2 of mean and standard deviation and a certainty of 95% with a relative estimation error of ± 35%. These sample size estimations have been used for sample planning in the German Food Monitoring Programme. This is in accordance with an EFSA guidance (2009c). It identifies a minimum number of 160 to support the P95 and of 800 to characterise the P99 appropriately. It must be stated that this number of samples, either for mean or for 95th percentiles, are often not reached in databases used for consumer exposure assessments.

3.4.9 Impact of Tiered Approach for Exposure Parameters, Shown by an Example for a Systematic Application of the Tiered Approach by Iteration of Exposure Parameters

To explain the steps for iterating the exposure parameters in a simple scenario and model, the exposure of eXaMPLe (XMPL) from eating fish is estimated by a strict and systematic approach. A data set derived from simulations based on measurements in the frame of food surveillance is used. The fish consumption data have been simulated for 150 exemplary adult individuals expressed in g per kilogram of bodyweight of each individual person. Exposure estimations for this scenario were calculated following a four-step tiered approach, starting with a conservative default-based estimate. The assumptions and results are presented in Table 3.11. The scenario and the model are not tiered in this example. Only the parameters (=model variables) are iteratively tiered by four consecutive steps as indicated.

First Tier

As the first step of this evaluation, the conservative estimate might be a default value of concentrations in fish of 0.1 µg/g. For consumption of fish, an amount of 1.54 g per kg of BW is assumed which represents a total amount of a fish meal of ~100 g, assuming that the person has a body weight of 60 kg. This body weight is slightly conservative and is used as a standard value in many assessments, although body weights have increased in both men and women during the last 20 years, as shown in food consumption studies (MRI 2008).

The exposure estimate reveals an amount of 0.165 µg/kg of BW per day.

It should be stated that this estimate represents an intended overestimation. It is generally accepted that an exposure assessment can be stopped at this point if the comparison with the ADI or TDI or any other relevant limit value reveals a ratio <1 in the risk characterisation. This example also shows that the choice of the default value, if not accepted as a standard, is due to the decision of the assessor and therefore an expert judgement. Maximum values from a series of measurements should normally not be chosen as defaults, because it is not certain that the next measurement will not present higher values. This is particularly important if the database is small. An alternative is to use higher percentiles for which the statistical validation is better.

Second Tier

The second iteration step follows an assumption that should be based on real data. In the data set used in this example, the mean concentration of XMPL in fish is 0.011 µg/g. This value will be multiplied with the mean consumption (0.246 g/kg of BW per day) for the normal consumer and with the 95th percentile of 0.7 g/kg of BW per day for the heavy consumer. The latter may be a person who eats larger quantities of fish at higher frequencies. For the normal eater, the daily intake yields an amount of 0.003 µg/kg of BW per day, and for the heavy consumer, the value is 0.008 µg/kg of BW per day.

This calculation can also be performed by using other statistical determinants. Table 3.11 shows that the median exposure estimate is about one third lower than the mean. This result indicates that the data used for the estimation are not following the Gaussian (normal) distribution but is skewed to the right.

Third Tier

If sufficient data are available, variation and range of the concentration and the consumption patterns can be assessed. Limit values (TDI, ADI) are normally related to a lifelong intake of substances. Therefore, it seems logical that one person will eat different amounts of meals that are differently contaminated with certain substances over longer times. The probabilistic estimation combines the whole ranges of both concentrations and consumption. In Table 3.11 the input data are shown as fitted distributions. The result of this estimate shows the variation and the range of the intake and the distribution can be expressed by a mean value, median and quantiles.

Table 3.11 Example for a tiered approach to calculate exposure from consumption of fish

Iteration step	Statistical	Fish consumption	Concentration in fish	Exposure estimate			
		g/kg BW per day	µg/g	µg/kg BW per day			
Tier 1	Default/ conservative	1.65	0.1	0.165			
Tier 2 deterministic	Mean	0.25	0.011	0.003			
	Median	0.17	0.011	0.002			
	P95	0.70	0.011	0.008			
Tier 3 Probabilistic	Distribution	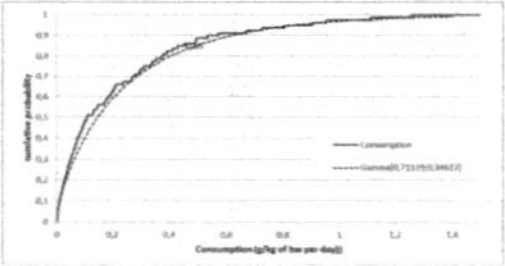	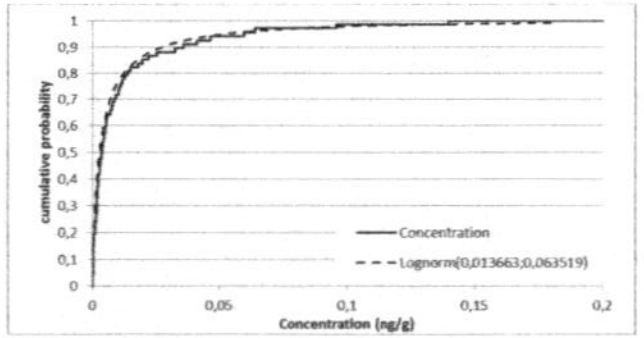	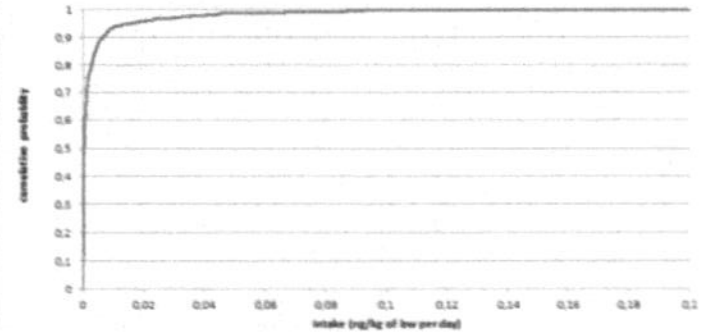			
	Distribution function	@RISK function: =gamma(0.71239;0.34617)	@RISK function: =(lognormal(0.014814;0.078994)	Output distribution			
				Mean	0.00322		
				Median	0.00034		
				P95	0.01167		
Tier 4 Probabilistic and individual	Individuals						
	Number of individuals	Individual estimate			Mean	Median	P95
	1	0.5280	@RISK function: = (lognormal(0.014814;0.078994)	P5	0.00003	0.00002	0.00009
	2	0.0620		Median	0.0025	0.0004	0.0092
	3	0.0042					
	. . .	. . .		Mean	0.0037	0.0007	0.0138
	152	0.5510		P95	0.0106	0.0019	0.0395

For illustration, 72 fish samples with fictional concentrations of XMPL were used and exemplary data of fish consumption of 150 individual persons, related to the individual body weights

In this assessment, the concentration of XMPS in fish is characterised by a lognormal distribution and the consumption by a gamma distribution. Both distributions can be generated by use of the best fit tool of the @RISK software. This tool fits a data set against selected distribution functions. The detailed description can be taken from various handbooks that guide users to work with the software.

The product of the two distribution functions results in the mean, median and P95 exposure estimate of 0.0032, 0.00034 and 0.0117 μg/kg of BW per day, respectively. These values are considerably higher than the deterministic estimates, but lower than the first conservative estimate. The mean value characterises the central tendency of the estimation, while the median separates the lower 50% of the population from the upper 50%. A median that is much smaller than the mean, like in the current example, indicates right-skewed data. Conversely a median larger than the mean indicates left-skewed data. The heavy consumers can be characterised by the 95th percentile (0.012 μg/kg of BW per day).

The use of distributions to estimate the exposure has the advantage that only one common calculation approach is used. In fact, there are thousands of calculations that are combined together in the final distribution. In contrast, the deterministic approach only combines a few numbers of single-separated calculation steps (mean × mean; mean × P95).

Fourth Tier

The fourth step of the estimation calculates the individual exposure of every single person taking each individual's daily amounts of fish consumption and multiplying it with the full range of concentrations in fish as given by the respective distribution. Taking individual consumption figures related to bodyweight considers that food consumption correlates with the body weight. This procedure generates a distribution for each individual person, and the statistical descriptors [e.g. P5, P50 (median), mean and P95] can be calculated on the individual level as well as on the distribution of individual intake estimates. In the last step of this calculation, the statistics for the latter values represent the interval (range) between the 5th and the 95th quantile for every individual and the whole sample of consumers (Table 3.11).

The mean estimate, given as the central tendency for exposure and representing the "normal" consumer, ranges from 0.00003 up to 0.0106 μg/kg of BW per day, with a mean value of 0.0037 and a median of 0.0025. For the P95 representing persons consuming bigger portions at higher frequencies, the estimate lies between 0.00009 and 0.0395, with a mean of 0.0138 and a median of 0.0092 μg/kg of BW per day. It should be noted that this example demonstrates the tiered approach and has its focus on the total population. There is no distinct separation of consumers and nonconsumers. Indeed, the calculation at the third tier using distributions only would not account for this, but calculating with individual values takes all the nonconsumers into consideration. From the individual results, the number of persons not eating fish can be identified. In the whole sample, they are visible by the low numerical results approaching zero of the lower percentiles of the final estimate.

The multiple iterations of the exposure estimation produce a frame of the exposure estimate. There is no distinct deterministic exposure estimate. It is also important to understand that the quality of the input data can broaden or narrow this window independently of the actual exposures. Besides, the default-based tier 1 estimate lies outside of that window.

It should also be considered that in many cases, comprehensive data are not available. For example, food exposure estimation is often based on data from several sources, having different quality and certainty. Therefore, a model can be fed with parameters of different kind, e.g. a combination of deterministic data, created by expert judgement and distributions from measurements in representative studies. Such estimations, however, need a thorough uncertainty analysis.

3.5 The Impact of Internal Exposure in Consumer Exposure Assessment

Gerhard Heinemeyer

Internal exposure assessment evaluates the amounts of substances in the human body. It is based on the principles of pharmacokinetics, which characterise the absorption, distribution and elimination (metabolism and excretion) of substances. The term "pharmacokinetics" is normally used to describe the concentrations of pharmaceuticals in the human body; the term "toxicokinetics" is used accordingly for toxic substances which means that the same principles are applied. Therefore, every pharmacokinetic model or model approach can be applied for toxicokinetics. This applies for the "classical" pharmacokinetics which focuses on the whole body burden as well as for the physiologically based pharmacokinetics (PBPK) which evaluates the distribution of agents and other biomarkers in the body compartments due to physiological parameters, e.g. the blood flow, and which focuses the exposure of particular organs and organ systems. The essential difference between pharmaco- and toxicokinetics is that the dose of the pharmaceutical is known (what the doctor is prescribing) but that of the toxic agent is an estimate.

It should be mentioned that external and internal exposure assessment cannot be done alternatively, but they complement each other. External exposure assessment evaluates the sources, exposure pathways and routes of exposure and characterises the dose of a chemical that is incorporated by humans. According to the scenario characterised in Sect. 3.2.7, internal exposure is representing a detailed characterisation of the target of exposure. The part of that external dose (the intake) that is absorbed characterises the amount of the substance that is actually in the body, internal dose, and which is systemically available. This dose can be characterised and estimated by the "classical" pharmacokinetic modelling. The part of this latter dose which acts at the target organ or site of action is called the target organ dose and can vary in the specific body compartments and can be (partly) estimated by the means of PBPK modelling and is hardly ever used in general exposure assessment. The internal dose does not provide any information about the source and (with some restrictions) the pathway of exposure. On the other hand, this information can be taken from the external exposure assessment.

It should be mentioned that the focus of internal exposure estimation is normally related to long-term exposure and chronic toxic effects.

3.5.1 Basic Principles for the Assessment of the Internal Dose

Internal exposure assessment is based on the measurement of substances circulating in the human body. This can be assessed by analyses of blood and urine and by application of particular pharmacokinetic approaches and resp. models. For estimations in particular organ systems, PBPK evaluations are needed. The elimination via the faeces is of interest if substances are not absorbed in the gut or if enterohepatic circulation happens.

For routine assessments, the excretion of the parent substance and its metabolites can be measured in the urine. Due to the many studies performed to elaborate the pharmacokinetics of therapeutic pharmaceuticals, there are clear rules that should be applied if the kinetics of a chemical is under consideration.

Table 3.12 lists the most important toxicokinetic parameters and information needed for reasonable human internal exposure estimation. The knowledge of the systemic availability is a prerequisite to estimate the part of the external dose that is absorbed into the body. If the systemic availability is 100%, then external exposure equals internal exposure. The systemic availability can be estimated from measurements of daily urinary excretion during steady-state conditions and from the knowledge of the exact external dose. Uncertainty of systemic availability has therefore at

Table 3.12 Important pharmaco(toxico)kinetic parameters to estimate the internal dose of substances

Pharmaco(toxico)kinetic parameter	Remarks
Systemic availability (bioavailability)	The part of the external dose which is entering the systemic circulation
Volume of distribution	The (apparent) volume in which the substance is distributed during steady state
Elimination half-life	The time in which the concentrations in blood reaches the half of the concentration Due to the compartmental concept of pharmacokinetics, several half-lives can be characterised. For risk assessment issues, always the terminal half-life is needed requiring a sufficient length of study Note: Half-life is not the loss of half of the amount in the body
Steady state	The situation where intake and loss of a substance is equal and distribution and redistribution are at a balance
Urinary excretion per time interval	The amount of a substance which is excreted in urine over a given time (parent substance and metabolites)

least two reasons: (1) the imperfect evaluation of daily excretion and (2) imperfect knowledge of the sources of exposure and the dose rate.

Toxic substances may exhibit extremely different toxicokinetics. This is primarily due to the physical characteristics of the chemical. As a rule of thumb, it can be assumed that substances having lipophilic characteristics, e.g. hydrocarbons, can be readily absorbed and they have a large (apparent) volume of distribution, in particular the halogenated ones (dioxins, PCBs). Lipophilic substances cannot be excreted directly by the kidneys: they must be transformed into more hydrophilic metabolites. Hydrophilic substances are normally excreted directly via the kidneys but can also be subject to further metabolism. Substances having large distribution volumes normally have long half-lives due to redistribution processes. For exposure assessment, the quantitative formation of the metabolites from their parent substances must be taken into consideration as quantitatively as possible. The formation of active metabolites must be accounted for in the toxicologic assessment.

Exposure assessment is normally done by measurements of substances in blood and/or urine. The toxicokinetic properties of the substances under concern should be considered for the planning of the sampling that should be done under steady-state conditions (Table 3.12). Figure 3.14 shows three concentration time curves in blood, after a series of the same daily doses of three different substances with different half-lives, 5, 10 and 24 h, from the bottom to the top. For the 5 h half-life, there is no accumulation of the concentration. For the 10 h half-life, the steady-state concentration is reached after about 50 h and for the 24 h half-life after 150 h with fluctuations due to the dosing interval. It can be concluded, as a general rule in pharmaco(toxico)kinetics, that steady-state conditions will be achieved after about five half-lives, with the major prerequisite that the frequency of intake and the dose rate are constant. For many substances this situation is hardly achievable due to

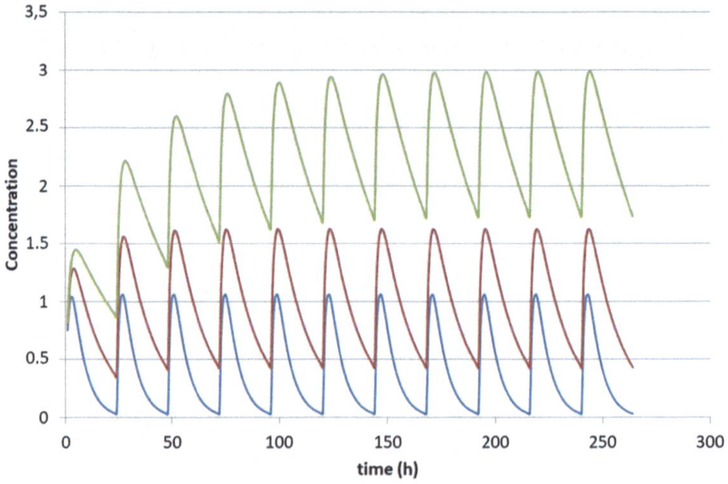

Fig. 3.14 Impact of half-life of 5, 10 and 24 h (from below to top) on accumulation of substances in blood

strong dose variability and to differences between distribution and redistribution processes due to their extremely long half-lives (e.g. dioxins). Substances having very short half-life (<10 h) will not accumulate, and therefore sampling strategies for urine and blood sampling must be adjusted according to the time of administration and dose rate.

Measurement of urinary excretion is the main instrument for human biomonitoring studies because the kidneys are the primary organ that eliminates the substances from the body. Substances which are not absorbed in the gastrointestinal tract or which undergo enterohepatic circulation are excreted via the faeces.

Measurement of urinary concentrations is therefore primarily applicable for substances which are water-soluble and for which the metabolites are completely known. The more a substance becomes fat-soluble and the metabolites that will be formed become unclear, the more uncertain is the urinary excretion in terms of exposure assessment.

Methods to Evaluate the Internal Dose of Substances

Urinary excretion of substances and their metabolites is the most preferred approach to estimate the body burden of a substance. The objective of these measurements is to cover the entire amount of the substance that is eliminated over a certain time period. This is based on the hypothesis of the presence of a steady-state situation. Therefore, the toxicokinetic properties of substances primarily govern sampling strategies.

The amount that is in the body can simply be calculated from measurements of the substance in urine by the formula

$$A_{\mathrm{U}_t} = C_{\mathrm{U}_t} * \frac{V_{\mathrm{U}_t}}{t} \tag{3.5}$$

where:

- $A_{\mathrm{U}t}$ is the amount of the substance to be estimated in urine collected over the time period
- $C_{\mathrm{U}t}$ is the concentration of the substance in the urine sample over the time period (t)
- $V_{\mathrm{U}t}$ is the volume of the urine sample collected over the time period (t)

$$A_{\mathrm{U}_{24\mathrm{h}}} = A_{\mathrm{U}_{24\mathrm{h}}} * \frac{V_{\mathrm{U}_{24\mathrm{h}}}}{V_{\mathrm{U}_t}} \tag{3.6}$$

The amount excreted over 24 h can be estimated by considering the ratio of the volume of urine excreted within the actual sampling period and the 24 h volume. Only a time of collection of 24 h guarantees that the amount of the substance measured in urine represents the daily excretion.

For practical reasons, urine sampling is often done by collecting spot samples, mostly in the morning at 6:00. This spot represents an interval between the last time of urination and the time of collection, e.g. for 10 h, but this time is normally not known.

This estimation approach of amounts of substances excreted in urine is based on the assumption that there is no circadian rhythm of urine production (glomerular filtration; tubular reabsorption) or of the excretion of the substances, but it is constant over the whole day and at steady state. In this case, an estimate can be made for spot urine samples. The longer the half-life of a substance, the more the concentrations in the spot, and the 24 h urine sample will approximate each other. Logically, this approach cannot be applied for substances with very short half-lives ($\ll$24 h). Steady-state conditions imply that spot urine concentrations and the urinary excretion rate are constant over at least 24 h. Therefore, they can be multiplied by each other to yield the daily excretion of the substance.

In this case, the V_{Ut} given in the equation above equals a mean of the daily urinary excretion in a collective of individuals, and the concentration C_{Ut} in the spot is assumed to represent also the concentration in the 24 h urine.

This approach represents an approximation of the mean excretion to the real excretion. The use of probabilistic calculations and thus consideration of variability may also estimate the range of the daily urine excretion respective the internal exposure. For individual estimation, however, the knowledge of the actual urine volumes produced over the time of collection is needed. The considerable variability of the exposure estimates should be mentioned.

Some methods exist to estimate the 24 h urine excretion from single urine samples. For example, the daily urine production in a healthy adult human accounts for about 2 L. If the spot urine in the morning is 500 mL, then the amount of a substance may be multiplied by 4. This approach assumes that the concentration in the urine over the whole day is constant and urinary production over the same time as well. Several choices can be made to extrapolate from spot urine measurements to a 24 h estimate. This however has different implications to the exposure estimation result and is discussed in the next chapter.

From the basic toxicokinetics principle, however, it should be assumed that a sample collected over 24 h or over longer periods reflects the daily intake of a substance from any sources, after subtraction of the dose which is not absorbed in the gastrointestinal tract and under the prerequisite that the metabolites are quantitatively known.

3.5.2 The Use of Human Biomonitoring Data for Exposure Assessment

According to WHO (2015), human biomonitoring is an approach that measures concentrations of chemicals and/or their metabolites in human matrices. The definition of the CDC (2005, 2009) also includes the assessment of effects into the HBM approach. In the German HBM Programme (UBA 2016), concentrations of substances are characterised which indicate a possible health risk for humans. This understanding of HBM implies a full risk assessment based on "internal" data.

HBM is seen as an important basis to estimate body burden and can therefore be understood as a comprehensive approach for exploring health risks in humans by assessments of substances and other biomarkers with respect to hazards (Batáriová et al. 2006).

Human biomonitoring is often used to estimate critical levels that might indicate health hazards. For example, the German Umweltbundesamt with its "Human Biomonitoring Commission" derives concentrations of substances of concern (Schulz et al. 2011). In this approach, the exposure data are derived from spot urine samples collected in the early morning. The commission uses these concentrations to derive the so-called HBM-values I and II. The HBM-I value covers a situation that indicates a "pay attention" scenario, where measurements and additional studies are needed. The HBM-II value indicates real health impairment that requires risk management measures. This concept can be accepted if it is seen as a closed concept where all measurements and assessments are based on the same approach. This means that exposure data derived by methods that are not matching the HBM concept (e.g. 24 h urine sampling) cannot be directly compared with the HBM-values. However, the experiences from many HBM studies have proven valuable in considering internal exposure estimations.

The UBA concept of HBM is clearly a risk-oriented approach. Therefore, taking results of the measurements of substances in spot urine in the morning can be considered to be a reasonable compromise. Indeed, with regard to preparing an estimate of the amount of substances in the body, only very few studies are available which examine the total excretion over time periods using 24 h urine sampling. In the INES study, Fromme et al. (2007a) evaluated about 400 urine samples collected from 50 individuals over 8 consecutive days. This approach does not follow the strict requirement of collecting the urine over the whole day, but can be seen as a compromise that might provide more realistic data than a single measurement.

To date, nearly all extensive national and international surveys (NHANES, UBA surveys) are using measurements in spot urine samples. This approach does not provide scientifically sound exposure information. These studies are not primarily focusing on quantitative assessment of exposure. They should be understood as a compromise, and all uncertainties involved in the collection approach should be covered in an uncertainty analysis.

Undoubtedly the population-based human biomonitoring activities (Angerer et al. 2007) have led to a tremendous increase of knowledge about concentrations of substances in urine and thus in the human body. The use of concentrations of substances from spot urine measurements, however, must be critically discussed.

As mentioned above, the steady state is reached after about five half-lives. This means that urine should be sampled after 5 days if the half-life is 1 day and the intake occurs every day to cover steady-state conditions. However, substances involved in consumer exposure are often characterised by irregular frequencies of intake.

Therefore, it can be argued that there is no steady state. This rule also implies that urine sampling for substances having long and extremely long half-lives never reach the steady state over the human lifetime. Assuming that intra-individual variability of concentrations over time is low, spot urine collection of urine for toxicokinetic analyses may be understood as a compromise. Therefore, this compromise between practicability and scientific requirements should be noticed and included into uncertainty considerations. It might be accepted that under these conditions, shorter periods may deliver data that are narrowing a steady-state elimination rate, but on the other hand, repeated measurements over several days would decrease the uncertainty that is involved in inappropriate collection periods.

The quantity of the substance in the 24 h urine is required for the estimation of internal exposure. Most accurate is the determination of the excretion of the substance in the entire quantity of urine that can be collected within 24 h. If the half-life of the substance of concern is very short, several fractions of the urine should be collected which should be evaluated by respective studies.

To adjust spot urine samples to 24 h estimates, some different approaches are used which are explained in the next (sub)chapters.

Volume-Adjusted Extrapolation of Spot Urine Data to 24 h

Taking spot urine for measuring the amounts of substances being eliminated represents a fraction of urine over some time. Therefore, the volume represents a certain fraction of the entire volume excreted over 24 h. If, for example, an amount of 30 μg of a substance is found in a urine volume of 400 mL, then the total amount excreted over the day might account to 150 μg assuming a daily urine volume of 2 L. It might however also reflect the fraction and thus an underestimate of the 24 h excretion. In addition this approach does not consider possible diurnal fluctuations of the excretion of the substance.

Extrapolations of Daily Excretion Based on Urinary Excretion per Time

Another approach is based on the hypothesis that the concentrations of substances in spot urine and 24 h urine are similar. Taking a urine production rate of 0.03 L/kg of BW per day (UBA 2016; Rudzok et al. 2016), the concentration can be multiplied with the urine excretion rate. Basically, this approach equals the volume-adjusted extrapolation. This approach was applied by Rudzok et al. (2016) and LANUV (2016) for estimating glyphosate exposure in children. In this study, the concentrations of glyphosate in spot and in the 24 h urine were approximately equal, and the multiplication by 0.03 L/kg of BW per day revealed similar exposure estimates. It should be mentioned, however, that the variation in urine production is large. In the respective collective of 350 children with an above-mentioned mean urine volume of 0.03 L/kg per day, the 25th percentile was 0.017 L/kg and the 95th percentile 0.047 L/kg. Individual values may therefore deviate considerably from an estimate focusing at a central tendency. This method cannot be applied if larger variability of the parameters in the model is expected.

Creatinine

Extrapolation of spot urine measurements to 24 h urine excretion can also be related to the excretion of creatinine. If measured in the spot urine sample, it was postulated that substances in that sample can be related to the concentration of creatinine. Creatinine concentration in serum follows a circadian rhythm. It is highest at the early afternoon and lowest in the morning (Knoll et al. 1978). The estimation of the daily elimination based on 6:00 morning urine samples might therefore lead to an underestimation. It should also be taken into consideration that creatinine excretion may be extremely variable with a reference range of 280–2170 mg/L in women and 390–2590 mg/L in men (Mazzachi et al. 2000).

$$E_{24\mathrm{h}}\left(g * 24\ \mathrm{h}^{-1}\right) = \frac{C_{\mathrm{urine}}\left(g * t^{-1}\right)}{\mathrm{Creat.}_{\mathrm{urine}}\left(\mathrm{nmol} * \mathrm{L}^{-1}\right)} * \mathrm{Creat}_{24\ \mathrm{h}}\left(\mathrm{nmol} * 24\ \mathrm{h}^{-1}\right) \tag{3.7}$$

Johner et al. (2015) found an excellent correlation of 24 h creatinine excretion and iodine and sodium excretion in a collective sample of 1463 individuals. The average daily creatinine excretion was 0.19 mmol/kg (0.021 g/kg) in men and 0.18 mmol/kg (0.02 g/kg) in women, 40–49 years of age. The authors recommend the use of creatinine excretion as a valuable tool to adjust spot urine measurements to 24 h urinary excretion. It must be noticed however that while this approach might be applicable for large collective population, samples of it might be inappropriate for estimations of daily excretion in individual cases. Finally, it is questionable if this conclusion regarding the excretion of sodium and iodine may be applied to the excretion of chemical substances with toxicokinetics significantly different from sodium and iodine.

Substance-Specific Approaches

Koch et al. (2012) evaluated samples from the environmental specimen bank for concentrations of bisphenol A. The median concentration of the urine collected over 24 h was 1.37 μg/L. Concentrations in Canadian spot urines were similar to these values (1.33 μg/L). Values from the NHANES study (1.99 μg/L) were slightly higher. The resulting total excretion was 0.037 μg/kg of BW per day, which resulted in a daily excretion of approximately 2.5 μg.

In a paper by Baez et al. (2000), partial collections of urine sampled for fluoride excretion over 24 h in 4–6-year-old children were investigated. Average excretion was 36.4 μg per hour in the morning, 45.6 μg per hour in the afternoon and 17.5 μg per hour overnight. The reason for the relatively low excretion overnight was explained by a decreased glomerular filtration rate overnight. A correction factor of 3 was assumed from the ratio of the excretion rate of the amount excreted over the day and the amount excreted in the spot urine. This factor has only been determined for fluoride. It is therefore not clear whether it applies to other substances as well. Furthermore, for the application of this factor, the absolute quantity is to be extrapolated over a period of time of 24 h. As a rule, however, only the concentrations are known.

In conclusion the extrapolation of amounts of substances in measured spot urine is an important reason for uncertainty and might provide misleading exposure assessment results. As extrapolation approaches work with constant extrapolation factors, they all refer to spot urine concentrations and thus might lead to erroneous quantitative estimations.

3.5.3 Balancing External Versus Internal Exposure

This chapter presents some examples that compare the results of external and internal exposure estimations. The substances DEHP, PFOS and PFOA, and the heavy metals cadmium and mercury are discussed (Table 3.13). There is no study in the literature that compares external and internal exposure directly and, indeed, such a study would be very challenging to execute. Duplicate diet studies have the intention to compare external and internal exposures directly, but they are focused primarily on the intake of substances via food consumption. Therefore, results from different studies were taken which have a focus on greater populations.

Table 3.13 Comparison of exposure estimates from external and internal assessment, shown for DEHP, PFOA, PFOS, cadmium and mercury

Substance name	External exposure estimate (reference)	Internal exposure estimate (reference)
DEHP	15–44 μg/kg of BW per day (Heinemeyer et al. 2012)	5.6–11.33 μg/kg of BW per day (Fromme et al. 2012)
PFOS	60 ng/kg of BW per day (EFSA 2008a)	0.22 ng/kg of BW per day (Zhang et al. 2015, Creat. Based) 0.29 ng/60 kg of BW per day (Harada and Koizumi 2009) 3.9–520 ng/kg of BW per day (Vestergren et al. 2008)
PFOA	2–6 ng/kg of BW per day (EFSA 2008a)	0.1 ng/kg of BW per day (Zhang et al. 2015, Creat. Based) 0.19 ng/60 kg of BW per day (Harada and Koizumi 2009) 0.3–140 ng/kg of BW per day (Vestergren et al. 2008)
Cadmium	0.15–0.5 μg/kg of BW per day (EFSA 2008a)	0.174/0.37–0.413 μg/kg of BW per day Table 3.14
Mercury	0.15–0.6 μg/kg of BW per day (EFSA 2004)	0.17 μg/kg of BW per day (Kolossa-Gehring et al. 2012)

DEHP

In a study evaluating the exposure of the German population with DEHP, Heinemeyer et al. (2012, 2013) prepared an external estimate based on German food consumption data and a collection of data of DEHP in various food items and in consumer products. The estimate of the total (external) exposure in children was 15–44 µg/kg of BW per day. Many external exposure approaches are more or less intended overestimations (see respective chapters), and the data and models are too raw to achieve a realistic estimate. An estimate of internal exposure in 600 children from the German Children's Environmental Survey revealed a mean of 5.6 and a P95 of 11.33 µg/kg of BW per day. Similar results were published by Fromme et al. (2012) in the LUPE III study. Interestingly, the external exposure of DEHP estimated for the house dust intake exceeded the estimate of exposure estimated by HBM. The authors explained this by an overestimation for DEHP by the house dust path due to overestimated default house dust intake assumptions. In a duplicate diet study, Fromme et al. (2007b) found comparable results for external and internal exposure of DEHP that demonstrates that a strict project plan is needed to get appropriate results. The real exposure might be overestimated by external exposure assessments, but it might also be underestimated by internal exposure assessment approaches, e.g. by inappropriate urine collection (spot) or by non-consideration of known or unknown metabolites.

Estimates of DEHP exposure based on internal exposure considerations revealed an average of about 5 µg/kg per day and an upper range (95th percentile of 11–30 µg/kg per day) (Bosgra et al. 2005; Wormuth et al. 2006; Becker et al. 2009; Koch 2006; Calafat and McKee 2006; Fromme et al. 2007b). All of these estimates are in a comparable but slightly lower range of exposure.

PFOS/PFOA

For PFOS and PFOA, internal exposure estimates revealed values of PFOS in a range of 0.2 ng/kg of BW per day and of PFOA of 0.1 ng/kg of BW per day which were calculated on the basis of data published by Zhang et al. (2015). In 24 h pooled samples of residents in Kyoto (Japan), Harada and Koizumi (2009) determined amounts of 17.6 (PFOA) and 11.3 (PFOS) ng/day. Related to a body weight of 60 kg, the resulting exposure accounts for 0.29 (PFOA) and 0.19 (PFOA) ng/kg of BW per day. These data were exceeded by far by the external estimate calculated by the EFSA expert panel of 60 ng/kg of BW per day (EFSA 2008b). The data show that the results may be hardly comparable due to possible regional differences (Europe vs. East Asia), although people in Japan eat much more fish than Europeans. Vestergren et al. (2008) reported exposure estimates in a general population in an industrialised area of 3.9–520 ng/kg of BW per day for PFOS and 0.3–140 ng/kg of BW per day for PFOA. The very large difference within the internal estimates as well as between internal and external exposure indicates several issues: They may express the large differences between regions, making comparisons difficult and indicates the need for systematic evaluation of exposure. As well, it might be assumed that external and internal exposures are hardly comparable and must be urgently correlated to each other.

Cadmium

The EFSA CONTAM (2009a) panel estimated a daily intake of cadmium of 0.15–0.5 μg/kg of BW per day. Slightly higher results were estimated by JECFA (FAO/IOC/WHO 2004) of 25 μg/kg per month, which is about 1 μg/kg of BW per day which is 60 μg/day for a person weighing 60 kg.

For comparison, Berglund et al. (2015) and Den Hond et al. (2015) in the DEMOCOPHES study measured concentrations of about 0.44 (GM) to 1.25 (P90) μg/l of urine. Related to creatinine, the excretion was 0.18 (GM) and 0.42 (P90) μg/g creatinine per kg of BW ($N = 1272$) in the non-smoking reference group. Approximately, this concentration would account for a daily excretion of ~0.9–2.5 μg of cadmium per day, assuming a daily urine volume of 2 L. Under consideration of a systemic availability of about 5% (EFSA 2009a), the daily intake accounts for about 9–25 μg per day.

Table 3.14 shows the results of a comparative calculation of the exposure from cadmium with the intention to demonstrate the sources of uncertainties and possible mistakes in the evaluation. External exposure estimates were taken from EFSA (2009a) and internal data from the WHO report on human biomonitoring (WHO 2015). A body weight of 60 kg is applied in all cases. The creatinine values were taken from Johner et al. (2015). The external exposure estimate was taken from the EFSA report. Interestingly, the internal exposure results conform relatively well with the external estimate at the first sight. However, the variability of the values should be mentioned. To demonstrate the practical extent impact on the numerical results, the urine production of 2.0 L over 1 day was chosen as an often-used default value for adults in clinical practice. The rate of urine production of 0.03 L/kg of BW per day has been derived by the HBM Commission (HBM-Kommission 2014) and is applied for all ages (*the non-linear relation of body weight to age should be kept in mind*). The example shows that the adjustment of the spot urine by a relation to volume as compared to creatinine might be erroneous. Taking a volume-related assumption, the estimate of

Table 3.14 Results obtained by the above characterised approaches for exposure estimation, taking cadmium as an example

	Measured conc in urine		Urine/creatinine production		Daily exposure	
	Cadmium: estimated external exposure: 0.15–0.5 μg/d/kg bw (EFSA 2009c) systemic availability: 5%					
Volume related	0.62	μg/l	2.0	l/day	0.413	μg/kg/d
Volume per time	0.62	μg/l	0.03	l/d/kg	0.372	μg/kg/d
Creatinine adjusted	0.42	μg/g Creat	0.021	Creat (g/d)	0.174	μg/kg/d

exposure is more conservative than the adjustment of the calculation to creatinine. A focus should also be paid to the choice of the value for bioavailability. This example was calculated by taking a bioavailability of 5%. Just changing this parameter slightly might lead to considerably different results. It must also be mentioned that this comparison cannot be taken to show that external and internal exposure are similar, due to uncertainties of the parameters and model choices.

Mercury

Kolossa-Gehring et al. (2012) and Schulz et al. (2007) reported mercury concentrations in urine of about 0.5 µg/L. Assuming a urine production of 2 L, the exposure will account to 1 µg per day. The systemic availability of mercury accounts for about 10%, and the exposure in an adult person weighing 60 kg is then 0.17 µg/kg of BW per day. An estimate of external exposure of mercury was calculated by the EFSA CONTAM panel (EFSA 2004) of 1–3.5 µg/kg of BW per week, which equals 0.15–0.6 µg/kg of BW per day.

In summary, short-term individual-level internal exposure estimates are in general rarely comparable to external exposure estimates. The degree of accordance of the two approaches, however, varies considerably, depending on the agent of interest and data collection procedures. The quality of the methodology used for the assessment has to be considered and is discussed in Sect. 3.7. This applies for both external and internal exposure assessments. The use of internal and external exposure assessment, in combination, can reduce the overall uncertainty. The need for improved data is stressed, with respect to excretion rates in urine, to frequency of samplings and length of sampling periods.

3.6 Results of Exposure Assessment

Gerhard Heinemeyer

The major purpose of the presentation of exposure evaluations is to support the risk assessment by adequate quantitative data to perform the risk characterisation and to enable the assessment of the risk management options.

The presentation of the results of the exposure assessment should include a characterisation of the sources and pathways and their quantitative contribution to the entire exposure.

In some regulations, the exposure assessment delivers the information to establish limit values. Exposure results are also used to establish registers of approved and non-approved substances in particular types of products. For example, in the cosmetics regulation, scientific committees and expert groups established lists of allowed substances based on an adequate exposure assessment.

3.6.1 Presentation of Exposure Assessment Results

First of all, the results of an exposure assessment should be presented in a transparent and well understandable manner, presented in a standardised way (EFSA 2009b).

As a general rule, the quality of exposure estimation results is always as good as the quality of the assumptions that are used in the assessment.

Different levels of exposure assessment characteristics should be taken into account, as presented in Fig. 3.15. Exposure assessment results can be presented as estimates characterising a realistic or a more or less fictive situation. The first one is characterised by using empirical data of sound statistical character. If such (high quality) data are available (compare Sect. 3.7), exposure can be estimated that well represents average exposure of the general population. This estimate can be presented with any type of mean value or median, including confidence intervals or other indicators of the range of the mean value. Accordingly, the same data also lead to respective values for higher percentiles of exposure estimates. These may correspond to the general population but also to particular subpopulations, e.g. children, or specific consumers or users of products. For example, consumers may use or consume the same type of product at all times, which might be highly contaminated with a certain compound.

On the other hand, if sound data are not available, the results will not fulfil scientific and statistical criteria. On the *Y*-axis in Fig. 3.15, the level of reality changes from the empirical to a more fictive (conservative) one, which, however, is needed for adequate protection of the population of concern. This is possible if the data are chosen as conservative or very conservative. Depending on the degree of the default used, the results may be of less quality, but due to the formulation of the exposure scenario, it is possible to perform an estimation that delivers appropriate data to perform a risk

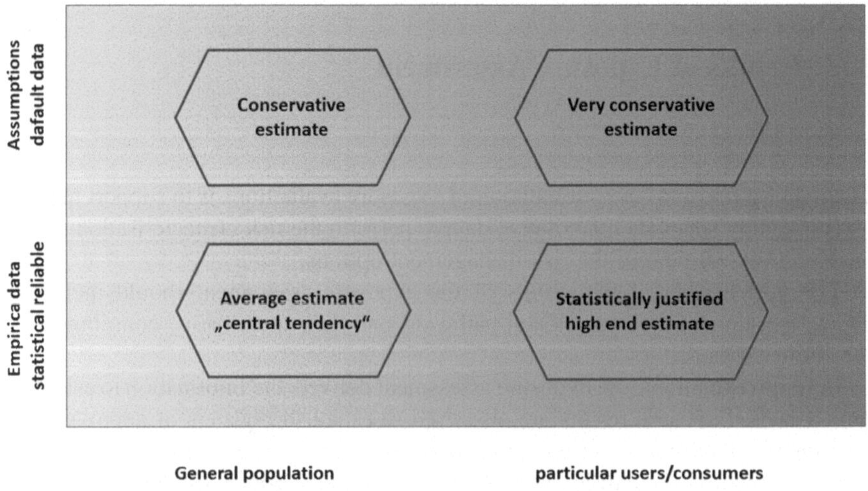

Fig. 3.15 Relationship of exposure assessment characteristics

characterisation as a conservative measure. This kind of assessment enables the risk assessor to decide whether or not additional assessments are needed and to start a tiered approach. Also, this type of assessment is typical for exposure evaluation under REACH which starts with an event-based assessment (Sect. 5.6.3). It should be noted that such default-based exposure estimates are often not only referred to acute toxicity corresponding to short events. It is often assumed that this use is repeated every day. The extrapolation of event-based exposures to long-term exposures is scientifically questionable (compare also Sects. 2.6.3 and 3.2.11).

According to estimations that cover the general population, an average result may be achieved that can be designated as "typical or normal". This means that the majority of the population may experience this exposure. For many regulatory purposes, this estimate is inadequate, because it might not secure the entire population. In this case, a conservative estimate is needed which covers every range of uses including the so-called "foreseeable" misuse. However, the aim of estimating extremes of exposure is similar, namely, to include subpopulations that deviate from the general population.

Subpopulations as characterised in Sect. 3.2.12 should be considered in the exposure results report.

Depending on the regulatory purpose of a risk assessment, an estimate characterising the central tendency and another characterising high exposure might be needed. The central tendency is used as a measure for the typical exposure of the general population and the high exposure estimate for consumers who are characterised by a high consumption. If the concern is not for acute toxicity, the frequency of use or consumption is the key determinant. If the substance of concern does not have a long persistence in the body, only frequent use or consumption lead to permanent presence of substances in the body and thus to chronic toxicity. A brand loyal consumer might use or consume frequently high quantities of a product.

Average consumption (typical exposure)	The average use or consumption of a product in the general population
High consumption	Frequent consumption, together with use or consumption of high amounts of a product, e.g. in combination with brand loyalty

In deterministic exposure assessments of long-term exposure, the average consumption is based on the use of mean or median values of exposure parameters in the model calculation. As pointed out in Sect. 3.4.8, average values can only be determined if adequate data are available. For many product types, these data are seldom available, and thus exposure estimates based on such assumptions are hardly characterising central tendencies. The correct characterisation of the exposure scenario is crucial in these cases, and therefore, many estimates tend towards conservative or high-end estimates. In food assessments, parameters describing average values are more often available, and most of the requirements for sound statistical

Table 3.15 Statistical descriptors that are a matter in a report about consumer exposure

Type of estimate	Statistical descriptors	Exposure situation	Population, consumer
Central tendency estimate	Average estimate (mean, median) It is appropriate to present mean *and* median	Average assumption (realistic approach) All combinations of contamination, consumption and frequencies of consumption/use	General (normal) population
Conservative estimate	High-end estimate (95th percentile)	High concentration and high consumption, average frequency of consumption/use	Particular population (to be characterised) Product loyal consumer
Very conservative estimate	Upper confidence interval of 95th percentile), or 99th percentile (to be justified)	High frequency of consumption	Brand loyal consumer

presentations of exposure data and results are formulated on the basis of food exposure assessment. The respective statistical descriptors and their characterisations are displayed in Table 3.15.

For this reason, the exposure characterising the central tendency and the conservative estimate must be transparently described in the exposure scenario, together with mentioning of the key parameters used.

In practice, exposure estimations for consumer products are normally only based on using assumptions (expert judgement) due to the lack of adequate data. Only in food exposure assessments, the assessor can present results that are based on consumption data having a representative background. Concentration data used in such assessments are representing a wide range, but are not necessarily characteristic or representative for the most important food groups. The advantages and disadvantages are described in detail in Sect. 4.8.

A special focus should be put on the tiered approach to present exposure estimation results and which makes it possible to shift from the more conservative to the realistic estimation. Deterministic estimations using default data and probabilistic estimations should be discussed as several iterative steps of the same approach.

Central Tendency and Conservative Estimates

For deterministic assessments, the following statistical determinants should be listed in a standard report: the average values (arithmetic mean, geometric mean, median) and the upper percentiles (90th, 95th percentile) of the estimate should be indicated to describe the range of the quantitative exposure. All values should be given and the respective confidence intervals at a level of 90% or 95%. This is only possible if adequate data are available. With the exception of body weights or surfaces, the low percentiles are normally not needed, because they indicate the lower levels of exposure, which are not needed for risk characterisation. This kind of assessment is only possible with access to relevant data. Probabilistic estimations will produce all the statistical determinants in one calculation step.

In a deterministic assessment, average estimates can be calculated if adequate exposure parameters are available that characterise mean values. To indicate inconsistencies as shown in the example in Sect. 3.4.9 (Table 3.11), mean and median intake estimates should be presented to provide an indication of the slope of the distribution.

To show the extremes of an exposure, conservative estimates should be presented. For practical purposes, the 95th percentile has become a good compromise. It marks a limit over which 5% of a population might be at risk; however, it also describes a population that might be characterised by special consumption behaviours and the consistent use of particular high concentrations of the substance of concern.

To estimate such exposure, an approach has been proposed by EFSA (2008c) for aggregated food exposures. This approach is based on a proposal of the SCOOP (EU 1998). First of all, in a series of disaggregated foods, the two foods showing the highest average exposure results should be identified. The total average exposure is calculated on the basis of mean values. Then, a second estimation should be performed in which the average estimates of the two foods having the highest consumption figures are exchanged by the 95th percentiles. This approach has a strong conservative character by taking the two foods most contributing to exposure. It leaves the others as they are. To date, in practice data that are adequate to perform this kind of assessments are only available for food exposure. However it should be discussed whether or not it could be modified and transferred to other products. This approach, however, requires adequate data.

Presentation of exposure results should consider age categories in relation to gender. In deterministic and probabilistic assessments, the differentiation according to age is essential due to the differences in body weight. Particular cases (pregnancy, nursing) should be indicated separately.

A template (Table 3.16) that can be used to present exposure results can be structured as follows. Due to the availability and the needs for differentiation, age groups may be expanded or constrained. Due to the availability of adequate data on food consumption and food contamination, this type of presentation is regularly used

Table 3.16 Template to present the results of an exposure assessment

	Mean (± CI)		Median		90th percentile		95th percentile	
	Male	Female	Male	Female	Male	Female	Male	Female
Newborn								
<2 months								
1–5 years								
6–12 years								
13–18 years								
Adults								
Elderly								

in opinions of EFSA and other agencies in the EU and worldwide. For other products, similar approaches can be applied if possible.

In deterministic estimations, average data and high-end estimates should be calculated in any case. In probabilistic estimations, the mean and median and the 90th and 95th percentiles should be presented, to show whether or not the slope of the distribution is small or even prolonged. An example to show results from deterministic and/or probabilistic exposure estimations is presented in Table 3.11.

3.7 Ensuring Quality of Exposure Data

Gerhard Heinemeyer

Exposure analysis requires approaches and information that need a high degree of quality. The choice of data for exposure assessment should include an evaluation of data quality, particularly for the establishment of guidance documents and fact sheets to be used as collections of standard parameters. US-EPA (2000) has reported an approach that was used for quality assessment of data to be considered in the US-EPA's *Exposure Factors Handbook* (2000, 2011) as well as WHO (2008) and the German XProb project (Mekel 2007). Quality assessment is also needed when data from new data sources are considered for use in an exposure assessment. Quality assessment should always be combined with the evaluation of uncertainty, due to possible overlaps.

This chapter reviews approaches to describe and characterise quality of data used for exposure assessment. Several reports have been published to describe criteria that can be used to assess quality. More or less, all the publications are focusing similar issues.

Table 3.17 lists some publications and guidance documents that deal with quality characterisation in exposure analysis. First of all, RIVM (Bremmer and Van Veen 2000, 2006) proposed a quality score of 1 (bad quality) to 4 (good quality) as the result of the quality assessment by expert judgement. Criteria to be used in a quality assessment have been developed by US-EPA (2000), in a working group of IPCS (WHO 2008) as well as in the XProb project (Mekel et al. 2007) for analysis of data that are foreseen to be put into the database RefXP. A suggestion made by Klimisch et al. (1997) for a reliable assessment, proposing to assign the results to one of four categories can be taken as a common basis of the development of quality criteria:

- Reliable without restrictions
- Reliable with restrictions
- Not reliable
- Not assignable (not sufficient details given for assessing reliability)

This approach has also been taken over by OECD (2005).

Table 3.17 Existing documents and approaches for quality assessment in exposure evaluation

Publisher	Characterization of quality	Remarks	References
RIVM	Quality score (1–5) from low to high	Quality characterization, not qualification	Bremmer and Van Veen (2000) and Bremmer et al. (2006)
X-Prob	List of quality criteria	Characteristics of criteria, recommendation	Mekel et al. (2007)
US-EPA-quality objectives	identify the type, quality, and quantity of data required	Step-by-step approach to characterize and narrow data quality	US-EPA (2000)
WHO-IPCS	List of quality criteria	Adressing four hallmarks of quality description similar to X-Prob and EPA approach	WHO (2008)
OECD	List of quality criteria		OECD (2005)
NCCMT (Canada)	Quality assessment tool	Practical tool which provides a score based quality result	NCCMT (2008) and EPHPP (2010)

Taking the IPCS guidance document, the following four quality descriptors should be evaluated for quality of the data as listed in Table 3.18(a–d).

3.7.1 Appropriateness

This descriptor characterises the applicability and utility of data for a particular exposure assessment. Good quality can be judged by the soundness of methods and procedures that were used in the exposure assessment.

3.7.2 Transparency

This descriptor assesses methods, data and processes used in the exposure assessment, which should be characterised in a clear and understandable way. This refers to the assumptions made in the assessment and the limitations underlying the process of the assessment, which should be explained and presented. Transparency enables an assessor to judge about the quality of the assessment for not only the intended purposes but other purposes as well.

Table 3.18 WHO criteria for quality assessment in exposure analysis [extracted from WHO (2008)]

Appropriateness
• Relevancy of a particular data set or method to a risk assessment or risk management question
• Making judgements about appropriateness of data sets for the purposes of the users of a data set
• Detailed description of the conditions and study design under which a data set was collected
• Thorough description of specifications, such as assumptions and simplifications
• Place, time and scale of the assessment (particular region, ethnic group or age group)
• Methods of analysis or study design
• Extrapolation of results
• Release of the chemical was collected for products that contained the chemical in a different matrix of materials
Transparency
• Awareness of all of the perspectives and viewpoints by which the quality of the data used in the assessment might be judged
• Replication or verification of data is possible independently by another group
• Someone having general knowledge of the area but is not intimately familiar with the particular methods can understand the methods/approaches being described
• Avoidance/elimination of "hidden" assumptions
• Sensitivity of the exposure assessment can be determined in relation to alternative values, assumptions or theories
• Description of the key assumptions and limitations of data and information
• Explicit description of elements that are either known or suspected to have been omitted
• Harmonised terminology
Accuracy
• Accuracy of the methods for sampling and analytics to evaluate the exposure to a contaminant
• Complete description of data
• Sampling design
• Calibration of instruments
• Analysis methods
• Reproducibility of data
• Step-by-step procedure
• Definition of the objectives
• Identification of the involved population
• Study design
• Type of sampling that was performed (e.g. randomly sampling)
• Collection of the samples and their analysis
• Accuracy for this step may include detailed protocols
• Description of the steps used to estimate any parameters or values from analysis of the samples
• Descriptions of the scientific principles used to calculate values and parameters from the observations
• Statistical mathematical methods or algorithms
Integrity (general considerations of assuring data quality)
• Assuring that data is properly recorded
• Provision of correct descriptions about the collection of the data

(continued)

Table 3.18 (continued)

Integrity (general considerations of assuring data quality)
• Quality assurance procedures during the collection of data
• Appropriate privacy safeguards have been followed
• Appropriate sampling
• Appropriate analysis
• Consideration of good laboratory practices
• Incorporation of the values from a data set in an appropriate manner to a risk assessment (here: exposure scenario)
• Extrapolation of other data if an appropriate data set is not available
• Consideration of uncertainty
• Proper protocols and review boards
• Protection of people during the collection of data
• Data collection that neglects appropriate safeguards for animals and people is unethical and erodes confidence that appropriate methods have been followed in other aspects of data collection

3.7.3 Accuracy

Parameters chosen for an assessment should be specified as close as possible to their true values. This includes laboratory analyses but also epidemiological and statistical methods. Accuracy is important, e.g. for understanding the representativeness of the data and the assessment.

3.7.4 Integrity

Data used in an exposure assessment should be as complete and accurate as possible. Appropriate quality control and quality assurance procedures should be taken into consideration. The data reported must be scientifically and ethically defensible.

Integrity is needed for high quality on an exposure assessment. The audience of the risk assessment report must have trust in the assessment. This can be approached if the report provides properly recorded data with correct descriptions about its collection. Quality assurance concepts are today a must for everybody who is involved in the process of risk assessment. The International Standards Organisation (ISO) (ISO 9000) has published general rules for quality management. Institutions and/or agencies involved in any laboratory, epidemiological, statistical and administrative work regarding risk assessment should be certified according to the ISO 9000 rules.

Table 3.18 lists the criteria extracted from the WHO/IPCS document and characterises the meaning of the four descriptors. They can be used directly to judge the quality of data used in the exposure assessment. Due to the above-mentioned overlap of the three concepts of WHO, US-EPA and XProb and to avoid double discussions, only the IPCS approach is discussed in detail; the other concepts can be understood by use of the tables.

The concept of quality assessment of exposure data was developed by the exposure planning group, an expert panel within the *IPCS harmonization programme*. The approach to assess quality was based on the principle published by Klimisch et al. (1997) and on the US-EPA (2000) recommendations for quality assessment. The concept of the XProb project was developed in parallel to the WHO by a German group (Mekel et al. 2007); however, the results were well comparable.

Data used for exposure assessment should be appropriate for the risk assessment question. In practice, this means that the purpose to choose a parameter for use in exposure estimation from another source should have an exposure assessment purpose. Many data are taken from studies having purposes other than exposure assessment. This should be made clear and transparent if they are chosen for an assessment. For example, an extrapolation of an estimate of the release of a chemical from a product but not relying on a comparable situation diminishes the quality of the particular parameter, in combination with an increased degree of uncertainty.

The data and methods used in the assessment should be described and characterised in a transparent manner. Transparency considerations have been pointed out as an important and particular extra issue of reporting data for exposure analysis by the EFSA scientific committee (ÊFSA 2009b). The IPCS expert panel characterises transparency analysis as an issue of quality assessment. Although quality evaluation is a particular issue, the three concepts, (i) quality assessment, (ii) uncertainty analysis and (iii) transparency characterisation are common approaches that should be presented and discussed in a common sense (WHO 2008).

Data used for exposure assessment should be accurate. This applies to laboratory analyses and for other methods to generate data for exposure evaluations such as questionnaires. Accuracy covers sampling, analysis, study design, statistical analysis and presentation of the data. The criteria listed in Table 3.18 were extracted from the guidance document published by WHO (2008).

The concepts published in the XProb project (Mekel et al. 2007) have been established to judge exposure parameters for input into the database that was developed in parallel.

For practical issues, the quality assessment tool, developed by the Canadian "Effective Public Health Practice Project" (EPHPP 2010) first published 1998 and updated 2010 by the "National Collaborating Centre for Methods and Tools (NCCMT 2008)", also allows a quick assessment of quality based on a scoring system and parameters which are very close to the approaches presented here by other institutions and authors.

3.8 Uncertainty Analysis

Michael Schümann

3.8.1 Introduction

The assessment of the population exposure is defined as the:

> process of estimating or measuring the magnitude, frequency and duration of contact to an agent, along with the number and characteristics of the population exposed. Ideally, it describes the sources, pathways, routes, and the uncertainties in the assessment. (Zartarian et al. 2005)

The promise given by an exposure assessment is that the estimated results would approximately reflect the real exposure situation for a defined population. Most exposure assessments rely on mathematical models to predict the level of individual, group or population exposure. Using this technique knowledge about models and input parameters should be sufficient for a valid prediction. Although the evaluation and documentation of uncertainty has been considered necessary since the 1980's (e.g. US-EPA 1985, 1992), this approach has become mandatory in exposure and risk assessment only within the last decade. A key document that is used as a framework in many regulatory approaches is the IPCS (WHO 2008) guidance document. It contains ten guidance principles for taking variability and uncertainty into account.

The main purpose of uncertainty analysis (UA) is to increase transparency regarding all elements of an exposure assessment and to support appropriate decision making. The results of this step of the risk analysis have a high impact on regulatory and enforcement action. Meanwhile the major regulatory organisations expect that assessments must state clearly and unambiguously what uncertainties have been identified and what is their possible impact on the overall assessment outcome.

Uncertainty is used as a general term referring to all types of limitations in available knowledge that affect the range and probability of possible answers to assessment questions. Exposure assessment conducted in parallel with an UA is characterised as an iterative process (WHO 2008), since the degree of the necessary stepwise refinement of the analysis depends directly on the overall safety level and the state of knowledge that is achieved. In the context of food safety, recent efforts by governmental bodies are focused to the develop guidance for assessors that deal with qualitative and quantitative methods for UA. Beside differences in detail, all recommendations for an informative assessment include (a) the analysis of the scope (the evaluation of the risk questions); (b) the selection and specification of relevant exposure scenarios; (c) the selection of the exposure models, the inherent parameters and the underlying data sources; and (d) the calculation rules (algorithms) to take into account the parameter variability and uncertainty (US-EPA 2011, 2016c) as

Table 3.19 Modified version of the EFSA (2006) and IPCS (WHO 2008) evaluation scheme for sources of uncertainty

Sources of uncertainty	Characteristics of uncertainty		
	Overall level of uncertainty	Appraisal of the knowledge base	Subjectivity of choices
Scope/assessment objectives			
Scenarios			
Conceptual model			
Mathematical model			
Parameters			
Result(s)			

well as population subgroup differences. A comparison of the mainly quantitative and qualitative approaches is published by EFSA (2018b, c, d), BfR (2015) and EFSA/BfR (2019).

Table 3.19 shows an abbreviated form of WHO (2008) scheme for the documentation of uncertainties. The rows reflect the steps of exposure and uncertainty assessment; the column headers might be used as a guide for the identification of the major sources of uncertainty: Each element of the matrix contains many aspects that might contribute to the overall uncertainty of exposure assessment results. For each element a classification of the uncertainty should be assigned. WHO (2008) recommends the terms (No, Medium, High or NA = "not applicable") for the quality and uncertainty assignments; EFSA (2006) proposed a ranking using two "+" and "−" signs indicating the direction and the magnitude of uncertainty for each subject of consideration. The last row contains an aggregated evaluation. In general, it is not a sum or maximum of the rows above, but an expert judgement for the overall qualitative valuation.

To this process UA adds a transparent estimation of the propagation of possible errors in parallel to the propagation of the empirical variance of influential factors (sensitivity analysis). In consequence, the sources and the different types of uncertainty that influence the exposure results should be characterised qualitatively and, if possible and necessary, described in a quantitative manner.

A properly conducted exposure assessment with an UA might clarify which answers to the research questions have a high degree of certainty, which are questionable and which additional research steps would have the highest impact to reduce remaining uncertainty. Relying on conventional defaults, assumptions or expert judgements might be appropriate as a starting point, but, if the exposure results point to high exposure levels, a refinement on the basis of empirical data, stratification by subgroups and the inclusion of new and reliable measurements should be mandatory to fulfil the requirements. Alternatively, the scope of the assessment might be changed, developing a model tailored for subgroups separately (e.g. "fish-eaters", e.g. EFSA 2012). It is the purpose of the exposure assessment to describe the expected exposure level of a study population. In consequence the exposure assessment must include information about the robustness and validity of the results for all subgroups of the study population, too. The amount of additional information and the degree of iteration to a higher-tier approach required in an UA will depend on the specific situation. The task is to choose a level of detail and refinement for an UA that is appropriate for the assessment objectives. It is based on available information and data and corresponds to the importance of the decision (EPA 2009a). In every case, the exposure, uncertainty and sensitivity analysis must be described in a transparent manner that allows re-examination and iteration. The amount of effort and detail of the uncertainty assessment should be proportionate to the scope and needs of the assessment, to the relevance and adversity of the possible health impact of exposure and to fulfil the regulatory requirements.

3.8.2 The Scope of the Assessment and Translation to Exposure Terminology

The objectives of the assessment might differ. It might aim to predict the exposure distribution of a population defined in a scenario (Sect. 3.2), it might describe the conditions under which a substance can be used safely, and risks are controlled (ECHA 2016a) in an exposure scenario under the REACH legislation (see Sect. 5.6.3). The exposure assessment might trigger the selection of necessary risk management measures, by comparing the scenario-specific exposure estimates with no effect levels determined in the hazard assessment.

The process of exposure estimation should start with simplified approaches, being repeated iteratively until the level of residual uncertainty can be tolerated with respect to the purpose (Sect. 3.2).

Each uncertainty may be analysed at one of three tiers: qualitative, deterministic or variance-based. The identification and evaluation of the different kinds and sources of inherent uncertainty is part of the overall analysis and documentation. Communicating the results of an exposure assessment that is based on model assumptions and numerical estimates is demanding, and communicating the inherent uncertainties at the same time makes the task complex.

The first step in exposure and UA is the translation of the assessment questions (e.g. formulated by governmental organisations, stakeholders) into a unified language (terminology) that is accepted in the scientific community and the participating organisations. Finding a common description language will simplify the communication and increase the acceptance of results.

In case of uncertainty about the scope of analysis, e.g. selection of relevant scenarios and target populations for which the analysis should be valid or with respect to the safety levels for the evaluation of results, the assessor should contact the customer (or awarding authority) before starting the project. It should be recognised that with very few exceptions, the data available for quantitative exposure assessment will not match the requirements of the exposure scenario that has given rise to the request. In consequence, there will be always a number of uncertainties in the quantitative risk estimate, most of which cannot be clearly defined or quantified. Exposure assessment is a prognostic technique based on knowledge, models and data. It does not have the quality of a specific empirical study that contains exposure measurement. It is important to clear up resulting uncertainties about the quality of results to avoid double work or conflicts (at the end of the project). Additionally, an agreement about the terminology to be used in the assessment document would be helpful

Different terminology is obviously used by the different players in the context of exposure assessment, dependent on the subject of interest. This is caused by sectoral legislation defining specific terminology and international standards for specific fields of risk assessment (FERA 2012). For example, ECHA (2013) published a table of terms and abbreviations for chemical safety assessments (see also Chap. 6). EFSA (2012) published a comparison of terminology with a focus on the Codex Alimentarius Commission (CAC, Codex 2003) in relation to food safety. Since exposure and risk assessment is in general a multidisciplinary task, scientists using different terminologies have to find an agreement that is accepted in the assessment team and by the scientific audience. Different ways of conceptualising exposure in administration, regulation, toxicology, nutritional sciences, exposure sciences, epidemiology and statistics with respect to the conceptual, mathematical and computational formulations require a strong effort of harmonisation in nearly every project. An alphabetical list of selected generic terms in hazard, exposure and risk assessment and their descriptions are provided together with a compilation of remarks and background notes in the IPCS harmonisation document (WHO 2004); Internet-based glossaries are published, e.g. by US-EPA (2015, 2017) or EFSA (2017). It is strongly recommended to use a unified terminology in the documentation of the assessments, at best with a reference to a published terminology standard. As pointed out later, the list of current terms to be harmonised however is much longer than these lists.

3.8.3 Planning the Assessment Strategy

Planning the exposure assessment strategy should include planning of the UA too. Done in parallel the amount of additional effort is minimised. The BfR-guidance document (BfR 2015) gives an example for this strategy by the formulation of a set of questions for each topic of the exposure assessment: goal and question formulation, exposure scenario to be modelled, choice of a conceptual and formula-based exposure model, qualitative and quantitative UA in relation to model parameters and qualitative evaluation of the algorithms for exposure calculation (see Fig. 3.16).

The questionnaire follows the three components of uncertainty specified in IPCS (WHO 2008):

1. The level of uncertainty of the exposure assessment that covers the potential deviation of the exposure assessment from the actual expected or empirical measured exposure situation

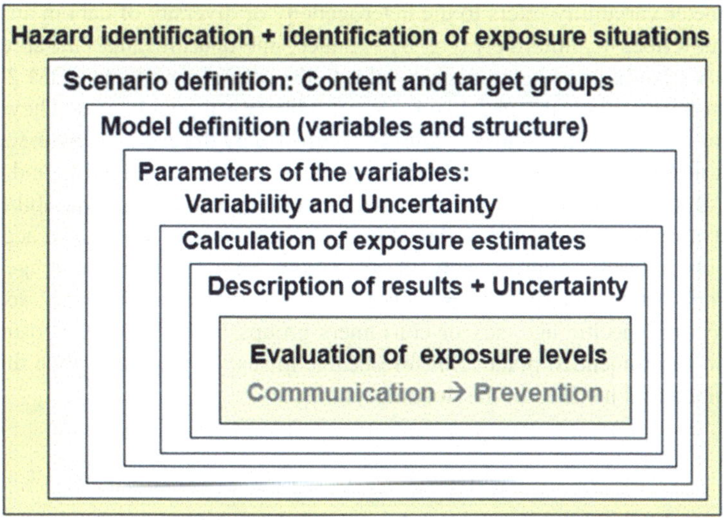

Fig. 3.16 Steps and hierarchy of an exposure assessment integrating an UA

2. The appraisal of the knowledge base of the exposure assessment that comprises the evaluation of degree of completeness of all available information that can be used for the exposure assessment
3. The subjectivity of choices inherent in an exposure assessment that includes the description of reasons for all decisions made between alternatives (selection of scenarios, models, parameters), with respect to deviant rules and recommendations within the scientific community and interest groups

The information needs for an exposure assessment are related to the completeness of the scenario characterisation and sufficient knowledge about the process of exposure and corresponding models. For describing the heterogeneity of the population exposure, a central tendency and a variance estimate are required for each influential factor. If defaults are used, it should be clarified what they stand for. If the information is based on empirical data, the statistical uncertainty (the standard error of estimates or confidence intervals) informs about the statistical precision that has been achieved. The variance of the influence factors (e.g. concentration, intake rates, exposure duration and frequency as well as body weights, surface and composition) must be included in the exposure parameter value selection: otherwise the population heterogeneity is not represented in the assessment. Using standard values of defaults might be systematically incorrect if the scenario population indicates to differences over subgroups (e.g. age/sex structure, nutritional preferences, regional or temporal deviation). Questions about possible measurement errors, bias and lack of reliability of measurement are related to the choice of the parameter values [for uncertainties in measurement see ISO (1993) and JCGM (2008)] for compliance evaluations (Eurachem/CITAG 2007). The degree of overall uncertainty due to imprecision and statistical uncertainty might be described on the same numerical scale as the exposure estimate.

Parameter variability refers to the heterogeneity or diversity of data in an assessment. It describes quantitatively (e.g. by variance, interquartile range) the spread of a data set. Variability and heterogeneity refer to the natural variations (IOM 2013) in the environment, exposure paths and susceptibility of subpopulations. They should be seen as characteristics, which cannot be controlled by the exposure assessor or the decision-makers. Collecting more information cannot reduce variability and heterogeneity. But, the statistical description error can be reduced. Since variability is an attribute of reality, it cannot be reduced in an assessment, but it has to be accounted for and characterised appropriately. If restrictions with regard to coverage of the population heterogeneity are associated with formulation of the scenarios (e.g. selecting specific age, sex or consumers groups, restriction to exposure with respect to recommended practice or foreseeable misuse of products), then this must be explained and justified in the documentation.

3.8.4 Taking Heterogeneity, Variability and Uncertainty into Account

An early approach to incorporate heterogeneity, variability and uncertainty into account was published by the US National Research Council (NRC 1996, 2009). It discusses mainly strategies to deal with variability. It proposes a stepwise refinement approach that starts with defaults for parameters and refines the degree of information used stepwise. Here, the recommendations of the IPCS (WHO 2008) document are used as an extension of this description for the stepwise refinement strategy by including uncertainty and variance-based analysis.

1. Ignoring variability (e.g. assuming a standard body weight of 70 kg) can be justified if the assessor has prior knowledge about the variability and if the consequences of ignoring variance are considered in detail (e.g. low variance of a parameter or minor influence on the results). The degree of resulting uncertainty using defaults or assumptions might be evaluated by repeating the calculation using constants that reflect the variability and additionally the uncertainty. Underestimation of the variability of parameters can have serious impacts on the results. Within this strategy more confidence in the level of conservatism of the expected exposure range estimates could only be maintained by the choice of more conservative input parameters.
2. Taking into account estimates of the low- and high-value ranges of the distribution (e.g. 5th and 95th percentiles) as a second step for each parameter gives an ample description of the expected ranges of exposure estimates (intervals). If this is done stepwise for each parameter of the model with a unified selection of the contained variance (e.g. %-percentiles, one standard deviation), the contribution of variance of each parameter on the variance of the exposure estimate might be compared roughly.

3. Using a strategy of disaggregating variability means that the interindividual variability is described for subgroups (stratification) that show lower within-group variance than the total population (e.g. stepwise selection by age, gender, consumption behaviour, season, specific subgroups like pregnant women). Accordingly, central tendency, variance and percentiles estimates must be available for all subgroups that should populate the selected scenarios and models. Remaining statistical parameter uncertainty can only be reduced with more and better data. Exposure factor handbooks that contain an appropriate stratified statistical description for many important variables are available (e.g. Vuori et al. 2006; Mekel et al. 2007; US-EPA 2008, 2011; JRC 2015; EFSA 2015a; Jang et al. 2014; see for an overview Zaleski et al. 2016) (see Sect. 7.3.2). These handbooks and data collections are intended to provide information on parameters (e.g. anthropometric, physiological, time budget, consumption-related behaviour) commonly used in assessing exposure to chemicals. And, it should be kept in mind that the model structure might be group-specific with respect to the pathways, sources and exposure events (e.g. playing and mouthing behaviour of children, specific user groups, usage and applications of cosmetics and articles for personal hygiene over age groups).
4. Using minimum and maximum of empirical data or consideration of the expected ranges (e.g. by expert judgement) or reference values is an approach that tries to translate the variability (the empirical range) in a population or measurement series into the model parameter. Using the terms "best" or "worst" case for this is mostly misleading since the empirical min/max estimates are very sensitive to sample size. There is almost a guarantee that a measurement series in a larger sample will show more extreme values, lower than "best" or higher than "worst". The author recommends instead the usage of low (e.g. 1% or 5% quantiles) and high quantiles (e.g. 95% or 99% quantiles); the choice of percentile should made mainly in dependency to the sample size and the level of intended conservatism. All parameters in the nominator of an exposure model equation should be selected from the upper area of the distribution (high percentiles, e.g. for consumption and concentration), all parameter values in the denominator of the model equation (e.g. body weight in ADI compatible exposure metrics) from the lower area (low percentiles). Nonetheless it should be considered that the width of confidence intervals of any statistical parameter estimate is proportional to the empirical standard deviation and the number of observation (roughly the square root of the sample size). This holds for percentiles too. Depending on the number of parameters, the range of the estimated exposure distribution might increase. The combination of moderate extreme values for the model parameters can reflect an extreme exposure situation that has a low probability for the combination. But, the resulting range of the estimated exposure interval covers a range of possible values. Using the total range is appropriate for a conservative safety evaluation. Since distributions that characterise variability in the input parameters (esp. concentration estimates) are often lognormally distributed, a high degree of skewness of the resulting exposure estimates must be anticipated in practice for the resulting exposure estimates.

5. Probabilistic techniques rely in general on Monte Carlo simulations (repeated sampling from the probability distributions) that calculate a distribution of the expected exposure distribution by sampling from defined distributions for every (or most influential) parameter. If the input distributions are well-characterised (e.g. by means, medians, variance or percentiles, type of distribution), the resulting exposure estimates will reflect the variability of input. Remaining uncertainty depends on the statistical errors of the input parameters and possible deficiencies of the model (e.g. ignorance of correlation, completeness of exposure sources and pathways, over-aggregation of variables over time or food categories). Bootstrapping techniques describe the resampling from empirical data sets. This sampling technique uses one data record per iteration (representing one individual); by this it is taking into account the inherent empirical correlation of variables (e.g. the interaction of the body weight with the intake or inhalation rate). It is not necessary to know the statistical distribution characteristics in advance. An adequate resampling technique gives in result information about the ranges, the statistical values and the confidence intervals of exposure estimates. But, conducting a Monte Carlo or a bootstrapping analysis on the basis of small samples does not generate "better" exposure results: the inherent uncertainty of the input propagates to the exposure estimates. Resampling using small data sets should be justified in the documentation, since the inherent statistical uncertainty remains. Small samples cover only a small range of the existing variability. Even a rough (e.g. lognormal) distribution approximation of parameters might be more appropriate in these situations, if the statistical uncertainty of parameters due to small sample size is described and taken into account in the UA. The location and scale of the (empirical) density function as well as the percentiles of the corresponding empirical cumulative function will have ranges of uncertainty for every statistical parameter (Fig. 3.17).

Using data from large surveys (e.g. national nutritional surveys) for a resampling approach is state of the art, but the restrictions due to subsample sizes taking into account the age and sex structure as well as the representativeness for the scenario population should be checked in advance and described in detail.

If quantitative estimates for uncertainty of the relevant parameters are available, this gives the opportunity to conduct a two-dimensional probabilistic analysis that includes variance and uncertainty. This requires that each of the selected parameters be modelled as the combination of sampled values from the defined distributions (or sample) and a separate term that reflects the contribution of uncertainty to the parameter value. As a result the impact of both components can be evaluated with respect to the variance and distribution of the resulting exposure estimates (see, e.g. EFSA 2006; Frey and Patil 2002). This type of analysis is recommended mainly for critical situations where the uncertainty should be separated from variability (e.g. for stimulating additional data sampling for influential but uncertain parameters). Additional 2D-uncertainties analysis is not necessary if it might be expected (see interval consideration above) that the influence of uncertainty of a parameter would makes no relevant (e.g. in relation to an ADI, TDI) difference to the range of the exposure.

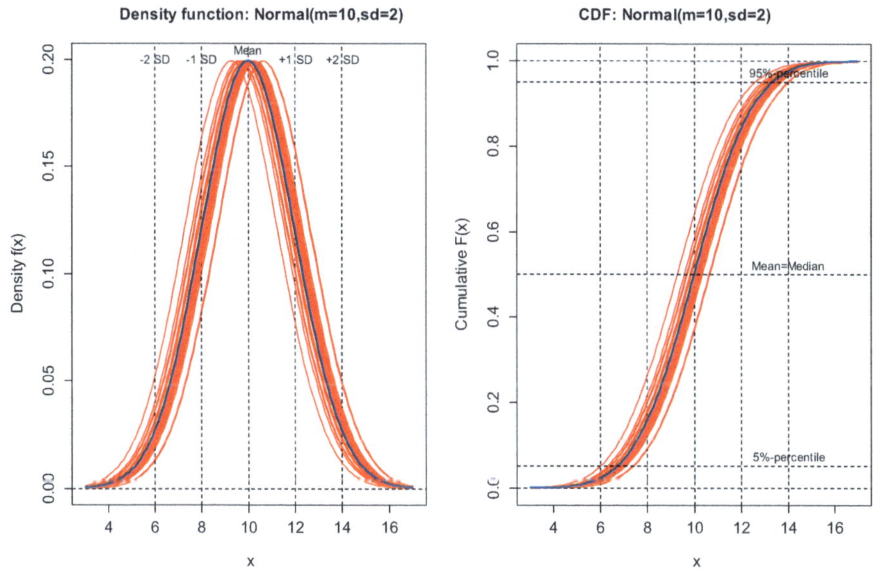

Fig. 3.17 Illustration of the uncertainty effect of the statistical error due to restricted sample size ($N = 50$) on the standard error of the mean (SEM): normal distribution with mean of 10 and a standard deviation of 2 units. Left: Density function with random shift around the mean (20 samples). Right: Cumulative density function with random shift around the mean (20 samples)

3.8.5 Evaluation of the Confidence in the Knowledge Base and the Subjectivity of Choices

Confidence in the knowledge base and subjectivity of choices are quality-related categories. These two dimensions can be described only qualitatively. The appraisal of the knowledge base comprises the evaluation of degree of completeness, suitability and representativeness with regard to all available information that can be used for the exposure assessment. If possible, the direction and the magnitude of the anticipated deviation of the results due to knowledge restrictions and possible systematic bias should be described for all criteria. Subjectivity of a choice might play a role if alternatives for the selection scenarios, models or parameters exist and preferences due to interests or values of the experts and the stakeholder might influence the choices. For the selected scenarios, models and parameters, the reliability, the consistency with other scientific concepts and the coherence of empirical data from different sources should be evaluated, by looking at the robustness of resulting exposure estimates. An evaluation of achieved quality by scientific review and, if possible, by validation and calibration studies is strongly recommended (WHO 2008).

3.8.6 *Identification of the Sources with High Contribution of Variability and Uncertainty to the Exposure Distribution*

The identification of those input variables that have a strong influence on the variance of the exposure will point to those variables that have (a) a high potential for possible exposure control (Frey et al. 2004) and (b) have in parallel a high sensitivity for influence of inherent errors and uncertainty. Investigating in influence and sensitivity analysis clarifies to which degree variance and error in the input and model parameters propagate to the exposure estimates. It provides information that is necessary to formulate answers to the questions (Saltelli 2002; Saltelli et al. 2004) by the customer and the public:

- How confident are you in the results?
- How much will the results change if the basic (input) data is slightly wrong or will change over time, over regions, over subgroups?
- Which impact any change of input data and assumptions will have?
- Which of the uncertain input factors is most important in determining the output?
- If we could invest in elimination of uncertainty in one of the input factors, which factor should we start with to reduce the uncertainty of results?
- Which are preventable?

3.8.7 *Exposure Models*

Exposure analysis relies on information on the concentrations of a pollutant in an exposure media, on the circumstances and the human behaviour and the activities that result in contact and exposure (exposure event) as well as on the transfer rates from the exposure media to the individual. An exposure model should integrate over all pathways (e.g. inhalation, ingestion, dermal) and all sources (e.g. water, food items and additional mouthing and dust intake for children) for all exposure events (e.g. drinking or eating something) in a defined time interval (e.g. 1 day). Using individual intake protocols, such an exposure measurement might describe the intake of an individual well on a high-resolution time scale. Repeated over some a certain period it would give information about the inter- and intra-individual variability.

$$\sum_{i=1}^{\text{Path}} \sum_{i=1}^{\text{Sources}} \sum_{i=1}^{\text{Events}} \text{Amount}_{\text{Source},k} * \text{Concentration}_{\text{Source}} * \text{Transfer}_{\text{Source}\rightarrow\text{Human}} \tag{3.8a}$$

$$\approx \sum_{i=1}^{\text{Path}} \sum_{i=1}^{\text{Sources}} \frac{n(\text{Events})}{\text{Time}} * \frac{\text{Amount}_{\text{Sources}}}{\text{Event}} * \text{Concentration}_{\text{Source}} \text{Transfer}_{\text{Source}\rightarrow\text{Human}} \tag{3.8b}$$

$$\approx \sum_{i=1}^{\text{Path}} \sum_{i=1}^{\text{Sources}} \text{Amount}_{\text{Source,Time}} * \text{Concentration}_{\text{Source}} \text{Transfer}_{\text{Source}\rightarrow\text{Human}} \quad (3.8c)$$

The conceptual exposure model, described within the scope of analysis, has to be translated into a mathematical formulation to allow calculation (e.g. within an Excel®-like table). Equation 3.8a–3.8c illustrates this translation, starting with a basic approach and ending with a time-aggregated model. The first formula describes the integration over all sources and all pathways for each exposure event over a defined period of time. Each element of the three-dimensional summation contains one exposure event (e.g. caffeine intake drinking a large morning cup of coffee = amount with an estimated caffeine concentration valid for a specific method of preparation = transfer from raw coffee). The sum over a day should take into account all sources of the substance of interest (e.g. different types of coffee and drinking coke, tea and other caffeine-containing beverages over the rest of the day). It would be almost impossible to estimate the exposure to a specified substance on an event-based level, due to a lack of appropriate data and since the intakes and the concentration in the sources will show variation from event to event (e.g. due to source factors such as origin of the brand, processing and preparation of coffee). The measurement efforts for an event-based assessment would be fairly high, and descriptive data is unavailable. In consequence, there would be a number of uncertainties in the quantitative event-based estimate, most of which cannot be clearly defined or quantified. Event-based exposure models are rarely conducted although they would reflect the reality best.

In general, exposure models are simplified by ignoring the time scale of the events by aggregating for a day by using day-related amounts of drinks and concentration estimates. If the distribution of exposure events per time unit and the intake per event is unknown, a second step of aggregation is commonly used. Only the intake per time unit and distribution of the substance concentration is taken into consideration (e.g. calculating the intake per day related to the body weight as the potential daily intake PDI). The transfer term used in the exposure model includes in general all factors like dilution (e.g. preparation, mixing), magnification (e.g. bacterial growth) and losses (e.g. dilution, peeling, washing) on the pathway from the source to the human body (e.g. preparation factors for food concentration, resorption factors for uptake). Very often this factor is chosen as a constant although a high degree of variability and uncertainty remains due to the inherent multiplicative nature of the combination of all inherent transfer factors should be expected.

The potential intake, described by formulae of the type above, represents only one component of often-applied evaluation targets, as an ADI as a long-term average or an ARfD as a short-term maximum daily intake, namely, the time-scaled nominator, e.g. (unit: mg potential intake per day). The denominator reflects the corresponding body weight (unit: kg^{-1} of BW). If the body weight is an influential factor on the amount of potential intake or the transfer factor (e.g. resorption rate in dependence to age), then this has to be taken into account, either by modelling correlation or dependency directly. In every case the parameters of nominator should be controlled for such an association, and the results should be reported. If the relationship indicates a proportion "the higher the body weight the higher the intake", then this might decrease the range of exposure estimates in comparison to an independency of variables model. This would support conservatism, but the interrelationship should be controlled for every model and every parameter separately. Depending on the targets of the analysis (e.g. for ARfD and ADI evaluation), the model output and the parameters must be chosen appropriately.

One disadvantage of exposure models of this type leading to systematic errors is the fact that all partial sums are assumed to be independent from each other, e.g. no correlation and no clustering due to other influences is considered. The same restriction to validity holds for independence of exposure events over time, ignoring constancy of exposure habits and periodicity like time structure of consumption (e.g. seasonality of intake, cultural habits like fish eating on Fridays, fixed consumption habits, avoiding a repeated consumption of food items on the following day/week). These dependency problems can be solved only by stratification or by selecting adjusted exposure scenarios (e.g. a describing typical Friday/weekend/workday). If the possible aggregation errors with respect to the time scale of measurement of the parameters and model structure are not considered, the resulting uncertainty in the exposure estimates by using a time-aggregated exposure model instead of measuring exposure on a high-resolution time scale might be high and hidden. Problems due to clustering of influential factors can only be solved by analysing or describing results for subgroups.

An uncertainty-related aspect inherent in the structure of all exposure models should be highlighted: The combinational operation of the parameters for each pathway and source is a multiplication. This implicates that also the variation, errors and uncertainty are multiplied for each exposure event, for each source and pathway related intake of a substance. This will result in right-skewed distributions of exposure even if each element (e.g. amount of intake, substance concentration and transfer factor) would be normally (or symmetrically) distributed. Outliers to the high end of exposure distribution have a higher chance to occur in right-skewed distributions. The skewed distributions for each item are summed up to one target exposure distribution, and in consequence, this distribution again will be right-skewed. In general, it will not follow a specific type of distribution but is a mixture of the sum of right-skewed distributions. The more sources and pathways have to be included in the analysis, the more complicated it will be to isolate the impact of variances and the statistical errors of a single source (or event) against the uncertainty impact of each component.

3.8.8 Type of Distributions and Inherent Uncertainties

Total exposure is the sum over all exposure conditions that are weighted by the amount of contact and substances intake. The parameters of the exposure model have some properties in common that influence the shape and range of the distribution of the resulting exposure.

First, nearly all parameters in an exposure assessment are constrained to values above zero (e.g. amount eaten, products applied to skin, rates of transfer). In consequence parameters and exposure estimates are left limited to zero values. If subgroups have systematically zero values for a group of nutritional parameters (e.g. vegetarians, non-alcoholics, non-smokers, nonconsumers of particular foods), this should be considered in the scenarios, in the data set and within the statistical analysis. A stratification for structural zeros avoids that possible secondary influence by within-group correlation of parameters (e.g. no meat product intake → higher amount of protein/fat intake from vegetable sources). Different associations for subgroups might disturb the results for the total population estimate. Assuming independence of contact to sources of interest might be wrong.

For substance concentration measurements in general, a limit of detection (LOD) and a level of quantification (LOQ) describe the lower bound of the numerical measurement values, due to the analytical technique and quality of the laboratory. Often, values below these limits are chosen as a constant value (e.g. half of the LOQ). Depending on the number of values below LOD, this decision might have no strong influence on the statistical central tendency descriptors of the parameter; it might influence strongly the total exposure estimates if a food item with high consumption rate is multiplied with an LOD/LOQ substitute. Using data sets with a high rate of values below the LOD (e.g. if the median correspond to the LOD) would amplify this distortion; sampling for a distribution for the LOQ range would reduce this (if no correlation has to be taken into account). However, considerable systematic errors might result if different laboratories are involved or if the LOQ or LOD has changed over time. The exposure estimate for such sources/products (and the estimated exposure contribution) might get remarkably high in contrast to a food item with a better analytical quality of substance measurement (lower LOD/LOQ values); a comparison of the relative contribution is hindered. The consequence of using LOD/LOQ values is that the exposure distribution might be left truncated too, although there might exist individuals with exposure below this limit. Conclusion like "all individuals of the selected scenario population are exposed at minimum with × (mg/kg of BW per day)" might be systematically wrong, due to the choice of algorithms for data handling. If the quality of laboratory measurement differs over the sources (e.g. food items), it is almost impossible to evaluate the lower range of the exposure distributions numerically.

If measurements do have a natural zero point (ratio scale), then the coefficient of variation (CV), also known as relative standard deviation (RSD), should give a valid characterisation of the variability in a data set.

For lognormal (and skewed) distributions, the coefficient of variation (CV) might be approximated without taking the mean into account. The corresponding CV is calculated on the basis of the logarithmic transformed measurement data as:

$$\mathrm{CV}_{\mathrm{standard}} = \frac{\text{standard deviation (SD)}}{\text{arithmetic mean}} \tag{3.9}$$

For lognormal (and skewed) distributions, the coefficient of variation (CV) might be approximated without taking the mean into account. The corresponding CV is calculated on the ground of the logarithmic transformed measurement data as:

$$CV_{\text{lognormal}} = \sqrt{e^{\text{Variance of } \ln(x)} - 1} \tag{3.10}$$

The coefficients of variation characterise the spread of data in a unit-less scale. This allows the comparison of parameters with respect to the heterogeneity over subgroups (without taking possible between-group differences in central tendency into account). The higher the CV, the more caution must be given to this variable with respect to influence of inherent variance and uncertainty on the expected spread of calculated exposure distribution.

Since the total exposure estimates contain variation and uncertainty due to variation and uncertainty in chemical concentration measurements for the food items as well as variation and uncertainty in the consumption estimates, the summative exposure estimate contains both. For a more specific discussion of the influence of analytical uncertainty on the overall uncertainty in exposure estimates, see, e.g. Bettencourt da Silva et al. (2012); for an evaluation of the impact due to uncertainty in food intake, see Boon et al. (2014), Kettler et al. (2015), Tennant et al. (2017), Ambrus et al. (2018) and Breysse et al. (2018).

Exposure distributions are normally skewed. For all right-skewed distributions (e.g. like the lognormal), the positions of the mode (highest frequency value, point of

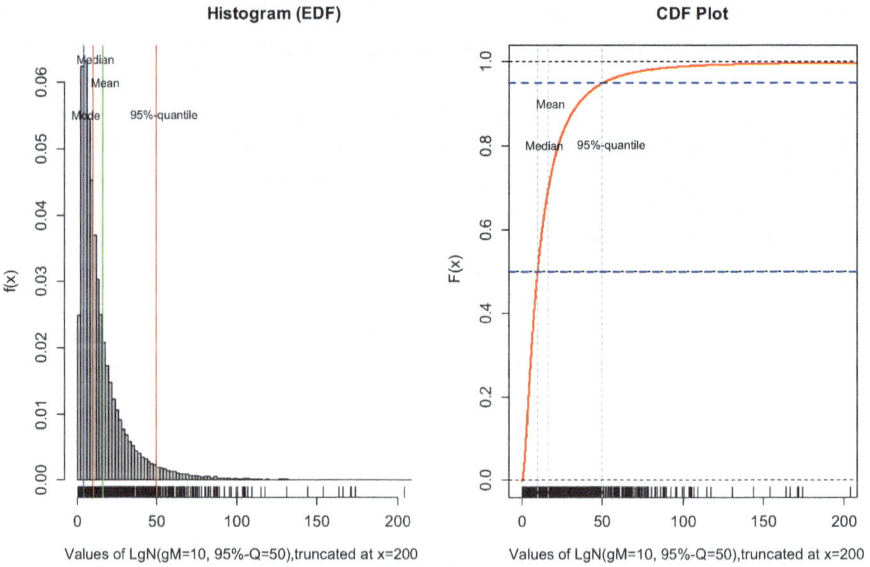

Fig. 3.18 Illustration of an empirical density function (left) and of the corresponding cumulative distribution function (CDF) of a lognormal distribution (geom. mean gM = 10, 95%-quantile Q95% = 50) with marked quantiles for 5%, 25%, 50% (median), 75% and 95%

global maximum of the probability/frequency density function), the median (50%-percentile corresponding to the geometric mean for lognormal distributions) and the arithmetic mean go from left to right (Fig. 3.18). The shift of the arithmetic mean in relation to the median (or geometric mean) is proportional to the variance of the distribution. This must be taken into account when using information about parameters from literature references. Accordingly the deviance of the upper percentiles to the median increases factorial esp. in skewed distributions.

The confidence intervals of the statistical location parameters (mean, median, percentiles) of skewed distributions are asymmetric too; the same holds in general for the uncertainty. If skewed distributions are used for probabilistic simulations, it should be controlled if the range of resulting values are within the limits of biological and physiological plausibility (e.g. for the amounts of food and water intake as well as for concentration estimates). In case of unrealistic ranges, a truncation of ranges might be reasonable, but this must be explained and described in the documentation in detail.

Evaluation of Combined Variation and Uncertainties: Identification of the Sources with High Contribution of Variability and Uncertainty on Exposure Estimates

An analysis of all possible outcomes for all ranges of the input variables (variability), together with a consideration of inherent quantitative (and numerical expressed qualitative) uncertainties, is a scientific task that will call for an involvement of mathematical, statistical and exposure science expertise. Cullen and Frey (1999) define sensitivity analysis as the assessment of the impact of changes in input values on model outputs. Sensitivity analysis is used to determine how different values of an input (the independent variables) will impact a particular output (the dependent variables), for a given model. Sensitivity analysis studies the relationships between information flowing in and out of the model (Saltelli et al. 2004). Frey and Patil (2002) and Frey et al. (2004) underline the use of sensitivity analysis in exposure assessment for the identification and comparison of sources of uncertainty that influence the target variables and the assessment conclusions. Since the multiplication of distributions always increases the variability and skewness of the resulting distribution, it should be anticipated that the distributions (and the propagated error) increase for every exposure source included in the analysis.

The estimates of the total exposure are calculated as the sum over time (time interval of the target exposure metric) for all sources and pathways. The process of calculating a sum over many components might level out some statistical errors, but only if the directions of errors in all elements of the calculation are independent. The total variance will increase for each item that contributes to the total exposure.

According to the recommendations of WHO/IPSC (2008):

> Sensitivity analysis should be an integral component of the uncertainty analysis in order to identify key sources of variability, uncertainty or both and to aid in iterative refinement of the exposure model

Incorporating variability and uncertainty into such comparative assessment is state of the art. Ignoring variability would mean neglecting existing differences in exposure conditions (e.g. over persons, over exposing situations, application/usage/ contact of hazards). Ignoring uncertainty would mean providing results as reliable, even if they were questionable.

Since the efforts for a statistical sensitivity analysis should be balanced with respect to cost and time versus the expected gain of information, any sensitivity and uncertainty analysis should start with a screening step, which uses defaults for all parameters, evaluating the change of the outcome by stepwise changing these values. The identification of those input variables that have a strong influence on the variance of the target variables will indicate to variables with a high potential for possible exposure control. The impact of variance of each variable on the model outcome might be controlled by a parameter-wise alternation of central tendency default to an upper-bound estimate (e.g. the 95th% or 5th quantiles). This procedure gives an overview about the 95% ranges of an influential variable, keeping all other influence variables on the mean or median (default for central tendency). If the data indicates a zero median, only the deviance for high percentiles can be evaluated. In a similar manner, the impact of statistical uncertainty might be controlled by a parameter-wise alternation of the central tendency and/or the upper-bound estimates using the confidence of these values: this describes the degree of uncertainty due to statistical stability of estimates. These "one-at-a-time" (OAT) methods (Murphy et al. 2004), changing always only one input parameter while keeping all other values constant, are recommended at the screening level. A "tornado diagram", ranking the variables by the outcome change, might illustrate the relative importance of each input parameter. For an overview of graphical methods for presenting quantitative uncertainty, see, e.g. Ibrekk and Morgan (1987) and Edwards (2012).

The main advantage of these OAT approaches is that the resulting changes in the model outcome are directly related to the change of input. An evaluation of results based on a OAT approach is in general understandable for risk managers and the public. But, it describes only the effects of variability or uncertainty for selected values. Possible associations (e.g. correlations, dependencies, interactions) between parameters are ignored. The behaviour of the model might deviate if all variables show variation (and uncertainty) in a multivariate setting with dependencies and interaction (Cacuci 2003; Murphy et al. 2004).

Taking into account the combined effects of many input variables requires simulation analysis. Looking at the variance impact is mostly done by Monte Carlo (MC) simulation. Software tools like "Crystal Ball®", "@RISK®" and "Analytica®" and "JRC/SimLab®", for instance, support the necessary calculations. If the uncertainty of the parameters might only be described by ranges (numerical intervals), these values might also be used (e.g. in a uniform distribution). If relevant association exists between the input variables, information about the correlation (covariance) structure should be used in the simulation model. Technically a Monte Carlo simulation consists of random combinations of random variables following the distribution function of each input variable. Repeating these random choices many times, the distribution of the exposure variable will represent the set of all possible combination of input variables, constraint by the distributions for the input. If uncertainty is included into the simulation, the analysis changes from a one-dimensional variance propagation model into a two-dimensional analysis.

Using Monte Carlo simulations, the dependency of the output values on the input variability might be evaluated by variance-based approaches. Ranking the correlation coefficients of all input variables to the resulting exposure values produces information that illustrates the positive/negative impact of the input variance and the degree uncertainty on the calculated exposure variance. Calculation of regression models with the input variables as independent and the exposure estimate as the dependent variable allows an integrative view. Using the standardised regression coefficients allows a direct comparison (Cacuci 2003).

Introducing quantitative estimates of uncertainty into a variation-based model results in a calculation that consists of a (in general an additive) mixture of variation and uncertainty. Uncertainty and variance components need to be represented by different variables within the model. If necessary due to model complexity or safety requirements, even more elaborate mathematical methods (e.g. Saltelli et al. 2004) might be appropriate. Due to the high efforts in practice, a full sensitivity analysis including variability and uncertainty components is rarely done.

If raw data sets from representative samples of the population (e.g. collected as national surveys) are available, then the original data set might be used as a calculation basis for exposure estimation. Using the individual consumption frequencies, the individual amounts eaten/used together with the individual anthropometric data (e.g. body weights), only the substance concentration distribution needs to be simulated according to the information about the type of distribution (see example in Table 3.11, Sect. 3.4.9). The calculation results in a population-based estimate of exposure. This approach avoids the problems of data dependency, correlation and interaction, and reduces data and model uncertainty. The techniques for a corresponding sensitivity analysis are the same as described above.

3.8.9 *Refining or Reducing the Model Complexity of the Exposure Assessment?*

If a sensitivity analysis identifies uncertainties in relation to knowledge about important input data, this should be seen as a prioritising argument for additional data collection and extended literature research for data references. If the quality of the input parameters is insufficient, the expected benefit of using a more sophisticated calculation technique is low. If time and urgency of the assessment have priority, using a high degree of conservatism for the choice of parameter values together with a lower tier of analysis is recommended. If relevant information for parameters is missing, informed assumptions, if possible supported by expert opinions, might be used as a placeholder, changing the character of the exposure assessment into a "what-if" analysis. The "what" reflects the results and the "if" a set of assumptions.

In a situation with poor information about the parameters of the model, a simplification of the scenario might be necessary. This should be done in cooperation with the customer, e.g. by reducing the pathways and the sources or by restriction to specific subgroups. Only with a moderate data quality an increase of information using a higher-tier analysis should be expected. In every case with minor quality, the assumption made and the consequences of the choices must be described in a transparent manner.

If different pathways (e.g. inhalation, oral ingestion and dermal) are relevant parts of the scenario description, it might be very informative to build as a first step the sum over all sources that contribute to each of the specific pathways before summing up to a total exposure estimate. For example, if the relative exposure contribution by inhalation is very low (with respect to the mean and the expected data ranges), the effort for literature and data retrieval might be saved by using a conservative default for breathing rates and substance concentrations (e.g. in air) in the assessment.

The same strategy of grouping parts of the sum of exposure sources and pathways might be applied for subgroups of sources, e.g. by aggregating over categories of food. This strategy for sources might reduce the complexity of the model and provide valuable information that is useful for communicating the results (see EFSA 2019). The calculation and evaluation of these sub-totals of the total exposure with respect to variability and inherent uncertainty might support the interpretation and the usefulness of the results even if for some subgroups the degree of uncertainty is evaluated as high.

3.8.10 Summary

The evaluation and description of uncertainty within an exposure assessment do not restrict the quality of the assessment; in contrary a documentation of information about what is known, what is reasonable to expect and what needs further clarification might have a high impact on the risk management process. It belongs to the obligation of the exposure assessment to inform the customer (or the risk management) about all inherent uncertainties and limitations. The development of approaches to assessing and taking these into account would support the transparency of the resulting scientific assessment and make it more robust for decision-making. Although an uncertainty analysis might restrict the utility of an exposure assessment for regulatory or prevention-directed purposes, it provides the necessary information for the risk management with regard to decision-making under uncertainty. Where appropriate to an assessment objective, exposure assessments should be iteratively refined over time to incorporate new data, additional information and higher-tier methods to better characterise uncertainty and variability.

Risk management actions cannot reduce variability for those variables that are not preventable. But the assessment can stratify for homogeneous subgroups to identify focused strategies for targeted prevention advice to reduce the expected impact of influential exposure sources. For this, reducing remaining uncertainty can be achieved by doing additional research, by the conceptualisation of more group-specific scenarios, by more comprehensive models and by collecting more data.

Obtaining measurements for individual exposure levels would be necessary for the validation of exposure models. Every evaluation should contain a reality check if the corresponding human biomonitoring data or empirical exposure measurements are published and in accordance with the assessment. An overall evaluation should include a statement whether the quality of the exposure prognosis is in accordance with the promises of this prognostic method. Where the level of uncertainty is too high, only doing additional research, collecting more information and/or obtaining better exposure measurements (e.g. together with human biomonitoring) will increase the quality of exposure assessments.

3.9 Exposure Calculation Strategies in Exposure Analysis

Christiaan Delmaar

3.9.1 Role of Exposure Assessment

Exposure assessments are conducted for different reasons, with different objectives and approaches. For example, an exposure assessment may be conducted for a product safety evaluation as part of the admission of a product or substance on the market. Or an exposure assessment may be conducted to assess the health risk of a population from the exposure to a chemical that is contained in multiple consumer products to inform restriction or intervention policies. Or the exposure from chemicals in a group of consumer products such as cosmetics or cleaning agents may be compared to that of other sources such as food or indoor air for a problematic substance to assess the potential effectiveness of restriction measures on the products in the product group considered.

3.9.2 Tiered Approach

In all these cases, it is usually good practice to follow a tiered approach (see also Sect. 3.2.10). Tiering exposure models can be done by starting with a simple model that refers to a simple scenario. This model may be expanded by introducing new model variables or by expanding an existing variable. Several models that present different consecutive iterations steps in the tiered approach are integrated in the ConsExpo tool (see Sect. 5.10.5).

In a first assessment (first tier), crude methods and assumptions are employed to screen the problem with comparatively limited effort. The large uncertainties that usually accompany a low-tier assessment are covered by making conservative assumptions (i.e. assumptions that will likely lead to an overestimation of the actual exposure) on scenario aspects, processes leading to exposure and exposure factors.

The results of a first-tier assessment are compared against some benchmark (e.g. an established safe level of chemical exposure in case of product safety assessments or the expected exposure levels of other sources). If exposure is sufficiently below the benchmark, the assessment ends; if not, the assessment is usually refined in a subsequent, higher tier. The refinement may concern both the exposure assessment and the benchmark (e.g. the hazard assessment).

A higher-tier assessment will attempt to reduce the uncertainty. This requires more detailed information and more advanced methods or models that provide a more realistic account of the exposure processes. It should be noted that new assumptions might be associated with new uncertainties.

Indispensable methods in a tiered approach are the use of sensitivity and uncertainty analysis. As in a tiered approach, the objective of every iteration step is to reduce the uncertainty in the evaluation; a proper assessment of this uncertainty in each iteration is required. An uncertainty analysis should identify the largest sources of error in the evaluation and, hence, the area where the largest potential gains in reducing this uncertainty lies in the quality of the exposure parameters.

Sensitivity analysis studies the changes in the results of the assessment with changes in the assumptions on which the assessment is based. For example, a sensitivity analysis may be conducted to demonstrate how much the outcome of the assessment changes with changes in one or more of its input parameters. In concert with an uncertainty analysis, a sensitivity analysis may help to identify which sources of uncertainty in the data on which the assessment is based may contribute to the largest extent to the overall uncertainty in the assessment. In this way, parameters or aspects of the assessment that require additional data collection can be identified.

3.9.3 Modelling of Exposure

Irrespective of the type of exposure assessment or the tier in the assessment, exposure analysis requires some sort of modelling: a method to quantitatively combine information, knowledge and assumptions into an estimate of the level of exposure. For example, an assessment may obtain disparate information on product composition, the usage of products and the populations using the products. This information will have to be combined with reasonable assumptions on the conditions under which the product is used and knowledge of physical-chemical and physiological processes which may determine the actual exposure during use.

For the modelling of exposure, different computer tools exist. These tools include both calculations for general-purpose modelling software and dedicated exposure assessment tools.

3.9.4 General-Purpose Modelling Software

Using general-purpose calculation software, the exposure assessor can implement algorithms that combine data and assumptions on exposure factors and exposure processes himself. Modelling tools vary widely in capabilities, expertise level needed to make efficient use of the tool and cost. Advanced modelling tools include, for example, MATLAB, Mathematica and the freely available R modelling language. These tools offer very advanced modelling capabilities and require a considerable expertise level of the modeller to use them efficiently. On the other side of the spectrum are spreadsheet tools such as MS Excel, which are fairly accessible and do not require much expertise to construct simple exposure models. Exposure models built in spreadsheet programmes are usually limited to very simple relations between parameters, which can be expressed in closed analytical form, and to simple representations of the data (e.g. by single, deterministic numbers). More advanced models, including numerical mathematical models or probabilistic models, can be built in a tool such as Excel. But this requires proficiency in programming languages such as Visual Basic for Applications (VBA).

Similarly, performing quantitative uncertainty or sensitivity analysis can be done in such simple spreadsheet models by varying the input over the uncertainty range of all parameters or excluding certain aspects of the model. But the options are usually limited, and such analyses quickly become unwieldy with increasing model complexity. So, also in these cases, to perform more advanced analyses of uncertainty and sensitivity requires proficiency in a programme language or the use of third-party add-ins such as @RISK or Crystal Ball, which provide functionality to use probability distributions for model parameters rather than single point values and perform probabilistic assessments of uncertainty or variability.

A list of dedicated exposure modelling tools and the detailed descriptions are given in Sect. 5.10.

References

Abraham K, Pfister M, Wohrlin F, Lampen A (2011) Relative bioavailability of coumarin from cinnamon and cinnamon-containing foods compared to isolated coumarin: a four-way crossover study in human volunteers. Mol Nutr Food Res 55:644–653. https://www.ncbi.nlm.nih.gov/pubmed/21462332

Ahlborg UG, Becking GC, Birnbaum LS, Brouwer A, Derks HJGM, Feeley M, Golog G, Hanberg A, Larsen JC, Liem AK, Safe SH, Schlatter C, Wvern F, Younes M, Yrjinheikki E (1994) Toxic equivalency factors for dioxin-like PCBs: report on WHO-ECEH and IPCS consultation. Chemosphere 28:1049–1067

Ambrus A, Horváth Z, Szenczi-Cseh J, Szabó IJ (2018) Factors affecting the quantitative uncertainty of the estimated short-term intake. Part I—calculation methods. J Environ Sci Health B 53 (6):394–403. https://doi.org/10.1080/03601234.2018.1439815

Angerer J, Ewers U, Wilhelm M (2007) Human biomonitoring: state of the art. Int J Hyg Environ Health 210:201–228

Asikainen A, Gotti A, Karakitsios S, Sarigiannis DA, Semple S, De Brouwere K, Galea K, van Tongeren M, Jantunen M (2011) INTERA: integrated exposure for risk assessment in indoor environments. Abstract nr. 387. The 12th international conference on indoor air quality and climate, Austin, TX, Jun 5–10, 2011

AUH (1995) Standards zur Expositionsabschätzung. Bericht des Ausschusses für Umwelthygiens. Arbeitsgemeinschaft der leitenden Medizinalbeamtinnen und – beamten der Länder. Herausgeber: Behörde für Arbeit, Gesundheit und Soziales (Hamburg) and Institut für Mathematik und Datenverarbeitung in der Medizin

Baez RJ, Baez MX, Marthaler TM (2000) Urinary fluoride excretion by children 4–6 years old in a south Texas community. Rev Panam Salud Publica/Pan Am J Public Health 7(4):242–247

Batáriová A, Spěváčková V, Beneš B, Čejchanová M, Šmíd J, Černá M (2006) Blood and urine levels of Pb, Cd and Hg in the general population of the Czech Republic and proposed reference values. Int J Hyg Environ Health 209(4):359–366

Beaugelin-Seiller K, Cessac B, Marot F, Chartier C, Deportes I, Metivier JM, Morin A, Mourlon C, Vincent G (2004) La banque de données CIBLEX, une compilation de paramètres d'exposition de la population Française au voisinage d'un site pollué. DÉCHETS – Revue francophone d'écologie industrielle – n° 35 – 3e trimestre 2004. http://lodel.irevues.inist.fr/dechets-sciences-techniques/docannexe/file/2125/5_beaugelin.pdf. Accessed 26 Oct 2019

Becker K, Seiwert M, Angerer J, Heger W, Koch H, Nagorka R, Roßkamp E, Schlüter C, Seifert B, Ullrich D (2004) DEHP metabolite in urine of children and DEHP in house dust. Int J Hyg Environ Health 207(5):409–417

Becker K, Göen T, Seiwert M, Conrad A, Pick-Fuß H, Müller J, Wittassek M, Schulz C, Kolossa-Gehring M (2009) GerES IV: Phthalate metabolites and bisphenol A in urine of German children. Int J Hyg Environ Health 212(6):685–692

Berglund M, Larsson K, Grandér M, Casteleyn L, Kolossa-Gehring M, Schwedler G et al (2015) Exposure determinants of cadmium in European mothers and their children. Environ Res 141:69–76

Bettencourt da Silva R, Bulska E, Godlewska-Żyłkiewicz B, Hedrich B, Majcen N, Magnusson B, Marinčić S, Papadakis I, Patriarca M, Vassileva E, Taylor P (2012) In Majcen N, Gegevičius V (eds) Analytical measurement: measurement uncertainty and statistics. Publications Office of the European Union, Luxembourg. https://doi.org/10.2787/58527

BfR (2012) Cassia cinnamon with high coumarin contents to be consumed in moderation. http://www.bfr.bund.de/en/press_information/2012/26/cassia_cinnamon_with_high_coumarin_contents_to_be_consumed_in_moderation-131836.html. Accessed 26 Oct 2019

BfR (2015) Guidelines on uncertainty analysis in exposure assessments. Recommendation of the Committee for Exposure Assessment and Exposure Standardisation of the Federal Institute for Risk Assessment (BfR). Available at: http://www.bfr.bund.de/cm/350/guidelines-on-uncertainty-analysis-in-exposure-assessments.pdf. Accessed 26 Oct 2019

Boon PE, van der Voet H, Ruprich J, Turrini A, Sand S, van Klaveren JD (2014) Computational tool for usual intake modelling workable at the European level. Food Chem Toxicol 74:279-288. issn:0278-6915. https://doi.org/10.1016/j.fct.2014.10.019; http://www.sciencedirect.com/science/article/pii/S0278691514004475

Bosgra S, Bos P, Vermeire T, Luit R, Slob W (2005) Probabilistic risk characterisation: an example with di(2-ethylhexyl) phthalate. Regul Toxicol Pharmacol 43:104–113

Bremmer HJ, Van Veen MP (2000) Factsheet algemeen, randvoorwaarden en betrouwbaarheid, ventilatie, kamergrootte, lichaamsoppervlak [General Factsheet, conditions and reliability, ventilation, room size, body surface]. RIVM report 612810 009, RIVM, Bilthoven. www.rivm.nl. Accessed 8 Mar 2018

Bremmer HJ, Prud'homme de Lodder LCH, van Engelen JGM (2006) General fact sheet. Limiting conditions and reliability, ventilation, room size, body surface area. RIVM report 320104002/2006; Updated version for ConsExpo 4. http://rivm.openrepository.com/rivm/bitstream/10029/3861/1/320104002.pdf. Accessed 26 Oct 2019

Breysse N, Vial G, Pattingre L, Ossendorp BC, Mahieu K, Reich H, Rietveld A, Sieke C, van der Velde-Koerts T and Sarda X (2018) Impact of a proposed revision of the IESTI equation on the acute risk assessment conducted when setting maximum residue levels (MRLs) in the European Union (EU): a case study, J Environ Sci Health B, 53:6, 352-365, DOI: https://doi.org/10.1080/03601234.2018.1439809

Bruinen de Bruin Y, Hakkinen PB, Lahaniatis M, Papameletiou D, Del Pozo C, Reina V, Van Engelen J, Heinemeyer G, Viso AC, Rodriguez C, Jantunen M (2007) Risk management measures for chemicals in consumer products: documentation, assessment, and communication across the supply chain. J Expo Sci Environ Epidemiol 17(Suppl 1):S55–S66

BVL (2002–2016) Berichte zur Lebensmittelsicherheit. Monitoring. Gemeinsamer Bericht des Bundes und der Länder. https://www.bvl.bund.de/DE/01_Lebensmittel/01_Aufgaben/02_AmtlicheLebensmittelueberwachung/04_Monitoring/lm_monitoring_node.html. Accessed 26 Oct 2019

Cacuci DG (2003) Sensitivity and uncertainty analysis: theory, vol 1. Chapman & Hall, CRC Press, Boca Raton

Calabrese EJ, Barnes RM, Stanek IIIEJ, Pastides H, Gilbert CE, Veneman P, Wang X, Lasztity A, Kostecki PT (1989) How much soil do young children ingest: an epidemiologic study. Regul Toxicol Pharmacol 10(2):123–137. https://doi.org/10.1016/0273-2300(89)90019-6

Calabrese EJ, Stanek EJ, Pekow P, Barnes RM (1997) Soil ingestion estimates for children residing on a superfund site. Ecotoxicol Envion Saf 36:258–268

Calafat A, McKee R (2006) Integrating biomonitoring exposure data into the risk assessment process: phthalates [diethyl phthalate and di(2-ethylhexyl) phthalate] as a case study. Environ Health Perspect 114:1783–1789

CDC (2005) Third national report on human exposure to environmental chemicals. Centers for Disease Control and Prevention. Atlanta, GA. https://www.jhsph.edu/research/centers-and-institutes/center-for-excellence-in-environmental-health-tracking/Third_Report.pdf. Accessed 26 Oct 2019

CDC (2009) Fourth national report on human exposure to environmental chemicals. Centers for Disease Control and Prevention, Atlanta, GA. https://www.hsdl.org/?view&did=784246. Accessed 26 Oct 2019

CDC (2016) National health and nutrition examination survey (NHANES). Page last updated: February 24, 2016. http://www.cdc.gov/nchs/nhanes.htm. Accessed 26 Oct 2019

Ciffroy P, Altenpohl A, Fait G, Fransman W, Paini A, Radovnikovic A, Simon-Cornu M, Suciu N, Verdonck F (2016) Development of a standard documentation protocol for communicating exposure models. Sci Total Environ 568:557–565. https://doi.org/10.1016/j.scitotenv.2016.01.134

Clausen PA, Spaan S, Brouwer DH, Marquart H, le Feber M, Engel R, Geerts L, Jensen KA, Kofoed-Sørensen V, Hansen B, De Brouwere K (2003) Working principles for risk analysis for application in the framework of the codex alimentarius. Technical report. Codex Alimentarius Commission: http://www.fao.org/docrep/006/y4800e/y4800e0o.htm. Accessed 27 Oct 2019

Clausen PA, Spaan S, Brouwer DH, Marquart H, le Feber M, Engel R, Geerts L, Jensen KA, Kofoed-Sørensen V, Hansen B, De Brouwere K (2016) Experimental estimation of migration and transfer of organic substances from consumer articles to cotton wipes: evaluation of underlying mechanisms. J Expo Sci Environ Epidemiol 26:104–112. https://doi.org/10.1038/jes.2015.35

CONCAWE (2015) Specific consumer exposure determinants (SCEDs) documents. https://www.concawe.eu/reach/specific-consumer-exposure-determinants-sceds-documents. Accessed 26 Oct 2019

Cullen AC, Frey CF (1999) Probabilistic techniques in exposure assessment. Plenum, New York

Danish EPA (2001–2016) Danish surveys on chemicals in consumer products. http://eng.mst.dk/chemicals/chemicals-in-products/. Accessed 26 Oct 2019

Danish EPA (2016) Determination of migration rates for certain phthalates. Survey of chemical substances in consumer products No. 149. http://mst.dk/service/publikationer/publikationsarkiv/2016/okt/determination-of-migration-rates-for-certain-phthalates/. Accessed 26 Oct 2019

Delmaar CJE, van der Zee Park M, van Engelen JGM (2005) ConsExpo 4.0 consumer exposure and uptake models program manual. RIVM Report 320104004/2005, 72. https://www.rivm.nl/bibliotheek/rapporten/320104004.pdf. Accessed 26 Oct 2019

Den Hond E, Govarts E, Willems H, Smolders R, Casteleyn L, Kolossa-Gehring M et al (2015) First steps toward harmonised human biomonitoring in Europe: demonstration project to perform human biomonitoring on a European scale. Environ Health Perspect 123:255–263

Dereumeaux C, Saoudi A, Pecheux M, Berat B, Crouy-Chanel P, Zaros C, Brunel S, Delamaire C, le Tertre A, Lefranc A, Vandentorren S, Guldner L (2016) Biomarkers of exposure to environmental contaminants in French pregnant women from the Elfe cohort in 2011. Environ Int 97:56–67. https://doi.org/10.1016/j.envint.2016.10.013

Dubois D, Dubois EF (1916) A formula to estimate the approximate surface area if height and weight be known. Arch Intern Med 17:863–871

Dubuisson C, Lioret S, Touvier M, Dufour A, Calamassi-Tran G, Volatier JL, Lafay L (2010) Trends in food and nutritional intakes of French adults from 1999 to 2007: results from the INCA surveys. Br J Nutr 103(7):1035–1048. https://doi.org/10.1017/S0007114509992625

Dudzina T, Delmaar CJE, Biesterbos JWH, Bakker MI, Scheepers PTJ, van Engelen JGM, Hungerbuehler K, von Goetz N (2015) The probabilistic aggregate consumer exposure model (PACEM): validation and comparison to a lower-tier assessment for the cyclic siloxane D5. Environ Int 79:8–16

ECETOC (2005) Targeted risk assessment, report 93. http://www.ecetoc.org/wp-content/uploads/2014/08/ECETOC-TR-093.pdf. Accessed 26 Oct 2019

ECETOC (2015) Advances in consumer exposure science: data, modelling and aggregate exposure assessment, 26 Jan 2016, Brussels, Workshop Report no 31. http://www.ecetoc.org/publication/workshop-report-no-31-advances-consumer-exposure-science-data-modelling-aggregate-exposure-assessment-26th-january-2016-brussels/. Accessed 26 Oct 2019

ECHA (2012) Guidance on information requirements and chemical safety assessment. Chapter R19: uncertainty analysis. ECHA-12-G-25-EN. https://echa.europa.eu/documents/10162/13632/information_requirements_r19_en.pdf. Accessed 26 Oct 2019

ECHA (2013) Guidance on information requirements and chemical safety assessment. Chapter R.20: Table of terms and abbreviations. ECHA12G26.1EN, European Chemicals Agency, Helsinki. https://echa.europa.eu/documents/10162/13632/information_requirements_r20_en.pdf. Accessed 26 Oct 2019

ECHA (2016a) Guidance on exposure scenario building (Part D). European Chemicals Agency, Helsinki. https://www.echa.europa.eu/documents/10162/13632/information_requirements_part_d_en.pdf. Accessed 26 Oct 2019

ECHA (2016b) Guidance on information requirements and chemical safety assessment Chapter R.15: Consumer exposure estimation. ECHA-10-G-03-EN. https://echa.europa.eu/documents/10162/13632/information_requirements_r15_en.pdf. Accessed 26 Oct 2019

Edwards D (2012) Introduction to graphical modelling (Springer texts in statistics, paperback). Springer, New York. ISBN 978-1-4612-6787-4, September 2012

EFSA (2004) Scientific opinion on the risk for public health related to the presence of mercury and methylmercury in food. EFSA J 34:1–14. http://onlinelibrary.wiley.com/doi/10.2903/j.efsa.2012.2985/abstract. Accessed 26 Oct 2019

EFSA (2006) Guidance of the scientific committee on a request from EFSA related to uncertainties in dietary exposure assessment. EFSA J 438:1–54. http://onlinelibrary.wiley.com/doi/10.2903/j.efsa.2007.438/full. Accessed 26 Oct 2019

EFSA (2008a) Marine biotoxins in shellfish – okadaic acid and analogues – Scientific opinion of the panel on contaminants in the food chain. EFSA J 589:1–62. http://onlinelibrary.wiley.com/doi/10.2903/j.efsa.2008.589/full. Accessed 26 Oct 2019

EFSA (2008b) Perfluorooctane sulfonate (PFOS), perfluorooctanoic acid (PFOA) and their salts scientific opinion of the panel on contaminants in the food chain. EFSA J 653:1–131. http://onlinelibrary.wiley.com/doi/10.2903/j.efsa.2008.653/abstract. Accessed 26 Oct 2019

EFSA (2008c) Guidance document for the use of the concise European food consumption database in exposure assessment. EFSA J 438:1–54. http://www.efsa.europa.eu/sites/default/files/efsa_rep/blobserver_assets/datexfooddbguidance.pdf. Accessed 26 Oct 2019

EFSA (2009a) Scientific opinion of the panel on contaminants in the food chain on a request from the European Commission on cadmium in food. EFSA J 980:1–139. http://www.efsa.europa.eu/en/efsajournal/pub/980. Accessed 26 Oct 2019

EFSA (2009b) Guidance of the scientific committee on Transparency in the scientific aspects of risk assessments carried out by EFSA. Part 2: General Principles. EFSA J 1051:1–22. http://onlinelibrary.wiley.com/doi/10.2903/j.efsa.2009.1051/full. Accessed 26 Oct 2019

EFSA (2009c) General principles for the collection of national food consumption data in the view of a pan-European dietary survey. EFSA J 7(12):1–51. http://onlinelibrary.wiley.com/doi/10.2903/j.efsa.2009.1435/abstract. Accessed 26 Oct 2019

EFSA (2012) Scientific opinion on risk assessment terminology. EFSA Scientific Committee. EFSA J 10 (5): 2664. http://onlinelibrary.wiley.com/doi/10.2903/j.efsa.2012.2664/full. Accessed 26 Oct 2019

EFSA (2015a) The EFSA data warehouse access rules. EFSA Support Publ 12(2):EN-768, 18 pp. http://onlinelibrary.wiley.com/doi/10.2903/sp.efsa.2015.EN-768/abstract. Accessed 26 Oct 2019

EFSA (2015b) The food classification and description system FoodEx2 (revision 2). EFSA Support Publ 12(5):EN-804, 90 pp. http://onlinelibrary.wiley.com/doi/10.2903/sp.efsa.2015.EN-804/epdf. Accessed 26 Oct 2019

EFSA (2015c) The EFSA comprehensive European food Consumption database. http://www.efsa.europa.eu/en/food-consumption/comprehensive-database. Last update 26 Apr 2018

EFSA (2017) Glossary. https://www.efsa.europa.eu/en/glossary-taxonomy-terms. Accessed 26 Oct 2019

EFSA (2018a) New uncertainty approach to make advice 'more transparent and robust'. https://www.efsa.europa.eu/en/press/news/180124-0 Accessed 26 Oct 2019

EFSA (2018b) Guidance on Uncertainty Analysis in Scientific Assessments. EFSA Journal 2018;16 (1):5123, 39 pp. https://doi.org/10.2903/j.efsa.2018.5123

EFSA (2018c) Scientific Opinion on the principles and methods behind EFSA's Guidance on Uncertainty Analysis in Scientific Assessment. EFSA J 16(1):5122, 235 pp. https://doi.org/10.2903/j.efsa.2018.5122

EFSA (2018d) Schendel T, Jung C, Lindtner O and Greiner M, 2018. Guidelines for uncertainty analysis: application of the respective documents of EFSA and BfR for exposure assessments. EFSA Support Publ 2018: EN-1472. 113 pp. https://doi.org/10.2903/sp.efsa.2018.EN-1472

EFSA (European Food Safety Authority), Hart A, Maxim L, Siegrist M, Von Goetz N, da Cruz C, Merten C, Mosbach-Schulz O, Lahaniatis M, Smith A, Hardy A (2019) Guidance on communication of uncertainty in scientific assessments. EFSA J 17(1):5520, 73 pp. doi:https://doi.org/10.2903/j.efsa.2019.5520

EFSA/BfR (2019) Guidance on communication of uncertainty in scientific assessments. EFSA supporting publication (Schendel T, Jung C, Lindtner O and Greiner M) Guidelines for uncertainty analysis: application of the respective documents of EFSA and BfR for exposure assessments. EFSA J 17(1):5520, 73 pp. https://doi.org/10.2903/j.efsa.2019.5520

Egsmose EL, Vaclavik Bräuner E, Frederiksen M, Mørcka TA, Siersma VD, Hansen PW, Nielsen F, Grandjean P, KnudsenLE (2016) Associations between plasma concentrations of PCB 28 and possible indoor exposure sources in Danish school children and mothers. Environ Int 87:13–19. http://www.sciencedirect.com/science/journal/01604120

ENNS (2006) Étude nationale nutrition santé ENNS, 2006: Situation nutritionelle en France en 2006 selon les indicateurs d'objectif et les repères du Programme national nutrition santé (PNNS). Premiers résultats. Colloque du Programme national nutrition santé (PNNS) 12 décembre 2007. INVS, 77 pp. http://www.ladocumentationfrancaise.fr/var/storage/rapports-publics/074000748.pdf. Accessed 26 Oct 2019

EPHPP (2010) Quality assessment tool for quantitative studies. Effective Public Health Practice Project, National Collaborating Centre for Methods and Tools (1998) updated 2010. http://www.nccmt.ca/knowledge-repositories/search/14. Accessed 26 Oct 2019

EU (1998) Report on methodologies for the monitoring of food additive intake across the European Union (Final report submitted by the Task Co-ordinator 16 January 1998). Reports of a Working Group on Scientific Co-operation on Questions Relating to Food. Task 4.2. SCOOP/INT/REPORT/2. European Commission Directorate General III Industry, Brussels. http://eur-lex.europa.eu/legal-content/EN/TXT/PDF/?uri¼CELEX:52001DC0542&from¼EN. Accessed 26 Oct 2019

EU (2003) Technical Guidance Document on Risk Assessment in support of Commission Directive 93/67/EEC on Risk Assessment for new notified substances Commission Regulation (EC) No 1488/94 on Risk Assessment for existing substances Directive 98/8/EC of the European Parliament and of the Council concerning the placing of biocidal products on the market. Part I. https://echa.europa.eu/documents/10162/16960216/tgdpart1_2ed_en.pdf. Accessed 26 Oct 2019

EU (2006) Regulation (EC) No 1907/2006 of the European Parliament and of the Council of 18 December 2006 concerning the Registration, Evaluation, Authorisation and Restriction of Chemicals (REACH), establishing a European Chemicals Agency, Amending Directive 1999/45/EC and repealing Council Regulation (EEC) No 793/93 and Commission Regulation (EC) No 1488/94 as well as Council Directive 76/769/EEC and Commission Directives 91/155/EEC, 93/67/EEC, 93/105/EC and 2000/21/EC. Official Journal L 396, 30.12.2006. https://eur-lex.europa.eu/legal-content/EN/ALL/?uri=CELEX%3A32006R1907. Accessed 18 Nov 2019

EU (2008) Regulation (EC) No 1272/2008 of the European Parliament and of the Council of 16 December 2008 on classification, labelling and packaging of substances and mixtures, amending and repealing Directives 67/548/EEC and 1999/45/EC, and amending Regulation (EC) No 1907/2006. https://eur-lex.europa.eu/legal-content/EN/TXT/?uri=CELEX%3A32008R1272. Accessed 18 Nov 2019

EU (2009) Regulation (EC) No 1223/2009 of the European Parliament and of the Council of 30 November 2009 on cosmetic products. https://eur-lex.europa.eu/legal-content/EN/ALL/?uri=CELEX%3A32009R1223. Accessed 18 Nov 2019

EU (2010) Directive 2010/63/EU of the European Parliament and of the Council of 22 September 2010 on the protection of animals used for scientific purposes. http://eur-lex.europa.eu/legal-content/EN/TXT/HTML/?uri=CELEX:32010L0063&from=EN. Accessed 26 Oct 2019

EU (2011) EURL ECVAM (European Union Reference Laboratory for alternatives to animal testing). https://eurl-ecvam.jrc.ec.europa.eu/glossary/glossary/european-union-reference-laboratories. Accessed 26 Oct 2019

Eurachem/CITAC Guide (2007) In Ellison SLR, Williams A (eds) Use of uncertainty information in compliance assessment, 1st edn. Available from https://www.eurachem.org/index.php/publications/leaflets/complianceleaflet. Accessed 26 Oct 2019

Fantke P, Ernstoff AS, Huang L, Csiszar SA, Jolliet O (2016) Coupled near-field and far-field exposure assessment framework for chemicals in consumer products. Environ Int 94:508–518. https://doi.org/10.1016/j.envint.2016.06.010

Fantke P, Aylward L, Bare J, Brown R, Chiu WA, Dodson R, Dwyer R, Ernstoff A, Howard B, Jantunen M, Jolliet O, Judson R, Kirchhübel N, Li D, Miller A, Paoli G, Price P, Rhomberg L, Shen B, Shin H-M, Teeguarden J, Vallero D, Wambaugh J, Wetmore BA, Zaleski R, McKone TE (2018) Advancements in life cycle human exposure and toxicity characterization. Environ Health Perspect 126(12):125001. https://doi.org/10.1289/EHP3871

FAO/IOC/WHO (2004) Report of the Joint FAO/IOC/WHO Ad hoc Expert Consultation on Biotoxins in Bivalve Molluscs, Oslo, Norway, 26–30 Sept 2004. Advance Pre- Publication Copy. Available at URL http://unesdoc.unesco.org/images/0013/001394/139421e.pdf. Accessed 8 Mar 2018; ftp://ftp.fao.org/es/esn/food/biotoxin_report_en.pdf. Accessed 26 Oct 2019

Flari V, Wilkinson D (2012) Terminology in risk assessments used by the scientific panels and scientific committee of EFSA. Food and Environment Research Agency (FERA), Sand Hutton

Frey HC, Patil SR (2002) Identification and review of sensitivity analysis methods. Risk Anal 22 (3):553–578

Frey HC, Mokhtari A, Zheng J (2004) Recommended practice regarding selection, application, and interpretation of sensitivity analysis methods applied to food safety process risk models. Office of Risk Assessment and Cost Benefit Analysis. U.S. Department of Agriculture, Washington, DC

Fromme H, Albrecht M, Angerer J, Drexler H, Gruber L, Schlummer M, Parlar H, Körner W, Wanner A, Heitmann D, Roscher E, Bolte G (2007a) Integrated Exposure Assessment Survey (INES). Exposure to persistent and bioaccumulative chemicals in Bavaria, Germany. Int J Hyg Environ Health 210:345–349

Fromme H, Gruber L, Schlummer M, Wolz G, Böhmer S, Angerer J, Mayer R, Liebl B, Bolte G (2007b) Intake of phthalates and di(2-ethylhexyl)adipate: Results of the integrated exposure assessment survey based on duplicate diet samples and biomonitoring data. Environ Int 33:1012–1020

Fromme H, Fembacher L, Schuster R, Dietrich S, Kiranoglu M, Meyer N, Kohlhuber M, Völkel W, Lahrz T, Burghardt R, Pfeiler P, Kraft M, Sievering S, Braukmann J, Buss M, van Houten U, Lorusso S, Haarmann H, Brinkmann C, Kadler D (2012) Länderuntersuchungsprogramm III (LUPE III). Gesundheitliche Bedeutung von Phthalaten in Kindertagesstätten ein integrativer Ansatz zur Risikoabschätzung. Bayerisches Landesamt für Gesundheit und Lebensmittelsicherheit, Landeslabor Berlin-Brandenburg, Landesamt für Natur, Umwelt und Verbraucherschutz NRW, Landesamt für Gesundheit und Soziales Berlin.

Gaitens JM. Dixon SL, Jacobs DE, Nagaraja J Strauss W, Wilson JW, Ashley PJ (2009) Exposure of U.S. children to residential dust lead, 1999–2004: I. Housing and demographic factors. Environ Health Perspect 117. https://doi.org/10.1289/ehp.11917

Garcia-Hidalgo E, Schneider D, von Goetz N, Delmaar C, Siegrist M, Hungerbühler K (2018) Aggregate consumer exposure to isothiazolinones via household care and personal care products: probabilistic modelling and benzisothiazolinone risk assessment. Environ Int 118:245–256

Harada K, Koizumi A (2009) Environmental and biological monitoring of persistent fluorinated compounds in Japan and their toxicities. Environ Health Prev Med 14:7–19

HBM – Kommission (2014) Grundsatzpapier zur Ableitung von HBM-Werten. Stellungnahme der Kommission Human-Biomonitoring des Umweltbundesamtes. Bundesgesundheitsbl Gesundheitsforsch Gesundheitsschutz 57:138–147

Health Canada (2015) The Canadian house dust study. http://hc-sc.gc.ca/ewh-semt/contaminants/dust-poussiere-eng.php. Accessed 26 Oct 2019

Heinemeyer G (2004) Expositionsszenarien. In: Reichl S (ed) Regulatorische Toxikologie. Springer, Berlin, pp 230–242

Heinemeyer G, Hahn A (2005) Use of product databases for risk assessment purposes. Toxicol Appl Pharmacol 207(2. Suppl):636–644

Heinemeyer G, Lindtner O, Scholz R, Rüdiger T (2006) Evaluation of behaviour of consumers for exposure assessment by questionnaire studies. International conference on environmental epidemiology and exposure, Paris, 2–6 Sept 2006

Heinemeyer G, Heiland A, Sommerfeld C, Springer A, Hausdörfer S, Treutz M, Lindtner O, Rüdiger T (2012) Phthalat-Belastung der Bevölkerung in Deutschland: Expositionsrelevante Quellen, Aufnahmepfade und Toxikokinetik am Beispiel von DEHP und DINP. Band I: Exposition durch Verzehr von Lebensmitteln und Anwendung von Verbraucherprodukten. http://www.uba.de/uba-info-medien/4326.html. Accessed 26 Oct 2019

Heinemeyer G, Sommerfeld C, Springer A, Heiland A, Lindtner O, Greiner M, Heuer T, Krems C, Conrad A (2013) Estimation of dietary intake of bis(2-ethylhexyl)phthalate (DEHP) by consumption of food in the German population. Int J Hyg Environ Health 216:472–480

Heuer T, Krems C, Moon K, Brombach C, Hoffmann I (2015) Food consumption of adults in Germany: results of the German National Nutrition Survey II based on diet history interviews. Br J Nutr 113(10):1603–1614

Huang L, Ernstoff A, Fantke P, Csiszar SA, Jolliet O (2017) A review of models for near-field exposure pathways of chemicals in consumer products. Sci Total Environ 574:1182–1208

Ibrekk H, Morgan MG (1987) Graphical communication of uncertain quantities to nontechnical people. Risk Anal 7(4):519–529

IOM (2013) Environmental decisions in the face of uncertainty. National Academies Press, Washington, DC

ISO (1993) Guide to the expression of uncertainty in measurement. International Organisation for Standardization, Geneva

Jang JY, Kim SY, Kim SJ, Lee KE, Cheong HK, Kim EH, Choi KH, Kim YH (2014) General factors of the Korean exposure factors handbook. J Prev Med Public Health 47(1):717

Jantunen MJ (2006) Expofacts, exposure factors sourcebook for Europe. Data sources and quality. http://cem.jrc.it/expofacts/docs/quality.pdf

JCGM (2008) Evaluation of measurement data. Guide to the Expression of Uncertainty in Measurement. Joint Committee for Guides in Metrology, 100:2008 (revised version 2010)

Jerrett A, Arain A, Kanaroglou P, Beckerman B, Potoglou D, Sahsuvaroglu T, Morrison J, Giovis C (2005) A review and evaluation of intraurban air pollution exposure models. J Exp Anal Environ Epidemiol 15:185–204. https://doi.org/10.1038/sj.jea.7500388

Johner SA, Boeing H, Thamm M, Remer T (2015) Urinary 24-h creatinine excretion in adults and its use as a simple tool for the estimation of daily urinary analyte excretion from analyte/creatinine ratios in populations. Eur J Clin Nutr 69:1335–1343

JRC (2015) European exposure factors (ExpoFacts) sourcebook. Joint Research Centre (JRC), Ispra/Italy; European Commission. http://expofacts.jrc.ec.europa.eu/. Accessed 26 Oct 2019

Kephalopoulos S, Bruinen de Bruin Y, Arvanitis A, Hakkinen P, Jantunen M. (2007) Issues in consumer exposure modeling: towards harmonisation at global scale. J Expo Sci Environ Epidemiol 17(Suppl.1):s90–s100. https://doi.org/10.1038/sj.jes.7500605. http://www.nature.com/jes/journal/v17/n1s/full/7500605a.html#bib3

Kettler S, Kennedy M, McNamara C, Oberdörfer R, O'Mahony C, Schnabel J, Smith B, Sprong C, Faludi R, Tennant D (2015) Assessing and reporting uncertainties in dietary exposure analysis: mapping of uncertainties in a tiered approach, Food Chem Toxicol 82:79–95. issn:0278-6915. https://doi.org/10.1016/j.fct.2015.04.007

Klimisch H-J, Andreae M, Tillmann U (1997) A systematic approach for evaluating the quality of experimental toxicological and ecotoxicological data. Regul Toxicol Pharmacol 25:S. 1–S. 5

Knoll E, Wisser H, Rebel FC (1978) Abhängigkeit der Konzentrationen von Kreatinin und Harnstoff im Serum von der Tageszeit bei normaler und eingeschränkter Nierenfunktion. J Clin Chem Clin Biochem 16:567–570

Koch HM (2006) Untersuchung der Di(-2-ethylhexyl)phthalat (DEHP)-Belastung der Allgemeinbevölkerung – Durchführung eines Human-Biomonitorings. Dissertation, Friedrich-Alexander-Universität Erlangen-Nürnberg

Koch HM, Kolossa-Gehring M, Schröter-Kermani C, Angerer J, Brüning T (2012) Bisphenol A in 24 h urine and plasma samples of the German Environmental Specimen Bank from 1995 to 2009: a retrospective exposure evaluation. J Expo Sci Environ Epidemiol 22(6):610–616. https://doi.org/10.1038/jes.2012.39. https://www.ncbi.nlm.nih.gov/pubmed/22617719

Koivusalo M, Jaakkola JJK, Vartiainen T, Hakulinen T, Karialainen S, Pukkala E, Tuomisto J (1994) Drinking water mutagenicity and gastrointestinal and urinary tract cancers: an ecological study in Finland. Am J Public Health 84:1223–1228

Kolossa-Gehring M, Becker K, Conrad A, Schröter-Kermani C, Schulz C, Seiwert M (2012) Environmental surveys, specimen bank and health related environmental monitoring in Germany. Int J Hyg Environ Health 215(2):120–126

Land Berlin (2005) Bauordnung für Berlin (BauO Bln) vom 29. September 2005 (GVBl. S. 495), zuletzt geändert durch Gesetz vom 29. Juni 2011 (GVBl. S. 315, in Kraft getreten am 10. Juli 2011) Senatsverwaltung für Stadtentwicklung, Referat VI D – Oberste Bauaufsicht; Württembergische Str. 6, 10707 Berlin-Wilmersdorf, http://www.bauordnungen.de/Berlin.pdf. Accessed 27 Oct 2019

LANUV (2016) Bestimmung von Schadstoffen und Schadstoffmetaboliten im Urin von 2- bis 6jährigen Kindern aus Nordrhein-Westfalen. Modul 2 Glyphosat. https://www.lanuv.nrw.de/umwelt/umweltmedizin/umwelt-und-epidemiologie/bestimmung-von-schadstoffen-und-schadstoffmetaboliten/. Accessed 26 Oct 2019

Lecomte J, Auburtin G (2006) Residential exposure assessment to household pesticides based on usage analysis. International conference on environmental epidemiology and exposure, Paris, 2–6 Sept 2006

Lindtner O, Ehlscheid N, Berg K, Blume K, Dusemund B, Ehlers A, Niemann B, Rüdiger T, Heinemeyer G, Greiner M, Hallerbach B, Thömmes O, Thier S (2014) Event-related survey of high consumers of energy drinks. BfR-Wissenschaft 05/2014. http://www.bfr.bund.de/de/suche.html?search%5Bquery%5D=energy+drinks. Accessed 26 Oct 2019

Lioy PJ, Smith KR (2013) Discussion of exposure science in the 21st century: a vision and a strategy. Environ Health Perspect 121:405–409

Mazzachi BC, Peake MJ, Ehrhardt V (2000) Reference range and method comparison studies for enzymatic and Jaffé creatinine assays in plasma and serum and early morning urine. Clin Lab 46:53–55

Meek ME, Boobis AR, Crofton KE, Heinemeyer G, Van Raaij M, Vickers C (2011) Risk assessment of combined exposure to multiple chemicals: a WHO/IPCS framework. Regul Toxicol Pharmacol 60(2011):S1–S14. https://doi.org/10.1016/j.yrtph.2011.03.010. https://www.ncbi.nlm.nih.gov/pubmed/21466831

Mekel O, Mosbach-Schulz O, Schümann M, Okken PK, Peters C, Herrmann J, Hehl O, Bubenheim M, Fehr R, Timm J. (2007) Evluation von Standards und Modellen zur probabilistischen Expositionsabschätzung. Forschungsbericht des Umweltbundesamtes 202 61 218/02 UBA-FN 00173/1 http://www.apug.de/risiken/forschungsprojekte/xprob/. Accessed 26 Oct 2019

Mohorovic L (2003) The level of maternal methemoglobin during pregnancy in an air-polluted environment. Environ Health Perspect 111:1902–1905. https://doi.org/10.1289/ehp.6055

MRI (2008) Nationale Verzehrs Studie II, Ergebnisbericht, Teil 1. Die bundesweite Befragung zur Ernährung von Jugendlichen und Erwachsenen. MRI, Karlsruhe. https://www.mri.bund.de/fileadmin/MRI/Institute/EV/NVS_II_Abschlussbericht_Teil_1_mit_Ergaenzungsbericht.pdf

Murphy J, Sexton D, Barnett D et al (2004) Quantification of modelling uncertainties in a large ensemble of climate change simulations. Nature 430:768–772. https://doi.org/10.1038/nature02771

NCCMT, National Collaborating Centre for Methods and Tools (2008). Quality assessment tool for quantitative studies. McMaster University, Hamilton, ON. Updated 3 Oct 2017. http://www.nccmt.ca/resources/search/14. Accessed 26 Oct 2019

Nielsen Company (2016) The dirt on cleaning. Home cleaning/laundry attitudes and trends around the world. http://www.nielsen.com/us/en/insights/reports/2016/the-dirt-on-cleaning.html. Accessed 26 Oct 2019

NRC (1996) Understanding risk: informing decisions in a democratic society. National Academy Press, Washington, DC; Framework for Environmental Health Risk Management – Final Report, vol 1

NRC (2007) Models in environmental regulatory decision making. National Academies Press, Washington, DC, 287 pp. https://www.nap.edu/download/11972

NRC (2009) Science and decisions: advancing risk assessment. Chapter 4 – Uncertainty and variability – the recurring and recalcitrant elements of risk assessment. National Research Council (NRC). Washington, DC. https://www.nap.edu/read/12209/chapter/6/. Accessed 26 Oct 2019

OECD (2005) Manual for investigation of HPV chemicals, chapter 3. Data evaluation, 3.1 Guidance for determining the quality of data for the SIDS dossier (Reliability, relevance and adequacy), December 2005, http://www.oecd.org/chemicalsafety/risk-assessment/36045203.pdf

Papameletiou D, Kephalopoulos S, Hakkinen P, Zenié A, Schwela D, Reina V (2005) European Information System on "Risks from chemicals released from consumer products/articles" "EIS-ChemRisks" (working paper) on behalf of DG SANCO (Directorate General for Health and Consumer Protection) Joint Research Centre, Physical & Chemical Exposure Unit Institute for Health & Consumer Protection, Ispra, 3 Feb 2005

Papameletiou, D; Lahaniatis, M; Bruinen de Bruin, Y; Hakkinen, P J.; del Pozo, C; Reina, V; Arvanitis, A; Zenié, A; Biavetti, I (2006) European exposure assessment toolbox. Proceedings of the ISEE/ISEA conference, November 2006, Epidemiology 17, Suppl. 6, p S65

Paustenbach DJ (2000) The practice of exposure assessment: a state-of-the-art review. J Toxicol Environ Health B 3:179–291

Ring CL, Arnot J, Bennett DH, Egeghy P, Fantke P, Huang L, Isaacs KK, Jolliet O, Phillips K, Price PS, Shin H-M, Westgate JN, Setzer W, Wambaugh JF (2019) Consensus modeling of median chemical intake for the U.S. population based on predictions of exposure pathways. Environ Sci Technol 53(2):719–732. https://doi.org/10.1021/acs.est.8b04056

Rudzok S, Sievering S, Chovolou Y, Kraft M (2016) Detection of Glyphosate and its metabolite AMPA in the urine of 2- to 6-year children from the German state North Rhine-Westphalia. Annual congress of the International Society of Exposure Science (ISES), Utrecht 2016, Contribution no. Tu-PL-H3.5

Ryan PB (1991) An overview of human exposure modeling. J Expo Anal Environ Epidemiol 1 (4):453–474

Saltelli, A (2002) Sensitivity analysis for importance assessment. Risk Anal, 22(3):579590. http://ipsc.jrc.ec.europa.eu/?id=752

Saltelli A, Tarantola S, Campolongo F, Ratto M (2004) Sensitivity analysis in practice: a guide to assessing scientific models. Wiley, Chichester

Schneider K, Recke S, Kaiser E, Götte S, Berkefeld H, Lässig J, Rüdiger T, Lindtner O, Oltmanns J (2019) Consumer behaviour survey for assessing exposure from consumer products: a feasibility study. J Expo Sci Environ Epidemiol 29(1):83–94. https://doi.org/10.1038/s41370-018-0040-2

Schulz C, Conrad A, Becker K, Kolossa-Gehring M, Seiwert M, Seifert B (2007) Twenty years of the German Environmental Survey (GerES): human biomonitoring—temporal and spatial (West Germany/East Germany) differences in population exposure. Int J Hyg Environ Health 210 (3–4):271–297

Schulz C, Wilhelm M, Heudorf U, Kolossa-Gehring M (2011) Update of the reference and HBM values derived from by the German Human Biomonitoring Commission. Int J Hyg Environ Health 215:20–35

Schwela D, Hakkinen P, Papameletiou D (2003) The new European information system (EIS-CHEMRISKS) for assessing exposures from consumer products in the indoor environment emission in the indoor environment. Healthy Buildings 2003 – Proceedings 7th International Conference, vol 1, National University of Singapore, 7–11 December 2003, pp 524–529

Shen J, Kromidas L, Schultz T, Bhatia S (2014) An in silico skin absorption model for fragrance materials. Food Chem Toxicol 74:164–176

Sieke C, Lindtner O, Banasiak U (2008) Pflanzenschutzmittelrückstände. Nationales Monitoring, Abschätzung der Verbraucherexposition: Teil 2. Dtsch Lebensmitt Rundsch 104:336–342

Silken Inc. (2003) Probabilistic methodology for improving solvent exposure assessment (PROM-ISE). American Chemistry Council

Tennant D, Bánáti D, Kennedy M, König J, O'Mahony C, Kettler S (2017) Assessing and reporting uncertainties in dietary exposure analysis—part II: application of the uncertainty template to a practical example of exposure assessment, Food Chem Toxicol 109(Pt 1):68–80. issn:0278-6915. https://doi.org/10.1016/j.fct.2017.07.061, http://www.sciencedirect.com/science/article/pii/S0278691517304490

UBA (2014) Datenbank für Expositionsfaktoren. Abschlussbericht des Xprob-Projektes und Datenbank RefXP für Expositionsfaktoren. http://www.umweltbundesamt.de/www.uba.de/xprob. Accessed 26 Oct 2019

UBA (2016) Human biomonitoring. http://www.umweltbundesamt.de/themen/gesundheit/belastung-des-menschen-ermitteln/human-biomonitoring. Accessed 26 Oct 2019

US-EPA (1985) Methodology for characterization of uncertainty in exposure assessments. EPA/600/S885/009, Sept 1985. United States Environmental Protection Agency, Office of Health and Environmental Assessment. Washington, DC

US-EPA (1991) MCCEM – Multi-chamber concentration and emission model. Users guide. Version 2.3. US-EPA, Office of Research and Development, Environmental System Laboratory, Las Vegas NV 89193-3478. https://www.epa.gov/tsca-screening-tools/multi-chamber-concentration-and-exposure-model-mccem-version-12. Accessed 26 Oct 2019

US-EPA (1992) Guidelines for exposure assessment (Chapter 6). U.S. Environmental Protection Agency, Risk Assessment Forum, Washington, DC, U.S.A., EPA/600/Z92/001. https://cfpub.epa.gov/ncea/risk/recordisplay.cfm?deid=15263. Accessed 26 Oct 2019

US-EPA (2000) Guidance for data quality assessment. EPA QA/G-9, Office of Environmental Information U.S. Environmental Protection Agency1200 Pennsylvania Avenue, NW, Washington, DC 20460. https://www.epa.gov/sites/production/files/2015-06/documents/g9-final.pdf. Accessed 26 Oct 2019

US-EPA (2008) Child-specific exposure factors handbook. National Center for Environmental Assessment, Washington, DC; EPA/600/R-06/096F. Available from the National Technical Information Service, Springfield, VA. https://ofmpub.epa.gov/eims/eimscomm.getfile?p_download_id=484738. Accessed 26 Oct 2019

US-EPA (2011) Exposure factors handbook: 2011 edition. National Center for Environmental Assessment, Washington, DC; EPA/600/R-09/052F. Available from the National Technical Information Service, Springfield, VA, and online at https://cfpub.epa.gov/si/si_public_record_report.cfm?Lab=NCEA&dirEntryId=236252. Accessed 27 Oct 2019

US-EPA (2015) Glossary of exposure science terms. Developed by the ES21 federal working group on exposure science. June 2015. https://view.officeapps.live.com/op/view.aspx?src=https%3A%2F%2Fwww.epa.gov%2Fsites%2Fproduction%2Ffiles%2F2015-06%2Fdraft_final_es21_glossary_june_2015_2.docx

US-EPA (2016a) Human exposure modeling – air pollutants exposure model APEX/TRIM.Expo Inhalation. https://www.epa.gov/fera/human-exposure-modeling-air-pollutants-exposure-model. Accessed 26 Oct 2019

US-EPA (2016b) Consolidated Human Activity Database (CHAD) for use in human exposure and health studies and predictive models. https://www.epa.gov/healthresearch/consolidated-human-activity-database-chad-use-human-exposure-and-health-studies-and. Accessed 26 Oct 2019

US-EPA. (2016c). EPA ExpoBox – uncertainty and variability. https://www.epa.gov/expobox/uncertainty-and-variability. Accessed 26 Oct 2019

US-EPA (2017) Terminology services terminology and acronyms report. Last updated on 1 Jun 2017, Version: release/4.6ADF. https://iaspub.epa.gov/sor_internet/registry/termreg/searchandretrieve/termsandacronyms/search.do?search=&term=Exposure&matchCriteria=Contains&checkedAcronym=true&checkedTerm=true&hasDefinitions=false. Accessed 26 Oct 2019

van den Berg M, Birnbaum LS, Denison M, De Vito M, Farland W, Feeley M, Fiedler H, Hakansson H, Hanberg A, Haws L, Rose M, Safe S, Schrenk D, Tohyama C, Tritscher A, Tuomisto J, Tysklind M, Walker N, Peterson RE (2006) The 2005 World Health Organization re-evaluation of human and mammalian toxic equivalency factors for dioxins and dioxin-like compounds. Toxicol Sci 2006 93, 223–241.

Van Engelen JG, Heinemeyer G, Rodriguez C (2007) Consumer exposure scenarios: development, challenges and possible solutions. J Expo Sci Environ Epidemiol 17(Suppl 1):S26–S33

van Veen MP (1999) Indoor air exposure to volatile compounds emitted by paints: experiment and model. J Expo Anal Environ Epidem 9:569–574

Vartiainen T, Liimatainen A (1988) Relation between drinking water mutagenicity and water quality parameters. Chemosphere 17(1):189–202

Vestergren R, Cousins IT, Trudel D, Wormuth M, Scheringer M (2008) Estimating the contribution of precursor compounds in consumer exposure to PFOS and PFOA. Chemoshere 73:1617–1624

Vuori V, Zaleski RT, Jantunen MJ (2006) ExpoFacts—an overview of European exposure factors data. Risk Anal 26:831–843

WHO (2004) IPCS risk assessment terminology. Part. 1, IPCS/OECD key generic terms used in chemical hazard/risk assessment and Part. 2, IPCS glossary of key exposure assessment terminology. IPCS harmonization project document no. 1. WHO, Geneva. http://www.who.int/ipcs/methods/harmonization/areas/terminology/en/. Accessed 26 Oct 2019

WHO (2005) Principles of characterizing and applying human exposure models. IPCS harmonization project document no. 3, WHO, Geneva. http://www.who.int/ipcs/publications/methods/harmonization/en/. Accessed 26 Oct 2019

WHO (2008) Uncertainty and data quality in exposure assessment. Part 1: guidance document on characterizing and communicating uncertainty in exposure assessment. IPCS harmonization project document no. 6. WHO, Geneva. http://www.who.int/ipcs/publications/methods/harmonization/en/. Accessed 26 Oct 2019

WHO (2014) Guidance document on evaluating and expressing uncertainty in hazard characterization. IPCS harmonization project document no. 11, WHO, Geneva. http://www.who.int/ipcs/publications/methods/harmonization/en/. Accessed 26 Oct 2019

WHO Regional Office for Europe (2015) Human biomonitoring: facts and figures. WHO Regional Office for Europe, Copenhagen. http://www.euro.who.int/__data/assets/pdf_file/0020/276311/Human-biomonitoring-facts-figures-en.pdf. Accessed 26 Oct 2019

Wormuth M, Scheringer M, Vollenweider M, Hungerbühler K (2006) What are the sources of exposure to eight frequently used phthalic acid esters in Europeans? Risk Anal 26(3):803–824

Young BM, Tulve NS, Egeghy PP, Driver JH, Zartarian VG, Johnston JE, Delmaar CJE, Evans JJ, Smith LA, Glen G, Lunchick C, Ross JH, Xue J, Barnekow DE (2012) Comparison of four probabilistic models (CARES, Calendex, ConsExpo, and SHEDS) to estimate aggregate residential exposures to pesticides. J Expo Sci Environ Epidemiol 22:522–532. http://www.nature.com/jes/journal/v22/n5/full/jes201254a.html

Zaleski RT, Egeghy PP, Hakkinen PJ (2016) Exploring global exposure factors resources for use in consumer exposure assessments. Int J Environ Res Public Health 13(5):E744

Zartarian V, Bahadori M, McKone T (2005) Adoption of an official ISEA glossary. J Expo Anal Environm Epidemiol 15:1–5

Zartarian VG, Glen G, SMITH L, XUE J (2008) SHEDS-multimedia model version 3 (a) technical manual; (b) user guide; and (c) executable file to launch SAS program and install model. U.S. Environmental Protection Agency, Washington, DC, EPA/600/R-08/118, 2008

Zhang T, Sun H, Qin X, Gan Z, Kannan K (2015) PFOS and PFOA in paired urine and blood from general adults and pregnant women: assessment of urinary elimination. Environ Sci Pollut Res 22:5572–5579

Chapter 4
Exposure to Substances via Food Consumption

Klaus Abraham, Davide Arcella, Katrin Blume, Polly E. Boon, Gianfranco Brambilla, Francesco Cubadda, Birgit Dusemund, Stefan Fabiansson, Rainer Gürtler, Gerhard Heinemeyer, Sven Knüppel, Oliver Lindtner, Birgit Niemann, Christian Sieke, and Andrea Straßburg

---

K. Abraham · B. Dusemund · R. Gürtler · B. Niemann
Department Food Safety, German Federal Institute for Risk Assessment, Berlin, Germany
e-mail: Klaus.Abraham@bfr.bund.de; birgit.dusemund@bfr.bund.de; rainer.guertler@bfr.bund.de; birgit.niemann@bfr.bund.de

D. Arcella
European Food Safety Authority (EFSA), Parma, Italy
e-mail: Davide.ARCELLA@efsa.europa.eu

K. Blume · O. Lindtner
Department of Exposure, German Federal Institute for Risk Assessment, Berlin, Germany
e-mail: katrin.blume@bfr.bund.de; oliver.lindtner@bfr.bund.de

P. E. Boon
Department Food Safety, Institute for Public Health and the Environment (RIVM), Bilthoven, The Netherlands
e-mail: polly.boon@rivm.nl

G. Brambilla · F. Cubadda
Department Food Safety, Nutrition and Veterinary Public Health, National Institute of Health (ISS), Rome, Italy
e-mail: gianfranco.brambilla@iss.it, francesco.cubadda@iss.it

S. Fabiansson
European Food Safety Authority (EFSA), Dietary and Chemical Monitoring (Retired), Parma, Italy

G. Heinemeyer (✉)
Department of Exposure, German Federal Institute for Risk Assessment (Retired), Berlin, Germany
e-mail: gerhardheinemeyer@o2mail.de

S. Knüppel
Department of Epidemiology, German Institute of Human Nutrition Potsdam-Rehbrücke (DIfE), Nuthetal, Germany
e-mail: sven.knueppel@dife.de

G. Heinemeyer et al. (eds.), *The Practice of Consumer Exposure Assessment*, https://doi.org/10.1007/978-3-319-96148-4_4

Many substances enter the human body via the consumption of food, and this can account for a large amount of the consumer exposures. According to the views of Lioy and co-authors (2013) and as pointed out in Sect. 3.2.5, food is not always the primary source of substances for consumers, because the origin is in the general environment. From this view, in the EU guidance documents for exposure assessment, food presents a pathway of exposure for the environmental exposure of humans. In standard food exposure evaluations, however, and from practical reasons, food is deemed as the (primary) source. The impact of the food chain for consumer exposure is pointed out in this Sect. 4.12. In addition, substances called process contaminants that may be formed from constituents in the food during the process of preparation of meals are having an increasing impact on food risk assessment.

4.1 Scenario of Exposure via Food Consumption

Gerhard Heinemeyer

Figure 4.1 shows a simplified scenario for exposure resulting from food consumption. It shows that exposure via the food chain is an environmentally driven exposure. Substances from the environment can come into water and plants, and via water and plants into animals. Water, plants, and animals form the important pathways of dietary exposure for humans. On this basis, several general food categories are formed, which can be differentiated into several levels of groups. Particular attention must be paid to industrial manufactured food in every food category. Manufactured and composite foods are a challenge for current and future considerations of studying food consumption and food contamination.

In case of focusing on a single person, it must be made clear whether there is an exposure to a particular substance. This situation might lead to concern if, for example somebody has eaten food with suspicious concentrations of a hazardous substance. In addition, this individual must be characterised in detail by the personal characteristics, which include gender, age, body weight, body height, location of living, season, and his/her individual eating habits. This type of exposure assessment might be relevant in cases of acute poisonings after intake of single meals. Some substances like shellfish toxins, pesticide residues or bacterial contamination that can cause such acute intoxications.

C. Sieke
Department of Pesticide Safety, German Federal Institute for Risk Assessment, Berlin, Germany
e-mail: christian.sieke@bfr.bund.de

A. Straßburg
Max Rubner-Institut (MRI), Federal Research Institute of Nutrition and Food, Karlsruhe, Germany
e-mail: andrea.strassburg@mri.bund.de

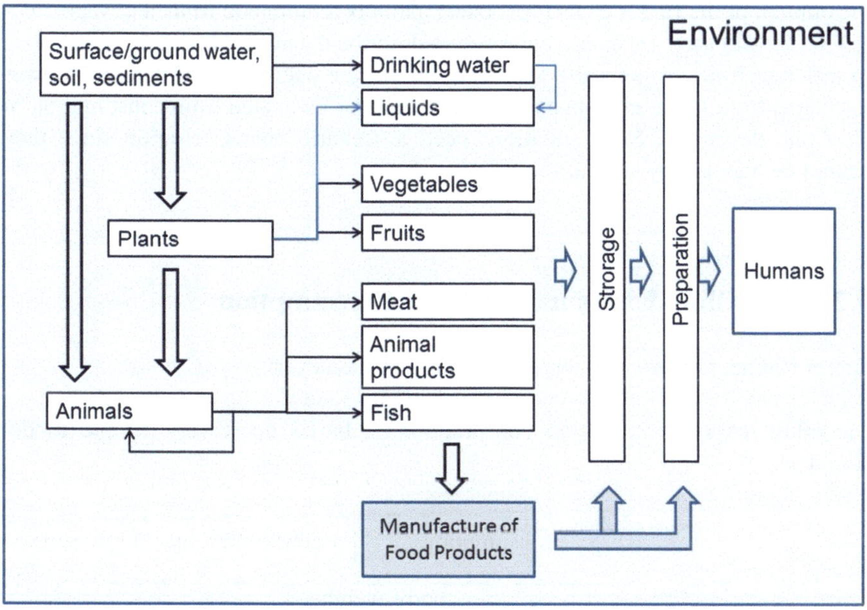

Fig. 4.1 Simplified Exposure Scenario "Food"

In most cases, an exposure assessment is focused on a particular population that should be described in the exposure scenario characterisation. At the beginning, an assessment can be focused on the general population. An important prerequisite for this assessment is that the data used are representative for that population, as described in detail in Sect. 3.4.

Standard procedures exist for assessments of exposure, for example the intake of a substance by food consumption if data from food consumption studies and concentrations of substances in food are available.

This includes the following as examples of exposure assessments:

- Population-based exposure assessment (whole population in a region or country)
- Consumer-based exposure assessment (normal, high food consumption)
- Risk-based exposure assessment (consideration of acute or chronic hazards)
- Methodology-based exposure assessment (different approaches of collecting data and use of statistical methods for exposure estimations).

For the types of exposure assessments listed above, the approaches for food-related exposures are more or less standardised, as is described in the following chapters. A particular scenario characterisation is not needed in any case.

In other cases, a scenario characterisation should be defined if non-representative data or defaults are used, which needs a detailed characterisation of the exposure situation. Also, a scenario should be formulated if special questions are to be clarified or if extrapolations from one food group to others are made. For example, the opinion of the EFSA CONTAM panel about perchlorate in fruits and vegetables (EFSA 2014e) used a scenario approach to describe the available data of perchlorate in different fruits. Among the approaches used, the occurrence of perchlorate was evaluated from literature data and the exposure was estimated from consumption of alcoholic beverages. Such situations need a scenario characterisation since they cannot be handled by standard tools.

4.2 Modelling Exposure via Food Consumption

Katrin Blume, Oliver Lindtner, and Gerhard Heinemeyer

Modelling exposure via food consumption is based on a very simple model (Eq. 4.1):

$$\text{Food} - \exp_{\text{pers}_i} = \frac{\sum_{j_{\text{food}}=1}^{n_{\text{food}}} \text{consumption}_{\text{pers}_i, j_{\text{food}}}\ x\ \text{concentration}_{\text{pers}_i, j_{\text{food}}}}{\text{body\_weight}_{\text{pers}_i}} \tag{4.1}$$

Consumption and concentration data (see next Sect. 4.3) are multiplied and normalized by body weight. All three parameters of the equation vary per individual as expressed by the index pers_i. This is self-explanatory for food consumption and body weight but individuals might show different probabilities of being in contact with low or high concentrations in consumed foods. Further, the exposure is normally summed-up covering different food categories. This is acknowledged in the equation by introducing the index j_{food} (Eq. 4.1). Beyond this, dermal and inhalative exposures and exposure via house dust have to be taken into account.

One reason for calculating exposure is to evaluate whether toxicological reference values are achieved or exceeded by one food category or from all dietary sources. In the same context, exposure assessments give insights into possible changes or the necessity of developing and installing maximum levels, for instance for environmental contaminants, food additives or pesticide residues in certain foods.

The objective of having maximum levels is to ensure food safety from the highest possible reduction of contamination. In the European Union, there are specified amounts for substances in food that are regulated and should not be exceeded. Surveillance programmes are responsible for controlling randomly whether foods on the market follow these levels.

To check whether maximum levels are at acceptable concentrations, data have to be collected in order to calculate total dietary exposure and short-term as well as long-term exposures via the respective food category or categories. Available food concentration data can also be used to derive proposals for maximum levels or to be compared to maximum levels already proposed.

4.3 Parameters for Estimation of Exposure via Food Consumption

Gerhard Heinemeyer

As shown in the previous chapter, information about (1) concentrations of substances in food, (2) consumption of food, (3) data on the influence of storage on concentrations in food, and (4) data characterising changes of concentrations from heating the food is needed.

In general, concentrations of substances in food can be found in scientific publications. It should be considered, however, that these data are normally evaluated for a non-exposure-related purposes and often are not representative. These data must be carefully evaluated for the exposure assessment in the scenario being characterised.

Food concentration data can also be gathered from official measurements of substances in food that can be made for several reasons. One important reason for measuring substances in food is the control of limits and to guarantee an unbiased trade of food commodities across different member states. These measurements are normally performed to identify levels of substances in certain foods that exceed the limit values (ADI, TDI, ARfD). These market control measurements are performed (1) according to official food control programmes or (2) in case of suspicion of exceeding the levels. A typical example for planned measurements is the control of pesticide residues in food and to control the correct application of pesticides. EU Member States are requested to perform measurements of substances in food and to develop a Multiannual National Control Plan (MANCP) according to Articles 41–43 of the EC regulation no 882/2004 (EU 2004). Similar measurements are also performed for other substances such as contaminants.

In Germany, the following actions are performed in the frame of the MANCP:

- The Federal Control Plan (FCP)
- National Food Monitoring (NFM)
- National Residue Control Plan (NRCP)
- Import Control Plan for food of animal origin (ICP)

Data about the concentrations of substances in food in Germany are available to the responsible federal office of consumer protection and food safety (BVL 1999–2017 in BVL 2018). Details about this are provided in the respective chapters.

It should be noted that the data used for exposure assessment should comply with degrees of quality and standards, as pointed out in Sect. 3.7. However, the data from market control are measured with the purpose to control the levels and not to identify the typical ranges of concentrations. Furthermore, the collection of the samples often does not follow a systematic and representative approach. This approach enables the assessor to characterise the upper extremes (the right or high side) of the concentration range, but the average values derived from that measurements often tend to higher values. However, it must be considered that this procedure may lead to a tendency to overestimate the exposure. Central tendencies can also be described, with a tendency for overestimation.

Data on the change of concentrations due to storage and transport of food is almost always lacking.

Obtaining adequate data for an extensive exposure assessment of substances in foods would lead to an overwhelming need of measurements due to the high number of substances in combination with many different foods. To overcome this problem, the concept of a total diet study (TDS) was developed as a special type of sample collection to gather adequate data, which can serve as a sound data basis for exposure assessment. The principle of the TDS is explained in Sect. 4.8.4. The basic approach is to pool the measurements of samples according to predefined food groups. The concentrations measured provide mean values for the group that is determining the pool. Another point is that the substances are measured in prepared food, including industrial manufactured food. The challenge of the TDS is to prepare an adequate plan for food sampling that fulfils the statistical requirements. This means that an adequate number of representative samples must be collected with respect to respective populations with different age and gender, seasons and regions, as described in detail in Sect. 3.4, and to prepare typical menus and meals.

Food consumption studies (FCS) deliver values based on representative questionnaire studies performed on the national level. Many states have performed FCS with many of them conducted in regular time intervals. Several methodologies exist to ask people about food eating habits by retrospective and prospective approaches, as described in detail in Sect. 4.6, and these have consequences for exposure assessment (see Sect. 4.6.5). The results of several FAO, European [DAFNE] and national [e.g., FACET] food consumption databases were compiled and are found in the European ExpoFacts on-line database (more in 7.4.2).

Preparation of food may lead to changes in concentrations of substances in food. As examples, substances can be decomposed by heat and there can be a shift of substances from the food into water associated with the food. However, it has also been observed that substances can be formed during a heating process, leading to the so-called "process contaminants". Examples of substances of concern as process contaminants include acrylamide, furane and 3-monochloropropane-1,2 diol esters (3-MCPD) (EFSA 2016a). These substances are the only ones for which food is the primary source of exposure.

Although food exposure assessment normally follows the fixed concept described above, it might be appropriate to prepare a description of the exposure that provides the formulation and usage of the equations, and to document the exposure parameters. This is especially important if special situations are to be assessed, and when different steps in a tiered approach are used.

4.4 Food Listing and Food Classification

Stefan Fabiansson

There are two primary purposes for employing classification systems in food exposure assessment: (1) to facilitate the comparison of foods from different sources (programs, surveys, countries, etc.) and (2) to support the accuracy of the exposure estimation by the use of aggregated levels of food categories.

Typically, estimating exposure from data on food component levels and data on foods consumed involves the comparison and matching of corresponding food records in two or more data sets. However, as descriptions of food are associated with large individual variability and uncertainty, the unique and unambiguous matching of foods by name from different data collections is often not possible. Foods, as purchased, can be described during data entry with as much information as is available, while the situation is more complicated for foods as consumed. Since food preparation has almost limitless variations, the final dish cannot usually be absolutely uniquely identified but has to be characterised with reasonable detail. The situation is further confounded by homonyms, synonyms, identical brand names for different food products, and culinary or technological terms (EFSA 2011f). Therefore, food records must most often be matched by aggregating food descriptions provided in each data set through a harmonised food classification system.

A hierarchical structure of the food classification system with individual food items aggregated into food groups and broader food categories providing different levels of detail is vital as the aggregation applied can be an important source of uncertainty in food exposure estimation. As will be shown later in this chapter, too broad-level aggregation (e.g. grains and grain products) can give an undue influence to rarely consumed foods sampled at high frequency, while too detailed food groups (e.g. wholemeal bread with nuts) might not provide the necessary level of matching.

To date, most food categorisation systems have been developed separately in the context of distinct applications such as food consumption, nutrient composition, or the monitoring of pesticides, contaminants, or additives. As a result, each system may address detail that is redundant in other areas and thus not covered by their system. Mapping between different individual food categorisation systems is time consuming and imprecise, since terms may differ in definition or level of specificity, making a one-to-one lookup difficult.

The main problem of food description is not the difficulty of finding the best terms or ways of describing food, but the fact that differing, inconsistent, and often incompatible terminologies are used in different areas. The main problem of food classification is the level of heterogeneity of foods included in each hierarchical group. On top of this, each area has its own classification and description language, based on different assumptions and designed to satisfy the immediate requirements of the initiator. Consequently, it is difficult to exchange data between countries, between organisations within the same country, between scientific disciplines, or even between staff in the same institution. A better approach would be to use a harmonised system in the various data sets that will need to be interlinked, using a common food classification and description system.

4.4.1 Systematic Classification Systems

Systems to characterise foods usually consist of two distinct steps, a description part and a classification part. A description system is a tool for the data originator, who wants to describe a reported food as precisely as possible. In contrast, the classification part of the system is a tool for the data end user, used to group or aggregate foods with similar characteristics. Most food classification systems consist of a thesaurus of up to three to four aggregation levels. Although it might initially seem sufficient to describe the food accurately, a hierarchical aggregation system is necessary to be better able to match data originating from different domains. This is partly due to consumption involving dishes or composite foods while laboratory testing commonly involves individual food items or raw ingredients. An exception is TDS where foods are prepared as consumed before being tested (see Sect. 4.8.4). By the use of facets covering further extrinsic details related to the food, like processing treatments, packaging materials, or preservation methods, individual food items can be aggregated along such aspects for exposure assessment.

Many different national and international systems to systematically define food items are available. Some are highly specialised, e.g. on trade, on crops, or on customs. Most of the more generalised are fit-for-purpose systems, mainly focusing on, for example food consumption as in DAFNE, Eurocode 2, and EPIC, or food composition as in BLS and EuroFIR. National food classification systems are often based on national criteria and the food groups can be very specific. Alternatives designed to cover a broader range of circumstances include the ADV catalogue system, the LanguaL classification system, and the EFSA's FoodEx system.

An exhaustive review of food classification and description systems, highlighting the advantages and disadvantages of the various systems, was published by Ireland and Møller (2010). Some of the more complete food classification systems are presented below.

Eurocode 2

The Eurocode 2 system (Table 4.1) was intended for use with food consumption surveys for nutritional epidemiology in Europe and related work. It consists of a three-level hierarchical classification system, and a descriptor system for coding supplementary information about the foods as consumed (Arab et al. 1987). The system aims to code foods in their consumed state, based on the following principles:

- Mixed foods integrated with natural foods or unprocessed foods.
- Food entries easily located without requiring practical food knowledge.
- A unique numeric code for each food or dish with all identifying detail of the food retained.
- Food grouping to enable international food comparisons.

Table 4.1 Two examples of the coding principle used in the Eurocode 2 system

Main group	Subgroup	Food item	Descriptor
6. Grain and grain products	630 Pasta and noodles	6.30.10 Dried main-dish pasta	T5 Dried
6. Grain and grain products	650 Fine bakery wares	6.50.20 Croissants	T1 Baked

- The core classification component consists of 13 main categories that are broken down into two further levels. For nutrient analysis or assessment of possible risks involving food processing, additional descriptions of foods as consumed may be appropriate. A separate descriptor system has therefore been developed as a semi-hierarchical system with 80 descriptors that can be added to the Eurocode 2 code if desired. The descriptors are divided into main groups denoted by a letter. These are:

T Thermal treatment at consumption
N Non-thermal treatment
P Preservation method/packing medium
A Component added
R Component removed

Development of and support for the Eurocode 2 system has ceased but descriptions of the system are still available on their website at http://www.ianunwin.demon.co.uk/eurocode/index.htm

BLS (Bundeslebensmittelschlüssel)

While the Eurocode 2 is a Europe wide system with its main focus on food consumption data capture, the classification of food as part of the German Nutrient Database (BLS) was developed as a standard instrument for the analysis of energy and nutrient intake data from epidemiological studies and nutritional surveys (MRI 2015) in Germany (Table 4.2).

As of 2017, close to 15,000 combinations of food items and further descriptors are contained in the BLS, from which analytical values of approximately 1200 primarily non-processed and primary processed food dishes are available.

The BLS was designed according to the following criteria:

- Hierarchical structure of food groups in such a way that foods having approximately the same analytical composition are combined into food groups.
- Clear and rapid coding with each food/food product having its own code number, whereby a quick, brief, and clearly arranged coding is possible.

Table 4.2 The more elaborate food description and classification system used by the German Nutrient Database

Main group	Subgroup	Food	Detail	Processing	Preparation	Weight
D. Cakes tart, pastries, and buscuits	D.1 Fruitcake	Strudel	Ingredients	Not applicable	Baked	As prepared
M. Milk, dairy products, and cheese	M.3 Hard cheese	Parmesan cheese	40% Fat	Commercially processed	Grated	As prepared

Each food item is connected to a food descriptor and a BLS code. The BLS code consists of seven digits representing the main food group, the single food item, the industrial process (e.g. canning and freezing), household preparation and the reference weight (e.g. drained) (Hartmann et al. 2006; Dehne et al. 1999).

Table 4.2: The principles of the BLS structure with two examples to show the more elaborate food description and classification system used.

The coding in the BLS uses a combination of alphanumeric digits. For example, at the first level, the letter B stands for Brot (bread and rolls), and F for Früchte (fruits). The second level (numerical) breaks the main group into subgroups, e.g. bread as wholegrain bread. The third and fourth levels provide a more detailed sequential classification of the individual food. The fifth and sixth digits characterise industrial processing and home preparation, respectively.

From the presentation in Table 4.2, the strudel would be coded as D10161 and the cheese as M30600, with the six for the latter indicating the fat content (>40%).

The BLS is an example of a national food classification system although attempts have been made to link the system to other European nutrientional databases through the EuroFir initiative (Roe et al. 2013).

Further information can be found at https://www.blsdb.de/bls

ADV Catalogues System

The ADV Catalogues System is an example of a national universal food coding system for data communication fulfilling the reporting requirements of the European Commission. It is maintained by the Bundesamt für Verbraucherschutz und Lebensmittelsicherheit (BVL) as it is responsible for collecting, managing, and reporting information received from the German federal states under various control programmes for food and other commodities (see also Sect. 4.8).

Data management of the information is based on a controlled language defined in the 65 so-called ADV (Automatisierte Datenverarbeitung) Catalogues. The ADV Catalogues form a faceted system covering not only food description (one facet of which provides a food classification), but also parameter (component and property) identification and value documentation (e.g. units and methods), together with administrative and regulatory aspects.

The following principles were used in developing the system:

- As detailed as necessary - as simple as possible.
- As compatible as possible with existing classifications.
- Expandable with the introduction of new reporting requirements.
- Wherever possible, use existing catalogues.
- Integrate all monitoring areas.

The online notification system will allow users to select the appropriate data for their application. The design of the catalogues provides flexibility with the matrix (food) catalogue broken down into basic foods and a catalogue of properties (facets). The number of valid facets is defined specifically for each food group, as reflected in the input screens that are defined for each group. The starting point for coding food products consists of a matrix of 310 possible basic food entries that are then further specified using any one of a range of 21 facets as appropriate for the specific food entry.

Further information of the ADV Catalogues can be found at http://www.bvl.bund.de/EN/01_Food/_01_tasks/02_OfficialFoodControl/11_Datamanagement/Datamanagement_node.html#doc5846170bodyText1

LanguaL

LanguaL ("Langua aLimentaria" or "language of food") is a multilingual, facetted thesaurus created to describe foods in a systematic manner. The thesaurus was initiated in the late 1970s by the Centre for Food Safety and Applied Nutrition of the United States Food and Drug Administration as an ongoing co-operative effort of specialists in food technology, information science, and nutrition. Since then, LanguaL has been developed in collaboration with the US National Cancer Institute (NCI), and, more recently, its European partners, notably in France, Denmark, Switzerland, and Hungary. Since 1996, the European LanguaL Technical Committee has administered the thesaurus.

When using the LanguaL thesaurus, each food in a database is described by a set of standard, controlled terms arranged in 14 facets characteristic of the food, such as the biological origin, technological treatments, packaging, as well as methods of cooking and preservation (Møller et al. 2008). The thesaurus provides a standardised language for describing foods, specifically for classifying food products for information retrieval.

LanguaL is based on the concept that:

- Any food (or food product) can be systematically described by a combination of characteristics.
- These characteristics can be categorised into viewpoints and coded for computer processing.
- The resulting viewpoint/characteristic codes can be used to retrieve data about the food from external databases.

As much detail of a food should be indexed as is available and can be described by LanguaL terms. More than 27,000 foods in European food composition databases are now LanguaL indexed. In addition, the USDA National Nutrient Database for Standard Reference is now partly LanguaL indexed. In total, more than 35,000 foods originating from Europe, North America, and other countries are now LanguaL indexed. Table 4.3 illustrates some of the characteristics captured for yoghurt in a few of the facet by Langual.

Table 4.3 An example of yoghurt systematically described by a combination of characteristics categorised in viewpoints and coded for computer processing

A. Product type	B. Food source	C. Part of source	E Physical state	F Heat treatment
Cultured milk	Cow	Milk	Semi-solid	Partial
A0101	B1201	C0235	E0134	F0018 T

As opposed to other food classification systems that use a hierarchical presentation of foods and food groupings with a possible linking of further facet descriptors, the sole focus of LanguaL is the elaboration of detailed facet descriptors for the foods. It was chosen to render the system language independent and therefore each descriptor is identified by a unique code pointing to equivalent terms in different languages. However, the complexity of the thesaurus with over 5000 descriptors in 14 facets is challenging for the user.

The Food Product Indexer (FPI) was developed to facilitate its use. The FPI software incorporates hierarchy (tree structure) by linking to foods in a number of external food classification systems, assists in assigning definitions, synonyms, and scope notes of descriptors in the LanguaL thesaurus, and includes a number of already indexed data sets as examples of food indexing.

Detailed information about the LanguaL system can be accessed on-line at http://www.langual.org

FoodEx

EFSA reviewed existing food classification systems in 2008 and concluded that because of its needs to cover a range of food safety domains including pesticides, food consumption, food composition, contaminants, additives, and food contact materials, none of the existing systems could fully cover their requirements. Attempting to improve on existing systems, EFSA started to develop a flexible food classification and description system called FoodEx.

FoodEx builds on different food description and classification systems including some of the systems described above. Legislative requirements were also considered, in particular the food classification presented in Commission Regulation (No 1881/2006 on (EU 2006c) setting maximum levels for certain contaminants in foodstuffs and Commission Regulation (EC) No 396/2005 (EU 2005a) on the maximum residue levels of pesticides in or on food and feed of plant and animal origin (EU 2005a).

FoodEx consists of descriptions of a large number of individual food items aggregated into food groups and broader food categories in a hierarchical parent–child relationship. The entire system is code based and each entry is identified by a unique code for the food item or food grouping, which in turn is associated with a proper description specifying which foods are included in the group. All terms in the food list are flagged with attributes defining their role (hierarchy, core list, or extended list) and their state (e.g. raw commodity, ingredient, simple, or composite food).

Hierarchies may be used to organise foods for data reporting (generation and storage of data in a database; these may be defined as input hierarchies) or for data analysis (grouping in meaningful classes when analysing the data and creating summary tables or documents; these may be defined as output hierarchies).

The new system was intensively tested with regard to the collection of food consumption and chemical occurrence data and this involved several European

Union member state organisations that operate in data collection. The testing phase highlighted the strengths and weaknesses of the classification system and provided suggestions for improvement. Based on the outcome of this testing phase, a major revision called FoodEx2 revision 2 was finally presented (EFSA 2015f).

Facets were introduced to add further detail to the information provided by the food list term. Through facets, collections of additional terms describing properties and aspects of foods from various perspectives provide an alternative to more elaborate food item definitions.

The terminology was significantly expanded, particularly in the area of raw commodities and natural sources. Furthermore, efforts were put into simplifying the coding aspect of the system by removing inconsistencies and ambiguities, and fixing the logic behind the relationship between the terms in the hierarchical tree of the system and the base terms and facets.

The major actions with regard to the logic of the system involved the creation of a specific reporting hierarchy, the revision of different facets, such as part-nature, process, source and source-commodities, and the creation of part-nature-based generic terms in order to discourage the use of hierarchy terms in cases where information is lacking. Figure 4.2 illustrates the principles of the FoodEx system capturing four hierarchies (after EFSA 2011a).

Overall, the eight hierarchies present in FoodEx2 revision 2 are:

- The master hierarchy for managing all the terms in the system (entire terminology, for technical use only).
- The reporting hierarchy, that is a data input-oriented hierarchy; it should facilitate the choice of the code and identify the minimum level needed for using the data.
- The exposure hierarchy that is designed to facilitate the grouping of food items for exposure calculations. It is also the preferred hierarchy for reporting consumption data.
- The pesticide hierarchy includes a parent group with the same pesticide code as the main product of each group as well as of the other products to which the same MRLs apply.
- The biological monitoring hierarchy (also referred to as the zoonoses hierarchy) reflects, within the first level group "meat and meat product", the animal species-based grouping defined in this domain. The remaining food groups are organised as in the reporting hierarchy.
- The feed hierarchy includes all the feed related entries and incorporates in the name the string "(feed)". For feed entries in the occurrence database, only a feed code may be used.
- The veterinary drug residues hierarchy encompasses the groups used for recording data on residues of veterinary medicinal products. In this domain, the data always refer to a specific animal, but may range between food samples and non-food animal-related samples, such as biological samples or even feed samples, environmental samples or water used in farming.
- The botanicals hierarchy supports EFSA's compendium of botanicals reported to contain naturally occurring substances of possible concern for human health when used in food and food supplements. Some of the plants included in this compendium are also food source plants, whereas some are not.

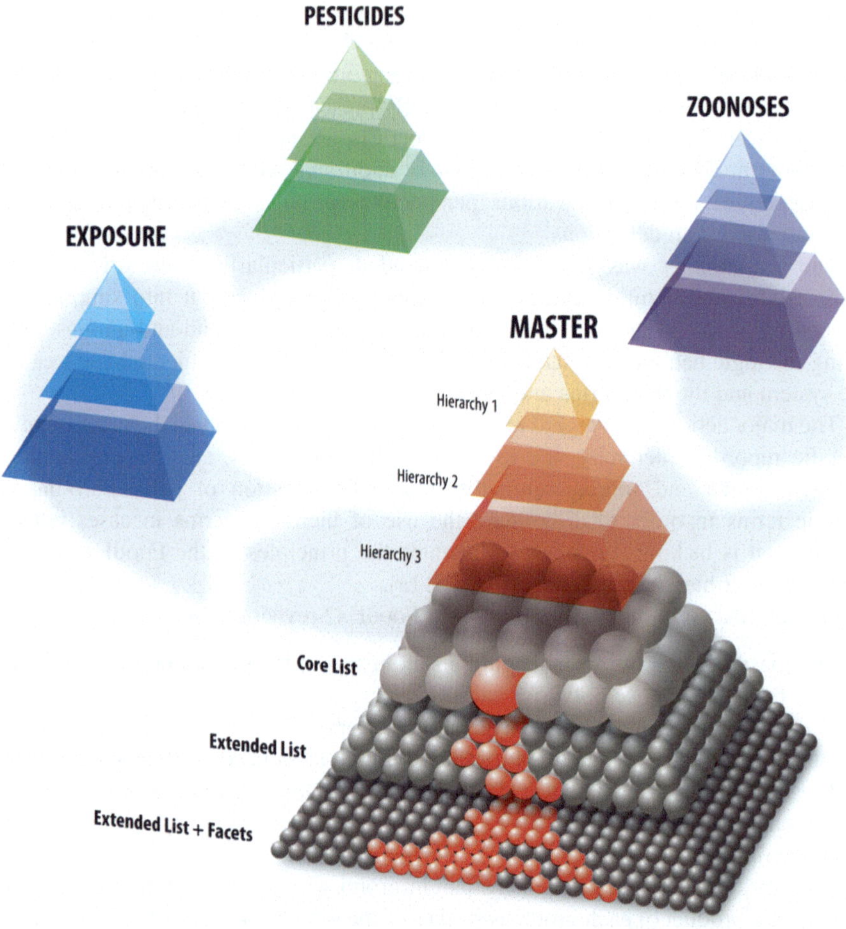

Fig. 4.2 The hierarchical structure of FoodEx with the commonly used core and extended list and facets, as well as three of the alternative three-level top structures graphically illustrated

Additives and food contact materials do not have individual hierarchies but are instead addressed through the use of specific facets.

In conjunction with the review of the system, a guide for the use of the system has also been prepared. The sections covering this guide present the rules for standardised coding and provide tips for the easier use of FoodEx2.

FoodEx2 is now the standard for the collection and transmission of data from the European Union member states to EFSA. FAO and WHO work together with EFSA to use the FoodEx2 as a basis for "a food classification system applicable at global level". This development is part of the framework for a pilot Global Individual Food consumption data Tool (FAO/WHO GIFT) that will facilitate access to data on food availability and produce food-based indicators comparable across the globe.

Further information on FoodEx2 can be found at https://www.efsa.europa.eu/en/data/data-standardisation

4.4.2 *Importance of Matching Food Entries at a Detailed Level*

The food description system that is ultimately chosen is less important than the requirement that the system be detailed enough to capture individual food items, provide a hierarchical aggregation system, and is uniform in its coverage of the different food domains involved in calculating exposure. The following example illustrates the importance of capturing foods at an appropriate level.

When EFSA was formed in 2002, it initially outsourced calculation of contaminant exposure to only a few European Union member states with access to adequate food consumption data. However, as this was not considered to sufficiently cover the European population, EFSA embarked on a collaborative project to collect food consumption data from a wider range of European Union member states to form the Concise Food Consumption database. Unfortunately, it was only possible to collect aggregated data in 15 broad food categories, with eight categories split into subcategories in some submissions providing a total of 28 possible categories (EFSA 2008c).

The Panel on Contaminants in the Food Chain (CONTAM) was among the first to use the Concise database for its opinion on cadmium (EFSA 2009a). At first, the cadmium concentrations and food consumption data were aggregated at the 15 food category levels (Concise Original) resulting in exposure varying between 4.49 and 6.56 μg/kg of BW per week across the participating member states as illustrated in Table 4.4.

However, as is common, cadmium hot spot food items like algal products, liver, and seaweed dominated in the reporting and skewed the aggregated data. When proportionate weighting based on a national detailed food consumption database was applied during aggregation of contaminant data to better reflect the contribution of

Table 4.4 Variations in cadmium exposure results using different levels of food aggregation for the calculations (Concise Original<Concise Adjusted<FoodExL2<FoodExL3)

Cadmium Exposure (μg/kg body weight per week)				
Country	Concise Original	Concise Adjusted	FoodExL2	FoodExL3
Belgium	5.13	2.11	1.87	2.10
Czech Republic	5.35	2.23	1.61	1.85
Denmark	4.98	2.19	1.57	2.00
Finland	4.49	1.88	1.51	1.82
France	5.98	2.24	1.79	2.13
Germany	6.56	2.66	1.50	2.11
Hungary	5.41	2.08	1.89	2.29
Ireland	6.03	2.62	1.97	3.21
Italy	5.35	2.02	2.20	2.32
The Netherlands	5.34	2.24	1.68	2.54
Sweden	5.08	2.20	1.76	2.02
UK	5.36	2.18	1.66	2.56

individual food items (Concise Adjusted), the exposure was reduced to between 1.88 and 2.66 μg/kg of BW per week, showing the considerable impact of better specifying the food items.

To address such shortcomings EFSA developed the FoodEx food classification system and embarked on assembling European Union member state food consumption information at the most detailed level possible. This resulted in the development of the Comprehensive Food Consumption database. With the new food consumption data and a slight expansion of the cadmium database, EFSA set out to compare the impact of using data at a more detailed level (EFSA 2012a). Matching data mainly at the FoodEx level 2 with some irregular foods at level 1 or level 3 (FoodEx L2), data comprising 154 food categories resulted in a cadmium exposure varying between 1.50 and 2.20 μg/kg of BW per week. Going one step further by matching data at FoodEx level 3 (FoodEx L3) comprising 553 food groups resulted in cadmium exposure varying between 1.82 and 3.21 μg/kg of BW per day.

Although results using the more refined models are in the same ballpark, it is always important to explore the reasons for the different outcomes of the exposure calculations. As mentioned above, matching data using only broad food categories allowed an undue influence when aggregating results of high frequency sampled food products that are highly contaminated. On the other hand, using too refined food categories left gaps in the matching process between consumption data and contaminant data with unpredictable results.

The use of facets could significantly increase the precision when matching food consumption and occurrence data and, consequently, reduce the uncertainty in the assessment of dietary exposure. The use of facets is fundamental to capturing as much detail as possible without exponentially increasing the list of foods and forcing the hierarchy of the food classification system in use. For example, the use of a facet for smoked foods could be fundamental to identify foods containing smoke flavourings from different food categories (e.g. meat, cheese, or snacks), the facet "sugar free" can be used to identify candies or beverages containing intense sweeteners, and facets for household processing (e.g. boiled, fried, and steamed) can be extremely useful for the assessment of exposure to processing contaminants (e.g. acrylamide or furan), and can also be used when a reduction in the levels is foreseen due to the processing of foods. For example, the latter facet is extremely useful when assessing the intake of minerals and vitamins since it is quite clear from the literature that some of these chemicals are significantly reduced when the food is processed.

4.4.3 Impact of the Aggregation Level on the Exposure Estimate

A similar result was reached when using different aggregation levels of foods in a German study calculating exposure to phthalates. Exposure of the German population aged 14–80 years to bis(2-ethylhexyl) phthalate (DEHP) from consumption of

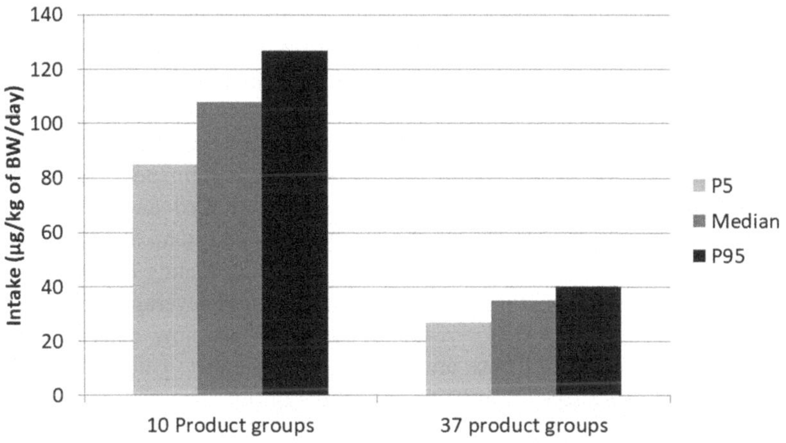

Fig. 4.3 Estimates of DEHP exposure using two different levels of food aggregation (Data were taken from Heinemeyer et al. 2012)

food was estimated by the use of probabilistic modelling (Heinemeyer et al. 2012, 2013). The study was performed on the basis of an extensive review of the literature from around the world reporting measured data on DEHP in food, as well as official German food control data at two different aggregation levels.

Figure 4.3 shows the result of DEHP exposure from food consumption. The first estimate is based on the use of 10 food groups characterising the upper level in a classification system (e.g. eggs, milk and milk products, grain and grain products). The disaggregated estimate is based on 37 food subgroups (e.g. cheese, bread, fine bakery).

The median exposure estimate (P50) is about twofold higher for the aggregated groups (P50 aggr), with exposure three- to fourfold higher for the 95th percentile (P95 aggr).

Both cadmium and DEHP examples show that it is necessary to carefully check the accuracy of the matching processs when calculating exposure. Exposure assessment is not an exact science but requires a lot of experience to get it right. The German language term "Fingerspitzengefühl" describes the intuitive instinct necessary to be a good exposure assessor.

4.4.4 Conclusions

Typically, estimating the exposure to a compound in food involves the comparison and matching of corresponding food records in two or more data sets. Although it might initially seem sufficient to use food descriptions in the respective data set, a hierarchical aggregation system is necessary to be better able to match data originating from different domains.

The aggregation part of the system is a tool for the data end user, used to group foods with similar characteristics when calculating exposure. In contrast, the description system is a tool for the data originator, who wants to describe a reported food as precisely as possible.

Many different national and international systems to systematically define food items are available. The food classification system that is ultimately chosen is less important than the requirement that the system be detailed enough to capture individual food items, provide a hierarchical aggregation system, and is uniform in its coverage of the different food domains involved in calculating exposure.

An important challenge in the classification of foods and beverages is the fact that more and more composite and/or ready to eat foods are now entering into the marketplace and consumed. These are particularly difficult to categorise and only a flexible system using facets can be used to sort out the increasing complexity of foods.

The data matching process is important. Matching data using only broad food categories can allow an undue influence of high-frequency sampled food products that are highly contaminated. Using too refined food categories can leave gaps in the matching process between consumption data and contaminant data with unpredictable results. It is thus necessary to carefully check the accuracy of the matching process when calculating exposure. A lot of experience is needed to get this process done correctly.

4.5 Population Aspects in Food Risk Assessment

Stefan Fabiansson

As has been mentioned throughout this book, the central tendency of dietary exposure is calculated by relating analytical data for the occurrence of chemical substances in a range of foods to the individual's consumption of the respective food in the population under consideration.

However, there can be considerable variations in food choice, consumption patterns, and sensitivity to possible toxic compounds. Consumption data at the individual level allow calculation not only of mean exposure for the population as a whole, but also high-level exposure due to the high-level consumption of food groups containing the hazardous compound at higher levels. The definition of high-level consumers is crucial to the outcome of the risk assessment because, in practice, it identifies the part of the population at the highest risk to be covered when considering whether action will be necessary to reduce dietary exposure to a safe (or less toxic) level.

4.5.1 *Calculating Dietary Exposure for "at Risk" Consumer Groups*

In order to calculate exposure for high-level consumers, the food consumption data should cover a representative sample of the population of an appropriate number of individuals, and allow individual matching between occurrence data and different dietary patterns. Individuals who consume large quantities of food in general or specific food items and thus could be "at risk" should be covered by considering higher percentiles of food consumption data (e.g. 90th, 95th, or 97.5th percentiles). Calculating high-level total population exposure might be sufficient in many cases to also cover subpopulations. However, groups considered by experts to be "at risk" on the basis of previous knowledge of dietary habits may not be well represented in national food consumption surveys (see Sect. 4.7). Therefore, separate exposure assessments may be required for particular subgroups of the population that have been deemed to be at higher risk than the general population for specific chemical substances of interest.

Additional risk may be the result of physiological or pathological factors, or an abnormal dietary intake (Verger et al. 1999). It is thus crucial to identify and evaluate subpopulations and individuals at potentially greater risk so that appropriate mitigation measures can be identified. On the one hand, individuals and groups can be at potentially higher risk if they are exposed to high concentrations of hazardous compounds (Sexton et al. 1993). On the other hand, they might be more susceptible to the adverse effects of a given exposure due to high sensitivity and/or vulnerability because of inherent genetic variability, age, gender, pre-existing disease (e.g. diabetes, asthma), inadequate diet, environmental, or lifestyle factors (e.g. smoking), stress, and inadequate access to health care (WHO 2000).

Subpopulations at particular risk are most often considered to include the very young, pregnant and nursing women, the elderly, and the immunocompromised, but might also encompass individuals with deviating food consumption habits due to social, religious, or ethical beliefs. There might also be seasonal variations to consider, as well as brand loyalty and hazardous compounds present in the local food supply.

As far as is possible, it is important to identify these susceptible individuals and groups to be able to understand the variability in exposure and to take account of this information in assessing and managing risks. Exposure and risk information for susceptible populations are critical since health standards and regulations are often developed with the intent of protecting such groups in particular.

As an example, in the case of chemical substances classified as endocrine disruptors, it would be important to specifically address exposure in infants and in women of childbearing age (Schug et al. 2011). If the survey is conducted at a regional level, exposure can be evaluated in each region separately to assess the contribution of local or regional food consumption and to issue specific local risk management measures. A typical example would be professional fishermen or recreational anglers and their families who regularly consume fish from the same region. There are several examples where localised contamination with heavy metals has influenced the health status in the local community, the most well-known being mercury poisoning in the Minamata City of Japan (Harada 1995).

4.5.2 Vulnerable and Sensitive Populations

The results from individual surveys typically contain food consumption data for different sex, age, ethnic, economic, and regional populations. Consumption data at individual level allow calculation not only of exposure patterns for the population as a whole, but also age-related exposure.

Children have a higher intake of food and fluid on a per kg body weight basis, and are therefore considered as an at risk group. On the other end of the age spectrum, the elderly are often considered more at risk because of deteriorating health. Figure 4.4 shows a typical example of an age-related distribution of dietary exposure to inorganic arsenic in the European population (modified after EFSA 2014c).

In this case, young children seem to be more at risk of the adverse effects of dietary exposure to inorganic arsenic compared to adults and the elderly. The contribution of rice to the dietary exposure of inorganic arsenic was noteworthy in all age classes as had been recognised previously (Fontcuberta et al. 2011). When examining existing food consumption data, the possible variation of food habits within subgroups of the population should be considered. Some subgroups within the population will show patterns of food consumption that differ widely from those of the population as a whole and include, for example, ethnic and cultural minority

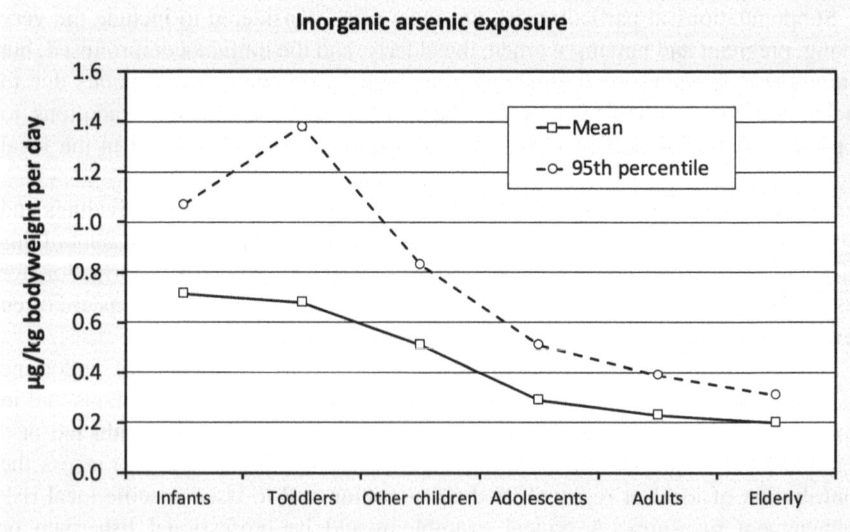

Fig. 4.4 Inorganic arsenic exposure during lifetime

groups within a community. As rice is a staple food for nearly one-half of the world's population and is consumed in high amounts by some specific populations, this is a typical situation where further food consumption data from ethnic subpopulations would be called for.

It is also important to consider endocrine development during childhood and adolescence, where the risks are based on the development of endocrine organs. The effects of endocrine disruptors should be given a special focus during this period of age even if the exposure is not changed. There are also age-dependent risks due to changes in metabolic capacity (quantitative and qualitative) in young children.

4.5.3 Rarely Consumed Foods

Some subgroups within the population consume large portions of specific food items rarely consumed by the general population. Some consumers may also be loyal to those foods or brands of food containing the highest concentrations of the hazardous compound, or may occasionally consume foods with very high concentrations of the compound. For certain chemical substances found only in such specific food items or food groups, calculation of the exposure for consumers of those food items or food groups can be of interest. This is done by identifying consumers of the specific food and calculating exposure by using only consumption information from this group.

An example is the consumption of honey in the European population with honey consumers varying from less than 5% to 50% of the total population across 16 different surveys. With less than 5% of the population consuming honey, no consumption would be picked up even by the 95th percentile exposure. As honey can be contaminated with pyrrolizidine alkaloids, a known carcinogen, it is important to calculate exposure for honey consumers separately. Table 4.5 shows habitual consumption of honey and the concomitant calculated exposure to pyrrolizidine alkaloids for all consumers and for honey consumers only as a range across the different surveys (modified after EFSA 2011d).

Mean dietary exposure to pyrrolizidine alkaloids was five-to-eight times higher for honey consumers only compared to calculations including all consumers, depending on the survey. Such exposure results for honey consumers only will make a big difference when assessing health risks. In the survey with less than 5% honey consumers, the 95th percentile exposure for all consumers did not indicate any exposure to pyrrolizidine alkaloids. Calculating the 95th percentile exposure might, thus, not provide sufficient protection for all subpopulations in all situations.

Table 4.5 Exposure to substances via consumption of honey

	Daily consumption (g/kg bodyweight		Dietary exposure (ng/kg body weight)	
	Mean	95th percentile	Mean	95th percentile
All consumers	0.002–0.034	0.000–0.261	0.02–1.22	0–9.41
Honey consumers only	0.008–0.251	0.026–0.722	0.10–9.03	0.33–26.0

Table 4.6 Exposure to lead via food consumption of conventional food vs. vegetarian diet

	Lead exposure (μg/kg body weight per day)	
	Mean	95th percentile
Conventional diet	0.49–0.97	0.83–1.53
Vegetarian diet	0.54–0.98	0.88–1.53

4.5.4 Special Considerations for Subpopulations

What people eat is formed and constrained by circumstances that are essentially social and cultural. Cultural influences can lead to differences in the habitual consumption of certain foods and in the traditions of preparation, and in certain cases can lead to restrictions such as vegetarians excluding all meat and milk from the diet, or strict vegans avoiding any animal-based products. There are about 375 million vegetarians in the world with the highest proportion of close to 30% found in India, with variations of about 2–10% in other countries (EXPO Net 2015).

It might be important to consider vegetarian exposure separately for compounds found particularly in plant foods because of their substantially different food habits. However, it might be difficult to find representative food consumption data for vegetarians as their diets differ greatly with respect to the consumption of many other food groups (Orlich et al. 2014). Some information about vegetarian consumption in broad food groups is available for a limited number of participants (Haddad and Tanzman 2003; Juan et al. 2015).

For the lead exposure assessment, EFSA calculated vegetarian exposure using the EFSA Concise European Food Consumption Database. Although the method was imprecise, the estimated exposure for 65 putative vegetarians in five surveys was compared with exposure for 8285 people with a conventional diet after weighting their contribution based on their proportional representation within each survey (EFSA 2010b; Table 4.6).

The difference in exposure between the diets varied by between 3% and 9%, with exposure via the vegetarian diet being slightly higher in most cases.

4.5.5 Local Consumption

The tendency of consumers to repeatedly purchase and consume the same food products, sometimes termed brand loyalty if associated with a specific manufacturer, may need to be considered and a range of concentrations may need to be used to generate dietary exposure estimates to cover various scenarios of consumer behaviour. Thus, if a specific brand of processed food contains a high level of a substance, regular consumers of that brand would have higher dietary exposure to the substance than those consuming the same amounts of brands with lower levels of the substance.

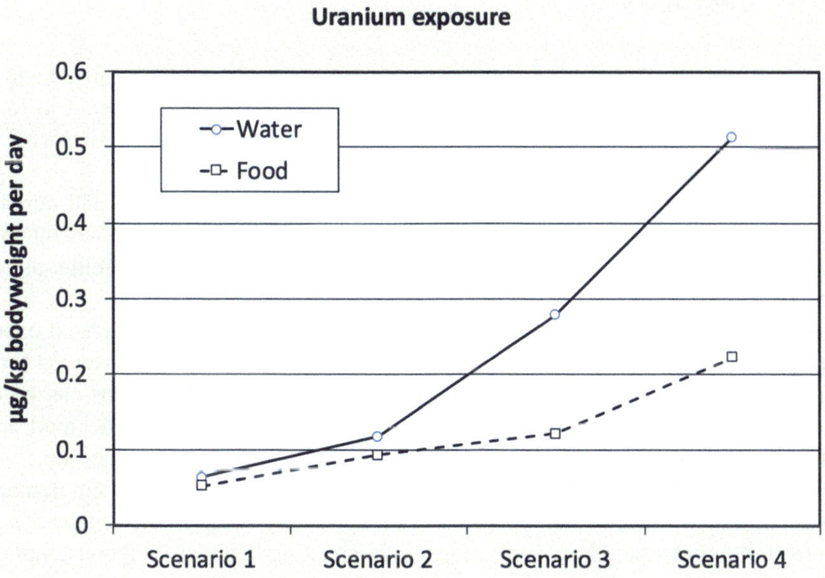

Fig. 4.5 Uranium exposure estimated for 4 scenarios

The impact of regular consumption of a certain locally produced food may also be important in the case of local contamination or residues of pesticides or veterinary drugs. Equally, consumer behaviour in relation to the selection of organic versus non-organic foods or the use of fortified and non-fortified foods may also need to be considered.

The use of water contaminated by uranium is a case in point as it is often very localised (Fig. 4.5). To get an impression of potential severe exposure to localised uranium contamination of water, EFSA constructed four different scenarios by multiplying the possible permutations of mean and 95th percentile uranium concentration and consumption (modified from EFSA 2009b).

The first and second scenarios included using mean consumption by mean occurrence values, and 95th percentile consumption by mean occurrence values, with the resulting exposure representing the usual general and high consumption situations, respectively. However, a third scenario was calculated using mean consumption by 95th percentile occurrence values to represent a local high contamination situation. An additional fourth scenario included high local contamination situations together with high consumption. It is clearly seen that water is more highly contaminated by uranium than food by the deviating lines across the four scenarios. Such scenario building could be used in the absence of matching consumption and concentration data for brand loyalty or local contamination.

4.5.6 *Conclusions*

It is often important when calculating dietary exposure to give special considerations to vulnerable and sensitive subpopulations with considerable discrepancies in food choice and consumption patterns. These factors could increase the sensitivity to possibly toxic compounds.

Subpopulations at particular risk include the very young, pregnant and nursing women, the elderly, and the immunocompromised, but might also encompass individuals with deviating food consumption habits due to social, religious, or ethical beliefs.

Consumption data at the individual level allow calculation of age-related exposure patterns. For certain chemical substances found only in rarely consumed food items, calculating exposure for only the consumers of those food items can be of interest. This is simply done by identifying consumers of the specific food and calculating exposure by using only consumption information from this group.

Calculating 95th percentile total population exposure might be sufficient in many cases to also cover minor subpopulations as long as they comprise more than 5% of the overall population. However, groups with deviating habitual food consumption considered to be "at risk" may not be captured in national food consumption surveys. Special dietary surveys might be necessary to capture food habits by such groups or exposures can be calculated by simulating expected consumption.

A tendency to be brand loyal by consistently selecting the same food product(s) from the same manufacturer(s) may need to be considered by generating dietary exposure estimates covering various expected scenarios of consumer behaviour. Similarly, the impact of regular consumption of locally produced food from "hot spot" areas can be addressed by simulating special exposure scenarios. In a hot spot scenario, the contribution to the overall human exposure from food intake may be affected by a greater variability of substance concentrations, related to the different pollution sources (air, water, soil, sediments, dust) and their direct and indirect impact on the biota and humans.

4.6 Activity Pattern Data: Assessing Food Consumption

Andrea Straßburg

The assessment of food consumption is a challenging task against the background of a complex and variable dietary activity pattern. The most appropriate assessment method for the specific purpose should be chosen because every method entails special strengths and limitations. Dietary assessment methods can be differentiated from methods that retrospectively assess the past food consumption (e.g. dietary recalls, diet history, food frequency questionnaire) and those that prospectively assess the present food consumption (e.g. dietary food records). Another classification comprises the current (dietary recalls) and usual food consumption (diet history, food frequency questionnaire) (Straßburg 2010). The sampling procedure and sample size of food consumption surveys are important to enable statistical sound data to represent the gender and age classes of the general population. Depending on the aims of the particular food consumption survey adaptations can be made, e.g. the level of detail of assessed foods or focus on special food groups.

4.6.1 Dietary Recall

The study participants in dietary recalls are asked about their food and beverage consumption during the previous 24 or 48 h. The interviews are conducted predominantly by trained interviewers who ask the participants personally in a study centre or by telephone. The first step following a multiple-pass approach is usually a so-called quick list, whereby the consumed foods are listed in a chronological order (Fig. 4.6). The background for this approach is that interviewees have fewer difficulties to recall all that they have eaten if they do not have to describe and quantify the consumed foods at once (Crispim et al. 2014). A detailed specification of the previously listed foods is carried out in the next steps.

For telephone interviews, the participants should get no prior notification concerning the day they will be interviewed to avoid changes in the usual diet. The general appointment times with regard to a good availability may increase the response. Food consumption differs between weekdays and weekends (Crispim et al. 2012; Rothausen et al. 2012; de Henauw et al. 2002; Tokudome et al. 2002; Mensink et al. 2001). Therefore, both weekdays and weekends should be considered in dietary recalls. The probability to assess rarely consumed foods rises with the number of the

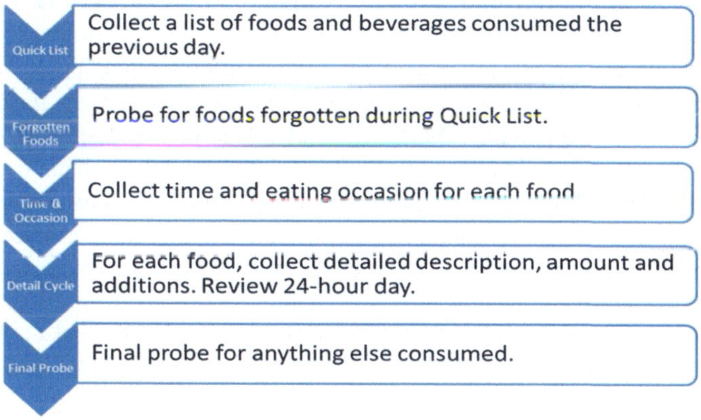

Fig. 4.6 Five-Step Multiple-Pass Approach of the Automated Multiple-Pass-Method (AMPM) used in What We Eat in America, the dietary interview component of the National Health and Nutrition Examination Survey (USDA-ARS 2016)

assessed days. Data collection on non-consecutive days is recommended because intra-individual variability can be estimated better than for consecutive days (EFSA 2009e). For calculation of energy and nutrient intake, the assessed foods must be linked with a nutrient database. If there is no prior linkage within the software program, the linkage has to be done after data collection and that might be very time consuming. However, the utilisation of the latest version of the nutrient database is possible. In any case, the classification of foods for this purpose could mean a great effort and needs expert knowledge.

Computerized Tools and Programs

There are different software programs that can be used for a standardised and efficient procedure (Conway et al. 2003; Slimani et al. 1999). Web-based self-administered versions can also be applied (Subar et al. 2012; Liu et al. 2011; Touvier et al. 2011; Arab et al. 2010). For national nutrition surveys in different European countries, the software program GloboDiet, previously known as EPIC-Soft, has been used. GloboDiet was developed for the European Prospective Investigation into Cancer and Nutrition (EPIC-Study) by the International Agency for Research on Cancer (IARC) in Lyon. The structure of GloboDiet is described in detail by Slimani et al. (1999, 2000). The program provides several quality control steps before, during, and after the interview (Crispim et al. 2014).

Dietary recalls were applied within the nutrition surveys of many European countries (Merten et al. 2011) and the US National Health and Nutrition Examination Survey (Dwyer et al. 2003).

- Example from national population-based surveys

24-h Recalls in the German National Nutrition Survey (NVS) II
The aim of the NVS II was to evaluate food consumption and other aspects of nutritional behaviour of a representative sample of the German population. In NVS II (2005–2006), the participants completed two 24-h recalls on non-consecutive days (MRI 2008). Trained interviewers in a specialized call centre asked the participants in a telephone interview about their food and beverage consumption of the previous day using the software program GloboDiet. In NVS II, the quantification of consumed foods was carried out with household measurements (e.g. cups, glasses, spoons, plates, bowls), direct recording of the weight in grams, standard portions, or units (e.g. one apple), and a picture booklet (Fig. 4.7), an excerpt of the original GloboDiet picture book (Slimani et al. 1999) modified for the NVS II. The assessment days covered weekdays and weekends.

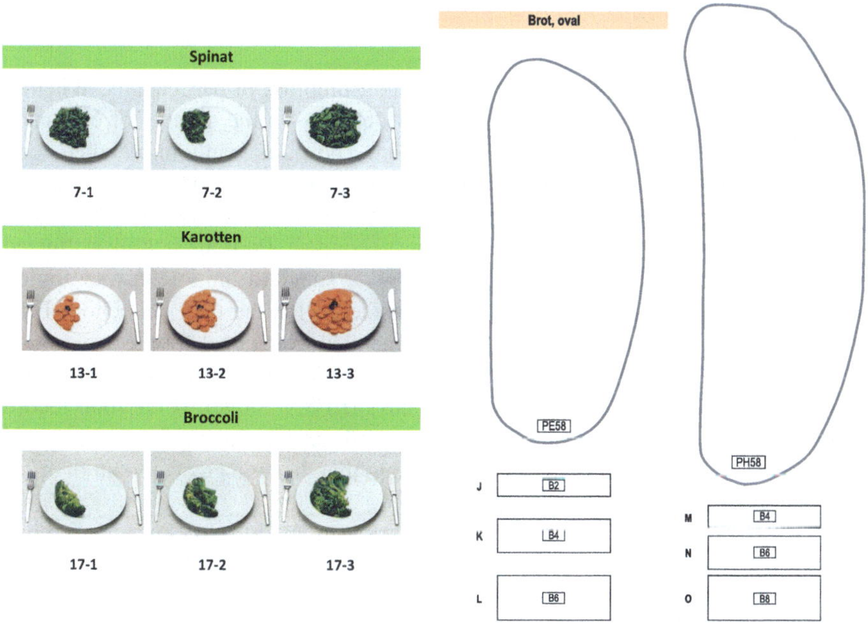

Fig. 4.7 Examples from the picture booklet (Reproduced with permission from EPIC-SOFT [renamed GLOBODIET in 2014] picture book for the estimation of food portion sizes. Van Kappel AL, Amoyel J, Slimani N, Vozar B, Riboli E. Lyon: International Agency for Research on Cancer. Copyright IARC 1995) used in the German National Nutrition Survey (NVS) II (on the left: different portion sizes of vegetables, on the right: shapes for estimating portion sizes of bread)

Strengths

An important advantage of dietary recalls is the low burden on participants. Participants have to spend only about 30 min on average for a 24-h recall. Dietary recalls provide very detailed information about, for example, the way of preparation, species or fat content, and are still suitable for large sample sizes. The food consumption of the participants is not affected; however, socially desirable answers can occur. In the case of a telephone or personal interview, the literacy of participants is not required.

Weaknesses

The accuracy of dietary recalls depends on the short term memory of the respondents. Further, one day does not represent usual intake and thus rarely consumed foods may not be adequately considered. However, statistical methods can be used in order to estimate the usual distribution of food consumption on the basis of at least two dietary recalls. In the case of a telephone or personal interview, the interviewer's influence may be possible. This does not occur for web-based interviews, but these may be more difficult to perform for participants and computer access is then a precondition. It should also be noted that telephone connections for all participants might not be available in all countries.

4.6.2 *Dietary Food Records*

Dietary food records assess the current food consumption of participants. Whether they assess the current or usual consumption depends on the number of evaluated days. Two kinds of food records are common. The participants in weighed food records are instructed to weigh all food consumed with a calibrated scale, and to record everything. Leftovers or waste should also be weighed and subtracted. The estimated food record is easier for participants to conduct. Portion sizes are estimated using household measures (e.g. cups, glasses) or standard units (e.g. one apple). Weekdays and weekend days should be considered for the data collection. Data collection on non-consecutive days is recommended in national food consumption surveys because intra-individual variability can be estimated better for non-consecutive days than for consecutive days (EFSA 2009e). If the food record is paper based, the calculation of the energy and nutrient intake for the assessed foods must be linked with a nutrient database and that means a great amount of effort for the scientific staff.

Computerized Tools and Programs

Mobile phones with integrated cameras or mobile phone apps may help to improve record keeping, especially for adolescents (Boushey et al. 2009). Because adolescents are supposed to have a great willingness to use technology, mobile phone-assisted methods are easy to implement into daily routines and may improve cooperation and accuracy (Boushey et al. 2016; Six et al. 2010). The inclusion of a fiducial marker is important for analysing portion sizes. Problems in capturing images may occur in snacking scenarios, e.g. on the bus or in the cinema (Six et al. 2010).

Dietary food records have been applied within the nutrition surveys of several European countries (Denmark, France, Hungary, Ireland, Italy, Spain, Sweden, and the United Kingdom) (Merten et al. 2011).

- Example from national population-based surveys

Food Records in the UK National Diet and Nutrition Survey (NDNS) (Public Health England 2016)
The aim of the NDNS is to assess diet, nutrient intake, and nutritional status in a sample representative for the UK population aged 1.5 years and older living in private households. Since 2008, the NDNS has been a rolling programme comprising between 1000 and 1500 individuals per year. Dietary food records (Fig. 4.8) are obtained on 4 consecutive days with the start date randomly allocated. An additional interview is conducted to collect background information on dietary habits, socio-demographic status, lifestyle, and physical activity (PHE, FSA 2014).

Time slot	Where? With Whom? At table? TV on?	Food/Drink description & preparation	Brand Name	Portion size or quantity eaten
		5pm to 8pm		
7.30 pm	*Kitchen/Diner* *With boyfriend* *At table* *No TV*	*Chicken in creamy mushroom and white wine sauce for 2, oven* *White rice, boiled* *Wine 13% alcohol*	*Sainsbury's, 370g (wrapper collected)* *Easy cook, Italian, Sainsbury's* *Sauvignon Blanc, New Zealand*	*½ pack* *2C* *1 small glass, 125ml*
		8pm to 10pm		
9.15 pm	*Sitting Room* *With boyfriend* *Not at table* *TV on*	*Squash, apple & blackcurrant, no added sugar,* *Crisps*	*Sainsbury's* *Pringles, sour cream and chives*	*1 average glass, 200ml* *5*
		10pm to 6am		
11.30 pm	*Bedroom* *Alone* *Not at table* *TV on*	*Water*	*tap*	*1 medium glass*

Fig. 4.8 Example of a weighed food record (contains public sector information licensed under the Open Government Licence v3.0 (PHE and FSA 2014) Available at https://www.gov.uk/government/statistics/national-diet-and-nutrition-survey-results-from-years-1-4-combined-of-the-rolling-program) used in the UK National Diet and Nutrition Survey (NDNS)

Strengths

Food records allow an accurate quantification of portion sizes if consumed quantities are really measured or weighed by participants. The method requires no memory effort and allows for a detailed description of the consumed foods.

Weaknesses

Food records require an immense workload of participants who must be highly cooperative and literate. The high burden may lead to low response rates, and to an overrepresentation of individuals who are interested in diet and health (EFSA 2009e). Furthermore, usual eating may be influenced by the recording process (Thompson and Byers 1994). The accuracy decreases with the number of reported days (Biro et al. 2002). Foods consumed away from home may be less detailed in reporting (van Staveren et al. 2012). Even if mobile phone cameras are used to capture images, this may be difficult in snacking scenarios, e.g. on the bus or in the cinema (Six et al. 2010). The workload is also enormous for scientific staff and therefore food records imply high costs, especially for the paper-pencil versions. Web-based versions lower the burden for the scientific staff but not for respondents.

4.6.3 *Diet History*

The diet history method assesses the usual food consumption during the recent weeks or months using an interview or questionnaire with detailed information about foods eaten at different times of the day, e.g. breakfast, lunch, dinner, and in between. The original method of Burke (Burke 1947) combined a 24-h recall, a food record of three days, and a food frequency questionnaire of the consumed foods of the previous month. This original method implies an extensive burden for respondents and scientific staff. Modified (Chinnock 2007; Sjöberg and Hulthen 2004; van Liere et al. 1997) and computer-assisted versions (Mensink et al. 2001) are now available.

Computerized Tools and Programs

The software program DISHES (Diet Interview Software for Health Examination Studies) was developed by the Robert Koch-Institut, Berlin, Germany, in collaboration with the software company "dato Denkwerkzeuge", Vienna, Austria. DISHES has integrated food composition and portion size databases (Mensink et al. 2001) and can be used to assess the usual food consumption of the previous 4 weeks in large-scale studies.

Diet history interviews were mainly applied in German national nutrition surveys: the German Nutrition Survey 1998 (Mensink et al. 1999), the nutrition module EsKiMo (2006) and EsKiMo II (2015–2017) of the German Health Interview and Examination Survey for Children and Adolescents (KiGGS) (Mensink et al. 2007), and the National Nutrition Survey (NVS) II from 2005–2006 (Heuer et al. 2015; Krems et al. 2006).

- Example from national population-based surveys

Diet History Interviews in the German National Nutrition Survey (NVS) II

Among other dietary assessment methods in NVS II, the food consumption of 15,371 participants was assessed with diet history interviews. Trained interviewers conducted the open-ended diet history interviews at study centres on the basis of the usual meal patterns of the participants (Fig. 4.9). Consumed quantities were quantified with the help of models of tableware (cups, glasses, spoons, plates, and bowls) and a 30-page picture booklet [an excerpt of the GloboDiet picture booklet (Slimani et al. 1999)] with different portion sizes and shapes of food items. Assessed food items were directly linked with the German nutrient database (BLS) (Max Rubner-Institut 2010). To increase data quality, a plausibility check to identify and to correct input errors was conducted (Heuer et al. 2015).

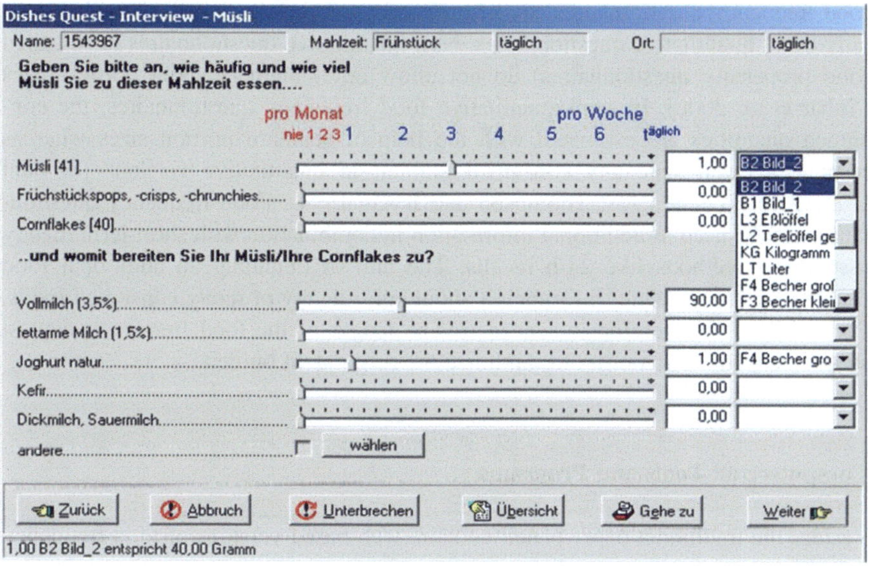

Fig. 4.9 Example from a Diet History Interview used in the German National Nutrition Survey (NVS) II (Reproduced with permission from Robert Koch Institut, Berlin)

Strengths

Diet history interviews are adequate to assess the usual dietary consumption and are not affected by the interview because food consumption of the past is determined. The literacy of respondents is not required for interviewer-administered versions. Computer-assisted versions are suitable for large-scale studies.

Weaknesses

A good long-term memory and regular dietary pattern of respondents are helpful for data assessment by diet history interviews. The estimation of portion sizes is difficult due to the long term timeframe, especially for inhomogeneous food groups and mixed dishes (Eisinger-Watzl et al. 2015). This method needs highly skilled and well-trained interviewers.

4.6.4 Food Frequency/Propensity Questionnaire

Food frequency questionnaires assess the usual frequency of consumed foods for a certain period in the past, e.g. the last month or the last year. The respondents are requested to state how often they usually consume each item of a predefined food list. The food list can be open ended in order to add their own foods that are not part of the original list. The food list should be adapted to the study population regarding eating habits, and the study aims should be taken into account. Three versions of food frequency questionnaires can be differentiated: non-quantitative, semi-quantitative, and quantitative questionnaires. Non-quantitative questionnaires (also called food propensity questionnaires) do not allow any statement about portion sizes (Subar et al. 2006). In semi-quantitative food frequency questionnaires, the consumed quantities are estimated with the help of standard portion sizes whereas respondents can state any consumed amount in quantitative versions (Slimani et al. 2015; Willett 1998; Thompson and Byers 1994). Food frequency questionnaires can be used as additional information in combination with short-term dietary assessment methods like 24-h recalls. The aim of obtaining an additional food frequency is to get better information about episodically or rarely consumed foods. Portion sizes are therefore not needed. The length of the food frequency request should be restricted in order to minimize the respondent burden.

Computerized Tools and Programs

Besides the traditional paper-pencil version, web-based versions of food frequency questionnaires are now available as well as computerized versions that can be filled-out in study centres (Garcia-Segovia et al. 2011). Computerized versions enhance data preparation but participants require appropriate computer skills.

Food frequency questionnaires were part of the US National Health and Nutrition Examination Survey (NHANES) (CDC 2016a, b; Fig. 4.10) and the German National Nutrition Monitoring (NEMONIT) (MRI 2012). Also, food frequency questionnaires have been used in the German Health Examination Survey (DEGS) (Haftenberger et al. 2010) and the European Prospective Investigation into Cancer and Nutrition (Kroke et al. 1999).

- Example from national population-based surveys

Food Requency Questionnaires in the US National Health and Nutrition Examination Survey (NHANES)
The aim of NHANES is to assess the health and nutritional status of adults and children in the USA. The NHANES program combines interviews and physical examinations of about 5000 persons each year and began in the early 1960s. Data from the nationally representative sample are released every 2 years. Food frequency questionnaires were used in NHANES I-III between 1971 and 1994. These frequency questionnaires varied regarding foods, food group specificity, the reference period, and the frequency response categories (Subar et al. 2006). In addition to 24-h recalls, a food frequency questionnaire was part of NHANES 2003–2004 and 2005–2006 (Fig. 4.10) (CDC 2017; USDA 2017).

Over the past 12 months...

73. How often did you eat **peanut butter** or **other nut butter**?

☐ NEVER

☐ 1–6 times per year	☐ 2 times per week
☐ 7–11 times per year	☐ 3–4 times per week
☐ 1 time per month	☐ 5–6 times per week
☐ 2–3 times per month	☐ 1 time per day
☐ 1 time per week	☐ 2 or more times per day

74. How often did you eat **roast beef** or **steak IN SANDWICHES**?

☐ NEVER

☐ 1–6 times per year	☐ 2 times per week
☐ 7–11 times per year	☐ 3–4 times per week
☐ 1 time per month	☐ 5–6 times per week
☐ 2–3 times per month	☐ 1 time per day
☐ 1 time per week	☐ 2 or more times per day

77. How often did you eat **other cold cuts** or **luncheon meats** (such as bologna, salami, corned beef, pastrami, or others, including low-fat)? *(Please do not include ham, turkey, or chicken cold cuts.)*

☐ NEVER (GO TO QUESTION 78)

☐ 1–6 times per year	☐ 2 times per week
☐ 7–11 times per year	☐ 3–4 times per week
☐ 1 time per month	☐ 5–6 times per week
☐ 2–3 times per month	☐ 1 time per day
☐ 1 time per week	☐ 2 or more times per day

77a. How often were the other cold cuts or luncheon meats you ate **light, low-fat**, or **fat-free**? *(Please do not include ham, turkey, or chicken cold cuts.)*

☐ Almost never or never
☐ About ¼ of the time
☐ About ½ of the time
☐ About ¾ of the time
☐ Almost always or always

Fig. 4.10 Example from a food frequency questionnaire (NCI 2017)

Strengths

Food frequency questionnaires are characterised by the low burden for the respondents and, as a result, by high response rates. The method does not affect the food consumption and is low in costs. Computer-assisted versions are suitable for large-scale studies.

Weaknesses

Food frequency questionnaires require complex cognitive tasks of respondents to estimate consumed foods (and amounts) over a long time span. As a result, it is possible to get only a few details about the consumed foods regarding the kind of foods, cooking methods, and portion sizes. Inhomogeneous dietary patterns are difficult to assess with food frequency questionnaires.

4.6.5 Which Dietary Assessment Methods Are Appropriate for National Food Consumption Surveys?

There is no best method for all purposes. The selection of the dietary assessment method depends on the objectives of the study, the foods, nutrients or hazardous chemicals of interest, the characteristics of the focussed study group (e.g. sex, age, sociocultural background), the need for data on an individual or group level, the time frame of interest, the level of detail in describing foods, and the financial and personal resources (Biro et al. 2002). For risk assessment purposes, this question is discussed extensively in Sect. 4.7. The EFCOSUM (European food consumption survey method) project concluded that both 24-h recalls and food records are appropriate to assess current and usual food consumption on an individual level by repeated short-term measurements, followed by statistical modelling in order to achieve usual intake distributions (Biro et al. 2002). A low respondent burden is important to minimize a selection bias and achieve high response rates. For adults, the 24-h recall method including two non-consecutive days is recommended by the EFCOSUM project (Biro et al. 2002) and EFSA (2009e) for national food consumption surveys. Diet history interviews and sole food frequency questionnaires are not recommended because of recall bias and insufficient detail.

Comparison of Different Dietary Assessment Methods

To achieve further insight into the specific strengths and limitations of different dietary assessment methods, the results of diet history interviews, 24-h recalls, and weighed food records were compared within the German National Nutrition Survey (NVS) II (Eisinger-Watzl et al. 2015; Straßburg et al. 2019). The greatest differences in food consumption were found between diet history interviews and 24-h recalls, and the least differences were between 24-h recalls and weighed food records. The ability to remember correctly is crucial for the accuracy of retrospective dietary assessment methods. Diet history interviews (as well as food frequency questionnaires) depend on long-term memory and require complex judgments about consumed quantities of a long time span. 24-h recalls stress only short-term memory whereas weighed food records do not require any memory effort. Besides the ability to remember as the crucial point, the estimation of quantities and frequencies especially in inhomogeneous food groups like vegetables or pastries may cause difficulties in diet history interviews. If these difficulties occur, social desirability possibly has an enhanced impact. This may explain the higher result for vegetables and the lower result for pastries in the diet history interviews in comparison to 24-h recalls (Fig. 4.11) (Eisinger-Watzl et al. 2015). Although the data on food consumption gained from weighed food records were close to 24-h recalls (Straßburg et al. 2019), 24-h recalls are more suitable for food consumption surveys due to a much lower respondent and interviewer burden in comparison to weighed food records. Thus, the results are in accordance with the recommendations for adults by the EFCOSUM project (Biro et al. 2002) and EFSA (2009e) to conduct at least two non-consecutive 24-h recalls in food consumption surveys.

Which Dietary Assessment Method Is Appropriate for Children?

Assessment of food consumption in children and adolescents is challenging because of the different abilities in different ages (Livingstone and Robson 2000). Pre-schoolers are not sufficiently aware of their food intake and are not able to conceptualize time (Livingstone et al. 2004). This ability starts approximately at the age of 7–8 years. However, children between 7 and 10 years often need help from parents or other adults for details about the consumed foods (Livingstone et al. 2004; Sobo et al. 2000). Parents can report children's food consumption at home but do not often know the children's food consumption elsewhere (Livingstone et al. 2004; Klesges et al. 1987). Because numerous children usually spend a part of the day in

Fig. 4.11 Comparison of food consumption assessed by diet history interviews and 24-h recalls within the German National Nutrition Survey (mean, 95% confidence intervals). This comparison entails food consumption data on an equal high aggregation level for both methods. A comparison on a lower aggregation level might show different results

childcare institutions like Kindergarten or school, information from child caregivers and parents are needed. EFSA recommends dietary food records, including two non-consecutive days, as being the most appropriate method for children up to 10 years because parents and/or caregivers can record foods if necessary (EFSA 2009e; Thompson and Byers 1994).

Which Aids to Quantify Portion Sizes Exist?

Weighing the consumed foods is, of course, the most accurate method to determine portion sizes but weighing each food item is a great burden for the study participants and may change their eating behaviour. Consumed quantities can be estimated with help of household measures (e.g. glasses), models of tableware (cups, spoons, plates, and bowls) or picture booklets with different portion sizes and/or shapes of food items (EFSA 2009e; Thompson and Byers 1994). In general, people tend to overestimate small portion sizes and to underestimate large portion sizes (Gersovitz et al. 1978). Portion size estimation aids should be country- and age-appropriate and validated.

4.7 Which Dietary Assessment Data Are Appropriate for Exposure Assessments?

Oliver Lindtner and Gerhard Heinemeyer

This chapter discusses what data from food consumption surveys should preferably be used for the different requirements of the assessment of chemical agent exposures. It presents the pros and cons of the data to be considered when using food consumption surveys for characterising the short- and long-term risks.

A systematic approach is needed for applying data from food consumption surveys for risk assessment. Figure 4.12 shows a scheme for how food consumption data can be used for risk assessments purposes on the basis of available survey results in Germany.

To perform an appropriate exposure assessment, the exposure scenario must link to the results of the toxicological testing. The substances of concern may occur in particular food. The survey methodology used to evaluate food consumption habits must be able to identify and characterise this food.

The type of food that is of concern (for the substance of interest in a particular food or food preparation) must be sufficiently characterised in the food list used in the food consumption survey, and for the food in which the concentration of the agent has been measured.

To characterise data needed for exposure assessment in the case of estimating the food intake, the point of departure is to differentiate the exposures due to the

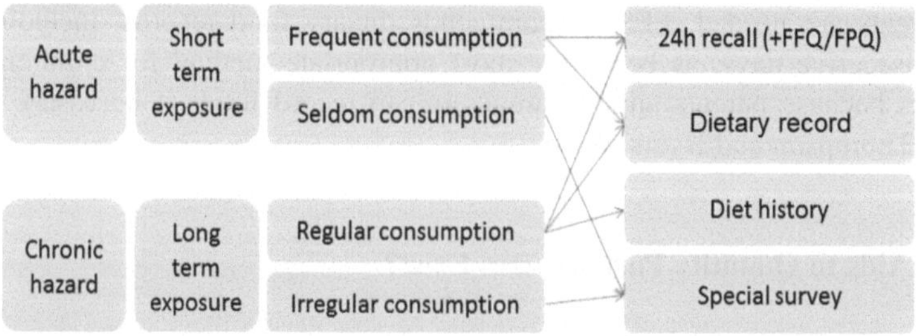

Fig. 4.12 Data needs for dietary exposure according to hazard

hazardous characteristics of the substance(s) of concern and the characteristics of the FCS to deliver adequate data.

The basic needs to derive the data for dietary exposure assessment are driven by the results of the hazard assessment. If the hazard is acute, a scenario characterising short-term exposure must be formulated. This means that the survey should report single days (24 h recalls, dietary records). If the hazard assessment reveals chronic toxicity, all exposure scenarios must be taken into consideration that lead to long-term exposures. As described in Sects. 2.6.2 and 3.5, exposures of substances having long persistence in the human body should be included, even if the exposure event is infrequent. For any exposure, the frequency may be high or low, and regular, rare, or irregular. The question arises whether or not the data from the available food consumption surveys are matching the needs for the exposure assessment. This applies not only for food exposure but also for other exposure situations.

4.7.1 Short-Term Exposures

Two issues must be distinguished when evaluating short-term exposure. These are the portion size and the incidence of the event of eating that portion.

Portion Size

The dose makes the poison. Acute toxic risks are characterised by the single dose that is involved. The portion size is therefore of central importance for short-term exposure. In practice, not only a typical portion size is of interest for short-term exposure but the knowledge of a single high dose that has been eaten might be sufficient. An anecdotal case or expert judgement might be included in the development of the estimate that characterises the portion size.

In the assessment of the risks of marine biotoxins, the CONTAM Panel of EFSA assumed a portion size of 485 g of mussels as a conservative estimate based on expert judgement (EFSA 2008b). To derive conservative estimates often only a very poor empirical basis is available. This availability problem of portion size data increases if the food is rarely eaten. Defaults relating to expert judgement or individual estimates are often used in these evaluations. Sometimes, only a handful of measurements are available having large ranges. High degrees of uncertainty become apparent. This estimate might be acceptable as a point of departure for a low level of the tiered approach. Higher tiers require appropriate statistical data that can be provided by FCS only if a sufficient number of individuals can be identified that eat that single portion to calculate statistical descriptors. The 95th percentile may be accepted as an appropriate descriptor. In special cases, the 99th percentile might be needed. The decision about what percentile is chosen must be based on the aim of the risk management. However, the choice of the percentile determines the number of individuals that need to be interviewed in the FCS (see Sect. 4.7). It has also to be questioned whether the methods used to report high percentiles of the portion size in FCS are appropriate to cover the entire variability even at the high-end tails of the population distribution. Variability might be described more appropriate using scales to weigh all foods eaten. When picture books are used, the quality of the portion size estimates is highly influenced by the number of portions per food, the scale used to report more or less than the exact portions shown in pictures and the number of foods that are not covered by pictures and similar foods needs to be used instead.

Incidence of Food Consumption

The incidence of the event "consumption of a particular food" characterises how often this event occurs in a population, for example per year. A FCS that looks at single events can answer this question. Therefore, diet history and FFQ surveys are not appropriate for short-term exposure evaluation. A sufficient size of the population is needed, and the number of persons who need to be interviewed is usually not reached in the actual food consumption surveys. Taking an adult population aged 18–80 years divided into 6 age decades, 1128 people (= 6 × 188; see Sect. 3.4.8) are needed to characterise the 95th percentile for every age group. If the incidence eating a specific food is 1%, a total of 120,000 individuals must be interviewed for an appropriate characteristic of the 95th percentile in a "consumers only"-scenario. This very rough estimate shows that sound statistical data to describe rarely consumed foods are hardly achievable in practice (e.g. by national surveys).

Dietary protocol/records and 24 h-recall studies deliver the adequate data to be used for short-term exposure assessments. Both types of food consumption interviews cover the information about the frequency of exposure and the portion size. Taking Germany as an example, exposure in children can be estimated by means of data from the VELS study (BfR 2006; Banasiak et al. 2005). For adults, exposure can be estimated from the national food consumption survey's 24 h-recall cohort and from the weighing protocol cohort.

4.7.2 Long-Term Exposures

The issues that must be differentiated when evaluating long-term exposures include the incidence and the affected population or individual, the mean and high consumption over a longer time period, the aggregation level in the food list, and the identification of seldom but regularly eaten foods in a small population.

Incidence of Exposure

Basically, the same rules as pointed out for short-term exposures are valid for long-term exposures.

Mean and High Consumption Over Long Periods of Time

The central tendency is in the focus of the assessment in long-term consumer exposure assessments. A person will eat portions of meals in varying amounts, every food with ranges of concentrations. Over long time periods, this combination of food and concentrations results in a mean intake estimate. Therefore, the mean exposure as the main estimate of long-term exposure (see Sect. 3.6.1) in the assessment of chronic risks has to take into account the day-to-day variability of food consumption and concentration. The following four main scenarios should be considered in dietary exposure assessment:

- Scenario 1: Average consumption with random changes of low and high concentration.
- Scenario 2: High consumption with random changes of low and high concentration.
- Scenario 3: Average consumption with long-term high concentration (e.g. brand loyalty).
- Scenario 4: High consumption with long-term high concentration (e.g. brand loyalty).

It can be seen that these scenarios combine eating average or high amounts of food with random changes and high concentrations of substances over long term. The relevance of the scenarios depends on the toxicity of the substance under consideration and on consumer behaviour. The mean, median, and 95th percentiles are normally available from every type of questionnaire approach used for FCS. Differences in the exposure result are only caused by the study design of the FCS (24 h recall vs. diet history vs. diary record). The particular characteristics of the food consumption questionnaires as pointed out in Sect. 4.6 and the respective statistical particularities of the data or statistical procedures to extrapolate (see Sect. 4.11) must be taken into consideration (compare Table 4.7).

Table 4.7 Some examples of substances in particular foods that need special characterisation

	Food	Remarks
Cadmium	Wheat and products made from wheat flour	High cadmium concentration in wheat bread as compared to other cereals
	Cacao	Occurrence of high amounts of Cd depending on the origin of cocoa
Ergot-Alkaloids	Rye and products made from rye grain	If the grain is not sufficiently controlled for ergot during milling
Dioxin in lamb liver	Liver from sheep	High concentration of dioxins in liver from sheep
Lead in game	Particular parts of meat	Applies for hunters as a special population eating other type of game meat as available in supermarkets or discounters and eating more frequently game than general population

Aggregation Level

The degree of disaggregation at a food-group level might be a limiting factor for appropriate estimations of dietary exposure. For example, cadmium primarily occurs in wheat. To evaluate cadmium exposure from food consumption, the foods being prepared from wheat must be identified and quantitatively described. In addition, the resulting list of food items should also be integrated into the food list of the food consumption survey. The latter must, therefore, ask for consumption of, for example, the bread made from wheat. In the official report from the German food consumption survey (MRI 2008), the category "grain and grain products" contains grain like wheat or rice and grain products such as cereals and pasta and other products produced from the grain and grain products. Those kinds of highly aggregated data in this and similar reports are of limited value for exposure assessment when there are differences of substance concentration in food is known, e.g. due to varying concentrations in different cereals or change of concentrations during manufacturing of the products. Therefore, for refined assessments more detailed food categories are needed and needs to be requested from data owners or used from available databases, as EFSA European Comprehensive Food Consumption Database. These EFSA data are based on the FoodEx2 (Sect. 4.4.1) classification system and considers 57 food categories for wheat and wheat products at different hierarchy levels. These are more appropriate to allow a sufficient estimation of, for example cadmium exposure from consumption of wheat and wheat products and to differentiate wheat from other grains under the prerequisite that the national data collected at EFSA matching the requirements of the FoodEx2 classification system are available.

In case of composite foods, food classification issues have to consider every variety of food and the different ingredients in the food. People normally eat prepared food, which may consist of a combination of several and different foods. Numerous variations exist regarding the composition of meals. Often composite foods are disaggregated to the ingredient level and not or only partly reported in the food lists of the FCS.

In particular exposure assessments, the degree of the disaggregated level is often not considered in the food list systems. For example, cadmium is a natural ingredient in cacao. Cadmium is a constituent of the earth's crust, and occurs at different levels in different regions of the world. Most of the cacao production in the world originates from West Africa (Infozentrum Schokolade 2016). Therefore, this cacao is going to be included in the majority of chocolate production. However, cacao is also grown in South America, where the cadmium concentrations in the ground are higher than elsewhere. Chocolate produced with cacao from South America, therefore, has high cadmium concentrations. However, food consumption surveys only ask for chocolate, or even for cacao-containing products. Normally they do not differentiate between chocolate of different origins. Estimating the risks from cadmium based on cadmium occurrence from South America and using consumption data from systematic food consumption surveys would lead to gross overestimations. This is due to the relatively small amounts of cacao from South America in the overall chocolate production.

Similar examples can be given for ergot alkaloids in rye bread. The information needed for exposure estimations is not handled by the existing food consumption studies. The ergot concentrations in rye flour depend on the efficacy of cleaning the grain from the ergot kernels during milling (LGL Bayern 2016), which will vary across manufacturers and is impossible to capture in food consumption studies. Equally, meat from different animals or from fish is not necessarily differentiated to the required degree of disaggregation in food consumption studies. In addition, the particular kinds of meat might not be represented appropriately in the occurrence samples. Therefore, further information sources must be made accessible. Systematic approaches are often inappropriate and a case-by-case approach is needed. This also requires a faceted food categorisation system (as described in Sect. 4.4) allowing different aggregations tailored to the specific case.

The frequent consumption of lamb liver may lead to health concern, due to possible high contamination with dioxins that may lead to health concerns, in particular in women in childbearing age and in children (EFSA 2011l).

Table 4.7 lists some examples where data in a high level of detail are needed from the food being classified to prepare appropriate estimates of dietary exposure. The EFSA FoodEx2 classification system fulfils these requirements. Of course, this requires data from national surveys that also classify the foods in a comparable depth.

Identification of Food Regularly Consumed in Small Populations

It should be taken into account that at least about 600 persons are needed to describe the central tendency of the consumption of a single food group across the six decades of age groups. This accounts for about 3% of the 20,000 persons involved in the German FCS. For example, a high risk has been identified by consumption of game due to high contamination with lead from gunshot ammunition. Game is regularly almost completely consumed by hunters. Due to the number of members of the German Hunters Association (2017) this population accounts to about 300,000 people in Germany, which, however, is only lower than 0.5% of the German population. It may also be assumed that this special subpopulation might not be homogeneously distributed and will not be representatively characterised even in the representative sample of the general population. This leads to the conclusion that the systematic FCS might not be appropriate to characterise eating behaviour of these people. This scenario is similar to the product loyal consumer scenario (see Sect. 3.2). However, the population of consumers can be relatively well characterised.

This example also shows that systematic FCS cannot identify occasionally consumed food and needs an alternative approach.

Sample Size

Taking into account detailed aggregated levels of foods, a population of men and women split across six decade long age groups, together with consideration of regional and seasonal food, vegetarians, vegans as well as pregnant and nursing women and the above-mentioned examples, a very raw estimate is that at least 100,000 participants in an exposure-oriented FCS would be needed. Food consumption surveys with the purpose of being for use in exposure assessments are still not available. On the other hand, it must be questioned whether the number of 5000 participants estimated in the EFSA guidance document (2009e) is appropriate. In the future, this issue should be discussed, ideally leading to a compromise that can be accepted between the statistical needs and the practical and financial limits.

To conclude, systematic food consumption surveys are appropriate to study average exposure in normally eaten food in a general population. They can also identify high-end exposures in that population; however, questionnaire studies are needed for detailed analyses.

4.7.3 Toxicokinetic Properties of the Substance Under Evaluation

To decide whether or not an exposure is short- or long-term, the knowledge of the toxicokinetic and hazardous properties of the agent of concern is crucial. According to Sect. 3.5, the question whether or not an exposure is short or long term is determined by the kinetic properties of the agent of concern, and is due to the relationship of the frequency of application/consumption to the clearance of the substance. For example, the exposure of dioxins having year-long half-lives is always long-term exposure, independent of the frequency of consuming food that might contain these agents. On the other hand, the consumption of very short-acting substances (e.g. nitrate that is rapidly metabolised) may only be attributed as long-term exposure if the frequency of the exposure events is high. Consequently, the characterisation of exposure of a substance where the toxicokinetics is not known as long-term results in conservative estimates that must be attributed as preliminary.

For practical reasons, the better way to deal with short- and long-term exposure in the absence of sound toxicokinetics data should be related only to the frequency of external exposure.

4.7.4 Needs for Food Consumption Data That Go Beyond Systematic Surveys

Gerhard Heinemeyer

The chapters above describe the approaches for gathering food consumption data on a systematic basis in the evaluations of dietary intakes of substances. The food consumption questionnaire studies, as pointed out in Sects. 4.6.1 to 4.6.4, use a systematic approach with a prospective methodology that includes predefined food lists and categories.

Small and Particular (Sub)populations

Sometimes particular consumption data are needed to answer special questions, e.g. in cases where the population of concern is very small, or the food of concern is only rarely or infrequently eaten. This has been noted in an earlier chapter (Sect. 4.5) of this book. Table 4.8 lists some examples of situations where special approaches are needed to evaluate food consumption.

Subpopulations that are not adequately covered in a systematic FCS are, for example pregnant and/or nursing women. This also applies for vegetarians and vegans, and to populations having other special eating habits, possibly due to their traditional or ethnic background. Their food consuming behaviours may not be sufficiently covered in the systematic consumption survey.

As shown in Sect. 3.4.8, at least 94 persons are needed to characterise the mean food consumption in a FCS for one food group. In Germany, about 450,000 pregnant women (= 60% of total births) in the age of 25–35 years were registered in the year 2012 (Stat. Bundesamt 2012), which accounts for less than 1% of the total female population. To describe a mean/median and 95th percentile as a representative estimate in a systematic FCS, a total of about 9400 female persons must be interviewed in the ages of 25–35 years to yield a mean estimate of food consumption, and 18,800 for the 95th percentile. However, the total German FCS only included a total of 7500 women in all age groups. This clearly shows that the

Table 4.8 Examples for special situations of food consumption that needs particular food consumption questionnaire methodology

Substance	Food	Special condition for occurence
Acrylamid	Starch containing foods (potatoes, Coffee, bakery)	Acrylamide is formed at temperatures exceeding 160–180 degrees.
5-hydoxy-methyl-furfural	Fruit beverages	Maillard reaction[a], occurring in many foods, e.g. beverages made from dried plums
Acrolein	Animal and vegetable fats, carbohydrates, and amino acids	Thermal treatment of food
Cumarine	Cinnamon	Traditional food eaten at special occasions
Caffeine, Glucuronolactone, taurine, vitamins	Energy drinks	Food consumed ar very special occasions by a very particular population
Lead	Game meat	Hunters and their families

[a]Maillard reaction: chemical reaction between amino acids and sugar, which gives food a characteristic flavour

systematic FCS is insufficient for studying food consumption in pregnant women. An extra study is needed that includes a representative sample of the pregnant population, possibly together with a control group.

Consumption of Particular Food Containing Substances of Concern

Table 4.8 lists some examples of substances in food that need a particular methodology to evaluate the consumption of the foods containing these substances due to special situations of occurrence (seasonal, preparation, or activity).

Coumarin

For example, toxicity studies revealed an acute hazard of coumarin due to liver dysfunction. Coumarin is an active ingredient in cinnamon that is used frequently in ginger bread and other baked goods that are particularly consumed during Christmas time. To create appropriate data, the BfR performed a small questionnaire and asked a number of, about 1000, persons representatively selected for consumption of ginger bread and other spiced baked goods (Abraham et al. 2010).

Lead

Systematic food consumption surveys may fail if particular populations are in the focus. Special questionnaires are needed if these populations are suspected to be at risk. This may apply for hunters and their families consuming game meat after shooting the animals with lead-containing ammunition. The consumption of game meat may account for daily meals that might lead to high lead exposures. The German food consumption survey does not include hunters particularly because this subpopulation is too small. About 300,000 hunters are living in Germany, which is about 0.3% of the total population. In the representative sample of the food consumption survey, a maximum of 50 persons eating meat could be expected. This number, however, is insufficient for a sound estimation of lead intake due to the above-mentioned consideration of minimum numbers of survey participants.

The example shown in Fig. 4.13 demonstrates an estimate of the influence of extreme and frequent consumption on the basis of the available data on lead concentrations in game meat (EFSA 2010b). About 12 consumptions per year (= 1 game meal per month) will lead to a range of about 0.3 (average case) to 1.4 (conservative case) and 2 μg/kg of BW per day (very conservative case). For frequent consumptions ($\geq$1 game meal per week) the respective intake estimate accounts for 1.3–6 to ~35 μg/kg of BW per day. It should be mentioned that in the

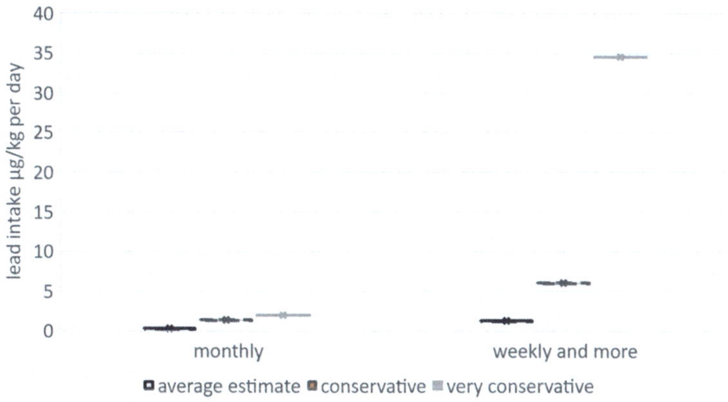

Fig. 4.13 Simulation of extreme lead input estimates from consumption of game meat

general population, the frequency of eating game meat is in a range of 0–5 meals per year.

Energy Drinks

The BfR and others have addressed risks related to the extensive consumption of energy drinks for several years. Active substances are caffeine and β-glucuronolactone. Detrimental health effects have been reported, e.g. dehydration, anxiety, insomnia, and tachycardia after consumption of excessive amounts (Nordqvist 2013). The effects can be aggravated in connection with alcohol.

When energy drinks are mixed with alcohol, the potential dangers for health of adolescents are much greater. In particular, young people are consuming energy drinks at parties, at rock festivals, or during Internet parties to stay awake. In this context, it is important that the evaluation of consumption behaviour cannot be performed by means of the regular approach of defining a representative population sample and using a systematic questionnaire. First of all, the typical consumption situations must be identified, and interviews should be focused on these events. This was done by BfR in a study to characterise the use of energy drinks in more detail (Lindtner et al. 2008) by identifying these special situations and the amounts of energy drinks that were consumed. The aim of this kind of study is to identify the amounts of consumed energy drinks during, for example a weekend or one night. Consumption of energy drinks may aggravate health effects due to an increased physical activity, fasting, lack of sleep, and the simultaneous consumption of alcoholic beverages. Between the ages of 25 and 35 years, energy drink consumption is higher at LAN parties and during gaming and sports, whereas those under 25 years consumed them at disco/dancing parties and music festivals.

For example, the highest intake was in a range of 2–4 L (upper fifth percentile of consumption). The highest estimates of caffeine intake may range from 1 to 1.3 g per day in comparison to the maximum permissible content of 320 mg per L.

Substances Formed During the Heating Process

New emerging risks have been found for substances that are formed during the heating process during cooking or frying starch containing vegetables, in particular, potatoes. Substances that can be formed in this way are acrylamide, 5-hydroxy-methylfurfural, and acrolein. These substances are characteristic of special scenarios where only a small number of foods are involved, which are not in the focus of food consumption surveys. Similarly, questionnaire studies were performed representatively regarding the formation of 5-HMF and acrolein in food (Abraham et al. 2011a, b). More details are presented in Sect. 4.18.

Conclusion

The examples show that systematic food consumption surveys may be inadequate to provide data for special populations and for foods eaten by those populations. Instead, studies that are individually focused to the exposure situation of interest are needed.

4.8 Chemical Occurrence Data: Substances (Agents) in Food

Katrin Blume and Oliver Lindtner

Concentration data of substances in certain foods in addition to consumption data for those foods are essential to provide refined and reliable estimates of exposure. Equation 4.1 notes the information that is needed to calculate dietary exposure. The main components are the frequency and amount of the food consumed (short-term food consumption) and the occurrence of substances of interest in the food.

Different statistical parameters are of specific interest when using measured concentration data. Most important is the mean of all concentration levels measured for a specific food to estimate the central tendency of exposure. This is for all long-term exposure assessments because it is often very likely that, over a lifetime, the probability of getting in contact with varying concentrations is very high and the arithmetic mean is the best descriptor to represent this situation (compare Sect. 4.6). Further, it is recommended to understand the underlying distribution as much as possible, for instance to identify possible multimodal distributions that indicate a need of stratification. Parameters of interest could be the percentage of detected values, the median compared to the mean, high percentiles, and the standard deviation. There are also cases in exposure assessment where using the mean is not appropriate. This applies for the case of brand loyalty in long-term assessments and which covers all situations where consumers prefer particular products that might lead to a higher probability to have only low or only high concentrations in the food over longer time periods. The use of mean data is also not appropriate in the case for short-term assessments involving acute risks where a conservative estimate is needed. High percentiles have to be assessed with good precision in both of these cases.

Besides the measured concentration data in food or legal limits like the maximum use levels, the reported use levels can be helpful for assessing dietary exposure. For example, this is done for food additives where the reported use levels submitted by industry are applied in the re-evaluation process of EFSA.

4.8.1 Classification and Aggregation

Important for the comparability of concentration data is knowledge about the characteristics of food classification structure since it provides details about the level of aggregation and about food descriptions. For example, the latter refers to information about packaging material, brand name, preservation method, and the organic or conventional origin.

The level of disaggregation of food groups has the main influence on the validity of an exposure assessment (Sect. 4.4.3). Usually three different levels of disaggregation are used. These are "intake level or food as eaten" (e.g. apple pie), "ingredient level" considering the processing of food (e.g. boiled apples), and finally the "raw agricultural commodity (RAC) level", which is the unprocessed state of food (e.g. apple, grains, or milk). A particular level of disaggregation is needed depending on the substance to be evaluated. Data for RAC products of plant origin, including fruits and vegetables, are needed for the regulation of maximum levels for pesticides concentration (Regulation (EC) No 396/2005 (EU 2005a; Valsta et al. 2017)). For environmental contaminants such as heavy metals or dioxins, partly unprocessed and processed foods such as infant formula are regulated (Commission regulation (EC) No 1881/2006 (EU 2006c)). In the German LExUKon-project, all consumption events reported in the German National Nutrition Survey II were recalculated to match the food categories of the above-mentioned legislation for environmental contaminants (Blume et al. 2010; Schwarz et al. 2014). Data on the intake level are important for substances that occur during processing of foods such as acrylamide, furan, or food additives (Sect. 4.18).

The more concentration data at a high disaggregated level of food categories are available, the better rarely eaten foods can be identified and interpreted. However, the overall disaggregation during statistical evaluation might lead to a loss of significance in connection with the number of samples and reliability of high percentiles.

Further, the season or region may have an influence on the variability of the concentration of a substance in a food item and thus needs to be considered when planning the data collection as well as in the exposure assessment. Elegbede et al. (2017) showed that in the case of the variation of food concentrations between different seasons, the yearly averaged concentration values would result in biased exposure estimates. Therefore, if necessary, exposure assessments should also consider seasonal or regional stratifications.

EFSA established a food coding system called FoodEx in order to standardise food classification and description (EFSA 2011a). It is a hierarchical system with facets to describe specific aspects such as the cooking method or fat content (See Sect. 4.4 for more details).

Since EFSA collects data on contaminants and residues as well as analytical data on food additives, it developed a guidance document on standard sample description (SSD) to define the requirements against European Member States (EFSA 2010d, 2013e). This was adapted within the TDS-Exposure projects to be able to also deliver results for a total diet study with necessary additional descriptions to allow for correct interpretation (ETHZ 2015).

4.8.2 Strategies for Evaluating Concentrations of Substances in Food

Scientific publications can provide concentration data only as summary statistics such as mean concentrations or as percentiles. The consequence is a lack of information about variability that limits the possibility of correctly interpreting the results. Therefore, the best option is to seek the availability of individual data.

There are various ways for obtaining individual data about substances in food. Depending on the purpose of data collection, the results may refer to more or less targeted or representative samples. According to EFSA, the following sampling strategies can be differentiated: objective, selective, suspect, or convenient sampling (EFSA 2013a). Objective sampling defines representative data for a specific population or nation where the food samples to be analysed are randomly selected. The strategy of selective sampling refers to a random sample from subpopulations that might be partly determined on a risk basis and can be higher represented by oversampling. If individual foods are selected in order to prove a suspicion of non-conformity, it is called suspect sampling. Finally, convenient sampling provides data for foods that are easily available without random selection.

For long-term exposure assessment, the intake over expanded time periods and therefore the central tendency (mean or median) concentration is needed. This is correlated with representative data for a defined market. In the case of short-term exposure assessment, high percentiles are of interest that can be also provided by targeted sampling.

While the majority of food concentration data are gathered according to the objective of surveillance programs, this is only the second choice to be used in exposure assessment because of the lack of representativeness. The first choice is to receive representative data from food monitoring programs or total diet studies. Both concepts will be introduced in the following chapters by examples using German data.

4.8.3 Food Monitoring Data

German food monitoring (BVL, 1999–2017) is part of the official food control. The official food control is organised by BVL in Germany. It consists of different concepts such as the Federal control plan (FCP) as a risk-based control programme and the National Monitoring as a systematic measuring programme. Besides the monitoring as a comprehensive long-term scheduled measurement program, project monitoring enables quick estimates relevant to questions that arise. Further, the National Residue Control Plan (NRCP) focuses on the detection of the misuse of veterinary drugs and collects data for animal foods under suspicion (see also Sect. 4.3).

According to this program, which is coordinated by the BVL, data from all federal states in Germany are collected and made available for exposure assessment since 1995.

In the scope of the German food monitoring program, individual foods are randomly and prospectively selected based on a representative food basket. These data are systematically analysed for certain substances such as pesticides and contaminants. The collection of data is based on the idea that, to be representative, the 98th percentiles of all foods available in the German market shall be a part of the sample (Sieke et al. 2008), representing at least 90% of the food eaten. Originally, the national monitoring was implemented to check foods due to the exceedance of maximum limits for pesticides. This was in order to ensure food safety and the analysed food was selected based on the First German National Food Survey (Schroeter et al. 1999; Adolf et al. 1995). Based on the VELS study (Banasiak et al. 2005) for the German pesticide monitoring, the food selection was revised by Sieke et al. (2008) to account for more than 90% of the whole food consumption. Further, foods mainly consumed by adult population such as alcoholic beverages have been added. Depending on how often maximum limits and toxicological levels are exceeded, the foods should be analysed every 3 or 6 years. Rarely consumed foods are per se not included in the food list. However, if foods have been often noted in the rapid alert system, proof of their concentrations are desired by including them in the monitoring program. Considering the assumption that pesticide residues can be partly extrapolated between similar foods, there should be analysed at least 100 foods with 94 or 188 samples resulting in about 3635 samples per year (Sieke

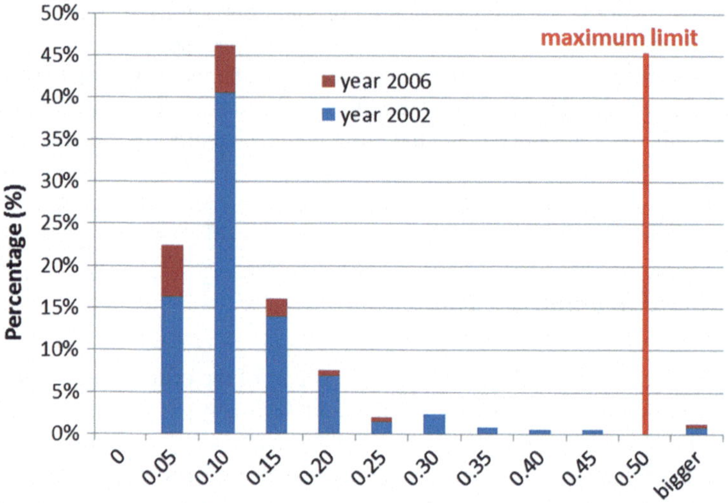

Fig. 4.14 Distribution of estimates for lead in beef kidney and exceedance of maximum levels in German monitoring data (© Bundesinstitut für Risikobewertung (BfR) 2018. All Rights Reserved)

et al. 2008). The concept was first derived for pesticide residues and was later adapted for other substances, especially environmental contaminants.

Figure 4.14 gives an overview regarding the distribution of estimates in the German monitoring for lead in beef kidney, and the percentage of results above the maximum limit of 0.5 mg/kg according to regulation (EG) 1881/2006 (EU 2006c). Beef kidney has been analysed in the scope of the German monitoring for lead content in the years 2002 and 2006. Three hundred and forty-two samples have been measured, resulting in a mean concentration of 0.10 mg/kg, which is one-fifth of the maximum limit. Exceedance of this limit can be observed for 1.2% of the results and with 94% of the samples being below half of the maximum limit for beef kidney. Thus, although single values might be above the maximum limit, the probability of being in contact with such food items for a longer time frame is quite low.

4.8.4 Total Diet Study Data

Total diet studies (TDS) are performed in a few countries in Europe. These include The Czech Republic, France, and UK. Also, TDS are performed outside Europe such as in Canada, New Zealand, and the USA (Moy and Vannoort 2013).

Fig. 4.15 Characteristics and steps of Total Diet Study (© Bundesinstitut für Risikobewertung (BfR) 2018. All Rights Reserved)

In contrast to monitoring data that refer mainly to raw foods, total diet studies give the opportunity to gain concentration data for prepared foods. This study design is based on three criteria (Fig. 4.15).

The first one regards the representativeness relative to the food consumption behaviour of the whole population. Therefore, the analyses in a total diet study cover at least 90% of the foods consumed in one country. The second is that substances are measured in foods as they are eaten. Last, but not least, the estimates are obtained for pooled samples instead of for single food items (EFSA/FAO/WHO 2011). This means that similar foods are pooled together to reduce the amount of samples to be analysed, which enables a very cost-effective data generation. However, mean values are the result and the variability in concentration gets lost. Consequently, the data from total diet studies can only be used for long-term exposure assessment. However, the concept of total diet studies is combined with higher analytical sensitivity of the laboratories meaning LODs and LOQs are set as low as possible and are not limited by evaluation of maximum values.

In addition to the refined TDS with pooling of different kinds of a very specific food in one sample (e.g. apples), there exists the possibility of using TDS as a screening tool. Therefore, fewer samples with foods of more aggregated groups such as fruits are pooled together for analysis (EFSA/FAO/WHO 2011).

In Germany, two main areas were identified where the TDS results are expected to fill important knowledge gaps. The first is to lower uncertainty in the assessment of environmental contaminants such as cadmium and lead that results in exposure levels around the health-based guidance values as currently assessed using food monitoring data (Schneider et al. 2014; Schwarz et al. 2014). The second relates to substances that are not at all covered in food monitoring programmes such as processing contaminants, food additives, and substances migrating from food packaging material.

4.8.5 Uncertainties

There is often the challenge to consider values below limit of detection (LOD) or limit of quantification (LOQ) when using monitoring data for exposure assessment. The data BVL received from the laboratories of each federal state show different LOD or LOQ for the same foods or substances according to the analytical sensitivity of the lab. Further, the LOD and LOQ are mainly set according to the maximum limits that rarely exhaust analytical sensitivity. This can lead to results mainly influenced by values below LOD and LOQ that will cause uncertainties in exposure assessment.

In order to deal with those values below LOD or LOQ, approaches such as the lower bound, medium bound, and upper bound can be applied, as explained in Sect. 4.9. Finally, the LOD or LOQ is rarely an issue when the interest is on the high-end percentages.

In the scope of the LExUKon project, the exposure of the German adult population to dioxin and dioxin-like PCB's due to food consumption has been estimated (Blume et al. 2010). To reduce uncertainties in the exposure results, the range between the lower and upper bounds for the whole population as well as for different subpopulations has been displayed (Fig. 4.16). The lower bound results are below the tolerable weekly intake (TWI) for PCDD/F-PCB of 14 WHO-TEQ/kg of BW, although adult men and adolescents from 14 to 18 years are very close to the TWI. In contrast to this, the upper bound approach leads to results above the TWI independently of the population group.

However, monitoring data are generated on the levels in individual foods and can give information about the whole distribution of the concentration of a substance. Therefore, monitoring data can be used for short-term as well as for long-term exposure assessment.

In TDS, the uncertainty may be related to inadequate planning of sampling of the foods to be pooled and analysed. This might be a limiting factor in particular for determining the aggregation level of the foods, which may represent an important source of uncertainties. From this reason, the TDS concept depends on the quality of the respective food consumption survey. It should be kept in mind that collecting a representative sample should follow the general rules as pointed out in Sect. 4.7. In the scope of a TDS foods are prepared as eaten. Exposure can be over- or

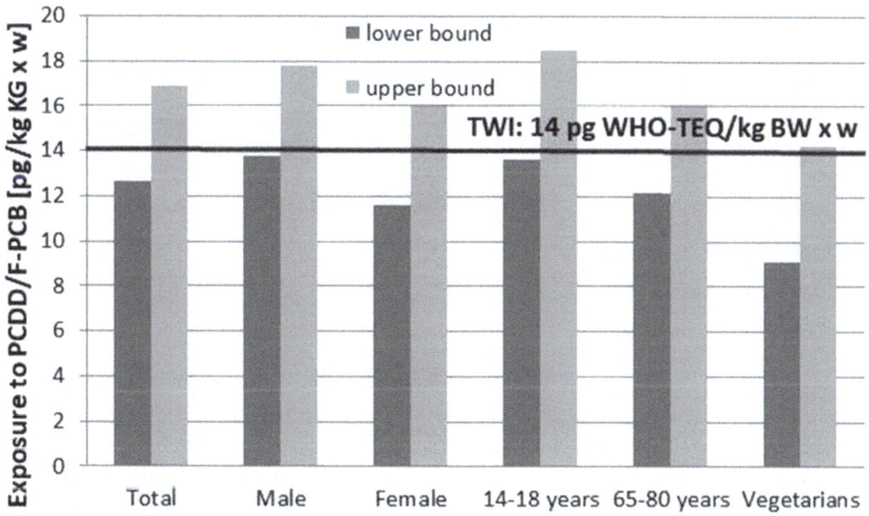

Fig. 4.16 Exposure to PCDD/F-PCB in German population depending on the lower or upper bound approach (Blume et al. 2010) (© Bundesinstitut für Risikobewertung (BfR) 2018. All Rights Reserved)

underestimated when using standard recipes instead of individual recipes. Total diet studies exist where only a small number of samples are collected. It should also be considered that TDS that are not based on similar sampling strategies are hardly comparable. Approaches that are TDS-like but not representative for the whole diet might underestimate exposure due to ignorance about certain foods. The results of inadequate sampling may therefore a source of uncertainty. Unknown hot spot samples may also result in overestimations of the central tendency exposure estimates. It should also be considered that the TDS has only limited value in estimations regarding to high-end exposure caused for instance by brand loyality. Further, a very high number of samples being pooled may bear the risk of "chemical dilution" even under the detection limit. Finally, it should also be mentioned that uncertainties from analytical reasons will be lowered by improved analyses due to improving the sensitivity of the measurement.

4.8.6 *Food Monitoring or Total Diet Study?*

Neither TDS nor food monitoring exclusively can answer all questions in exposure assessment. The decision tree shown in Fig. 4.17 developed in the scope of the "TDS_Exposure" project can help in choosing the best suitable data source if both are available. In the scope of work package 7 ("variation and trends") in this project, a literature research was performed to identify the objectives when collecting food monitoring or TDS data (BfR 2016b). Regarding "TDS", fewer publications

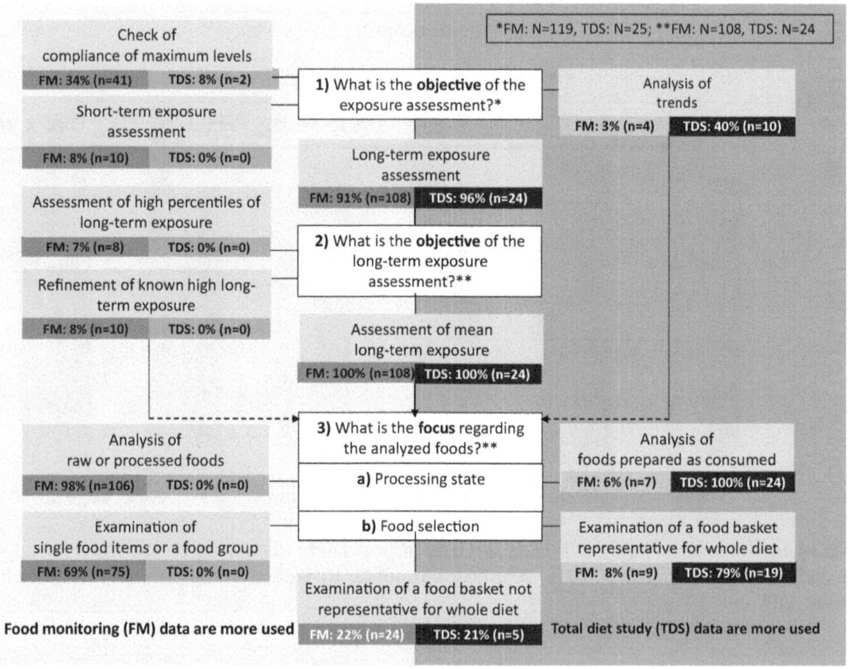

Fig. 4.17 Decision tree for use of food monitoring and TDS data in exposure assessment - literature review in work package 7 of TDS_Exposure project (BfR 2016a) (© Bundesinstitut für Risikobewertung (BfR) 2018. All Rights Reserved)

($n = 25$) could be taken into account than for "food monitoring and other data" ($n = 119$). A decision about the use of food monitoring or TDS data is in a first step, and is dependent on whether the data is used for short- or long-term exposure assessment or for trend analysis. Both approaches can be applied if the central tendency of exposure should be the result for estimating long-term exposure. However, TDS is a more cost-effective method and can cover the whole diet in a shorter period. Further, uncertainties related to the use of concentration data in raw foods instead of processed foods can be avoided. Food monitoring is the method of choice if high percentiles of exposure should be used since individual data points are generated instead of the pooled samples in a TDS approach. Another case is if the high exposure has already been identified by population screening assessments and the mean contribution of single food items should be assessed.

Besides long-term exposure it might be necessary to assess short-term exposure or to observe concentration data according to exceedance of maximum levels. Therefore, food monitoring is the best approach since individual samples are measured.

Finally, Fig. 4.17 shows that TDS is mostly used for trend analyses. The reason is that data about the whole diet available within a short timeframe are of interest. Food monitoring data are collected within a longer timeframe but can be used for trend analyses if only single food groups are desired.

Regarding the German situation, Lindtner et al. (2013) discussed the advantages and disadvantages of the German monitoring program versus the TDS approach. TDS gives the opportunity to measure the whole range of substances including environmental and processing contaminants. However, it cannot be neglected that financial resources often limit the size of a TDS. This is also the reason that TDS can mostly consider only standard recipes. The food monitoring can only be applied for contaminants that are not influenced by heat, and can be further refined by processing factors for mixing, dilution, and drying. Nevertheless there is the theoretically but rarely used possibility that individual recipes can be used to calculate data for composite foods from the RAC level of the individual foods.

Finally, the approaches of food monitoring and TDS described above can be applied depending on the objectives and requirements regarding the specific purpose of exposure assessment. It should be added that special questions can need special studies to measure concentrations of substances in food, for example for particular populations, substances or regional, and seasonal issues.

After finalisation of the first German TDS, it will be possible to compare its results with those from the German food monitoring. So far there are results from the pilot TDS in Germany obtained in the scope of TDS Exposure project (NIPH/SZU 2015). Among other food substance combinations, lead (Pb) in mussels, mercury (Hg) in tuna (canned), and manganese (Mn) in oilseeds have been analysed in the pilot TDS between 2013 and 2015. Therefore, 12 single food items of each of these foods have been pooled together before the content of Pb, Hg, or Mn has been measured. In that way, the mean concentrations have been received that are displayed in Table 4.9. The same table gives results from food monitoring that have been obtained earlier (2004 for Pb; 1999 for Hg; 2004–2013 for Mn) from the German food monitoring program.

It can be seen that number of samples collected in the frame of the German food monitoring is much higher than in the German pilot TDS (Table 4.9). In the latter, a mean concentration has been generated out of two or four measurements for a pooled sample. The food monitoring provides individual data and gives additional information about variability, for example via the percentiles or maximum values. The examples displayed in Table 4.9 partly show considerable differences in mean concentrations that might be caused by different years of analysis. Variations in LOD or LOQ can be neglected in this case since number of results below LOD/LOQ is very low in this food monitoring data.

Regarding manganese in oilseeds, it has to be considered that the pooled sample in the pilot TDS includes only linseed and pumpkin seed. These are the most consumed oilseeds in Germany. This might be an explanation for lower mean concentration compared to the monitoring data that also consider poppy seed, pine nut, sesame, rapeseed, and sunflower seed.

Table 4.9 Comparison of concentration data in the pilot TDS (2013–2015) (NIPH/SZU 2015) and food monitoring data (BVL 1999–2017) in Germany

		German pilot TDS		German food monitoring			
Substance	Food	N	Mean (mg/kg)	N	N < LOD/ LOQ	Mean (mg/kg)	P95 (mg/kg)
Lead (Pb)	Mussels	12	0.13	201	2 (1%)	0.32	0.88
Mercury (Hg)	Tuna(canned)	12	0.11	262	4 (2%)	0.15	0.44
Manganese (Mn)	Oilseeds	12	26.9	352	0 (0%)	45.4	101.2

4.9 Censoring of Food Concentration Data

Stefan Fabiansson

Occurrence data is assembled from the results of chemical analyses of the substance of interest in a number of foods expected to contain the substance. It is important to have basic knowledge about the limitations of analytical chemistry to handle the occurrence data appropriately. In particular, it is essential to know how to handle sample results where substance concentrations are below the sensitivity of the analytical method but not necessarily zero, i.e. what is called left-censored data.

4.9.1 Testing Food and Environmental Samples for Residues and Contaminants

Analytical chemistry is the discipline that deals with separating chemical substances from a range of different substrates and determining their identity and concentration using a variety of chemical and physical methods. Analytical methodologies have evolved dramatically over the past 100 years, from basic assays based on solution chemistry to those relying on complex instrumental platforms. Advancements in the analytical methods have incorporated bioassays, fluorescence spectroscopy, gas chromatography (GC), GC coupled with mass spectrometry (GC-MS), liquid chromatography (LC), and ultimately LC-MS-MS with progressively increasing specificity and sensitivity.

As an example, detection limits for techniques used to analyse lead in food samples have evolved from the simplicity of X-ray fluorescence spectroscopy at 10 mg/kg, which only requires homogenisation of the samples, to flame atomic absorption spectrometry at 0.5 μg/L and inductively coupled plasma mass spectrometry at 0.05 μg/L, which both require more elaborate high temperature acid digestion of the sample (EFSA CONTAM Panel 2010c).

Each analytical method has its particular specificity and sensitivity. While previously, the concentrations of a substance could be detected at milligram or microgram levels, modern instrumentation allows detection at nanogram, picogram, or even femtogram levels. This has resulted in significantly higher sensitivity and vastly improved specificity of detection. Specificity is a qualitative measure and gives an indication of the identity of the chemical species in the sample, while sensitivity will determine the possible quantification of the chemical species expressed as the concentration of the component in the substrate (Currie 1997).

Despite substantial advances there is still a significant degree of difficulty in testing food and environmental samples for residues and contaminants as they comprise complex matrices and this will influence sensitivity. Thus, one of the most important tasks of analytical chemistry is to maximise the desired signal from the chemical substance while minimising the associated noise. However, even the most sensitive method cannot approach zero and this is a problem for exposure assessment.

Left-Censored Data

Analytical chemists use the limit of detection (LOD) and the limit of quantification (LOQ) to express the lower limit for identifying the presence of a substance and the lower limit where the substance can actually be quantified, respectively. The LOD represents the minimum concentration or mass of an analyte that can be detected with a given confidence for a given analytical procedure. Similarly, the LOQ is the minimum concentration or mass of the analyte that can be quantified with acceptable accuracy and precision (Keith et al. 1983).

Most analytical instruments produce a signal even when a blank is analysed. This signal is referred to as the noise level. Most commonly the LOD is defined as the analyte concentration that is required to produce a signal greater than three times the standard deviation of the noise level and the LOQ is the analyte concentration corresponding to the noise level plus ten standard deviations (Keith et al. 1983) as shown in the following equations:

Definition of LOD and LOQ
LOD = 3 * (standard deviation of the blank)
LOQ = 10 * (standard deviation of the blank)

This is a simple and quick method to characterise the analytical method and matrix combination applicable for most substances in food and environmental samples. It follows that LOQ/LOD = 3.3. However, there are some exceptions. The European Community heavy metal legislation deviates from the norm in that it specifies that LOQ/LOD = 2 for the instrumental results. Also for radioactive compounds, there is an anomaly in that radioactive emission can be quantified before the substance can be identified and thus LOQ < LOD.

By definition, results below the LOQ and LOD cannot be expressed quantitatively with a necessary level of credibility. These results are categorised as not detected (ND) or not quantified (NQ) or <LOD and <LOQ, while it is important that the actual LOD and LOQ concentration levels be specified for each method/matrix combination. Unless there is reason to assume that a food does not contain a chemical of interest (e.g. foods for which a pesticide is not registered for use or for which a food additive is not permitted, or foods that undergo extensive processing during which a chemical is likely to be completely removed), it should be assumed that samples without detectable or quantifiable concentrations may contain the chemical below the LOD or LOQ (WHO/FAO 2009). In statistical terminology, such results are defined as left censored. Left censoring occurs when a data point is below a certain value, but it is not known by how much. Right-censored data are also possible and would occur when high concentrations saturate the detector. However, this can be overcome by further dilution of the sample and repeat testing and should not be of practical concern for exposure assessments.

4.9.2 What to Do About Left-Censored Data?

Several techniques can be used to handle the left-censored values. However, it is important to understand that the protocol used for assigning concentration values to left-censored results will be critical to the dietary exposure estimate. There are no international guidelines on the need to report both the LOD and LOQ in a standardised manner. Thus, inconsistent reporting of LODs or LOQs may lead to differences in the numerical value assigned to left-censored results for use in dietary exposure estimates (WHO/FAO 2009).

Occurrence data most often consist of analytical results collected at different times and for different purposes by different laboratories, and measured by different analytical methods. The results in many cases would reflect the ultimate sensitivity of the relevant analytical method. However, most data delivered from surveillance programs are for market control and are not analysed with the purpose of risk assessment. For many substances, regulations prescribe that certain limit levels in food must not be exceeded before marketing. In such cases, the analytical methodology is aimed at checking for compliance with the respective legislated limit value. Hence, in practice, it might happen that the analytical sensitivity is often only set to this limit, although the methodology would allow a lower limit to be used, but at a higher cost and effort. Consequently, one of the challenges of the risk assessment process is to describe in a comprehensive way the available information and to highlight the relevant parameters for the later estimation of dietary exposure.

The risk assessor must decide what values to assign to left-censored results with a variety of statistical methods available. These include simple deletion and substitution methods, or the more statistically advanced maximum likelihood estimation

(MLE), log-probit regression and non-parametric methods. All left-censored data are simply disregarded when applying the deletion method, which is likely to result in a sizeable overestimation of levels of contamination. This approach is not recommended.

The Substitution Method

The substitution of non-reported results with imputed fixed values is a widely used method in food risk assessment (Hornung and Reed 1990; Glass and Gray 2001). It customarily uses three different imputation strategies (US-EPA 2000; WHO 1995):

Upper bound
values <LOD are set at LOD
values <LOQ are set at LOQ
Middle bound (half-way between upper bound and lower bound)
values <LOD are set to LOD/2
values <LOQ are set to LOQ/2
Lower bound
values <LOD are set to 0
values <LOQ are set to 0

Although, based on the definitions of LOD and LOQ, reporting a value as "<LOQ" (not quantifiable) implies that the value is also ">LOD" (detected). This reasoning suggests that for the imputation of values reported as less than the limit of quantification, the LOD and (LOD + LOQ)/2 for the lower and middle bound should be used, respectively. However, this is best avoided given the inconsistencies in laboratory reporting.

Some guidance has been provided for the deterministic presentation of left-censored values (Helsel 1990; WHO 1995). For substances unlikely to be present unless specifically used for producing the agricultural commodity or the food, like pesticides or additives, using a lower bound mean concentration only is usually the norm. Typically, for substances like contaminants and natural toxicants that are likely to or can possibly be present in the food, both lower and upper bounds should be calculated for the mean food concentration. However, GEMS/Food Europe has suggested that if fewer than 60% of the results are less than the LOD or LOQ, then a reasonable estimate of the mean can probably be obtained by setting all left-censored results to LOD/2 or LOQ/2, respectively (WHO 1995).

As the best practice for data sets that have a large number of samples that are less than the LOD or LOQ, it may be advisable to perform sensitivity analyses by first assigning all non-reported results to zero, setting these values to the full LOD or LOQ and then evaluating how the exposure estimates change.

Table 4.10 Results in μg/kg of concentrations of glycidyl fatty acid found in some selected foods illustrated using the substitution method (N = number of samples)

Food item	N	% LC	Mean MB (LB–UB)	Median MB (LB–UB)	P95 MB (LB–UB)
Margarine	170	6	361 (358–364)	175 (175–176)	1280
Soya oil	189	30	171 (157–186)	120 (120–140)	560
Rapeseed oil	290	49	166 (144–188)	100 (100–110)	560
Smoked fish	60	57	5.8 (4.7–6.8)	3 (0–4.8)	18

The assignment of different values to left-censored results may have a significant impact on estimated dietary exposures, the effect being greater for less sensitive analytical methods with a higher LOD or LOQ. A conservative approach using the upper bound for exposure calculations is recommended when the difference between the lower and upper bounds are large. However, with a small difference between the two estimates, use of the middle bound to calculate exposure is sufficient.

The information in Table 4.10 is an example of the presentation method adopted by the European Food Safety Authority when using the substitution method. It is an extract of the presence of glycidyl fatty acid concentrations expressed as μg/kg in some food items (EFSA CONTAM Panel 2016a).

For full transparency it might be beneficial to report the full set of calculations regardless of the results of the sensitivity analysis as the Authority has selected to do. Here the Middle Bound (MB) results are given with the variation between Lower Bound (LB) and Upper Bound (UB) indicated within brackets. The spread in values increases with the proportion of left-censored data (%LC). There is a difference between mean LB and UB for margarine being 2%, soya oil 18%, rapeseed oil 31%, and smoked fish 45%.

It has been widely recognised that the substitution method is biased, the bias being a function of true variability in the data, the percentage of censored observations, and the sample size (El-Shaarawi and Esterby 1992). Another disadvantage of the substitution method is that it does not consider the overall distribution of the concentrations of the substance found in quantified samples. In other words, whether a sample contains 10% or 90% of quantified samples, it does not drive the way left-censored results are treated. In practice, it is likely that the proportion of quantified results also influences the underlying distribution of the left-censored data (Helsel 2004).

The robustness of the substitution approach can also be limited by the computational difficulties in calculating percentiles and/or basic statistics. Examples include the mean and standard deviation, if the percentage of censored observations is substantial and in the application of basic statistical techniques, e.g. regression methods, where log-normality of the variable of interest is a desirable prerequisite.

4.9.3 Advanced Statistical Methods

There are more sophisticated statistical alternatives available to evaluate the impact of the left-censored tail of the distribution, including log-probit regression (Jahn et al. 2015; Mulhausen and Damiano 1998), maximum likelihood estimation (Finkelstein and Verma 2001), and non-parametric methods (Schmoyer et al. 1996; She 1997). Estimation of the full distribution set of occurrence data would also allow the computation of the uncertainty in the exposure.

In the log-probit regression method (also called regression on order statistics), the data are sorted and a linear relationship is assumed between the logarithm of occurrence values and the inverse cumulative normal distribution of the observations. A number of variations have been suggested, mainly to address departures from the lognormal distribution (Hewett and Ganser 2007). However, it has been suggested that this method is not appropriate for complex data sets with multiple reported left-censoring limits that are typical for food contaminate data (Hewett and Ganser 2007).

On the contrary, the consensus view is that the maximum likelihood estimation (MLE) method is a good approach for imputing left-censored data from a methodological perspective. The sample parameters are those estimates that maximise the likelihood function after the definition of a parametric distribution to best fit the data. This method is very versatile as it can be applied to data that do not necessarily follow the well-known normal distribution (Helsel 2004).

MLE uses three pieces of information:

1. Values above detection limits.
2. Proportion of data below the detection limit.
3. Assumption of the shape of the distribution of positive values.

Data below and above the detection limit are assumed to follow the same statistical distribution. The parameters of the chosen distribution are estimated to best fit the distribution of the observed values above the detection limit, compatible with the percentage of data below the limit. The estimated parameters are the ones that maximise the likelihood function. However, there is a major drawback in the use of parametric procedures in that the tail behaviour is strongly influenced by the choice of underlying distribution.

As an alternative, non-parametric procedures can be used that will allow the asymptotic behaviour of the left tail to be determined and allow the construction of asymptotic confidence intervals (Tressou 2006). Non-parametric methods do not involve computing "parameters" such as the mean or standard deviation of a given distribution. Instead they use the relative magnitude (ranks) of data. The standard non-parametric technique for censored data is the Kaplan–Meier method based on the empirical cumulative distribution function (Kaplan and Meier 1958). It was originally developed for estimating the mean of right-censored survival data and was first suggested to analyse left-censored concentration data in environmental studies (Schmoyer et al. 1996; She 1997). The advantage of such an approach is the possibility of estimating the mean, together with the median and other quantiles, in the presence of non-detect values, without relying upon distributional assumptions (Tressou et al. 2004). The weight of the censored data is distributed over the different observed values below the censoring values, i.e. LODs and LOQs, and zero. This method is available in many statistical packages, but because it was originally intended for right-censored data sets, the concentration data must be "flipped" before analysis (i.e. the left-censored data set must be converted to a right-censored data set).

Comparing Left-Censored Imputation Methods

In 2010, the European Food Safety Authority undertook a project to evaluate the effects of different methods of handling left-censored data. Table 4.11 illustrates an example of the experimental comparison of two substitution methods (left-censored results set to LOD in Substitution 1 and to LOD/2 in Substitution 2), and four curve-fitting models. It is an extract of results for the presence of cadmium expressed as mg/kg in more than 19,000 meat product samples containing 26% left-censored data (EFSA 2010c).

The outcome of the different simulations varies considerably. In this particular case, all three parametric simulations (Log-normal, Weibull, and Gamma) have to be rejected due to the results of a lack-of-fit test (not shown). This would leave only the use of the non-parametric Kaplan–Meier model as an alternative to the substitution methods.

Different comparisons showed that Kaplan–Meier performed well, both for the estimation of the mean (up to percentages of censored values of 50%) and for the 95th percentile. In other simulation scenarios, it appeared that the maximum likelihood estimation can be a very valuable solution when data can be accurately fitted with one of the many distributions implemented in any of the available statistical packages. In the settings explored in the simulation study, the gamma and particularly the Weibull appeared to be very flexible in order to accommodate a variety of data distributions.

In summary, it should be acknowledged that the analysis of the distribution of occurrence data relates to the objectives of the risk assessment and to the results of the hazard characterisation. When the resulting exposure is below the threshold for safety concern, a more refined analysis is not necessary. Conversely, in cases where the exposure distribution overlaps or exceeds the toxicological reference doses, more advanced statistical methods should be applied.

Table 4.11 Cadmium concentrations in meat product samples (mg/kg) calculated by using two substitution methods or four curve fitting models, respectively

Statistics	Substitution 1	Substitution 2	Log-normal	Weibull	Gamma	Kaplan–Meier
Mean	0.1134	0.1130	0.7040	0.1271	0.1128	0.1128
Median	0.0290	0.0289	0.0126	0.0181	0.0205	0.0286
P95	0.3700	0.3700	1.3376	0.5931	0.5428	0.3700

Conclusions

It is fully acknowledged that exposure assessment calculations are often context specific, thus likely to depend, among others, on:

- The type of chemical
- The nature of data
- The statistical knowledge at hand, and
- The availability of high-performing computational resources.

Despite its drawbacks, the substitution method for replacing left-censored data is still widely used, mainly with the justification that it is easy to implement. It is widely understood and that the upper bound practice leads to conservative estimates for exposure assessment calculations, i.e. overestimation of the mean and underestimation of the variability.

When the difference between lower and upper bound is not negligible, and the upper bound estimates are in the range of toxicological reference doses, it is recommended that exposure assessment calculations become more refined, possibly by a full probabilistic exposure assessment. However, the feasibility of adopting such a refined methodology will depend on access to advanced statistical expertise and computational resources.

4.10 Modelling of Food Exposure and Computerized Tools for Food Exposure Assessment

Polly E. Boon

Typically, short- and long-term exposure estimates are the two types of dietary exposure estimates of adverse substances in food. They are distinguished based on the toxicity profile of the substance of interest. Short-term exposure refers to exposure with a duration of 24 h or less, and long-term exposure refers to exposure during a longer period of time, often a couple of months up to several years. A tiered approach is usually used to assess these types of exposures and the accuracy of the exposure estimate will increase when going from a lower to a higher tier. However, also the resources and the input data needed will increase when going from a lower to a higher tier (Fig. 4.18), making higher tier assessments often more laborious, time

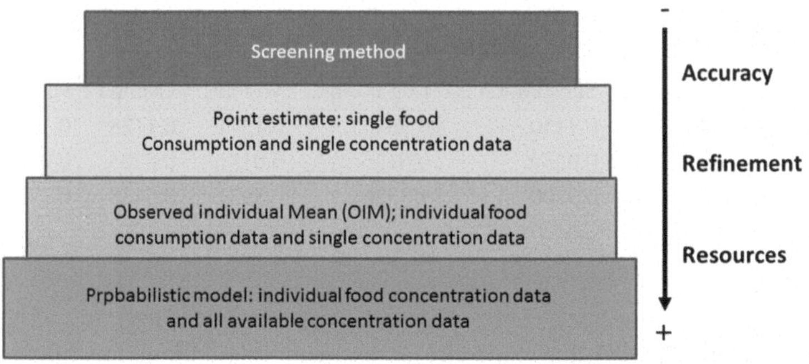

Fig. 4.18 Tiered exposure models to calculate the exposure to substances via food

consuming, and costly. In practice, higher tier assessments are therefore often only performed if exposure estimates obtained in a lower tier cannot exclude the possibility of a health risk and the input data needed for such an assessment are available. On the other hand, there are exposure situations that can only be dealt with by higher tier assessments. Examples are the assessment of the dietary exposure to multiple compounds or assessing the exposure via all relevant exposure routes (dermal, oral, and inhalatory) and from different sources (such as air, food, and consumer products): cumulative and aggregate exposure, respectively. Apart from using a higher tier assessment to increase the accuracy of an exposure estimate, an exposure estimate can also be refined within a tier via the use of more realistic input data. The more refined an exposure estimate is, the closer it will be to the true exposure.

This chapter briefly addresses the input data needed to assess the short- and long-term exposure to (an) adverse substance(s) via food(s) and a more elaborate discussion is provided in the respective sections of this book. This chapter also describes the different models that are available to assess the exposure. This description will follow the tiered approach as depicted in Fig. 4.18.

4.10.1 Input Data

Food consumption and concentration data of substances in the relevant food(s) are needed to assess the dietary exposure to adverse substances. In the lowest tier assessment (screening method), maximum physiological levels of consumption and upper concentration limits as set in legislation are used as proxies of these two input values. Food consumption data can be used to refine this, with an increasing level of accuracy derived from food balance sheets, household budget surveys, or individual food consumption surveys. An example of food consumption data based on food balance sheets is the GEMS/Food cluster diets (Sy et al. 2013). These diets are used by the Joint FAO/WHO Expert Committee on Food Additives (JECFA) to assess the exposure to contaminants (e.g. JECFA 2016, 2017). An example of individual food consumption data is the Comprehensive database (EFSA 2011a). This database contains individual food consumption data from different European countries and is used by EFSA to assess the long-term exposure to different types of substances, such as contaminants (e.g. EFSA 2015c, 2016e) and food additives (e.g. EFSA 2016d, 2017a, b). The exposure assessment can also be refined by using measured concentrations of a substance instead of upper concentration limits. Further, use levels provided by food industry can be considered to refine the exposure assessment of authorised food additives (e.g. EFSA 2017a, b; Sprong et al. 2015).

Individual food consumption data can cover the general population (e.g. the mean consumption of apples in a population) or specific subpopulations such as males, children, elderly people, and women of childbearing age. The (sub-)population of interest depends on the toxicological profile of the substance (Sects. 4.5 and 4.7). For example, different health-based guidance values (HBGVs) have been derived for the environmental contaminant lead, which are relevant for children up to age seven, pregnant women and adults (EFSA 2010b). Preferably, each of these three sub-populations should be addressed when assessing the health risk of lead exposure. In addition, possible vulnerable groups may need to be addressed separately. Examples of such groups are young children due to their higher food consumption levels per kg body weight. Another relevant subpopulation consists of those individuals that actually consume the food(s) that may contain the substance of interest, the so-called "consumers-only". This subpopulation is especially relevant for substances that are present in rarely consumed foods. A general population exposure assessment [including both consumers and non-consumers of the food(s)] would underestimate the health risk of those consuming the relevant food(s). Food consumption data based on food balance sheets refer to a general population, and do not allow the identification of subpopulations. Household budget surveys provide information about food availability at the household level; however, information about the division of the foods among the members of the household is not available. Data from household budget surveys are therefore also not suitable for assessing the exposure to subpopulations.

Depending on the substance of interest, additional input variables can be included to optimize the exposure assessment. Examples are information about the market share of relevant foods, the effects of processing (such as washing, peeling, cooking, frying, baking, and oil extraction) on the presence of the substance and unit variability. This last input variable refers to potential variability in the concentrations from commodity unit to commodity unit, when the concentrations are derived from samples containing multiple units that are homogenized before analysis. This factor is important in short-term exposure assessments and addresses the possibility of exposure to the highest possible concentrations to be expected for a single unit (FAO 2016). These additional input data can be included at each tier and can result in more refined exposure estimates within a tier. For example, an exposure estimate obtained with the point estimate approach can be refined by correcting the concentration for the effects of processing.

4.10.2 Lower Tier Models to Estimate Exposure via Food

Figure 4.18 depicts the different exposure models as part of a tiered approach. A screening method, like the budget method, is a low-tier model that aims to overestimate the exposure of high-end consumers to avoid an erroneous indication of no safety. For this, a maximum physiological level of consumption is combined with an upper concentration limit that is set in legislation (Douglass et al. 1997). The budget method has been used to assess the exposure to food additives by JECFA (FAO/WHO 2001), and within the EU (EU 2001). In the EU, this conservative method has been replaced by more realistic long-term exposure assessment models based on individual food consumption data (EFSA 2012c, 2014a).

It is a particular characteristic of low-tier models that they use point estimates. In this model, a single numerical input parameter for food consumption (often a mean or high consumption level) is combined with a single concentration value (e.g. upper concentration limit, use level or a mean or high concentration value based on measured concentrations). The difference with the screening method is that the food consumption data used are based on actual food consumption data.

The point estimate is used to assess both short- and long-term exposures. To assess the short-term exposure, a high consumption level is combined with a high concentration. For example, for residues of plant protection products (PPPs) a large portion (97.5th percentile of consumption based on individual food consumption data) of the relevant commodity is used (EFSA 2007). For the concentration, the highest residue level derived from supervised residue trials or monitoring programmes is used and corrected for unit variability and possibly, for the effect of processing (Sect. 4.10.1). The resulting short-term exposure estimates per commodity are not summed to estimate possible overall short-term exposure via all commodities that may contain the residue. It is considered unlikely that a consumer will eat two or more different commodities in quantities equal to the large portion size during a period of 24 h or less, and that those commodities will all contain the highest likely residues of the same PPP. When a short-term exposure assessment should cover all commodities, a higher tier assessment should be performed (Sect. 4.10.3).

To assess the long-term exposure with the point estimate, single mean or median input variables are often used for both consumption and concentration. In long-term assessments, consumption and concentration are assumed to level out to a mean or median level over time. When assessing long-term exposure, calculated exposures per food may be summed to estimate the total exposure. Several options are available to ensure that long-term exposure assessments are also sufficiently protective for possible high-level consumers. For example, a high consumption level (e.g. 95th percentile) of one food may be linked to the mean concentration of the chemical in the relevant food(s). For the remainder of the foods that contain the substance, the mean consumption may be linked to the mean concentration. If only mean consumption levels are available for the exposure assessment, a high concentration level of the food (or multiple foods) can be used to arrive at a high exposure level.

EFSA uses the point estimate approach to estimate short- and long-term exposure to residues of PPPs (e.g. EFSA 2015b, 2016b). The method used for this purpose is the Pesticide Residue Intake Model (PRIMO) (EFSA 2007). The Joint FAO/WHO Meeting on Pesticide Residues (JMPR) uses a similar approach to assess the short- and long-term exposure to these residues (FAO 2016). JECFA uses the point estimate approach to assess the short- and long-term exposure to food contaminants (e.g. JECFA 2016, 2017). Another example of a point estimate approach is the Food Additives Intake Model (FAIM) (EFSA 2012e). This model has been developed by EFSA to support applicants in the estimation of exposure to food additives, and has replaced the budget method that was previously used within the EU for this purpose (see above). FAIM can provide potential conservative estimates of long-term intake for average and high-level consumers.

4.10.3 Higher Tier Models to Estimate Exposure via Food

Lower tier exposure estimates are either ultraconservative (screening methods) or high-end estimations (point estimate) of the true dietary exposure to substances within a (sub-)population. When mean input variables are used in the point estimate, the resulting long-term exposure estimate provides information about the mean exposure within a (sub-)population. If the resulting exposure estimates in comparison with an HBGV show that a health risk cannot be excluded, more refined exposure estimates may be needed. Also, a higher tier assessment may be performed if there are concerns regarding the upper tail of the exposure distribution within a (sub-)population. A higher tier assessment may also be performed when the goal of an assessment is to obtain a realistic exposure estimate, i.e. as close to the true intake as possible. These more refined assessments are based on whole food consumption and/or concentration databases instead of only using single input variables. Long-term exposure estimates can be refined using both higher tier models (observed individual means (OIM) model and probabilistic models), whereas short-term exposure estimates can only be refined with probabilistic models (Fig. 4.18). Both models are described below in relation to the two types of exposure estimates. Also, some additional features of higher tier models are addressed.

Long-Term Exposure

Long-term dietary exposure can be estimated using different higher tier models: the observed individual means (OIM) model and a probabilistic model (Fig. 4.18), also referred to as statistical models in case of long-term exposure (EFSA 2012c). The challenge with these models is to estimate the long-term exposure within a (sub-) population based on food consumption data that typically cover only a limited number of days (2 up to 7) per individual. Food consumption data that cover only one day per individual are not suitable to assess the long-term exposure using higher tier models (Kroes et al. 2002).

The OIM model is a simple, distribution-free method to estimate long-term exposure, similar to the simple distributions approach as described by Kroes et al. (2002). It is known to estimate the long-term exposure relatively accurately at the mean level, but to overestimate the upper tail of the long-term exposure distribution (EFSA 2012c; Goedhart et al. 2012). The OIM model is currently used by EFSA to estimate the long-term exposure to contaminants (e.g. EFSA 2015d, 2016a) and food additives as part of the re-evaluation of authorised food additives (e.g. EFSA 2015e, 2016d, 2017a, b). In the OIM model, long-term exposure is calculated as follows:

1. All relevant foods consumed on a person-day present in the food consumption database are multiplied with the mean concentration of the relevant substance in that food.
2. Exposures per food are summed to obtain the overall exposure on that person-day and adjusted for that person's body weight.
3. To obtain the mean exposure per person, the exposure per person-day is averaged over the person-days of that person resulting in a distribution of mean daily exposures per person (OIMs). The number of mean daily exposure levels equals the number of respondents present in the food consumption database.

Note that the availability of the whole concentration database is not necessary when using the OIM model, and that the mean concentrations of relevant foods suffice. However, in this case the uncertainty in the concentration data cannot be reflected in the exposure estimate (see below). The availability of the whole food consumption database is essential when using the OIM model.

In the OIM model, the distribution of mean exposures over the person-days per person is assumed to equal the "true" long-term exposure distribution of a given (sub-)population. Given the limited number of person-days present in a typical food consumption database per person (typically 2 days) and the variation in daily food consumption patterns within a person, the distribution of mean exposures over persons obtained with the OIM model will often be too wide in comparison to distributions of "true" long-term exposures [compare the green ("true" intake) line with the blue line in Fig. 4.19] (EFSA 2012c; Goedhart et al. 2012).

To address this limitation of the OIM model, several sophisticated statistical long-term intake models have been developed. Examples are the Iowa State model for Foods (ISUF) (Nusser et al. 1996, 1997), the BetaBinomial Normal (BBN) (de Boer et al. 2009), LogisticNormal-Normal (LNN) (Goedhart et al. 2012) and the National Cancer Institute (NCI) model (Tooze et al. 2006), see also Sect. 4.11. These models remove the within-person (between days) variation from the person-day exposure distribution (the result of step 2 of the OIM model), thus leaving the interindividual (between person) variation. The intraindividual (within-person) variation is not important by definition (the long-term exposure distribution is the interindividual variation). Removal of this source of variation results therefore in a more accurate estimation of the long-term exposure. This variation is removed in the following way. The exposure frequencies and exposure amounts are modelled separately,

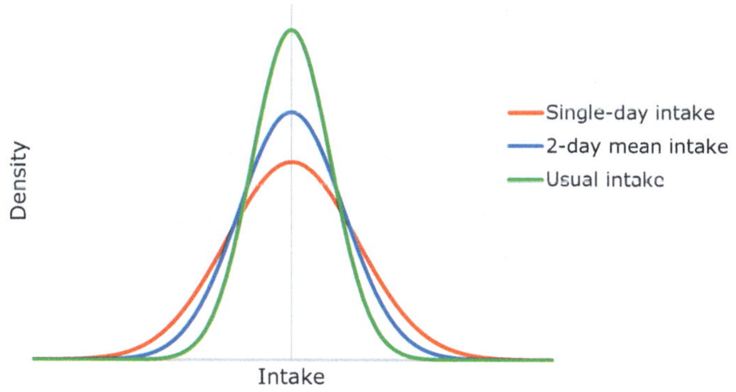

Fig. 4.19 The effect of the within-individual variation on long-term exposure distributions

followed by an integration step. By modelling intake frequencies, the model accounts for the fact that there is no daily intake for some substances, and that intake occurs only on a fraction of the days. For example, the consumption of beer could determine the exposure to a certain substance. If a respondent reports beer consumption on all its person-days, the model estimates the probability that this person will consume beer on all days over a longer period of time or just a fraction based on information on beer consumption in the whole food consumption database. This probability will be estimated as low if the number of person-days on which beer consumption is recorded in the whole database is low.

For the modelling of the positive exposure amounts, the positive daily exposure distribution is transformed into a more symmetric distribution. Then a normal distribution is fitted to estimate the between person variation while removing the within-person variation. The resulting between person normal distribution is then back-transformed and combined with the exposure frequency distribution to obtain the long-term dietary exposure distribution. This is achieved by sampling a large number of times from both the exposure frequency distribution and the back-transformed positive exposure distribution (Monte Carlo integration). The different statistical models differ in the way they transform the positive daily exposure distribution in a normal distribution. For example, the LNN uses either a logarithmic or power distribution, and the ISUF model can follow this with an additional spline transformation. In the European Tool Usual Intake (ETUI) project, different models were evaluated to assess long-term exposure (Goedhart et al. 2012; van Klaveren et al. 2012), and the LNN was recommended when the normality criterion is met and a realistic estimate of high exposure is preferred. The OIM model would be a good choice when a conservative estimate of high exposure suffices or the normality criterion is not met (van Klaveren et al. 2012). The LNN is similar to the NCI model.

The long-term exposure models described above first add the exposure of foods per person-day and then apply statistical modelling to derive a long-term exposure distribution, i.e. the Add-Then-Model approach. Another approach is to model the long-term exposure via single foods or food groups and to subsequently add these estimates to obtain the overall exposure, i.e. the Model-Then-Add approach. The exposure per food(group) can be modelled using OIM or a statistical model. The Model-Then-Add approach is especially useful in those situations where the normality criterion is not met and when this is due to one or more distinct foods or food groups (van der Voet et al. 2014). In the field of nutrition, a comparable approach has been developed to address the total intake of nutrients via the diet and other sources such as food supplements (Verkaik-Kloosterman et al. 2011).

For a more detailed description of statistical long-term exposure models, see Sect. 4.11.

Short-Term Exposure

The Monte Carlo approach should be used when assessing the short-term exposure according to a probabilistic methodology (EFSA 2012c). In this approach, person-days are randomly drawn from the food consumption database. The consumed amount of each relevant single food item on that particular day is multiplied by a randomly selected concentration available for that food item from the concentration database. The exposures per food are subsequently added to obtain the total intake for that particular person-day and divided by the individual's body weight to obtain the exposure per kilogram body weight. This procedure is repeated many times, resulting in a frequency distribution that reflects all plausible combinations of daily consumptions and concentrations. The procedure is repeated until a stable frequency distribution is obtained. The upper part of the distribution represents consumers with a high intake of the compound (high-level consumers), which is important for assessing the short-term risk. Examples of such assessments can be found in (Boon et al. 2009; Caldas et al. 2006; Ruprich et al. 2009).

Variability and Uncertainty

Probabilistic models can provide information about the variability and uncertainty in the exposure estimates, due to the variability and uncertainty in the input data. In deterministic approaches, variability and uncertainty are addressed by selecting conservative inputs for the assessment. In the resulting exposure estimates, i.e. a single mean and/or high exposure level, the variability, and uncertainty are no longer visible.

Variability is the natural variation occurring in the real world (e.g. individual differences in food consumption habits), whereas uncertainty refers to lack or imperfection of knowledge of the real world (e.g. concentration data). Both variability and uncertainty are related to all input variables of an exposure assessment, including the food consumption and concentration data, processing factors, and unit variability. By using the whole distribution of an input variable, variability is accounted for in a probabilistic assessment and is translated to a distribution of possible exposures within a (sub-)population. For this, the true variability in the data should be covered by the available data. Uncertainty is more difficult to quantify, and the uncertainty related to the sampling size of the input databases can be quantified using the bootstrap approach (Efron and Tibshirani 1993). With this technique, multiple input databases are created from the original input databases and used to assess the exposure, resulting in a 95% confidence interval around the exposure estimates. The more limited the sampling size of the databases, the broader the confidence interval will be and subsequently the larger the uncertainty around the estimated exposure estimates will be. In practice, the bootstrap technique is used to assess the uncertainty due to the sampling size of the food consumption and concentration databases, as the sampling size of the other input databases is often too limited.

Apart from bootstrapping, uncertainty can also be quantified using parametric methods. With these methods, the uncertainty in processing factors and concentration data can be modelled, resulting also in confidence intervals around the exposure estimates. For more details, see van der Voet et al. (2015) and Kennedy et al. (2015b). Quantification of uncertainty using parametric models is, however, not yet a common practice. Typically, uncertainty is assessed via sensitivity analyses. In sensitivity analyses, different values of an input variable are used to study how the exposure outcome is affected. Sensitivity analyses are, for example, used to address the uncertainty related to the concentrations to be assigned to samples reported to contain the substance at a concentration below a certain limit value (i.e. non-detect samples), such as the limit of quantification (LOQ), detection (LOD), or reporting (LOR). Another example of a sensitivity analysis is the proposal in the EFSA guidance on probabilistic modelling to calculate the exposure via two runs: an optimistic and a pessimistic model run (EFSA 2012c). The major uncertainties are treated using assumptions that are expected to result in underestimates of exposure in the optimistic model run, whereas these uncertainties are treated in such a way that it is expected to result in overestimates of exposure in the pessimistic model run. Sensitivity analyses are not exclusively applicable to higher tier assessments, but can also be applied in the point estimate approach.

Another option to address the uncertainty is to qualitatively evaluate the contribution of the different uncertainties to the overall exposure estimate in a tabular form. The EFSA guidance on probabilistic modelling recommends that for every probabilistic assessment, assessors should systematically examine all parts of the assessment for unquantified uncertainties and evaluate them using the tabular approach as recommended by (EFSA 2006b, 2012c). An example of such a tabular approach can be found in Boon et al. (2015, 2017), and is used by EFSA to evaluate the uncertainties related to the exposure assessment of authorised food additives (e.g. EFSA 2015e, h, 2016d, 2017a, b) and contaminants (e.g. EFSA 2015d, 2016c). If the uncertainty is too large to conclude about the exposure or whether a health risk can be excluded, additional input data are needed to improve the exposure estimation or it may be decided that an intake assessment is not feasible.

Empirical Versus Parametric Modelling

When modelling the exposure to substances present in food using higher tier models, concentrations can either be selected randomly from a concentration database, i.e. the empirical approach, or the concentrations can be assumed to follow some kind of theoretical distribution, i.e. using parametric modelling (EFSA 2012c). Parametric modelling of the concentration data is an alternative approach when the available concentration data do not cover the whole range of possible concentrations to which consumers may be exposed via food. In this way, concentration data beyond the range of observed data may be modelled, which is especially relevant when assessing the short-term exposure. However, a distribution assumes knowledge about its shape, which may be uncertain. Furthermore, distributions of single foods do not take into account possible correlations between foods. If these correlations are known, certain probabilistic modelling tools can take them into account. This is especially important for food consumption data because dietary patterns contain many correlations such as the unlikely consumption of spinach and endive on one day. Food consumption data are therefore preferentially modelled empirically, including all potential correlations between consumption of different foods (EFSA 2012c). An example of parametric modelling of concentration data to assess the long-term exposure to lead can be found in Boon et al. (2017).

4.10.4 Cumulative and Aggregate Dietary Exposure

The interest in the use of probabilistic models to assess dietary exposure and especially cumulative exposure has increased since the publication of Regulation 396/2005 (EU 2005a) of the European Parliament, and of the Council of 23 February 2005 on maximum residue levels of plant protection products (PPPs) in or on food and feed of plant and animal origin (EU 2005a). Cumulative exposure is the exposure to a group of substances belonging to a Cumulative Assessment Group (CAG) "that could plausibly act by a common mode of action, not all of which will necessarily do so"(EFSA 2008a). Regulation 396/2005 (EU 2005a) includes a requirement that when suitable methods are available, cumulative exposure from multiple PPP residues should be assessed, as well as exposure to PPP residues considered individually. Probabilistic models are needed to assess cumulative exposure. The EFSA guidance (EFSA 2012c) contains guidance on how to conduct these assessments for PPP residues. As part of the EU project ACROPOLIS, software was developed that is in line with these requirements (van der Voet et al. 2015) and a cumulative exposure assessment to selected groups of PPPs of the triazole group was performed (Boon et al. 2015).

Regulation 396/2005 (EU 2005a) also states that aggregate exposure should be addressed when setting maximum residue levels of PPPs as soon as the methodology for this is available. Aggregate exposure combines dietary and non-dietary sources of exposure. As part of the EU project ACROPOLIS, a conceptual framework of aggregate exposure was developed in which calculations for each of the contributing sources can range from simple deterministic estimates to fully probabilistic models that account for variability and uncertainty. The potential of this model was tested with four different aggregate exposure scenarios to PPP residues (Kennedy et al. 2015a). The developed framework was considered as a first prototype of how an aggregate exposure assessment can be performed using probabilistic techniques. To develop aggregate exposure modelling further, more experience is needed with this type of exposure modelling, and robust data to perform these assessments (i.e. pesticide usage data, and activity patterns of operator, worker, bystander, and resident) should be generated (van Klaveren et al. 2015).

Cumulative and aggregate exposure models are currently developed further in the ongoing EU project EuroMix. Results of this project are expected in 2019.

4.10.5 Computerized Tools for Dietary Exposure Assessment

Several automated programs have been developed to calculate the exposure using higher tier models (Sect. 4.10.3). This section briefly describes a selection of these programs. This description is not exhaustive and is meant to illustrate the types of programs available. For more details, see the references or websites provided.

Crème

Crème Global is an organisation that develops, among other software, cloud-based software in the area of predictive intake modelling. Different software packages are available of which Crème Food Safety™ is relevant within the context of this book. This model covers all models to assess the short- and long-term exposure to single substances in a probabilistic manner as outlined in the EFSA guidance on probabilistic modelling (EFSA 2012c). To assess the long term exposure, Crème uses models comparable to those of the USA's National Cancer Institute (Tooze et al. 2006). Crème contains food consumption data from the EU, USA, South America, and Asia. However, users can also enter their own input data for the exposure assessment. Crème is regularly updated to include new developments regarding the modelling of dietary exposure, and is available by license as a software service at www.cremeglobal.com.

DEEM

The Dietary Exposure Evaluation Model - Food Commodity Intake Database (DEEM-FCID) software is the standard program used by the Environmental Protection Agency in the USA (US-EPA) to "conduct dietary risk assessments to support the establishment of tolerances for residues of pesticides in/on raw agricultural and processed commodities. It also allows the Agency to look at exposures at each eating occasion rather than grouping the entire day's food and drink intake at once. The agency can thus conduct more refined risk assessments that will help inform regulatory decisions for pesticides". The software can be used to estimate short- and long-term dietary exposure for the population and for subpopulations of the USA. The US DEEM model incorporates food consumption data from the National Health and Nutrition Examination Survey/"What We Eat in America" (NHANES/WWEIA) dietary survey. It is not possible for the user to use other food consumption data with this software. The user can provide other input data needed for the assessment, such as concentration data and processing information. DEEM-FCID is freely available from EPA at https://www.epa.gov/pesticide-science-and-assessing-pesticide-risks/deem-fcidcalendex-software-installer

A comparison in exposure results between MCRA (described below) and DEEM-FCID v. 3.16, using the same NHANES/WWEIA data, showed that, in general, the outcomes were comparable (van der Voet et al. 2015). However, there were some differences that were related to technical differences in the algorithms used to calculate the exposure. These differences related to the binning algorithm of DEEM. MCRA stores all exposure estimates for the simulated person-days, whereas DEEM applies a binning algorithm to reduce storage requirements and gain speed, reducing the accuracy of the estimated lower (below 70th) percentiles of exposure (van der Voet et al. 2015).

FACET

The FACET (Flavourings, Additives, and Food Contact materials Exposure Tool) exposure tool was developed as part of the EU project FACET "in response to a call by the European Commission to produce a risk management tool consisting of a database containing information on concentrations of food additives and food flavourings, potential migrants in food contact materials and food consumption data". The FACET tool contains fixed food consumption data at the individual level derived from different countries, including Finland, France, Hungary, Ireland, Italy, Poland, Portugal, and the UK (Oldring et al. 2014). These data are coded in a harmonised way according to a three-tier food coding system. The tool contains features that are specifically useful for the assessment of flavourings, food additives, and migrants of food contact materials, such as consumer loyalty, the probability that a substance is added to a food, and market share. The tool only supports long-term exposure assessments for which it uses the OIM methodology. The FACET tool is freely available for download from Joint Research Centre (JRC) website (expofacts.jrc.ec.europa.eu/facet).

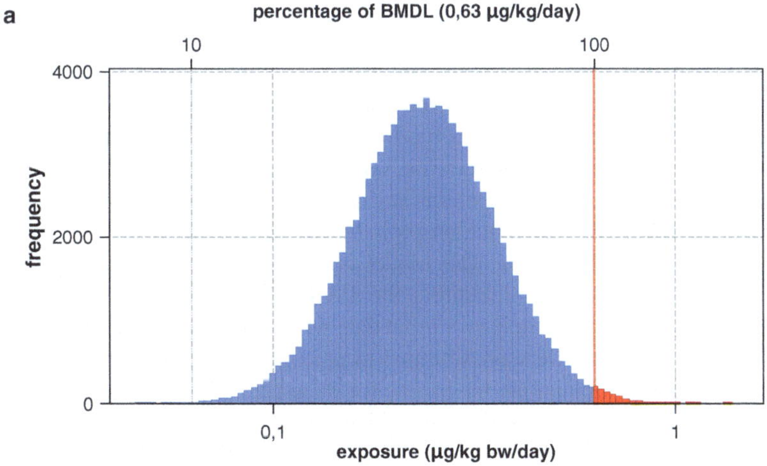

b

Mean exposure: 0.43 (0.4097, 0.4743) (µg/kg bw/day)

Percentage	Exposure (µg/kg bw/day)	Lower bound (p2,5)	Upper bound (p97,5)	Percentage of reference dose
50.00	0.408	0.388	0.4474	64.76
90.00	0.6191	0.5895	0.6969	98.27
95.00	0.6969	0.6602	0.7962	110.62
99.00	0.867	0.8207	1.012	137.62

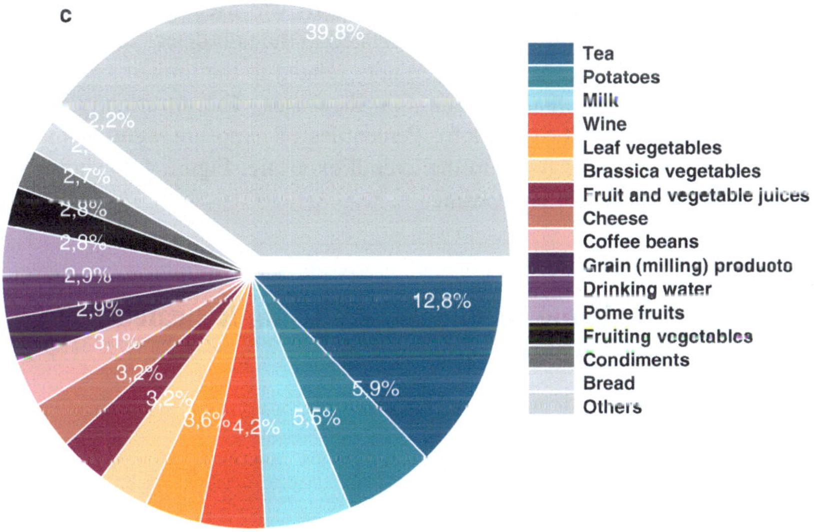

Fig. 4.20 The typical output of an MCRA long-term exposure assessment includes the long-term exposure distribution, percentiles of exposure (as specified by the assessor) with lower (2.5) and upper (97.5) confidence limits of the 95% confidence interval, and the contribution of foods to the overall exposure (BMDL = lower confidence limit of the benchmark dose)

MCRA

The Monte Carlo Risk Assessment (MCRA) is a computational tool from RIVM, WUR, and FERA that has been developed to assess both the short- and long-term exposure to substances present in food. Furthermore, both the exposure to single substances or groups of substances (i.e. cumulative exposure) can be assessed (van der Voet et al. 2015). MCRA contains standard procedures to calculate the exposure according to the EFSA guidance on probabilistic modelling of dietary exposure to PPP residues (EFSA 2012c). The model contains different statistical models to assess long-term exposure, including the ISUF, BBN, and LNN modelling (see Sect. 4.11). Users can use their own input data when formatted in the correct manner. Furthermore, MCRA offers possibilities to calculate the exposure for the general population or subpopulations based on, for example, age, gender, education, and consumption patterns (e.g. only people that consume apple). MCRA is regularly updated to include new developments regarding the modelling of dietary exposure, and is available for registered users at http://mcra.rivm.nl

A typical estimate of exposure obtained with MCRA was published for lead in the Netherlands (Boon et al. 2017). In this assessment, the long-term exposure to lead was estimated in young children aged 2–6 years and people aged 7–69 years. Recent national Dutch food consumption data for these two subpopulations were available. Given the toxicological profile of lead (EFSA 2010b), the exposure of women of childbearing age (as a proxy for pregnant women) and adults aged 18–69 years were also estimated. The long-term statistical model LNN was used. Sensitivity analyses were performed to address the uncertainty related to the non-detect samples and the bootstrap was used to quantify the uncertainty related to the limited sampling size of the food consumption and concentration databases. The uncertainty was also evaluated qualitatively (EFSA 2006b). Percentiles of exposure were estimated as well as the contribution of foods to the overall exposure. Figure 4.20 shows some typical MCRA outputs of this exposure.

4.11 Modelling Usual Dietary Intake Using Repeated Short-Term Measurements

Sven Knüppel

4.11.1 Measurement Error in Dietary Assessment

In this section, we emphasize key concepts of modelling usual (habitual) dietary intake using repeated short-term measurements to characterise (long-term) dietary exposure. Usual dietary intake is defined as long-term daily average intake of a substance, food, or food group. The interest in assessing usual dietary intake can be to evaluate the nutritional status of a population, to provide a basis for dietary recommendations, to find associations between dietary intake and health outcomes or the occurrence of a disease, or to perform an estimate of long-term exposure for potentially hazardous substances. As the long-term average of dietary intake cannot be directly observed, the dietary assessment is based on self-report instruments such as food records (also known as food diaries), 24-hour dietary recalls (24hDR), food frequency questionnaires (FFQ), dietary screeners, and diet histories (Thompson et al. 2015). The drawback of all self-reporting dietary assessment instruments is that these are not perfect and are prone to different types of measurement error (Thompson and Subar 2013; Willett 2013). Measurement error is the result of the difference between the reported and the (true) usual dietary intakes over a given time period, and originates from random and systematic error, also known as bias. Findings from an error-prone self-reporting dietary assessment instrument can be misleading and loss of statistical power will be present (Carroll et al. 2012; Freedman et al. 2011; Kipnis et al. 2003). The use of adequate study designs and more sophisticated statistical methods are necessary to remedy the bias and loss of statistical power.

The daily dietary intake of individuals depends on many factors such as individual dietary and life habits, food preferences, dietary recommendations, residential area, availability of foods in different seasons, and the day of consumption (e.g. weekday/weekend/working day). Individuals can vary their diet each day, resulting in day-to-day variation in dietary intake. Therefore, the dietary intake reported on a given day may differ to the usual dietary intake. The difference is called random error, also known as day-to-day variation or within-person random error. A dietary assessment instrument that only suffers from random error is called accurate or unbiased. The sample mean is not influenced by the random error but the variability is increased compared to the usual dietary intake distribution. Averaging the reported intakes of repeated short-term dietary intakes approximates the usual dietary intake.

Averaging reported intakes does not help when a dietary assessment instrument suffers from systematic error because a systematic difference between the reported and (true) usual dietary intake would be observed. The sample mean is also affected and may result in misleading findings. The systematic error may occur because individuals may under- or overreport their dietary intakes. Reported intakes may depend on personal characteristics or the social/cultural background, e.g. individuals may report higher amounts of fruit and vegetables compared to the true dietary intakes because fruit and vegetables are believed to be healthy. Validation studies with an unbiased reference measurement are needed to correct for systematic error (Freedman et al. 2011). Recovery biomarkers, which are directly related to dietary intake in a specific time period, are considered to be proper reference measurements (Willett 2013). For example, urinary nitrogen is "recovered" as protein intake (Bingham 2002). Another example is the doubly labelled water technique as a recovery biomarker for energy expenditure, measuring energy intake assuming that energy expenditure, and intake are in balance (Freedman et al. 2011). Only a few proper recovery biomarkers to characterise (long-term) dietary exposure are known and they are seldom available because the assessment is cost-intensive with a high burden for the study participants, or the measurement is invasive. In the absence of a proper biomarker, a good available approximation to estimate (long-term) usual dietary intake can be obtained from repeated 24hDRs (Kipnis et al. 2009). An optimal dietary assessment instrument should only suffer from random error.

Multiple-day food records, repeated administrations of 24hDR and FFQs are often applied in nutritional studies. A multiple-day food record is a short-term instrument in which study participants record their type and total consumption amount of food and beverages over more days. A 24hDR is also a short-term measurement asking for detailed quantified dietary intake of the previous day. In contrast to 24hDRs, FFQs are long-term measurements asking for the past intake of a long time period, e.g. one year. FFQs are cognitively complex to answer, leading to large amount of systematic error. Thus, FFQs are limited in their feasibility to estimate usual dietary intake (Freedman et al. 2011). Multiple-day food records and 24hDRs are less prone to systematic error because they are less cognitively complex to conduct. To estimate usual dietary intake (long-term dietary exposure), multiple-day food records or repeated administration of 24hDR per study participant are necessary to capture random error (Carroll et al. 2012; Kipnis et al. 2003, Thompson and Subar 2013; Willett 2013).

4.11.2 Statistical Methods to Estimate Usual Dietary Intake

This section focuses on statistical methods to estimate (long-term) usual dietary intake from repeated short-term measurements (e.g. repeated administration of 24hDRs) for a population, or on the individual level. These methods serve as the basis to relate dietary intake to health outcomes and diseases, or to estimate long-term exposure for potentially hazardous substances.

One 24hDR per study participant is sufficient to estimate the mean usual dietary intake in a population (Dodd et al. 2006). Repeated administration of 24hDRs is required to estimate the usual intake distribution of a population or an individuals' usual dietary intake. The naïve mean values do not represent the usual dietary intake due to day-to-day variation in human diets (Dodd et al. 2006; Goedhart et al. 2012). The usual intake distribution estimated by a naïve mean from a few 24hDRs is generally wider than the true usual intake distribution, resulting in over- or underestimating the proportion of the population above or below a certain cut point. Thus, more sophisticated statistical methods were developed.

One of the first statistical method estimating usual dietary intake was developed at the Iowa State University (Nusser et al. 1996). Four recent statistical methods are:

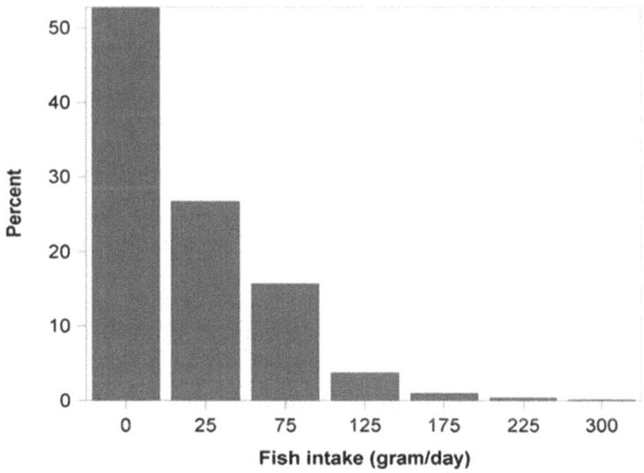

Fig. 4.21 Distribution of Fish intake (gram/day) in EPIC-Potsdam substudy (2010–2012)

- National Cancer Institute (NCI) Method (Tooze et al. 2006, 2010; Kipnis et al. 2009).
- Multiple Source Method (MSM) (Haubrock et al. 2011; Harttig et al. 2011).
- Statistical Program to Assess Dietary Exposure (SPADE) (Dekkers et al. 2014).
- BetaBinomial-Normal model (BBN) (de Boer et al. 2009).

The web address of each method can be found in the Appendix A. Some examples for tools are given in Table 4.14.

While the four recent statistical methods share some common structure, they show some differences. Each of these statistical methods can be applied to daily consumed (consumed nearly every day by nearly everyone) and episodically consumed (consumed not daily and not by everyone) substances, foods, and food groups assessed by a short-term dietary assessment instrument. For simplification in the text below, we will assume that a repeated administration of 24hDR is applied.

Fish intake data from the German EPIC-Potsdam substudy can be used to illustrate the principles of modelling repeated short-term measurements (2010–2012; Gottschald et al. 2016; Wientzek et al. 2014). Eight hundred and fourteen study participants answered up to three 24hDRs. Figure 4.21 shows the fish intake distribution. Two common issues can be seen in the figure:

1. Episodically consumed foods may be overproportioned by reported days without consumption, referred to as a spike at zero, or excess to zero, or zero-inflated, and
2. Reported intakes on days with positive consumption amounts may not be symmetric, referred to as a skewed distribution. Dietary intake distributions are often positive skewed, meaning that many individuals consumed little and only a few individuals consumed much of the substances or foods under investigation.

Therefore, statistical methods should address the following challenges to estimate usual dietary intake from repeated short-term measurements (NHANES Dietary Web Tutorial):

- Consider reported days without consumption
- Distinguish within-person variation (day-to-day variation, random error) from between-person variation and
- Consider consumption-day amounts that are positively skewed.

Further, the statistical methods should consider the inclusion of covariates related to the probability of consumption, or the consumption-day amount, or both.

To model dietary components episodically, the four recent statistical methods include the following three steps:

1. The probability of consumption is estimated for each individual.
2. The person-specific consumption-day amount (estimated consumption amount, when a person would consume the specific dietary component) is estimated, and
3. The usual dietary intake is estimated by multiplying the probability of consumption with the person-specific consumption-day amount.

   In the case of daily consumed dietary components, only the amount needs to be applied because the probability of consumption is assumed to be one. Statistical methods should address a potential correlation between both the probability and the amount part.

In the first step, the reported intakes are dichotomised (as consumed or not consumed) on a given day. To estimate the probability of consumption, a logistic regression with person-specific random effects (NCI) or a Beta-Binomial model (SPADE and BBN) are applied on the dichotomized intakes. MSM uses the residuals from a classical logistic regression to estimate the probability of consumption for each individual.

The second part is much more complicated. The statistical methods have to address day-to-day variation of non-zero dietary intakes and skewed distributions to estimate the person-specific consumption-day amounts.

A minimum of two 24hDRs are needed at least in a subsample of the study sample to capture the day-to-day variation. The distribution of the dietary components is often skewed with a long right tail (see Fish example in Fig. 4.22a). Therefore, the NCI method, SPADE, and BBN transform the raw intake data to normality because modelling of skewed distributions could be difficult (see Fish example in Fig. 4.22b). The Box-Cox transformation is an often used method that is a scaled version of a power transformation including logarithmic transformation (Box and Cox 1964). After transformation, linear regression with random effects is applied to separate within-person variation from between-person variation. Using a shrinkage technique, the within-person variation is adjusted out and only between-person variation remains (see Fish example in Fig. 4.22c). In the end, the shrunken values have to be backtransformed to the original scale. The backtransformation can be done by approximation (Tooze et al. 2010), numerical integration (Dekkers and Slob 2012), or exactly for special cases (Hoffmann et al. 2002). For a more detailed

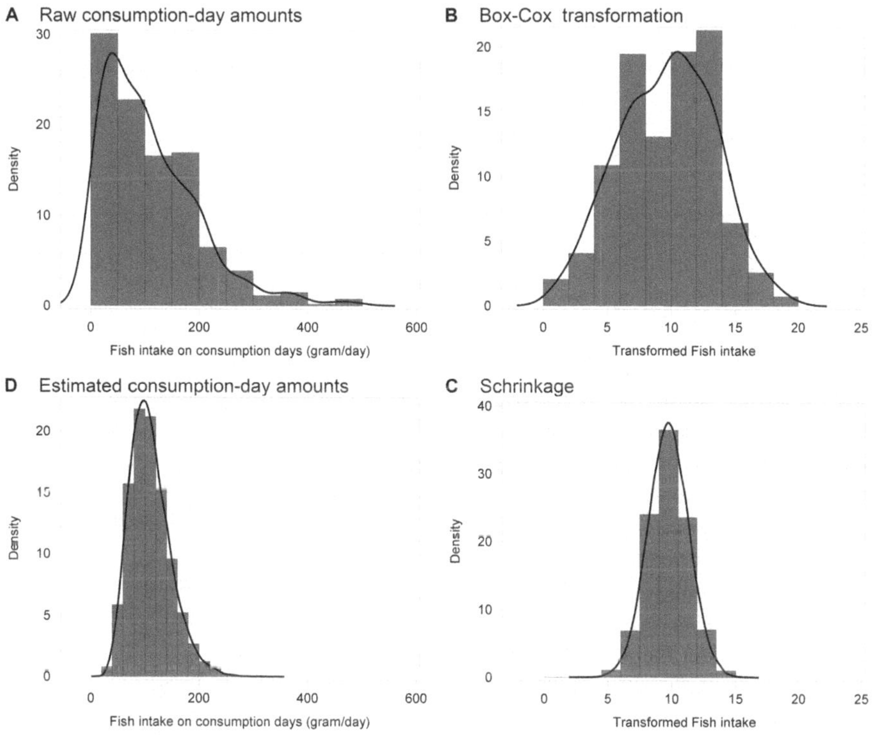

Fig. 4.22 Histograms including kernel density estimates showing steps to estimate person-specific consumption-day amounts of Fish intake from repeated 24hDRs in EPIC-Potsdam substudy (2010–2012)

discussion of backtransformation, see Dekkers and Slob (2012). The backtransformed values are the estimated person-specific consumption-day amounts on the original scale (see Fish example in Fig. 4.22d). The final estimated distribution is narrower than the original distribution because the day-to-day variation is removed.

The MSM estimates the person-specific consumption-day amounts slightly differently. After applying a classic linear regression on the non-zero dietary intakes, the residuals are transformed to normality applying two-parameter Box-Cox transformation, shrunken to remove within-person variation, and backtransformed by an exact calculation (Haubrock et al. 2011).

In the third step, multiplication of the consumption probability with the person-specific consumption-day amounts gives the estimation of individuals' usual intake. The model parameters can be used to approximate the usual intake distribution of a population by simulating a given number (e.g. 100) of estimated values per person on the transformed scale assuming normal distributed errors (Tooze et al. 2010). The simulated values are backtransformed to approximate the usual intake distribution.

Table 4.12 Estimation of usual Fish intake (gram/day) distribution in EPIC-Potsdam substudy (2010–2012) by naïve averaging the reported intakes of each individual over three 24hDRs, and by applying the NCI method ignoring a correlation between consumption probability and consumption-day amount and allowing a correlation between both parts

	Mean	Percentile						
Model		5	10	25	50	75	90	95
	gram/day							
Observed individual mean	23.3	0.0	0.0	0.0	0.0	35.1	69.0	98.9
NCI method (ignoring correlation)	23.6	3.6	5.2	9.5	18.1	31.9	49.5	62.1
NCI method (allowing correlation)	23.4	5.7	7.6	12.4	20.5	31.4	43.2	50.9

Table 4.12 shows the estimation of usual Fish intake (gram/day) distribution in EPIC-Potsdam substudy (2010–2012; Gottschald et al. 2016; Wientzek et al. 2014) by observed individual mean intake over three 24hDRs, and by applying the NCI method. This ignores a correlation between consumption probability and consumption-day amount, which is equal to assuming a correlation of zero, and by modelling a correlation between both parts (Kipnis et al. 2009; Tooze et al. 2006). A positive correlation refers to a population in which persons tend to eat more on a given day, when they eat the specific dietary component more often. Otherwise, a negative correlation refers to a population in which people tend to eat less on a given day, when they eat the specific dietary component more often. The estimated mean intakes are nearly the same for all three approaches, and this is expected because the repeated 24hDRs are assumed to be unbiased. The estimated distribution of the usual Fish intake applying the NCI method is much narrower compared to the distribution estimated by the observed individual mean because the day-to-day variation was adjusted out. The correlation between consumption probability and consumption-day amount was estimated to be −0.60. The effect of this negative correlation coefficient is that the resulting distribution was narrower than the estimated distribution ignoring this correlation. In the case of a positive correlation, the resulting distribution would be wider compared to the distribution ignoring the correlation.

In the Fish example, unadjusted models were used to illustrate the principles of modelling short-term measurements; however, dietary intakes often depend on personal characteristics, the timing of the 24hDR (e.g. day of the week and season) or on a nuisance parameter (e.g. mode of assessment telephone interview vs. web-based questionnaire). Such covariate information can be considered in the two modelling parts. Adjusting for frequency information (e.g. taken from a FFQ) can be helpful to improve the estimation on the individual level but not for estimation of the usual intake distribution in a population (Carroll et al. 2012; Goedhart et al. 2012; Kipnis et al. 2009; Verly et al. 2012).

The modelling of rare and low frequent consumed foods can be problematic because of the low number of consumers. In the EPIC-Potsdam substudy, the overall empirical consumption frequency of Fish intake was 22.4%. Different types of fish and shellfish compose the food group Fish. The four most reported types were Salmon, Herring, Tuna, and Trout with consumption probability of 7.3%, 7.0%,

1.2%, and 1.1%, respectively. The subdivision of food groups into their underlying foods leads to decreasing consumption frequencies until they reach nearly zero. If the consumption frequency is too low, the number of individuals with at least two different non-zero consumption days could be very low or non-existent. To estimate day-to-day variation, some individuals should have consumed the dietary component on at least 2 observation days. For example, in the EPIC-Potsdam substudy (Gottschald et al. 2016; Wientzek et al. 2014) 669 out of 814 (82.2%) individuals reported no consumption of Salmon at all, 118 (14.5%) reported Salmon consumption on one day, 21 (2.6%) on 2 days, and 6 (0.7%) on all 3 days. Tuna intake was only from one individual reported on 2 days and none on all 3 days. Trout consumption was only reported from 27 (3.3%) individuals on one day and no others had Tuna intake.

As an orientation, Table 4.13 shows the expected consumption probability depending on the number of repetitions of a short-term measurement assuming a binomial distribution. Addressing the day-to-day variation of a dietary component (e.g. Tuna and Trout consumption in the EPIC-Potsdam substudy) with consumption

Table 4.13 Expected number of consumption days with positive consumption amounts

Consumption probability per day p	Expected number of consumption days D within one year $D = 365 * p$	Number of repetitions of a short-term measurement n	Expected probability of consumption days x with positive consumption-day amount assuming a binomial distribution X~Binomial (n, p) $x = 0$	$x = 1$	$x \geq 2$
0.5%	1.83 (on average around 2 days a year)	2	99.00%	0.99%	0.00%
		3	98.51%	1.49%	0.01%
		4	98.01%	1.97%	0.01%
		12	94.16%	5.68%	0.16%
1%	3.65 (on average around 1 day a quarter)	2	98.01%	1.98%	0.01%
		3	97.03%	2.94%	0.03%
		4	96.06%	3.88%	0.06%
		12	88.64%	10.74%	0.62%
5%	18.25 (on average around 1 or 2 days a month)	2	90.25%	9.50%	0.25%
		3	85.74%	13.54%	0.72%
		4	81.45%	17.15%	1.40%
		12	54.04%	34.13%	11.81%
10%	36.50 (on average around 3 days a month)	2	81.00%	18.00%	1.00%
		3	72.90%	24.30%	2.80%
		4	65.61%	29.16%	5.23%
		12	28.24%	37.66%	34.10%
20%	73 (on average around 1 or 2 days a week)	2	64.00%	32.00%	4.00%
		3	51.20%	38.40%	10.40%
		4	40.96%	40.96%	18.08%
		12	6.87%	20.62%	72.51%

Table 4.14 Some examples for tools that calculate usual dietary exposure

Method	Developer	Available at
NCI method (SAS macros)	National Cancer Institute (NCI)	https://epi.grants.cancer.gov/diet/usualintakes/method.html?&url=/diet/usualintakes/method.html/
MSM Multiple source method	DIFE (German Institute of Human Nutrition)	https://msm.dife.de/ Web application using R
SPADE Statistical Program to Assess Dietary Exposure	RIVM, National Institute for Public Health and the Environmen	https://www.rivm.nl/en/spade (R package)
MCRA Monte Carlo Risk Assessment	RIVM, National Institute for Public Health and the Environment, and Wageningen University	https://mcra.rivm.nl/

probability of around 1% is problematic. This is because the expected and the observed numbers of individuals with two or more consumption days are quite low if the number of repetitions is low.

Reporting zero dietary intake in all 24hDRs does not mean that the participant never consumes that food. It only shows no consumption on the reported days. To decide if a person is a “never consumer” or an “occasional consumer”, an additional source of information can be used (e.g. frequency information from a FFQ). MSM and SPADE provide the option to consider never consumer in the analysis. Keogh and White (2011) extended the NCI model to consider never consumers by extending the model with an extended part that estimated the probability of being a consumer vs. a never consumer.

Comparisons of the NCI method, MSM, SPADE, and BBN based on empirical and simulated data showed that no one statistical method performs best in all situations (Goedhart et al. 2012; Laureano et al. 2016; Souverein et al. 2011; Verly et al. 2012, 2016; Table 4.14). The performance and applicability of the statistical methods depends on the research question, study design, and available nutritional data. Choosing a suitable statistical method may not only depend on the statistical properties but also on the availability of the algorithms, their features, and the required programming and statistical skills.

Foods are not consumed apart from each other and thus the extension to multivariate statistical approaches to model more foods together is a necessary step forward. Also, the simultaneous modelling of dietary intake along with energy intake and supplements may be of interest. Zhang et al. (2011) extended the NCI method to apply a multivariate model including daily and episodically consumed foods as well as energy intake.

Agogo (2016) proposed a flexible zero-augmented calibration model to correct for measurement error in intake reported in the FFQ, and to handle excess zeroes, skewness, and heteroscedasticity simultaneously without transforming the 24hDR intakes. The authors showed with empirical and simulated data that this new approach gives similar results to the NCI method.

The principle structure of the two modelling parts (probability of consumption and person-specific consumption-day amount) can be used to simplify the dietary assessment for large-scale epidemiological studies. The consumption frequency seems to contribute more to the variation between persons in substance and food intake than the variation in portion sizes (Noethlings et al. 2003). In this regard, one solution for large-scale studies is to separate the probability from the amount part. To achieve this, the 24 h food list was developed for repeated applications that allow the estimation of the probability of consumption and it does not require information about portion sizes (Freese et al. 2014). The 24 h food list used in a web-based tool only ask for a yes or no about the consumption of the foods and the food groups of the previous day. The person-specific consumption amounts were estimated by using a reference source (e.g. a representative national nutrition survey). This promising new approach has been applied in a large-scale cohort study in Germany (Knüppel et al. 2019).

Modelling the usual dietary intake using repeated short-term measurements comes with many statistical challenges. Many issues of the statistical challenges are solved but some need further investigation. The statistical methods can help to improve our understanding of the relationship of diet and health.

A list with useful web resources (Appendix A) about the measurement error and estimating usual intake, and more information about dietary assessment instruments can be found in Appendix A.

4.12 Impact of Downstream Use for the Food Chain and Downstream Use for Consumer Exposure

Gianfranco Brambilla

Headline hitting food safety scares in the 1990s were about "mad cow" disease from Bovine Spongiform Encephalopathy (BSE) agent present in ruminant rendered proteins administered to ruminants (OIE 2017), and about the dioxin crisis from industrial dielectric fluids that entered fats and oils in the feed administered to food-producing animals (Bernard et al. 2002). These incidents clearly led the feed and food safety legislation towards a comprehensive strategy based on the overall traceability (verification) of the place(s) of origin and history, and on the safety of the food process.

Such strategy, reflected in the "from farm to fork" or "from stable to table" slogans, aims for consumers to have as much information as possible about potential risks and what is being done to minimise them. Keeping risks to a minimum requires the application of modern food and hygiene standards drawn up to reflect the most advanced scientific knowledge. Food safety starts on the farm, where the food is produced (EU Commission, the white paper on food safety (EU 2000a)).

The challenge of the identification of the chain of exposure (downstream use) from the origin to the contact with people has been covered the most in the animal food chain. Ideally, when a hazard is introduced in the food chain, the overall system will not allow it to propagate and amplify following its introduction, thus determining an acceptable or unacceptable exposure for consumers. Information about the downstream use is based on the records of administrative checks on traceability (the origin and provenience of feed and food through online platforms for data exchange), on animal tagging, on the geo-referenced inventory of farms, on the authorisation of feed and food production plants, and on the records of internal and official monitoring activities carried out on a regular basis by Food Business Operators, and Food Safety Authorities. The strategy has been proven to be effective at reducing to a minimum the risk for chronic toxicity in the general population as the sources of contamination are usually identified and removed/reduced from the food chain within a reasonable time, thus preventing long-term exposures (EU Commission, RASFF (EU 2017) preliminary report 2016; EFSA 2008d).

Of course, the downstream use should also take into account emerging risks not fully framed into feed and food checks. This could be the case for contaminants associated to the consumption of novel foods such as seaweeds, or to the pollution from anthropogenic waste such as fertiliser in feed and food, when there is evidence that an increased consumption/use of such items could result in a meaningful exposure to contaminant(s). In this case, Food Safety Regulatory Authorities ask for the collection of evidence about the occurrence of contaminants and time trends in feeds, livestock, and food. This information can be used to verify if there are potential concerns for intake. If there is a concern, a risk management option could be the setup of maximum residue limits and the insertion of such matrices/substances into the monitoring activity. This approach depends on the availability of pertinent toxicological guidance values for alimentary intakes (EFSA 2012g).

The possible benefits from the native presence of nutrients and detoxifying substances in the food/diet that can mitigate/counterbalance the toxicity of a contaminant could be assessed. The main question is whether the health risks clearly outweigh the health benefits or vice versa, as in the case on the benefits of fish/seafood consumption compared to the risks of methylmercury in fish/seafood (EFSA 2010f, 2015j).

For toxicity linked short-term exposures, the acute and developmental toxicity can be framed (Boobis et al. 2017) even if the exposures mostly cause delayed effects. Information about downstream uses can lead to awareness about case reports of collective intoxication. For example, several collective intoxications around the world have been reported for cardiovascular impairment related to the presence of residues of the beta-adrenergic drug Clenbuterol in liver and meat of illegally treated cattle and pigs (Estrada-Montoya et al. 2008). The reported onset of gynecomastia in pre-pubertal children related to the presence of injection sites in the second choice cuts of meat used for the preparation of baby foods (Loizzo et al. 1984). In these cases, the evidence addressed the non-controlled/non-authorized administration of drugs to food production animals. The safety of the downstream use of foods by consumers relies on the empowerment to purchase food from official and traceable channels.

From a toxicological point of view, a residue may differ in bioavailability and toxicity than those of the parent compound, e.g. as a consequence of its metabolic transformation in food-producing biota (Palazzesi et al. 1995). Examples include the presence of covalently bound residues to proteins (Pietraforte et al. 2008), and the bioselection of more pharmacological active L-forms in the case of enantiomers (Parr et al. 2017). In these cases, appropriate in silico and in vitro experiments could be used to refine the risk assessment.

For consumer exposures, the downstream use may be more relevant for susceptible and vulnerable groups (Consumers only, Children, Pregnant, and Breast-feeding women) and those groups that for socio-economic, cultural, and ethnological reasons are likely to frequently or exclusively eat peculiar food items of local origin. For example, this could be the case for shooters of wild game, and the anglers of feral fish. In such scenarios, the downstream use could result in target actions to empower such a consumer group to make a resilient food choice (EFSA 2015a).

4.13 Hazard Assessment and Derivation of Health-Based Guidance Values

Rainer Gürtler

Briefly, hazard assessment in food risk assessment has the goal of deriving the limit levels on the basis of toxic doses obtained from chronic and acute toxicity testing (see also Sect. 2.2.1). For chronic hazards and in the case of additives or residues, an ADI is the estimated maximum amount to which individuals may be exposed daily over their lifetimes without an appreciable health risk. For substances unintentionally occurring in food, the corresponding dose is the TDI. Based on the most sensitive endpoint from a range of toxicological hazards and their dose–response relationships, a no-observed-adverse-effect Level (NOAEL) or a benchmark dose lower confidence limit (BMDL) is established and used as a point of departure for deriving an ADI (EFSA 2009d, 2012d). This point of departure is divided by an uncertainty factor that covers uncertainties due to the extrapolation of data from animal studies to the human situation as well as individual variabilities. Generally, a factor of 100 is applied as a default value. A smaller factor could be applied if human data are available, e.g. toxicokinetic data, that allow for comparison of internal doses in experimental animals and humans. A factor that is larger than 100 would be required in the case of additional uncertainties. The ADI/TDI is expressed in mg per kg body weight per day, and is established for compounds for which a threshold mechanism of toxicity can be demonstrated.

An acute reference dose (ARfD) is derived based on adverse effects observed after a single dose or short time exposure.

4.14 Risk Characterisation in Food Risk Assessment

Katrin Blume

In addition to food consumption data, data about occurrences of a substance in certain foods is essential to provide refined and reliable risk characterisation. Detailed description and characterisation of these data are given in Sects. 4.6 and 4.8. Beyond this, dermal, inhalative, and oral exposure from sources other than food, e.g. via house dust, should be taken into account, too.

Although a detailed description of risk characterisation goes beyond the scope of this book, risk characterisation combines hazard and exposure data. As pointed out in Sect. 2.2.5, risk characterisation evaluates the ratio of the exposure estimate with the TDI/ADI (in case of chronic hazards) or the ARfD (for acute hazards).

One further reason for calculating exposure and performing risk characterisation is to evaluate whether toxicological reference values are achieved or exceeded. These values are derived from exposure assessment, taking into consideration all known (food) sources and their respective consumption data. In the same context, exposure assessment gives insight into the possible adoption, or the necessity to install maximum levels for environmental contaminants, food additives, or pesticide residues in certain foods.

The objective of maximum levels is to ensure food safety by the highest possible reduction of contamination determined to be needed. In the European Union, there are specified amounts for some substances in regulated food that should not be exceeded. These maximum levels follow the principle of ALARA meaning they are set at a level that is as low as reasonably achievable (Regulation (EC) No 1881/2006b). Surveillance programmes are responsible for controlling randomly whether foods on the market follow these levels.

Risk characterisation should include all known food sources and other sources of exposure.

4.15 Food Regulations

Gianfranco Brambilla

4.15.1 Introduction

Food regulations can be framed as the food safety and quality actions needed to cover the entire food chain, from production to consumption, ensuring that all food is as safe as possible (UNECE 2012) (Fig. 4.23). In almost all countries, the food legislation regulates the production, trade, and handling of food. Thus, it considers

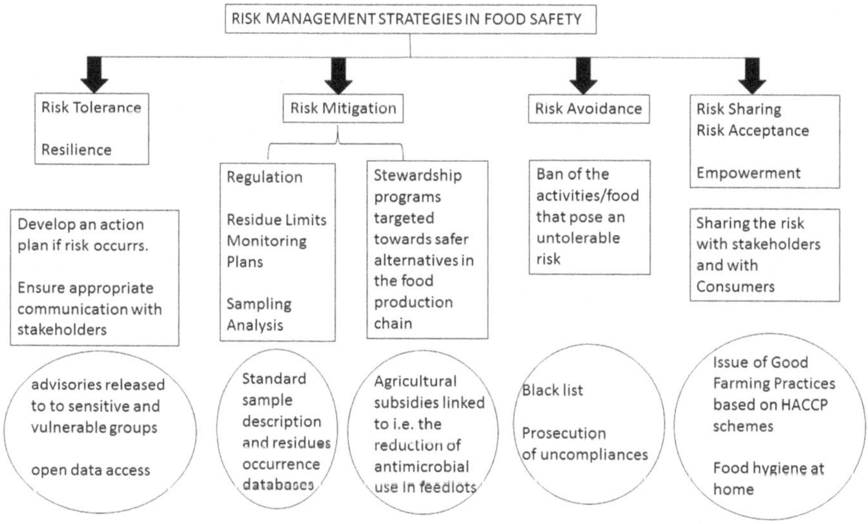

Fig. 4.23 Regulatory limits framed within the overall food safety risk management options: boxed, the actions to be taken; circled, the products (From UNECE 2012, modified)

all the steps, beginning with provisions for animal feed, on-farm controls, and early processing through to final distribution and use by the consumer. At the individual country level, most of such a regulatory framework is harmonised with international agreements such as the World Trade Organization (WTO) and Codex Alimentarius Commission (CAC) standards, and/or centralized at a transnational level, as in the case of the European Union via the European Commission. As part of the risk management, the regulatory aspects are generally based on risk/benefit evaluations (WHO/FAO 2009; Vapnek and Spreij 2005).

4.15.2 *Risk vs. Benefit Evaluations Behind Food Safety Risk Management Option*

The risk is usually represented by unacceptably high food intakes with respect to toxicological guidance values. For chronic alimentary exposures, it is considered to be the estimated amount of a substance in food and drinking water, expressed on a body weight basis, that can be consumed every day for a lifetime by humans without presenting an appreciable risk to their health. Such guidance values are defined as Acceptable Daily Intake (ADI) for a) licensed veterinary drugs and pesticides administered/used under good veterinary and farming practices, and b) as Tolerable Intakes on Daily (TDI), Weekly (TWI), or Monthly (TMI) basis, or as Benchmark Dose Levels (BMDL) (EFSA 2016i) for those contaminants whose presence in the food commodities could be mainly ascribed to involuntary factors.

A guidance value for acute toxicity, the Acute Reference Dose (ARfD) is the amount of a substance in food and/or drinking water that can be ingested in a period of 24 h or less without appreciable health risk to the consumer is also considered, rarely for veterinary drugs but more often for pesticide residues (WHO/FAO 2009).

It should be noted that the above-mentioned limit values apply not only to food, but also to total exposure from all sources. If there is a risk from food intake, additional exposures from, for example, consumer products might not be acceptable.

While the food safety risk is clearly framed within the discipline of toxicology, the benefits rely on social, economic, and environmental aspects. These include to facilitate the fair trade of food products and the related business sector, and to authorize the use of veterinary drugs and additives able to improve the animal health and welfare, the quality and quantity of food production, and of the shelf-life of processed items. They also include guaranteeing the intake of nutrients from certain food able to satisfy the recommended daily allowance, as in the case of polyunsaturated fatty acids related to fish consumption, and to support the food production systems and the social and economic relevance of family farming activities. Further, they include being a tool to preserve local heritage and landscape, and to help build consumer confidence associated with food consumption (CAC 2018; UNECE 2012).

To this respect, most of the benefits may be summarized within the definition of food security: when all people, at all times, have physical, social, and economic access to sufficient, safe, and nutritious food, which meets their dietary needs and food preferences for active and healthy life (FAO 2003a).

4.15.3 From Risk Assessment to Risk Management: The Regulatory Levels in Food

One of the more sensitive risk management options is represented by the setting of the Maximum Residues Level (MRL), defined as the highest level of a chemical residue that is legally tolerated in or on food or feed according to good farming practices. Owing to the above, the regulatory levels are almost always harmonised within a transnational or international context, leaving to local competent authorities the issue of specific decisions in the case of emergency contingency plans (i.e. restriction in food farming/production in a contaminated area).

Due to the very sensitive aspects related to the issue of a regulatory measure relevant to food safety, the risk management decision process for the people involved in the risk analysis process should be supported by the scientific quality and representativeness of the experimental and observational data, scientific excellence, independence, transparency, and accountability, with a clear separation and independence between the assessment and management parts. Under the pressures of the food safety crisis related to Bovine Spongiform Encephalopathy (BSE) and dioxin in feed and food, it appeared mandatory to recover the accountability of public opinion in the Competent Authorities and to keep the bodies in charge of the risk evaluation separate and independent from the risk management tasks.

On the first years of the twentieth century, it has been taken place the constitution both of national and transnational independent risk assessment Agencies, for example the Food Standards Agency in the UK, the French Agency for Food, Environmental and Occupational Health & Safety (ANSES), the Federal Institute for Risk Assessment (BfR) in Germany, and the European Food Safety Authority (EFSA) of the European Union; For some aspects, i.e. those related to pharmaceuticals and pesticides, the food safety agencies are supported by other Agencies with duties on veterinary medicines, chemical substances, and environment [i.e. European Medicine Agency (EMA), European Chemicals Agency (ECHA), European Environment Agency (EEA)]. Also within the Codex Alimentarius, the risk assessment of food additives, pesticides, contaminants, and drug residues is carried out by selective Joint Experts Groups, who report to the risk management body represented by the Codex Alimentarius Commission (CAC) (WHO/FAO & CAC 2016).

As part of a risk analysis process, it is worthwhile to illustrate how the risk management is fed by risk assessment, and how risk management could produce evidence for the reassessment of risk on the basis of data produced under regular monitoring plans. For exemplification, these are the steps for the setting of a maximum residue level (WHO/FAO 2012).

1. Knowledge of the use (amount, frequency, time-window of administration, target animals/crops) of the product, under good practices.
2. Knowledge of the residues expected in food items, according to the *P*harmaco-*D*ynamics/*P*harmaco*K*inetics (PDPK), or ToxicoDynamics/ToxicoKinetics (TDTK) for drugs and pesticides, respectively, and definition of an MRL and/or of a median value of residue occurrence along with the proposition of an appropriate withdrawal time.
3. Toxicological guidance values for the active drugs and its metabolites. Chronic toxicity is measured with the Acceptable Daily Intake (ADI), and acute toxicity with the Acute Reference Dose (ARfD).
4. Availability of a standard global diet or of different country diets (such as GEMs Cluster diets), to combine food consumption data with residues occurrence in food commodities.
5. Risk characterisation by comparing intake estimates, corrected for the residue bioavailability, with the considered guidance levels, and if the case, to reconsider the proposed MRLs.

The main determinants in such an MRL assessment are represented by the residue occurrence descriptors, and by the adopted food consumption database. It is usually considered to be a conservative approach: the putative concentration in all the food commodities interested by residue(s) may be set at the proposed MRL by default, thus calculating a Theoretical Maximum Daily Intake (TMDI). However, the median residues concentration, when supported by an adequate data set, has been adopted instead of the MRL (WHO/FAO 2006) as the best point estimate of a central tendency over a prolonged period. The resultant dietary exposure estimate using the median residue concentration is termed the estimated daily intake (EDI), restricted to long-term dietary exposure estimation. For instance, the EDI calculation, where "muscle" refers to either meat or fish, will be as follows:

$$\begin{aligned}\text{EDI} = {} & (300\ \text{g} \times \text{median residue}_{\text{muscle}}) + (100\ \text{g} \times \text{median residue}_{\text{liver}}) \\ & + (50\ \text{g} \times \text{median residue}_{\text{kidney}}) + (50\ \text{g} \times \text{median residue}_{\text{fat}}) \\ & + (100\ \text{g} \times \text{median residue}_{\text{eggs}}) + (1500\ \text{g} \times \text{median residue}_{\text{milk}}) \\ & + (50\ \text{g} \times \text{median residue}_{\text{honey}}).\end{aligned}$$

In Table 4.15, the comparison in the computing of the TDMI and the EDI for the same veterinary drug residue is reported using the set of global standard food consumption figures. How the consumption figures were derived is described a few paragraphs below.

In the case of regulatory levels targeted on ARfD, the residues concentration considered is the MRL or high residue concentration derived from depletion studies, such as the upper one-sided 95% confidence limit over the 95th percentile residue concentration.

The other sensitive point where risk assessment supports the risk management in the MRL setting relies on the food consumption database considered compared to food intake estimates.

Within the Codex Alimentarius activities, the Joint Expert Meeting in charge to evaluate MRL setting on veterinary drug residues and more in general on chemicals (JECFA) historically considered a global standard food consumption database until 2005. Such database reported the daily per capita food consumption estimates for each macro-commodity based on the five standard regional diets (Middle Eastern, Far Eastern, African, Latin American, and European). Under a conservative approach, the highest consumption of a certain food category was accounted for (see Table 4.15). From 2006, the five regional diets have been progressively replaced by the Global Environment Monitoring System/Food Contamination Monitoring and Assessment Programme consumption cluster diets (WHO 2012). Currently, the FAO, in partnership WHO and other international partners, is building a harmonised pilot Global Individual Food consumption data Tool (FAO/WHO GIFT) so that the data can be comparable between different population groups and geographical areas. In this connection, a food classification system applicable at global level is also under construction in collaboration with the European Food Safety Authority (EFSA).

Within this frame, the consistency of regulatory residues levels in food with acute and chronic toxicology guidance values has been proposed to be assessed through the global estimates of chronic and acute dietary exposure (GECDE and GEADE). The GECDE to a veterinary drug residue for the general population is the highest exposure calculated using the 97.5th percentile consumption figure for a single food selected from all the foods, plus the mean dietary exposure from all the other relevant foods. In most cases, the food with the highest estimate of exposure using the 97.5th

Table 4.15 TMDI and EDI assessment accounting for the FAO global standard diet, for a veterinary drug residue with an Acceptable Daily Intake (ADI) of 0.420 mg/person and with different Maximum Residue Limits (MRL) set for the different food items

Food category	Consumption (g/person/day)	MRL (ng/g)	Occurrence P50 (ng/g)	Intake TMDI (ng/person)	Intake EDI
Muscle	300	150	38	45,000	11,400
Liver	100	150	38	15,000	3800
Kidney	50	200	145	10,000	7250
Fat	50	150	82	7500	4100
Milk	1500	50	11	75,000	16,500
Egg	100	300	24	30,000	2400
Intake estimates	2100			182,500	45,450
ADI				420,000	420,000
MOS				0.43	0.11

The Margin of safety (MOS) is the ratio of the intake estimates and the ADI for a 60 kg BW person (from FAO/WHO 2012, modified)

percentile consumption value drives the resulting dietary exposure estimate. In the rare case where two foods have similar 97.5th percentile exposure values, the calculation should be undertaken for each one to determine the higher GECDE.

As a more conservative approach, the highest reported mean food consumption values (or the highest 97.5th percentile value derived from surveys with individual records) of those available for consumers in different countries can be selected. When necessary, the considered food consumption values are targeted on the most sensitive group, such as children aged 2–6 years, as a consequence of their higher food intakes with respect to their body mass. Particular attention is paid to those foods most frequently consumed (i.e. milk and dairy products against liver and offals).

On the other hand, the global estimate of acute dietary exposure, or GEADE, considers the high daily consumption (97.5th percentile) of a single food item (meat, offal, milk, or other items) for consumers, accounting for the occurrence of the highest residue concentration observed/expected (WHO/FAO 2012).

Owing to the above, the MRL setting is affected by different uncertainty factors, mainly related to the quality and representativeness of the food consumption databases, and also accounting for the variability in human body weights. A high-level aggregation of food consumption data (e.g. one consumption factor for total muscle consumption, as opposed to beef muscle + pork muscle + chicken muscle, etc.) on one side simplifies the exposure assessment targeted to evaluate the consistency of the MRL. However, there might be situations where exposure assessment at the individual animal species and/or edible tissue level would be required or preferred, using specific data at a much lower level. This could be the case for huge differences in the contamination level of pollutants in the different seafood species, for the pharmacodynamics of certain antimicrobials in the lung, and for a food item that at the local level may represent a relevant food source.

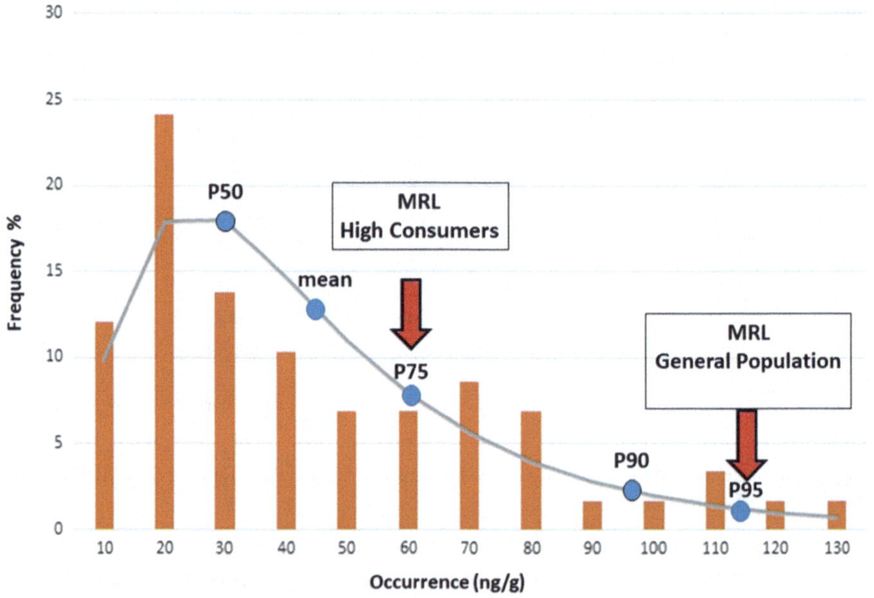

Fig. 4.24 Exemplification of food safety/food security issues in the presence of log–normal distribution of the occurrence of a residue in a food commodity

Data shown in Fig. 4.24 are based on a scenario of a TDI of 18,000 ng/person day, and a mean and P95 food consumption figures, and an averaged contamination of most consumed seafood species. Groups above the dotted line reach the TWI for methylmercury at a lower number of servings per week than needed to attain the DRV. When computed on 95th percentile consumption figures, almost all the groups considered exceeded the TWI (from EFSA 2015a).

There are situations where the regulatory levels are not able to protect all of the population. This is the case of those contaminants showing a log-normal distribution skewed towards higher levels in some food commodities (persistent organic pollutants and methylmercury in fish and seafood), with great differences in food consumption patterns (sensitivity) among the population (i.e. coastal population vs. inland population), and in the time-framed physiological susceptibility to the hazard (vulnerability), i.e. prenatal exposures (EFSA 2015j) (Fig. 4.26). The setting of the most conservative levels, even if broken down for each animal species within the same food commodity would hamper the access and trade to a valuable food resource: under such a situation, the management options include the release of advisories addressed to vulnerable groups, along with the specification of the food commodity, its geographical origin, and suggested consumption figures/frequency of the servings.

Figure 4.25 shows an example drawn from the Swedish Food Agency, addressed to women of reproductive age. Other risk management options rely on the setting of dedicated regulatory limits in food for infants, such in the case of

Advice about food for you who are breast feeding. The Swedish National Food Adminsitration, 2008

Fish and shellfish is good for you

Fish and shellfish are rich in vitamin D, iodine and selenium. All of which are important when you are breastfeeding. Oily fish, such as salmon and mackarel, also contain omega 3-fat. So eat 2-3 times a week and choose different kindes, both oily and non-oily Just as when you wer pregnant there are some kinds of fish that may contain raised levels of mercury or dioxins and PCB. You shoud not eat these as often, maximum 2-3 times a year.
This is particular relevant if you eat fish that have not been caugh commercially since many of the kinds in question are caught by anglers.

Eat 2-3 times a week, choose different kinds
Examples of ordinary fish, fish products and that are save to eat

All farmed fish
Alaska pollock
Anchovies
Blue mussels
Canned zuna
Catfish
Crab, with white flesh
Crayfish
Fishballs
Flounders/dabs
Haddock
Hake
Herring, produced pickled
Hoki
Lobster
Mackerel
Plaice
Prawns
Saithe
Salmon and trout
Sardines
Scallopw
Stockfish
Tilapia
Whitefish

The list is not exhaustive and there are many other kind that are good to eat

A couple of times a year

Eat maximum 2-3 times a year – because of mercury

- Atlantic halibut (Hippoglossus hippoglossus)
- Burbot
- Perch
- Pike
- Pikeperch
- Ray
- Shark
- Swordfish
- Tuna, freshly frozen

Eat maximum 2-3 times a year – because of dioxin, PCB

- Baltic herring, fermented Baltic herring

Salmon and salmon trout from the Baltic, Lake Vättern and char from lake Vättern. The advice primarily conserns woman who eat fish zhau have been caught non-commercially, since those fish are seldom to be found in normal shopw.

Certain fish are less good to eat for environmental reasons. Read more about fish and the enviroenment on www.livsmedelsverket.se

Fig. 4.25 Advisories released to pregnant and breastfeeding women, to prevent unacceptable exposures from methylmercury and persistent organic pollutants associated with the consumption of geo-referenced fish and seafood species

pesticides, heavy metals, and mycotoxins, the adoption of all the measures able to prevent or minimize the entry of the hazard in the food chain, such as the implementation of the good agriculture and animal farming practices (FAO and OIE 2009), the setting of environmental quality standards tailored on food safety,

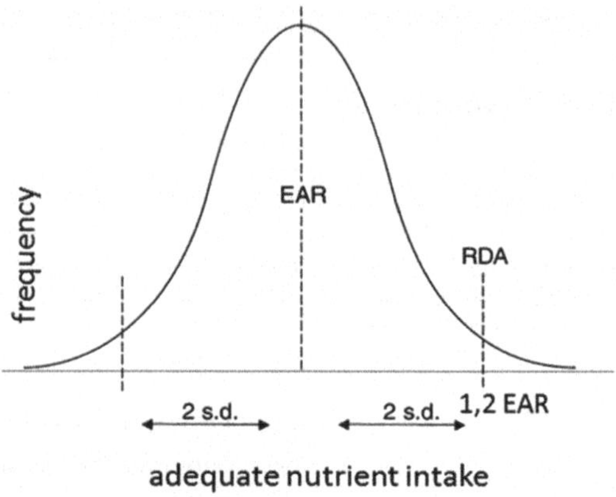

Fig. 4.26 Distribution of the individual nutrient requirements in a population assuming that the requirements are normally distributed. The *R*ecommended *D*aily *A*llowance (RDA) is a sum of the Average Requirement (AR) and two standard deviations (SD), if the requirements and their variation between individuals are known. Otherwise, the RDA will be calculated assuming a coefficient of variation of 10% of the estimated AR leading to an RDA of 1.2 EAR (modified to EFSA 2010)

and the definition of food safety-based end-of-waste criteria within the circular economy (Saveyn and Eder 2014).

The influence of risk assessment on risk management options can be expected when unintended/unmonitored chemicals are detected in food and drinking water, with a probability of human exposure. The risk managers In such cases should be supported by a risk prioritisation approach. EFSA and WHO proposed a tiered approach based on the Threshold of Toxicological Concern (TTC) for the safety assessment of those chemicals when hazard data are incomplete and human exposure can be estimated. For genotoxic compounds, a TTC of 0.0025 μg/kg of BW per day has been proposed as sufficiently protective, based on linear extrapolation for known genotoxic carcinogens. High potency carcinogens, i.e. aflatoxin-like, azoxy- or N-nitroso-compounds, and benzidines, are excluded from this approach (EFSA 2016e). A TTC is not applicable when compound-specific assessment and toxicity data are available, or are required under existing regulations. From such a TTC, it could be possible to derive residue limits in food and drinking water.

On the other hand, very strict regulations based on the lowest achievable analytical performances in residue analysis, also supported by a toxicology assessment, have been issued for illegally administered drugs to food-producing animals, such abuse has been proved to provoke acute intoxication outbreaks and concern in the public opinion about the safety of all the meat in the food chain. In the past, gynecomastia in prepubertal age children due to the presence of the injection sites of anabolic steroids in the meat used for baby food production was described (Loizzo et al. 1984); more recently, several cases of collective poisoning with tachycardia and muscular tremors have been reported in different countries after the ingestion of pork or beef liver tainted with clenbuterol, a bronchodilator drug illegally administered to feedlot as a meat leaning agent (Yan et al. 2014). EFSA was asked by the European Commission to deliver Reference Points for Action (RPAs) for non-allowed pharmacologically active substances in food of animal origin (EFSA 2013b). Such RPAs are expressed as the concentration for a non-allowed pharmacologically active substance that can be determined by official control laboratories and is low enough to adequately protect the consumers of food commodities that contain that substance. Such RPAs can not be applied when the analytical controls are targeted mainly on living animals and on their non-edible matrices (urine, blood, hair, faeces), or as an anti-fraud action, or to guarantee and protect those farmers working under good practices, or to prevent the rise of concern in the public opinion, and when there is no toxicological threshold for residues in the cases of malachite green and nitrofuran metabolites. In such cases, the regulatory level is based on the minimum required performance limit (MRPL) on analytical basis.

4.15.4 From Risk Management to Risk Assessment

The regulatory core of the food safety management is represented by the set-up and the implementation of National residue monitoring plans (EU Council Dir 93/26/EC), with qualifying aspects targeted on criteria for sampling and analysis with respect to the different food items and substances, the appointment of official personnel and structure in charge of the duties, the Quality Assurance and Quality Control requirements of the analytical labs, and the fluxes of data and information among risk managers, risk assessors, and stakeholders (EU Decisions 98/179/EC (EC 1998) and 2002/657/EC).

The open question is how data from regular monitoring activities can feed the risk assessment for alimentary exposures. To this respect, the following particularities in each step of the food monitoring strategy should be considered.

Sampling The samples figures and the representativeness of the samples drawn are usually tailored to the number of farms at the national level and on the amount of raw food produced. Therefore, the samples are more representative of the agriculture farms/production at the national level, and not necessarily of the food items most consumed and/or most contributing to the alimentary exposure for a determined substance or group of substances. In most of the cases, the analysis is carried out on the raw food commodity: this implies potential variation in the food contamination due to the ability of processing/cooking treatments to reduce or increase the contaminant burden. For instance, during grill cooking, the lipids loss may reduce the burden of lipophilic contaminants, but you can expect an increase of Polycyclic Aromatic Hydrocarbons (PAH) released from the combustion process. This implies that occurrence data in raw food maybe not directly transposed into intake estimates. However, the added value of data achieved under official monitoring plan relies on non-biased sampling performed by independent personnel.

Analysis The requested practicability and feasibility of residues monitoring plans, and the need to give a feedback in a short time-frame about the compliance or non-compliance of the food item, encourage the use of highthroughput and cost-saving qualitative/semi-quantitative screening methods, that do not provide quantitative data for the risk assessment. The dynamic range of the quantitative analysis of residues driven by the fixed MRL (usually a validated and accredited method should span within the MRL/5 – MRL x 2 range) could affect the statistical descriptors of the occurrence of contamination, i.e. due to the high rate of left-censored data, and/or due to an indicative quantification in those samples with an outlier contamination that undoubtedly trespasses the regulatory limit.

The strength is the quality of the analysis carried out by a network of officially appointed labs, working under Quality Assurance and Quality Control Systems, under the supervision of National Reference Labs, and in the case of the European Union, of European Reference Labs.

4.15.5 Towards an Improvement of the Regulatory Approach to Support Risk Assessment

The open discussion about a more risk-oriented planning of residue monitoring plans would help to improve the overall quality of data for risk management activities.

To this purpose, the following aspects could be envisaged as considerations to reduce the uncertainties brought into the risk assessment from risk management activities:

- To target the sampling activities towards more consumed, more susceptible to contamination, and more contributing to the overall intake of food items, along with appropriate sample standard descriptors, on which base the results can be reported in a harmonised way.
- To improve the dynamic range of the analysis towards lower levels in a way that uncertainties related to left-censored data should not bring into the intake assessment an uncertainty $>10\%$ of the target guidance level (Table 4.15). Therefore, the performance of the method with respect to the limit of determination should be tailored for each food and also accounting for its potential contribution to the intake and accounting for the guidance levels: higher consumption figures in milk would correspond to lower quantification limits, under the consideration that a low contaminated/high consumed food may contribute to the alimentary intake as or more than a high contaminated/low consumed food.
- To improve multi-target/not target analysis in a way to allow the evaluation of associated exposure to different residues for the evaluation of combined exposures, as in the case of pesticides.
- To set a biobank of food samples and their analytical extracts, in a way to allow retrospective studies, e.g. in the case of the onset of emerging risks.

Table 4.16 Example of the uncertainties (U) brought to the intake estimates from different Limit of Determinations of residues in food commodities

	Maximum intake		U as % on the overall intake (1800 ng/person/day)	
Food categor	g/person/day mL/ person/day	LOD 1 ng/g	LOD 0.1 ng/g	LOD 0.01 ng/g
1 Cereals	316	17.5	1.75	0.18
2 Sweets	82	4.55	0.46	0.05
3 Vegetables	332	18.4	1.84	0.18
4 Potatoes	195	10.8	1.08	0.11
5 Fruit	438	24.3	2.43	0.24
6 Meat	233	12.9	1.29	0.13
7 Fish	45	2.5	0.25	0.03
8 Eggs	38	2.11	0.21	0.02
9 Milk	391	21.7	2.17	0.22
10 Fats	71	3.94	0.39	0.04
11 Miscellaneous	50	2.78	0.27	0.08
12 + 13 Water and coffee. etc	1469	81.6	8.16	0.92
14 Alcoholic beverages	319	17.7	1.77	0.18
SUM	3979[a] g	221%	22.1%	2.23%

Remark: The maximisation of food consumption in general population is not realistic, as far as larger intakes in a single food category are usually antagonist towards other foods category (i.e., fish against meat intake); however, such maximisation better highlights the U% associated to the selected LOQs

- To channel when appropriate the results of internal checks performed by food producers and food retailers in the occurrence database, when achieved under QA/QC schemes.

Last but not least, the regular activation of Total Diet Studies (EFSA/FAO/WHO 2011) and Duplicate Diets (EFSA 2011g), targeted on general population and sensitive groups, would help to refine the uncertainties from the food processing/cooking, and from the consumption habits, when such habits are not fully intercepted or represented in the food frequency questionnaires. The simulation shown in Table 4.16 accounts for a daily reference dose for intakes of 30 ng/kg of BW and for an average body weight of 60 kg (maximum allowed overall intake = 1.800 ng/person/day). The selected categories of the food items reported account for a theoretical highest amount ingested (worst-case assumption in the general population).

4.16 Contaminants

Gerhard Heinemeyer

The term "contaminant" covers substances that are entering the food chain from the environment and from other sources than food, mostly without direct human influence. To get an overview about contaminants, the list of opinions prepared by the

EFSA expert panel for contaminants in feed and food (CONTAM[1] Sect. 7.1.2) offers a compilation of the main substances in feed and food. The focus here shall be on food, although there is an impact of contaminants in the feed which will reach food, e.g. via milk, meat, or other animal products. The main groups of substances that are addressed by the panel are toxins, metals, and hydrocarbons. The latter represent a heterogeneous group that can be classified in order of their chemical structure and their use. They include non-halogenated and halogenated (i.e. brominated, chlorinated, and fluorinated) hydrocarbons, e.g. PBB, PCB, dioxins and dioxin-like PCB's, and perfluoro compounds. Many of these substances are undesired chemicals, which are not used anywhere (dioxins), while some are part of industrial production processes and some are used as pesticides, insecticides, flame retardants, or plasticisers (phthalates and bisphenol A). Some substances have hormonal disruption activity in humans.

A large group in the list of the CONTAM panel is for mycotoxins and other toxins. Metals represent the third large group.

In general, the risk assessment for contaminants follows the principles that are described in this book. Risk assessment for contaminants has the objective to characterise and control the maximum limit values for substances in particular foods, and to adjust them to the actual data. Risk assessment opinions are prepared mostly on behalf of the European Commission and partly for agencies of member states. There is no special regulation as for pesticides or additives. The risk characterisation of contaminants follows the general strategy of hazard identification, exposure assessment, and risk characterisation as described in Sect. 2.2.5. Uncertainty analyses are regularly included in the opinions.

For chronic hazards, a TDI has to be derived, and the long-term exposure is evaluated. All known sources of food are included in the evaluation. If the substance has a long persistence time in the human body, short-term exposures must also be considered in the evaluation. Food consumption data are presented by national surveys in the comprehensive food consumption database. The requirements for choosing and compiling the food consumption data are pointed out in Sect. 4.6.5. Food occurrence data are available from national sources as described in Sect. 4.8. A particular focus is put on heavy metals (lead, cadmium, mercury) and mycotoxins (e.g. aflatoxin, ochratoxin, nivalenol, and zearalenone).

For acute hazards, the ARfD will be derived, and the short-term exposure is evaluated. This needs data that will inform about the portion size and provide an appropriate range of concentration data, including the high-end levels of concentrations. Although the list of substances leading to considerable acute risks is short, prominent examples are represented by marine biotoxins. The CONTAM panel of EFSA offered, for example a series of opinions for different toxins formed primarily

[1]http://www.efsa.europa.eu/en/search/site/contam%20panel?f[0]=sm_hierarchy_type%3Aopinion

by shellfish, all of them producing severe acute intoxications. The panel derived acute hazard levels and re-evaluated methods for analytical measurements of shellfish toxins to establish reliable concentration data.

In general, exposure estimation results in an average and a high-end estimate to characterise the average risk and the risk for consumers having particular habits leading to high exposure. How these results should be presented is pointed out in Sect. 3.6.

4.17 Food Additives

Rainer Gürtler and Davide Arcella

4.17.1 Introduction

Substances added intentionally to food such as flavouring substances and processing aids might be considered to be food additives. However, according to European legislation, the use of food additives and flavourings is regulated by separate regulations. These regulations also address the risk assessment of food additives and flavourings that is to be performed by the European Food Safety Authority (EFSA). Different approaches are used for the risk assessment of food additives and flavourings. Accordingly, the approaches used for exposure assessment also differ. This chapter focuses on specific aspects of the exposure assessment of food additives that are used for technological functions such as preservation or food colouring (but not on flavouring substances and processing aids). It provides the legal background and delineates the principles and requirements for the risk assessment, with a special focus on exposure. The approaches used for exposure assessment of food additives are described and the challenges relevant for monitoring the use and consumption of food additives are addressed. A major uncertainty is that the food categories as defined in food legislations are not fully identical with the food categories for which food consumption data are available.

4.17.2 Legal Background

The regulations, definitions, and approval processes for food additives may vary among different regulatory authorities and jurisdictions.

Generally, any substances added intentionally to food including flavouring substances, food enzymes, and processing aids might be considered to be food additives. According to European legislation, the term food additive is, however, restricted to those substances that are used for technological purposes only. According to Article 3 of Regulation (EC) No. 1333/2008 (EU 2008c) of the European Parliament and of the Council of 16 December 2008 on food additives, a food additive

> shall mean any substance not normally consumed as a food in itself and not normally used as a characteristic ingredient of food, whether or not it has nutritive value, the intentional addition of which to food for a technological purpose in the manufacture, processing, preparation, treatment, packaging, transport or storage of such food results, or may be reasonably expected to result, in it or its by-products becoming directly or indirectly a component of such foods.

Thus, substances that have a technological function, such as preservatives, antioxidants, food colours, sweeteners, emulsifiers, stabilisers, or gelling agents, comply with the definition of a food additive, while flavouring substances, food enzymes, and processing aids are regulated separately in the European Union.[2] According to Article 6 of Regulation (EC) No. 1333/2008 (EU 2008c), a permission for the use of food additives can only be given for a substance provided that

> (a) it does not, on the basis of the scientific evidence available, pose a safety concern to the health of the consumer at the level of use proposed; (b) there is a reasonable technological need that cannot be achieved by other economically and technologically practicable means; and (c) its use does not mislead the consumer.

Consequently, the safety of food additives is to be assessed by EFSA prior consideration by the European Commission for market authorisation. Applicants seeking such an authorisation are required to provide all of the relevant data.

Globally, food additive standards were established by the Codex Committee on Food Additives (CCFA)[3] of the Codex Alimentarius Commission (CAC)[4] in the framework of the Joint Food and Agriculture Organization (FAO) of the United Nations/World Health Organization (WHO) food standards programme. In this respect, the CCFA established a General Standard for Food Additives (GSFA)[5] in which provisions for the use of food additives in certain food categories were established. The task of the Codex Alimentarius Commission is, among other things, to develop and harmonise worldwide health standards. Another task of this Commission is to issue guidelines and recommendations on agricultural and fishery products, foodstuffs, food additives and other substances with the aim to protect the health of consumers and to ensure fair practices in international trade.

---

[2]Flavouring substances are regulated according to Regulation (EC) No 1334/2008 (EU 2008d) and food enzymes according to Regulation (EC) No 1332/2008 (EU 2008b); Regulation (EC) No 1331/2008 (EU 2008a) established a common authorisation procedure for food additives, food enzymes and food flavourings. While processing aids are defined in Article 3 of Regulation (EC) No 1333/2008 (EU 2008c), their use does, apart from extraction solvents, not require an approval in the European Union. For extraction solvents used in the production of foodstuffs and food ingredients, the Directive 2009/32/EC (EU 2009) of the European Parliament and of the Council is applicable.

[3]http://www.fao.org/fao-who-codexalimentarius/committees/committee-detail/en/?committee=CCFA

[4]http://www.fao.org/fao-who-codexalimentarius/en/

[5]http://www.fao.org/fao-who-codexalimentarius/codex-texts/dbs/gsfa/en/

Accordingly, while the GSFA is not legally binding in the European Union, it may have relevance for trade issues between the European Union and third countries when it comes to negotiations at the World Trade Organization (WTO).

4.17.3 International Scientific Expert Committees Evaluating the Safety of the Use of Food Additives

The risk assessment for the use of food additives was established at a global level in the middle of the twentieth century. In 1955, the Food and Agriculture Organization (FAO) and the World Health Organization (WHO) of the United Nations installed the Joint FAO/WHO Expert Committee on Food Additives (JECFA).[6] In its first report, JECFA provided general principles for the use of food additives and pointed out that

> critically designed animal tests of the physiological, pharmacological and biochemical behaviour of a proposed additive can provide a reasonable basis for evaluating the safety of use of a food additive at a specified level of intake (WHO 1957).

Procedures for the testing of food additives were first published by JECFA in 1958 (WHO 1958). After establishing the EFSA in 2003, food additives in Europe have originally been evaluated by the respective EFSA Panels (AFC until 2008, ANS from 2008 to 2018 and FAF since 2018, see Sect. 7.1).

4.17.4 Data Required for the Risk Assessment of Food Additives

JECFA defined principles for the safety assessment of food additives and contaminants in 1987 (WHO 1987). Later, it published principles and methods for the risk assessment of chemicals in food (WHO/FAO 2009). JECFA currently performs risk assessments of food additives according to these principles and methods.

The requirements applicable in the European Union for the risk assessment of food additives are described in a guidance document adopted by the EFSA ANS Panel in 2012 (EFSA 2012d). According to this guidance, a tiered approach that balances toxicity data requirements against the risk is applicable for the risk assessment of food additives. The approach was established to evaluate the following core areas: toxicokinetics, genotoxicity, toxicity (encompassing subchronic toxicity, chronic toxicity, and carcinogenicity), and reproductive and developmental toxicity. Additional studies on other toxicological endpoints may be required on a case-by-case basis. The approach consists of three tiers. It provides guidance to applicants for

[6]http://www.fao.org/fao-who-codexalimentarius/scientific-basis/jecfa/en/http://www.who.int/foodsafety/areas_work/chemical-risks/jecfa/en/

designing their toxicity testing strategy in which the decision on the requirement for further testing can be based on the results of certain initial studies. While maintaining the high level of consumer safety, the application of this tiered approach could result in a smaller number of animal tests or more refined animal studies compared to the requirements applied before, and thus benefit animal welfare.

According to EFSA's guidance for submission of food additive applications (EFSA 2012d), a minimum set of data is required for all food additives at tier 1. It covers data on absorption (i.e. systemic availability), genotoxicity in vitro, and subchronic toxicity. Depending on the results, further toxicity studies are required at tier 2 and 3. Tier 2 studies will be required in order to generate more extensive data for substances that are absorbed and demonstrate toxicity or genotoxicity in tier 1 tests. If higher tier testing is required based on results in one of the core areas, such testing would only be required in this core area (but not in the other areas). For example, if results from the subchronic toxicity study require further tier 2 studies but the tier 1 in vitro genotoxicity is clearly negative, there would be no need for tier 2 follow-up of genotoxicity. Tier 3 testing should be performed on a case-by-case basis. Results at higher tiers will in principle supersede the results observed at lower tiers. The risk assessment approach applied by EFSA is described in more detail by Gürtler (2014).

4.17.5 Derivation of an Acceptable Daily Intake

The main purpose of the risk assessment of food additives is the derivation of an Acceptable Daily Intake (ADI) as a health-based guidance value. The ADI is applicable to the general population except for infants below 12–16 weeks (WHO 1978; SCF 1998; EFSA 2017c). Based on a numerical ADI and an exposure assessment, the conditions of use, e.g. maximum levels for certain food categories, can be derived for food additives by risk managers. The current conditions of use applicable in the European Union are defined in Regulation (EC) No. 1333/2008 (EU 2008c) in combination with more than 100 Commission Regulations amending its annexes.[7] There are no numerical ADIs for some food additives. For substances of very low toxicity that do not represent a hazard to health on the basis of the available data on toxicity and intake, e.g. citric acid and citrates (E 330–E 333), the outcome of the evaluation by JECFA or SCF was "ADI not specified". Such substances may be used at "quantum satis" level. According to Regulation (EC) No. 1333/2008 (EU 2008c),

> *quantum satis shall mean that no maximum numerical level is specified and substances shall be used in accordance with good manufacturing practice, at a level not higher than is necessary to achieve the intended purpose and provided the consumer is not misled.*

[7]Regulation (EC) No 1333/2008, all Commission Regulations amending its annexes as well as the current consolidated version of this regulation are available at http://eur-lex.europa.eu/legal-content/en/ALL/?uri=CELEX:32008R1333

An occasional exceedance of the ADI might be tolerated provided that (1) there is still a tolerable margin of safety between the NOAEL and the exposure, (2) the effect on which the ADI was derived is not observed after acute exposure, and (3) the exceedance does not occur frequently so that the long-term exposure is not significantly affected (Gürtler 2010). However, the provisions of use should ensure that the ADI is generally not exceeded. Therefore, an exposure assessment is required that takes into account not only the reported uses but also the provisions of use, e.g. the maximum levels for certain food categories.

4.17.6 Requirement for Data on Exposure Assessment

In the framework of the Codex Alimentarius, the endorsement of the proposed use by the Codex Committee on Food Additives should take into account the ADI and the probable daily dietary exposure to the additive from all food sources (FAO 2014).

In the European Union, it is stated in Article 13 of the Commission Regulation (EU) No 234/2011 of 10 March 2011 implementing Regulation (EC) No 1331/2008 of the European Parliament and of the Council establishing a common authorisation procedure for food additives, food enzymes, and food flavourings (EU 2008a, 2011c) that the risk assessment shall include

> a dietary exposure assessment for the European population taking into account other possible sources of dietary exposure; (...) when the dietary exposure exceeds the health-based guidance value identified in the overall risk assessment, the dietary exposure assessment of the substance shall be detailed, providing where possible the contribution to the total exposure of each food category or foodstuff for which the use is authorised or has been requested.

Commission Regulation (EU) No. 257/2010 (EU 2010) set up a programme for the re-evaluation of approved food additives in accordance with Regulation (EC) No. 1333/2008 (EU 2008c) within the European Union. In the framework of the re-evaluation programme, EFSA is currently evaluating the safety of all food additives authorised for use in the EU prior to 20 January 2009. This evaluation also includes an assessment of exposure under different scenarios.

To enable the risk assessment, information should be provided on known or anticipated human exposure (including data from epidemiological or biomonitoring studies) to the proposed additive from food (including natural dietary sources) or toxicologically relevant components of the additive, and any other potential non-dietary sources (e.g. from drinking water, consumer products such as cosmetics and pharmaceuticals), when information on exposure from these sources is available. Exposure estimates are also to be provided on any potential exposure to residues or contaminants present because of the use of the additive.

4.17.7 Approaches for the Exposure Assessment of Food Additives

There are several possible approaches to evaluate the intake of a food additive. Ideally, the intake can be assessed by combining data on concentration in all food items with their consumption. However, it is considered to be neither cost effective nor necessary to collect detailed data for every additive (Lawrie and Rees 1996; Nutriscan 1992; Petersen et al. 1994). In this context, the adoption of a tiered approach seems to be the most suitable. The first step provides potential intakes of additives using highly aggregated data. If the intake levels exceed the ADI, a new step is performed by including more detailed information on food consumption and additive occurrence. The process is stopped if the ADI is not exceeded (Gibney and Lambe 1996). Such a prioritisation system avoids the wasting of resources to collect large amounts of unnecessary data. A tiered approach for the exposure estimation for food additives is also recommended by FAO and WHO (EHC 240; WHO/FAO 2009), and is used by JECFA for the evaluation of food additives. According to these FAO/WHO recommendations, the initial steps rely on conservative screening methods such as the budget method or models based on poundage data.

Different methods including indirect (household or national level) and direct (retrospective or prospective at the individual level) methods have been discussed for the estimation of food additive intake (Chaisson and Petersen 1996; Gibney and Lambe 1996; Gibney 1999; Gilsenan et al. 2003a, b; Howlett 1996; Langlais 1996; Lawrie and Rees 1996; Löwik 1996; Penttilä 1996; Verger 1996; Wagstaffe 1996).

The "budget method", a conservative technique that was originally developed to determine food additive use limits has now been used as a screening tool for food additive intake in order to establish monitoring priorities (Hansen 1966, 1979, 1990; Douglass et al. 1997; EU 1998, 2001; WHO 2008). The budget method is based on assumptions regarding the maximum intakes of food and beverages that potentially contain the additives (Hansen 1966, 1979). In this context, conservative assumptions were made on (1) the level of consumption of foods and non-milk beverages, (2) the level of presence of the additives in foods and non-milk beverages, and (3) the proportion of foods and non-milk beverages that may contain the additives. More specifically, the levels of consumption of foods and beverages considered are the maximum physiological levels of consumption, i.e. the daily consumption of 0.1 L/kg body weight of non-milk beverages and the daily consumption of 100 kcal/kg body weight from foods (equivalent to 0.05 kg/kg body weight based on an estimated energy density of 2 kcal/g) (Hansen 1979). In a 60-kg person, these levels correspond to the daily consumption of 6 litres of non-milk beverages and 3 kg of food (WHO 2008; WHO/FAO 2009). The budget method has also been used within a tiered approach of food additive monitoring at the first tier (tier 1) in the European Union (EU 1998, 2001).

In 2012, EFSA developed the Food Additive Intake Model (FAIM) template (EFSA 2012e, 2013f) to support the calculation of exposure estimates to food additives. The FAIM template is a screening tool for estimating long term exposure to food additives and can be used as a first step in the dietary exposure assessment process by applicants, risk assessors, and risk managers. It is based on summary food consumption statistics (in grams per kg body weight) from the EFSA Comprehensive European Food Consumption Database tailored to food additive categories. FAIM allows the user to estimate the mean and high-level exposure to food additives for different population groups throughout several European countries and age groups. According to the "Guidance for submission for food additive evaluations" (EFSA 2012d), two scenarios can be applied, taking into account:

- The Maximum Permitted Levels (MPLs) of use, as set in the current EU legislation and
- The levels of use as reported for existing food additives, if available, or as proposed, in the case of new applications.

The model also provides information on the food groups contributing to the total mean exposure. The FAIM template can be used for the estimation of exposure to a new food additive or exposure resulting from new uses of an already authorised food additive.[8] Recently, a model for food additives similar to the FAIM template was suggested by Tennant (2016). It is the Comprehensive European Dietary Exposure Model (CEDEM).

More refined methods should be applied when screening methods cannot rule out a safety concern. With access to food consumption survey information at an individual level, chemical concentrations and food consumption can be directly matched for each subject and the respective food or food category and a distribution of total exposure estimates can be produced and any percentile of interest can be calculated. The average contribution of each food or food category to the total exposure can also be estimated. EFSA started to use the individual food consumption data and body weight information from the Concise Database for the assessment of exposure to contaminants in 2010. Data from the EFSA Comprehensive European Food Consumption Database (Comprehensive Database) (EFSA 2011a) are currently used and this is now EFSA's preferred model for refined exposure assessment. This database is used to estimate exposure for the six following population groups: infants (from 12 weeks of age), toddlers, children, adolescents, adults, and the elderly. Dietary exposure estimates are in the majority of cases produced by considering the consumption of all foods and all subjects involved in the dietary surveys, but estimates for "consumers only" are also calculated under specific circumstances.

A central aspect in the assessment of exposure is the concentration of the additive in food as consumed. Depending on the purpose of the exposure assessment (pre-regulation or post-regulation), food additive concentration data can originate from:

- Legislated Maximum Permitted Levels (MPLs)
- Usage levels reported by manufacturers
- Monitoring and surveillance programs
- Total Diet Study (TDS).

[8]An updated FAIM version is available at https://www.efsa.europa.eu/en/applications/foodingredients/tools

Typically, very limited information is provided about the representativeness of the usage levels submitted with respect to the food market share. In the case that a reported usage level refers to a niche product that belongs to a food category for which additional reported usage levels are available, this reported use level is excluded from the exposure assessment. The representativeness of the data will vary according to the measurement method, whether it is based on estimated levels or actual analytical results, the sampling strategy, and the market coverage.

The availability of data per food additive and food category can vary, ranging from no data up to very large amounts of data. A significant part of these analytical results may be left censored (i.e. either below the LOD or the LOQ) or qualitative to indicate the presence or absence of the food additive in the sample. In addition, analytical results may be provided for food categories in which a given food additive is not authorised. The presence of the food additive in those food categories could, for instance, be due to carry-over or natural occurrence. To consider left-censored analytical data, the substitution method as recommended in the "Principles and Methods for the Risk Assessment of Chemicals in Food" (WHO/FAO 2009) and the EFSA scientific report "Management of left-censored data in dietary exposure assessment of chemical substances" (EFSA 2010c) is used and the mean or median, whichever is highest, middle-bound concentration is calculated. In the refined dietary exposure estimates, only authorised uses with analytical data at or below the MPL are considered because results above the MPL are part of risk management measures, e.g. for non-compliance purposes. For this reason, such results are not taken into account, except in specific cases (e.g. in the case of natural occurrence).

In the framework of the re-evaluation programme, EFSA regularly issues public calls for usage and/or concentration data (analytical data) on food additives to be re-evaluated. The food industry, food industry associations, and food additive (s) producers provide data on the usage levels of food additives. Analytical data are provided mainly by national food authorities of the Member States (MSs).

Exposure assessments of food additives under re-evaluation are carried out by the EFSA FAF Panel.[9] They are based on two different sets of concentration data: (a) maximum permitted levels (MPLs) of use set down in the EU legislation (defined as a regulatory maximum level exposure assessment scenario) and (b) usage levels and/or analytical data provided through the calls for data (defined as a refined exposure assessment scenario).

Based on the available data set, refined exposure estimates are calculated based on different model populations. In the brand-loyal consumer scenario, it is assumed that a consumer is exposed long-term to the food additive present at the maximum reported use/analytical level for one food category and at the typical reported use levels or the mean of analytical results, whichever was highest or available, for the remaining food categories. In the non-brand-loyal consumer scenario, it is assumed that a consumer is exposed long term to the food additive present at the mean reported use/analytical levels in food.

[9]Food additives were evaluated by the ANS Panel until June 2018. Since then, they are evaluated by the FAF Panel.

Additional exposure scenarios might be contemplated in order to consider the presence of food additives due to carry-over or due to natural occurrence. Also, ad hoc exposure scenarios on consumers only are considered for food supplements in all age groups, and for foods for special medical purposes (FSMP) in infants and toddlers.

This approach is described, for example in the scientific opinion on Indigo Carmine (EFSA 2014b).

Since 2015, the Mintel's Global New Products Database (GNPD)[10] has been used by EFSA for checking the labelling of products containing food additives. The Mintel GNPD is an online database that monitors product introductions in consumer packaged goods markets worldwide. It allows users to search for the compulsory ingredient information presented in the labelling of products. The Mintel GNPD currently contains information for over two million food and beverage products, of which more than 900,000 are or have been available on the European food market.

Additional databases used for the assessment of exposure to food additives are the Irish National Food Ingredient Database (INFID) (Gilsenan et al. 2002) and a German database (Diouf et al. 2014). The German database was created to enable the tier 2b approach while accessing the food additive intake for infants, toddlers, and children in Germany. In this database, the consumption data of two dietary surveys were combined with the permitted use levels of food additives of Council Regulation (EC) No. 1129/2011 (EU 2011a) to enable a tier 2 approach. The connection of these data with packaging information on the occurrence of food additives in the foods consumed allowed the tier 2b approach (Diouf et al. 2014). Data on the occurrence of food additives in food are also available for a limited number of food additives investigated as part of the Flavourings, Additives, and Food Contact materials Exposure Tool (FACET) project (Vin et al. 2013). This tool facilitated the investigation of exposure levels to 13 food additives in four European countries, at both tier 2 and tier 3 levels (Vin et al. 2013).

4.17.8 Guidance Documents of International Expert Committees on the Estimation of Food Additive Intake

JECFA noted in its principles for the safety assessment of food additives and contaminants in 1987 (WHO 1987), that

> it is not possible to describe specific procedures for estimating exposure for all food additive and contaminant situations.

However, JECFA described the three general types of estimation procedures that have been accepted by previous Committees: per capita estimates, estimates from dietary food intake surveys, and analytical values from market-basket/total-diet surveys. While JECFA at that time described these types of exposure estimation

[10]http://www.mintel.com/global-new-products-database

procedures only briefly, referring to guidelines for the study of dietary intakes of chemical contaminants published in the framework of the Global Environmental Monitoring System (WHO 1985), more detailed attention was given to the estimation of exposure to food additives in its Principles and Methods for the Risk Assessment of Chemicals in Food (EHC 240) published in 2009 (WHO/FAO 2009). According to these guidelines (EHC 240), the mean concentration of food additives is often used to reflect the expected concentration in food over time. It may be derived from manufacturers' use data and monitoring or surveillance data (WHO/FAO 2009).

The FAO/WHO published guidelines for the preparation of working papers on the intake of food additives for the Joint FAO/WHO Expert Committee on Food Additives (FAO 2001).

According to the call for data for the 84th JECFA meeting (WHO 2016), the following data were required for the intake assessment for food additives:

All data relevant to:

1. Technical levels of use of the additive in the foods in which it may be used.
2. Annual poundage of the additive introduced into the food supply.
3. Estimation of additive intakes based on food consumption data for foods in which the additive may be used.
4. Food consumption patterns; also considering different (age-)population groups.

The Codex Alimentarius Commission established guidelines for the simple evaluation of dietary exposure to food additives in 1989 that were revised in 2014 (FAO 2014). This document describes a stepwise approach to estimate exposure to additives in order to check if an ADI is potentially exceeded. These guidelines are intended to facilitate the work of governments, particularly for countries with limited resources, on the assessment of dietary exposure to food additives by reflecting current procedures in place to carry out such work in a simple way. It is, however, also pointed out that these guidelines are not intended to provide support to CCFA on the work on the GSFA, as JECFA is the international expert scientific advisory body to provide such advice to the committee based on the Principles and Methods for The Risk Assessment of Chemicals in Food - Environmental Health Criteria (EHC 240; WHO/FAO 2009).

The EFSA Scientific Committee (EFSA 2005a) published a scientific opinion related to exposure assessments and made reference to the FAO/WHO guidelines for the preparation of working papers on the intake of food additives for the Joint FAO/WHO Expert Committee on Food Additives (FAO 2001). The guidance for submission for food additive evaluations of the EFSA ANS Panel also contains a section on exposure assessment (EFSA 2012d).

Recent Exposure Assessments for Food Additives

The results of the evaluations performed by the EFSA are published on its website.[11] There are also a number of studies in which the intake of food additives has been investigated, e.g. Leclercq et al. (1999, 2000, 2003), Arcella et al. (2003, 2004), Husain et al. (2006), Soubra et al. (2007), Bemrah et al. (2008), Tennant (2008), Connolly et al. (2010), Dixit et al. (2011), Mischek and Krapfenbauer-Cermak (2012), Diogo et al. (2013), Ha et al. (2013a, b), Vin et al. (2013), Mancini et al. (2015), Doell et al. (2016), Martyn et al. (2016), O'Sullivan et al. (2016) and Gajda-Wyrębek et al. (2017).

A stepwise approach [using EFSA's Food Additives intake model (FAIM), followed by tier 2 and refined tier 2] was used to assess the exposure of the Belgian adult population to 70 food additives (Van Loco et al. 2015).

In several cases in which the EFSA identified a safety concern because the estimated exposures exceeded the respective ADI, a refined exposure assessment was performed based on new data, e.g. new suggested or reported use levels or new analytical data (EFSA 2013h, i, 2014f, g, 2015h, k, l, 2016h). In all of these cases, the refined exposure estimates were lower than the previous estimates.

One example is the refined exposure assessment for Ponceau 4R (cited from EFSA 2015l):

> In 2009, the EFSA ANS Panel "concluded that, at the maximum levels of use, intake estimates for adults at the high percentile (97.5th) and for 1–10-year old children at the mean and high percentiles (95th/97.5th) were generally above the ADI, even for the refined intake estimates. Consequently, Annex II to Regulation (EC) No 1333/2008 was amended by the European Commission regarding the conditions of use of Ponceau 4R (E 124), applicable as from 1 June 2013. Many Maximum Permitted Levels (MPLs) of Ponceau 4R (E 124) were withdrawn ($n = 24$) or decreased by a factor 1.2 to 200 ($n = 29$), depending on the food category. Following this, EFSA performed a refined exposure assessment for this food colour, using new usage data from industry, as well as analytical data submitted to EFSA by Member States, combined with food consumption data of the EFSA Comprehensive European Food Consumption Database. Usage data from industry were provided for three out of the 31 food categories for which Ponceau 4R (E 124) is authorised as a food additive; while analytical data from Member States were provided to EFSA for 18 food categories. Using MPLs and these new data, none of the exposure estimates exceeded the ADI of 0.7 mg/kg BW per day in any population" (EFSA 2015l).

4.17.9 *Challenges Related to the Exposure Estimation for Food Additives and the Monitoring of Food Additive Intake*

The links between food consumption information on the one hand and food additive concentration data on the other are mostly not compatible with each other. The

[11] http://www.efsa.europa.eu/en/topics/topic/food-additives

national and international food description and classification systems are basically designed for raw food commodities (see Sect. 4.4) and focus, for example on food consumption, food composition, legislation, or trade. National food classification systems are often based on national criteria and the food groups can be very specific. However, the systems are not necessarily compatible and the user needs to get a clear understanding of the way the systems are designed and how they are used for specific applications.

Consumption data in the Comprehensive Database are codified according to the FoodEx classification system (EFSA 2011f, i). Nomenclature from the FoodEx classification system has been linked to the Food Classification System (FCS) as presented in Annex II of Regulation (EC) No 1333/2008 (EU 2008c), part D, to perform exposure estimates to food additives. In practice, FoodEx food codes were matched to the FCS food categories. Furthermore, the food categories defined in Regulation (EC) No 1333/2008 (EU 2008c) are not fully compatible with the food categories defined in the Codex Alimentarius General Standard for Food Additives.

A major uncertainty related with the exposure estimation to food additives is due to the fact that the food categories, as defined in Regulation (EC) No 1333/2008 (EU 2008c), are not fully identical with the food categories for which food consumption data are available. Some food categories or their restrictions/exceptions are, for example not referenced in the consumption data and therefore could not be taken into account in the assessment. This is the case, for example for "Pre-cooked or processed cereals, only pre-cooked cereals" and "Soups and broths, only dehydrated soups and broths", which are not present in the EFSA Comprehensive Database. In certain circumstances, the restrictions/exceptions that apply to the use of a food additive could not be taken into account, and therefore the whole food category is considered in the exposure assessment. This may result in an overestimation of the exposure.

Another uncertainty is related to the use of a food additive as a carrier of other food additives, or as a food additive in other food additives, food enzymes, flavourings, and nutrients in accordance with Annex III to Regulation (EC) No 1333/2008 (EU 2008c). This source of exposure is difficult to assess because data on such uses as well as analytical data are lacking in many cases and, because according to Article 20 of Regulation (EU) No 1169/2011 (EU 2011b), such uses do not require to be labelled on the product (provided that the additive does not have a technological function in the final product). This can be illustrated using the example of sorbitan esters (E 491–E 495). These food additives are authorised according to Annex II (use for technological purposes in different food categories at certain maximum permitted levels) and Annex III of Regulation (EC) No 1333/2008 (EU 2008c). According to Annex III, sorbitan esters (E 491–495) are authorised as a carrier in food additives (colours, antifoaming agents, and glazing agents for fruit) at *quantum satis*[12] the amount which is enough, as food additives other than carriers

[12]According to Article 3 of Regulation (EC) No 1333/2008, 'quantum satis' shall mean that no maximum numerical level is specified and substances shall be used in accordance with good manufacturing practice, at a level not higher than is necessary to achieve the intended purpose and provided the consumer is not misled.

in food additives (preparations of colours, anti-foaming agents, and glazing agents for fruit) at *quantum satis*, and as food additives in nutrients (in beta carotene, lutein, lycopene, and vitamin E preparations at *quantum satis*, and in vitamins A and D preparations at 2 mg/kg in final food). The EFSA ANS Panel noted in its scientific opinion on the re-evaluation of sorbitane esters (E 491–495) that these uses were not considered in the exposure assessment scenarios because no data were available (EFSA 2017d). Accordingly, the Panel pointed out that analytical data on the actual presence of sorbitan esters (E 491–495) in food are needed to reduce these sources of uncertainty. However, the provisions according to Annex III of Regulation (EC) No 1333/2008 are not very specific. Sorbitane esters may be used, for example in colours. Since there are about forty different permitted colours that could be used in many different food categories, it would be difficult for official food control agencies to decide what food categories should primarily be analysed for a potential presence of sorbitane esters. Therefore, appropriate information from the food industry is required in such cases.

Another challenge is related to the monitoring of the use of food additives. According to Article 27 of Regulation (EC) No 1333/2008 (EU 2008c):

> member States shall maintain systems to monitor the consumption and use of food additives on a risk-based approach and report their findings with appropriate frequency to the Commission and the Authority. After the Authority [EFSA] has been consulted, a common methodology for the gathering of information by the Member States on dietary intake of food additives in the Community shall be adopted in accordance with the regulatory procedure referred to in Article 28(2).

Such a common methodology is not yet established. However, there was a study commissioned by the European Commission on (1) the ways this is currently done by the different EU Member States, (2) identifying the needs in this respect by EFSA, and (3) considering several options on how this monitoring could be done in the future (EU 2015a). The report of this study (EU 2015b) is available to the European Commission, the EU member states, and the EFSA.

The following nine key points have emerged as short statements from the study (EU 2015b):

1. The obligation to monitor the use of food additives needs a clear definition and mandate that is currently not available.
2. Monitoring of food additives should be given a high priority.
3. Exchange of data is essential and needs harmonisation and cost-effective approaches.
4. Better understanding of terminology is needed.
5. The collection of reliable and representative chemical concentration data sets is needed.
6. Within the tiered approach, the classification of substances per tier is already a prioritisation risk-based mechanism that is completed by a system to prioritize substances within individual tiers.

7. Specific data collections mechanisms have to be implemented that fulfil the requirements for exposure assessment of food additives.
8. A common understanding about data collection mechanisms (the "options") addressing presence data, use level data and analytical data is needed.
9. A preliminary assessment is needed that focuses on the advantages and disadvantages, feasibility, and stakeholders' acceptance, with the aim of supporting forthcoming discussions and analyses.

Validated analytical methods are needed to fulfil the monitoring requirements according to Regulation (EC) No 1333/2008 (EU 2008c). There are many publications on the chemical analysis of food additives. As an example, a review published by Scotter (2011) is on the available literature on the methods of extraction for approved natural colours in food and drink. In this review, the available analytical methods for the determination of European Union permitted natural food colour additives in foods and beverages were assessed for their fitness for purpose in terms of their key extraction and analysis procedures, selectivity, and sensitivity, especially with regard to maximum permitted levels, and their applicability for use in surveillance and in an enforcement role. Remarkably, there is a European reference laboratory for feed additives but not for food additives. Thus, it is still a challenge to harmonise the analytical methods that are required for the monitoring of food additives.

4.18 Process Contaminants

Klaus Abraham

4.18.1 Introduction

Foods were already being heated more than 700,000 years ago. This was done in order to preserve them, to enhance their taste and flavour, and to improve their digestibility. Since the late twentieth century, industrialized processes have increasingly replaced home cooking, with the initial focus on preservation and later on microbial safety and quality issues (van Boekel et al. 2010).

During the heating process, aromatics and numerous other compounds are formed by the so-called Maillard reaction or by lipid oxidation. Particularly at the higher temperatures occurring during baking, frying, grilling, deep frying, or roasting, undesired compounds (the so-called heat-induced contaminants) can also be formed. Some of them have been shown to be harmful to health in animal experiments, at least at high doses. This was discovered in the early 1970s when polycyclic aromatic hydrocarbons (PAH) and heterocyclic aromatic amines were identified, in particular in grilled meat and fish. Since then, several hundred different heating-related contaminants have been identified in foodstuffs. For most of these contaminants, very little is known about their occurrence in food and their toxicological properties (Frenzel et al. 2017). Nowadays, a few important compounds or groups of compounds have been identified as having relevance regarding occurrence and toxicology; they are described in the following sections. The main challenges for a comprehensive exposure assessment of these substances are the identification of all relevant food sources, the availability of a representative set of occurrence data for all of these foods (also considering home cooking and a possible trend over time), and the availability of the corresponding consumption data.

4.18.2 Acylamide

The occurrence of relevant levels of acrylamide was discovered in 2002 when it was found in potato and cereal-based products such as potato chips, "French" fries, roasted and baked potatoes, bread (especially crispbread), coffee, and pastry products (Tareke et al. 2002). Acrylamide is formed in carbohydrate-rich foods that are subjected to high temperatures (above 120 degrees Celsius, e.g. during frying or grilling). Therefore, it is not found in raw or cooked food. Acrylamide is a product of the Maillard reaction between certain amino acids (primarily aspartic acid) and reducing sugars (e.g. glucose or fructose) (Lineback et al. 2012). Another important source of exposure is cigarette smoke.

In mammals, acrylamide is mainly converted by CYP2E1 to glycidamide (epoxidation), which has a significantly higher genotoxic potential than acrylamide. The main metabolic pathway of acrylamide and glycidamide in humans is conjugation with glutathione; the ultimately generated mercapturic acids are detectable as short-term biomarkers in the urine. In rodent experiments with long-term administration of high doses of acrylamide, neurotoxicity, reproductive and developmental toxicity, as well as carcinogenicity with tumours in a variety of tissues were found to be the most important effects. According to the International Agency for Research on Cancer (IARC), acrylamide was classified as probably carcinogenic to humans (group 2A). As a reference value for the effective dose for carcinogenicity, a BMDL10 value (benchmark dose lower confidence limit 10%) of 0.17 mg/kg of BW per day was calculated for the tumour formation in the Harder's glands of male mice (EFSA 2015d). According to the current concept for the risk assessment of genotoxic carcinogens, the exposure to acrylamide has to be below 0.017 μg/kg of BW per day (Margin of Exposure = 10,000) in order to be of low concern from a public health point of view (EFSA 2005b). The Margin of Exposure (MoE) is a dimensionless number, which indicates the ratio of an effective dose in the animal experiment (e.g. the BMDL10 value) to the exposure dose in humans.

EFSA (2015d) performed a comprehensive exposure assessment using more than 40,000 analytical results of foods from 24 European countries analyzed since 2010. In the following, the acrylamide levels calculated as average middle bound levels are given in brackets (Table 4.17). The highest levels were found in "Coffee substitutes (dry)" (1499 μg/kg) and "Coffee (dry)" (522 μg/kg). High levels were found in

Table 4.17 Formation particularities and sources of process contaminants

Process Contaminant	Acrylamide	Furan	MCPD	5-HMF	Acrolein
Formation characteristics	Carbohydrate-rich foods subjected to high temperatures (above 120 degrees Celsius, e.g., during frying or grilling). Product of the Maillard reaction	Heating of carbohydrates, amino acids, ascorbic acid and polyunsaturated fatty acids Roasting of foodstuffs or heating processes in closed vessels	Refining of vegetable fats and oils at high temperatures above 200 degrees Celsius	Maillard reaction during thermal treatment of carbohydrate-containing foods	Thermal treatment of animal and vegetable fats, carbohydrates and amino acids
Major food sources	Potato chips French fries Roasted and baked potatoes Bread (especially crispbread) Coffee Pastry products Cigarette smoke	Roasted coffee Instant coffee Jarred baby food Soups and sauces Meat Cereal products	Refined fats and oils Margarine Infant formula Baked or roasted cereal Potato products Soy sauce Hydrolyzed vegetable protein	Beverages made from dried plums	Lack of reliable content measurements
Exposure estimates	Children: 0.5 to 1.9 (average) 1.4 to 3.4 (P95) Adolescents, adults: 0.4 to 0.9 (average). 0.6 to 2.0 (P95) depending on the survey and age group) All values in μg/kg BW per day	Children 0.04 to 0.22 (average) 0.09 to 0.46 to 1.4 (P95) Adults: 0.03 to 0.59 (average) 0.09 to 1.29 (P95), different population groups and studies All values in μg/kg BW per day	Children: 0.5 to 1.5 (average) 1.1 to 2.6 (P95) Adults 0.2 to 0.7 (average) 0.3 to 1.3 (P95) different population groups and studies All values in μg/kg BW per day	Up to 350 mg/kg BW per day	uncertain

"Potato crisps and snacks" (389 µg/kg) and "Potato fried products (except potato crisps and snacks)" (308 µg/kg), and lower levels in "Processed cereal-based baby foods" (73 µg/kg), "Soft bread" (42 µg/kg), and "Baby foods, other than cereal-based" (24 µg/kg). A data set of manufacturers' measurements of acrylamide in freshly sliced potato crisps from 20 European countries showed a substantial downward trend for the mean levels from 763 µg/kg in 2002 to 358 µg/kg in 2011. Such a trend was not observed for other food categories, despite the efforts made by manufacturers (e.g. the acrylamide reduction tools developed by FoodDrinkEurope). EFSA's estimation of long term dietary exposure to acrylamide revealed infants, toddlers, and other children to be the most exposed groups, and was estimated to be on average 0.5–1.9 µg/kg of BW per day, depending on the survey and age group (95th percentile: 1.4–3.4 µg/kg of BW per day). For adolescents, adults, elderly and very elderly, long-term dietary exposure was estimated to be on average 0.4–0.9 µg/kg of BW per day, depending on the survey and age group (95th percentile: 0.6–2.0 µg/kg of BW per day). In general, the main contributor was found to be the category "Potato fried products (except potato crisps and snacks)". Thus, preferences in home cooking can have a substantial impact on the dietary exposure to acrylamide (EFSA 2015d).

In its risk assessment report, EFSA concluded that the current dietary exposure to acrylamide is not of concern with regard to non-neoplastic effects. However, EFSA stated that although the epidemiological associations have not proven acrylamide to be a carcinogen in humans, the low MoE values indicate, based on animal evidence, a concern with regard to neoplastic effects. Across all surveys and age groups, the MoE values were calculated in the range from 425 (minimum lower bound) to 89 (maximum upper bound) for the mean exposure estimates, and from 283 to 50 for the respective 95th percentile exposure estimates (EFSA 2015d).

In general, the exposure to genotoxic carcinogens should be "As Low As Reasonably Achievable" (ALARA principle), and, especially in view of the low MoE values, the efforts to reduce acrylamide concentrations in food have to continue. In order to limit the levels in foods, indicative values are set by the EU Commission Recommendation 2013/647/EU "on investigations into the levels of acrylamide in food" (e.g. 600 µg/kg for ready-to-eat French fries and 1000 µg/kg for potato chips from fresh potatoes and potato dough). In cases where these values are exceeded, Member States should carry out further investigations into the production and processing methods used by food producers.

4.18.3 *Furan*

Furan is a heterocyclic, aromatic compound with high volatility that is generated during heating of carbohydrates, amino acids, ascorbic acid, and polyunsaturated fatty acids (Crews and Castle 2007). It is formed, especially during roasting of foodstuffs, or during heating processes in closed vessels. Food categories with high levels of furan are primarily roasted coffee, instant coffee, jarred baby food, soups, sauces, meat, and cereal products.

Furan is rapidly absorbed in the gastrointestinal tract and metabolized by liver CYP2E1 to its primary metabolite *cis*-2-butene-1,4-dial, an unsaturated dialdehyde with high reactivity and genotoxic and cytotoxic properties. It binds to thiols (e.g. glutathione) and amino groups of proteins and was found to be mutagenic. However, in vivo studies on the genotoxicity of furan gave no definite results. Long-term studies in rodents revealed furan to be carcinogenic (Von Tungeln et al. 2017). Overall, the hazard potential including the underlying mechanisms of furan carcinogenicity has not yet been fully clarified (Moro et al. 2012). Currently, no scientific body has derived a health-based guidance value or a risk assessment using the MoE concept. IARC has classified furan as possibly carcinogenic to humans (group 2B).

In 2011, EFSA performed an exposure assessment using more than 5000 analytical results of foods from 20 European countries analysed between 2004 and 2010. The highest furan levels were found in coffee; the mean values ranged from 45 μg/kg (brewed coffee) to 3660 μg/kg (roasted coffee beans, with the highest 95th percentile of 6407 μg/kg). In the non-coffee categories, the mean values ranged from 3.2 μg/kg (infant formula) to 49 μg/kg (jarred baby food with vegetable, with the highest 95th percentile of 123 μg/kg). EFSA estimated the mean furan exposure across surveys to range from 0.03 to 0.59 μg/kg of BW per day for adults (95th percentile 0.09–1.29 μg/kg of BW per day). Brewed coffee was major contributor to furan exposure in adults with a mean contribution of 88% across the surveys. Mean furan intake for adolescents was estimated to range from 0.02 to 0.13 μg/kg of BW per day (95th percentile 0.06 and 0.31 μg/kg of BW per day), with coffee being still a major contributor. Other important contributors were cereal-based products, ready-to-eat soups, sauces, and fruit juice. Mean furan intake of other children was estimated to range from 0.04 to 0.22 μg/kg of BW per day (95th percentile 0.09–0.46 μg/kg of BW per day), and of toddlers from 0.05 to 0.31 μg/kg of BW per day (95th percentile between 0.2 and 1.4 μg/kg of BW per day). Major contributors to furan exposure in these two population groups were fruit juice, milk-based products, and cereal-based products. In addition, jarred baby foods accounted for an average of 40–86% to furan exposure, depending on age group and survey. Based on the limited food consumption data available for infants (two surveys only), the mean exposure for infants was estimated to range between 0.09 and 0.22 μg/kg of BW per day (95th percentile 0.97 μg/kg of BW per day). Jarred baby food was the main contributor in one country and ready-to-eat soups were in the other country (EFSA 2011e).

A meaningful risk assessment of these exposure estimates is currently not possible, as the hazard potential of furan is not conclusively known. Furan exposure could be reduced by consuming mainly freshly prepared food and less ready-to-eat and canned foods. Since furan is a volatile compound, it is also recommended to heat jarred baby food in an open container and to stir regularly. Furthermore, it should be kept in mind that the consumption of coffee is the most important source of furan in adults.

4.18.4 MCPD and Glycidyl Fatty Acid Esters

Fatty acid esters of 2- and 3-monochloropropandiol (2- and 3-MCPD) and of glycidol are generated, in particular, during the refining of vegetable fats and oils in the so-called deodorisation process that removes unwanted odours and flavours at high temperatures above 200 °C (Craft et al. 2013). Refining vegetable oils in this way has a long history, but the problem was only quite recently discovered in 2006 for MCPD and in 2009 for glycidol. Analytical quantification is a complex and matrix-dependent issue, but methods have been developed in recent years and the knowledge about the occurrence of these compounds in different foodstuffs has significantly increased. The largest sources of exposure are refined fats and oils, or the foodstuffs produced from this vegetable raw material, such as margarine or infant formula. Other foods like baked or roasted cereal and potato products also contain these fatty acid esters. Unesterified 3-MCPD and other chloropropanols had already been identified as relevant contaminants in soy sauce and hydrolysed vegetable protein in the late 1970s. Technological measures have led to a significant reduction in these levels, and this source of exposure is currently no longer relevant (Lee and Khor 2015).

Kinetic studies in laboratory animals have shown a more or less complete hydrolysis of these esters during intestinal passage, leading to an internal exposure to unbound MCPD or glycidol. As these are the substances relevant for risk assessment, occurrence data are evaluated as ester bound MCPD or glycidol.

For 2-MCPD, no sufficient toxicological data are available, which would allow the derivation of a health-based guidance value.

Free 3-MCPD has been shown to be carcinogenic in animal experiments, probably due to non-genotoxic modes of action, and was classified by IARC as possibly carcinogenic to humans (group 2B). Renal toxicity (tubular hyperplasia) in male rats was identified as the most sensitive effect in long-term studies. Using the same key study, different scientific bodies have currently derived different values for a TDI between 2 μg/kg of BW per day (EFSA 2018b) and 4 μg/kg of BW per day (FAO/WHO 2016). The differences mainly result from the use of different mathematical models for the calculation of BMDL10 values.

The long-term administration of free glycidol increased the incidence of tumours in several tissues of rats and mice, likely via a genotoxic mode of action. According to the IARC, glycidol was classified as probably carcinogenic to humans (group 2A). As a reference value for the effective dose for carcinogenicity, a T25 value of 10.2 mg/kg of BW per day was calculated for peritoneal mesothelioma in male rats (EFSA 2016a). According to the MoE concept of EFSA, an exposure to glycidol has to be below 0.4 μg per kg of BW per day in order to be of low concern from a public health point of view (EFSA 2005b).

The most comprehensive exposure assessment was done by EFSA in 2016, and considered a total of more than 7000 occurrence data on 2-MCPD, 3-MCPD, and glycidol. These data were mainly from oils and fats. For food categories not represented in the data set, the occurrence of the compounds was calculated, where possible, using a model based on the available data on fats and oils. The highest occurrence values were found in the food group "Fats and oils", with "Palm oil/fat" showing a mean middle bound level of 1565, 2912, and 3955 μg/kg for bound 2-MCPD, 3-MCPD, and glycidol, respectively. Lower mean levels were calculated for other oils, ranging from 48 to 608 μg/kg for bound 3-MCPD, from 86 to 270 μg/kg for bound 2-MCPD, and from 15 to 650 μg/kg for bound glycidol. "Margarine, normal fat" had mean levels of 236, 668, and 582 μg/kg for bound 3-MCPD, 2-MCPD, and glycidol, respectively. In other food categories, the highest levels were observed in "Potato crisps" (mean levels of 135, 216, and 110 μg/kg for total 2-MCPD, total 3-MCPD, and bound glycidol, respectively), "Hot surface cooked pastries" (mean levels of 123, 247, and 137 μg/kg for total 2-MCPD, total 3-MCPD, and bound glycidol, respectively), "Cookies" (mean levels of 103, 200, and 134 μg/kg for total 2-MCPD, total 3-MCPD, and bound glycidol, respectively), and "Short crusts" (mean levels of 79, 154, and 149 μg/kg for total 2-MCPD, total 3-MCPD, and bound glycidol, respectively) (EFSA 2016a).

EFSA estimated the mean exposure to 3-MCPD as 0.5–1.5 μg/kg of BW per day across the dietary surveys for the age groups "Infants", "Toddlers", and "Other children" (95th percentile: 1.1–2.6 μg/kg of BW per day). In adolescents and adult population groups, the mean exposure to 3-MCPD ranged from 0.2 to 0.7 μg/kg of BW per day (95th percentile: 0.3–1.3 μg/kg of BW per day) (EFSA 2016a). These values in part exceed the EFSA TDI of 0.8 μg/kg of BW per day, but not that TDI of 4 μg/kg of BW per day recently adjusted by JECFA (FAO/WHO 2016). According to the occurrence data, the estimated mean exposure to 2-MCPD was roughly half as much as that of 3-MCPD (EFSA 2016a); currently, no health-based guidance value is available for 2-MCPD allowing for a risk assessment of the estimated exposure.

The mean exposure to glycidol was estimated by EFSA to range from 0.3 to 0.9 μg/kg of BW per day across the dietary surveys for the age groups "Infants", "Toddlers", and "Other children" (95th percentile: 0.8–2.1 μg/kg of BW day). In adolescents and adult population groups, the mean exposure to glycidol ranged from 0.1 to 0.5 μg/kg of BW per day (95th percentile: 0.4–1.1 μg/kg of BW per day in adolescents, and 0.2–0.7 μg/kg of BW per day in adult population groups) (EFSA 2016a). Based on the T25 value recalculated as a T10 value of 4.08 μg/kg of BW per day (corresponding to a BMDL10), the MoE estimates for the mean exposure in "Infants", "Toddlers", and "Other children" ranged from 13,600 to 4500 (95th percentile: 5100–2000). For "Adolescents" and adult population groups, the MoE for the mean exposure ranged from 40,800 to 8200 (95th percentile: 20,400–3700). Values below 10,000 indicate a public health concern.

A separate consideration has to be made for formula-fed infants, as their fat requirement is the highest in life (up to 6 g fat/kg BW per day), and the fats in infant formula are composed of refined vegetable fats. Exposure scenarios of infants receiving formula only, based on mean consumption and mean occurrence in the formula, resulted in an intake of 2.4–1.9 μg/kg of BW per day for 3-MCPD and glycidol, respectively (95th percentile occurrence data: 3.2–4.9 μg/kg of BW per day, respectively) (EFSA 2016a). Using a reference dose based on the T10 value mentioned above, the calculated MoE for glycidol is about 2200 for the mean occurrence (95th percentile: 840), much lower than 10,000 and thus indicating a public health concern. Therefore, a most urgent need is to reduce bound glycidol in infant formula and maximum levels are under discussion in the European Commission.

4.18.5 5-Hydroxymethylfurfural

5-Hydroxymethylfurfural (5-HMF) is generated as a product of the Maillard reaction during thermal treatment of carbohydrate-containing foods. In vitro genotoxicity was found to be positive in the case of metabolic conditions allowing the formation of the reactive metabolite 5-sulphoxymethylfurfural. However, in vivo genotoxicity has thus far been negative. The results obtained in short-term rodent studies on the potential induction of neoplastic changes in the intestinal tract were negative or cannot be reliably interpreted as "carcinogenic". In the only long-term carcinogenicity study in rats and mice, no relevant tumours or their precursory stages were induced by 5-HMF. Therefore, no relevance for humans concerning the carcinogenic and genotoxic effects can be derived. Various animal experiments failed to reveal the adverse effect in the range 80–100 mg/kg of BW per day. The estimated intakes are between 4 and 30 mg per person per day, while intake of up to 350 mg may result from beverages made from dried plums. In general, the safety margins were assessed to be sufficient (Abraham et al. 2011b).

4.18.6 Acrolein

Acrolein is an α, β-unsaturated aldehyde formed during thermal treatment of animal and vegetable fats, carbohydrates, and amino acids. Further, inhalational exposure (e.g. from smoking) and endogenous formation (e.g. through lipid peroxidation and polyamine metabolism) may be relevant. Acrolein forms adducts with glutathione and other cellular components and is therefore cytotoxic. Mutagenicity was shown in some in vitro tests. Acrolein forms different DNA adducts in vivo. In subchronic oral studies, local lesions were detected in the stomach of rats, but mutagenic and carcinogenic effects have not been demonstrated for oral exposure. Systemic effects have not been reported from basic studies. A tolerable oral acrolein intake of 7.5 μg/kg of BW per day was established by a WHO working group. Acrolein exposure via food is hard to assess due to the lack of reliable content measurements. Human biomonitoring of an acrolein urinary metabolite allows rough estimates of acrolein exposure in the range of a few μg/kg of BW per day; high exposure could be ten times higher after the consumption of certain foods. Although the estimation of dietary acrolein exposure is associated with uncertainties, a health risk seems to be unlikely (Abraham et al. 2011a).

4.18.7 Polycyclic Aromatic Hydrocarbons (PAH)

In contrast to the small molecular substances mentioned above, polycyclic aromatic hydrocarbons (PAH) are a large and complex group of about 250 aromatic substances containing 2–7 rings of hydrocarbons. The lead compound is benzo(a)pyrene and it is used to evaluate the PAH content of food. The substances are process contaminants, especially in smoked and grilled meat, and sausages. However, many other sources are relevant for human exposure via inhalation (from combustion processes including traffic and cigarette smoking) and via contact with consumer products (polluted rubber or plastic articles, and wood preservatives). The assessment of the total exposure is very difficult not only due to the different routes needing to be considered, but also due to the high number of different compounds with different toxicological properties. In this respect, cancer is the main concern. While benzo(a)pyrene is classified as human carcinogen (IARC group 1), many other PAHs are considered probable or possible human carcinogens. For certain foodstuffs, the EU Commission Regulation (EC) No. 1881/2006 (EU 2006c) sets maximum levels for benzo(a)pyrene between 1 and 10 μg/kg.

4.18.8 Implications for Exposure Estimation of Process Contaminants

The exposure assessment of PC's has to take into account the particularities of their formation and their various sources. This has consequences for at least the following issues:

- The classification rules of the food involved in process contamination.
- Discussion of the questions whether the food sources of process contaminants are covered in the systematic food consumption study.
- The identification of unknown food in which process contamination takes place.
- Consideration of the conditions of the formation of PCs during processing.

As shown in Table 4.17, the food sources of PC's vary considerably, but the listing is not complete. Performing an estimation of exposure for acrolein seems impossible due to lack of sound analytical measurements.

Food category systems as described in Sect. 4.4.1 seem inappropriate because a high level of detail is needed to classify the food. This detail of food classification is not offered for example in the officially available part of EFSA's comprehensive food consumption database. Further, food consumption studies will not always provide adequate consumption data. In addition, there could be possible unidentified food sources of PC's and thus the knowledge about exposures may be incomplete and therefore uncertain.

PC's can only be identified quantitatively in processed food. Therefore, the TDS concept (Sect. 4.8.4) offers an ideal source of information. To improve the knowledge about process contaminants, it might be promising to start with TDS data to identify foods in which PC's are formed and to develop concepts for food consumption evaluation and classification.

4.19 Natural Substances in Food

Birgit Dusemund

4.19.1 Introduction

Food from botanical origin forms an intrinsic part of traditional diets all over the world, and diets rich in vegetables and fruits are usually associated with health benefits. Further, the increasing trend of consuming natural, "organic", herbal, and plant-based products, especially food supplements and related dietary preparations (slimming products, special food products for sporting people) based on botanicals,[13] and their preparations[14] obtained by all kind of processing, e.g. pressing, extraction, fractionation, concentration, drying, and/or fermentation, has led to increased sales of these products in the European and North American markets. In addition, there are tendencies to replace synthetic food additives or flavourings in food products with additives and flavourings derived from plants. Novel foods or genetically modified foods represent another source of intentional dietary intake of botanicals; however, they are not included in this chapter. The unintentional exposure of consumers to botanicals and their components is occurring via contamination of food with whole plants, plant parts, or their ingredients. This may be the consequence of inappropriate harvesting or cleaning methods, adulteration, or due to carry over from animal feed through animal products, or via mistaking edible versus non-edible plants (Speijers et al. 2010; Koleva et al. 2012; Dusemund et al. 2012).

A large variety of botanicals, including certain medicinal plants, is used in food supplements and related products. The plants especially used for therapeutic purposes may contain biologically active substances such as mono-, di-, tri-, and sesquiterpenes, alkaloids, cardiac glycosides, and anthraquinones. Dose–effect relationships for these substances are often well-known from human data on medicinal uses. The need for an EU-wide harmonised approach based on appropriate systematic safety evaluations of botanicals and botanical preparations used in food

[13]This term refers to plants, including algae, fungi and lichens, and parts of plants as a whole or cut.

[14]This term refers to preparations obtained by all kind of processing, e.g. pressing, extraction, fractionation, concentration, drying and/or fermentation.

supplements is seen to assure that intake doses of biologically active plant components in these products are within safe limits (EFSA 2009c, f; Silano et al. 2011).

General procedures and guidelines for exposure assessment are followed for food additives and food flavourings of botanical origin (EFSA 2010e, 2012e). For botanical contaminants, exposure assessments are developed from a combination of analytical occurrence data of the substances of concern (which may be grouped in the case of chemically related substances with similar toxicological profile) and from consumption data of relevant food categories (EFSA 2011a). Exposure estimates for botanicals and botanical preparations used in food supplements are based on the available dosage recommendations.

4.19.2 Specifics in the Risk Assessment of Botanicals: Some Examples

Substance of Concern

The risk assessment of botanicals and their preparations is frequently based on a constituent (or a group of constituents) that provide reasons for health concern due to their chemical, physiological, or toxicological properties. This substance or group of substances of concern has often been identified in the past as an "active principle" responsible for toxic, addictive, psychotropic, or other adverse health effects in case reports, experimental studies, or clinical trials. The EFSA "Compendium of botanicals reported to contain naturally occurring substances of possible concern for human health when used in food and food supplements" has the aim of flagging these substances of possible concern that may require specific attention (EFSA 2012f). An exposure calculated for the substance/group of substances of concern is then compared to derived health-based guidance values (HBGV), or used for a Margin of Exposure (MOE) or Margin of Safety (MOS) approach. For example, the risk assessment of hemp food is based on exposure estimates and HBGVs of delta-9-tetrahydrocannabinol (THC) as the main representative of the cannabinoids occurring in *Cannabis sativa* L. (BgVV 1997, 2000; EFSA 2015f). Another example is the risk assessment of food based on lupin seeds, for which a MOS for quinolizidine alkaloids occurring in certain species of the genus *Lupinus* is calculated (BfR 2017).

Complex Mixture: Importance of the Definition of Identity and Specifications

Since botanicals and botanical preparations generally consist of a complex mixture of phytochemicals, the risk assessments of botanicals and their preparations differ from those for clearly defined single substances as a component of food. For the active principle (i.e. the substance of interest under technological or health aspects), the purity is often rather low and the remaining substances in the preparation may not be characterised sufficiently. A botanical preparation of a certain species may be represented by different extracts varying in composition, due to the use of

- Extracts varying in composition, due to the use of source materials (e.g. differences in botanical subspecies
- Geographical origin and conditions of growth and harvesting
- Parts of a plant and
- Extraction methods and solvents.

Enrichments of natural ingredients of concern or contaminants such as heavy metals or pesticides may occur in the case of selective extractions. These parameters have to be taken into account when defining the identity and specifications of botanical preparations. It has to be ascertained for risk assessment that the identity and specifications of the botanicals in the products of trade, those investigated in the toxicity testing, and those considered in the exposure assessments are the same, allow for the use of read-across approaches, or are based on the same relevant substance(s) of concern.

This may be illustrated by an example belonging to the category of food additives. For the natural botanical food colour lutein (E 161b), the content of total carotenoids/saponified carotenoids as colouring matter may be as low as 5% when extracted from grass, nettle, lucerne, or spinach, but amounts to ≥60% when *Tagetes erecta* flowers are used as a source. The EFSA Panel on Food Additives and Nutrient Sources added to Food (ANS) in the process of re-evaluation of food additives established an ADI of 1 mg/kg of BW per day for the more concentrated extracts (total carotenoids ≥60%) from *Tagetes erecta* flowers based on the existing data on the chemical composition and toxicity testing. However, EFSA's Panel on Food Additives and Nutrient Sources decided that the data available is too limited to conclude that the ADI also applies to lutein preparations of lower purity or from other sources (EFSA 2010a, 2011b).

Interactions, Matrix Effects, and Cumulative Approaches

Biologically active substances usually occur in a botanical as a mixture of chemically related compounds that have been formed, e.g. as precursors or by-products of biosynthesis pathways. They may have the same structural alerts for specific toxicity endpoints due to their similarity in structure, may react as agonists or antagonists in receptor-mediated reactions, or they may compete for binding sites in metabolising enzymes responsible for activation, deactivation, or elimination. As a consequence, matrix effects (interactions of the active substance with accompanying ingredients) may weaken or enhance the toxic effects of a known substance of concern. Basing the risk assessment of botanicals, botanical preparations, or botanical contaminants on the exposure and toxicity data of only one active ingredient and ignoring the accompanying ones may be a cause of uncertainties in the risk assessment.

Possible additive acute effects have been addressed, e.g. in the risk assessments of botanical contaminants in food performed by national authorities and EFSA. One example is the risk assessment of ergot alkaloids occurring in the sclerotia of *Claviceps purpurea* that are grain contaminants. Reference is made to the exposure to a sum of ergot alkaloids including ergometrine, ergotamine, ergosine, ergocristine, ergocryptine, and ergocornine. They have in common the tetracyclic ergoline ring system that is associated with their activity as ligands for adrenergic, serotonergic, and dopaminergic receptors (EFSA 2012b). Also, EFSA took exposure to the sum of (−)-hyoscyamine and (−)-scopolamine into account, assuming equivalent potency as antimuscarinic agents, in the risk assessment of tropane alkaloids (EFSA 2013g; Mulder et al. 2016).

A cumulative assessment approach is also generally accepted for 1,2-unsaturated pyrrolizidine alkaloids (PA), the double bound being a prerequisite for acute liver toxicity and metabolic activation to genotoxic carcinogens (COT 2008; EFSA 2011d; BfR 2007, 2011a, 2013, 2016b). Based on available occurrence data (Mulder et al. 2015), EFSA estimated the exposure to PA in the European population via the consumption of plant-derived foods. This resulted in the highest estimates of mean long-term dietary exposure of 34.5–48.4 ng/kg of BW per day in toddlers, and 154–214 ng/kg of BW per day in the highly exposed population (also in toddlers). Following a rather conservative scenario, the highest estimates of acute mean exposure and 95th percentile exposures were also for toddlers. The mean exposure was up to 311 ng/kg of BW per day and the 95th percentile exposure was up to 821 ng/kg of BW per day. Tea and herbal infusions were by far the main average contributors to the total exposure to PAs (EFSA 2016g).

In evaluating the risks due to the contamination of poppy seeds, derived from *Papaver somniferum*, with opium alkaloids, the focus was on morphine since it showed the highest concentrations in the vast majority of seed samples (BfR 2004; EFSA 2011c). The concentration profiles of opium alkaloids in poppy seeds were from analytical data generated in European countries and Australia. The mean concentrations in seeds in Europe ranged from 33.3 to 33.6 mg/kg of BW for morphine, from 2.85 to 3.09 mg/kg of BW for codeine, from 3.23 to 3.40 mg/kg of BW for thebaine, from 0.09 to 0.59 mg/kg of BW for papaverine, and from 1.02 to 1.70 mg/kg of BW for noscapine (EFSA 2011c). Based on the average poppy seed consumption recorded in the Comprehensive European Food Consumption Database (EFSA 2011a), the exposure to morphine was the highest for adults and ranged from 3.11 to 90.9 μg/kg of BW per day. Based on hypothetical single portion scenarios for consumers of foods with high content of poppy seeds, the exposure to morphine was the highest for children within the age range of 3–10 years at 47.8–252 μg/kg of BW (EFSA 2011c).

Interactions may also play a role in the toxicokinetics, as assumed for the polyphenols contained in green tea extracts derived from the leaves of *Camellia sinensis*, inter alia used in food supplements. It has been hypothesised that the principal ingredient (−)-epigallocatechin-3-O-gallate (EGCG), taken as part of a green tea extract, shows slower elimination than when used as an isolated compound due to competition with other accompanying polyphenols in the extract for binding sites in metabolising enzymes (EFSA 2009f).

Experiences and Data from Human Exposure via Herbal Medicinal Products

Knowledge about the adverse effects of botanicals is often derived from experiences and data associated with the therapeutic use as herbal medicinal products for humans. This information is available in the form of epidemiological and clinical studies, reports of adverse health effects, and case reports of intoxications. In contrast, data from validated toxicological studies in experimental animals are often sparse and this is the case for the above-mentioned alkaloids. The risk assessment of *Ephedra* herb in food supplements is another example that was performed using dose–effect relationships from the medicinal use of individual ephedra alkaloids (EFSA 2013c).

Low Intake in the Traditional Diet Versus High Intake with Food Supplements

Botanical components that commonly occur at low levels in the diet may be extracted and reintroduced in concentrated form for certain food supplements or related health products, resulting in comparatively high exposure levels (EFSA 2004). In some cases, serious adverse effects have been assigned to this type of products. For example, capsules containing high-dose dried green tea extracts taken for weight loss purposes have been associated with severe liver toxicity while these adverse effects have not been reported to the same extent for traditional consumption of green tea infusions (EFSA 2009f; Speijers et al. 2010). Certain case reports on adverse cardiovascular effects show a possible association with intake of food supplements advertised for weight loss or improvement of physical fitness containing high doses of (−)-synephrine as part of bitter orange (*Citrus aurantium* ssp. *aurantium*) extracts partly in combination with caffeine. In contrast, no adverse effects are known from (−)-synephrine exposure via traditional foods such as bitter orange marmalades or orange juice (NTP/NIEHS 2004, EFSA 2009f, Health Canada 2011; BfR 2012a).

4.19.3 *Guidance for Safety Evaluation: An Approach by EFSA*

In view of the specific aspects in the safety evaluation of plant-based food supplements and their expanding market with numerous claims, EFSA saw a need for a better characterisation of the botanicals in use and for harmonisation of their risk assessments (EFSA 2004). To suggest a science-based approach for the safety assessment, EFSA developed the "Guidance on safety assessment of botanicals and botanical preparations intended for use as ingredients in food supplements"

(EFSA 2009c). Even though the guidance is focussed on the use of botanicals in food supplements, EFSA notes that the principles of the approach chosen are also applicable to other uses of botanicals and botanical preparations in the areas of food and feed.

The Data Set Needed

The necessary information for a botanical or botanical preparation comprises technical, exposure, and toxicological data. The technical data comprises details on (1) the identity and nature of the source material, (2) the manufacturing process of the botanical (preparation), (3) its chemical composition, (4) its specifications, (5) its stability in food (supplements), (6) the proposed uses and use levels, and (7) the information on existing assessments. Regarding exposure, information is required on (1) the anticipated exposure via the food supplement or food, (2) the total exposure via different categories of food, including food supplements, and via medicinal products, (3) the modality of use, and (4) the information on historical use of different categories of food, including food supplements, and medicinal products. Regarding the toxicological data, studies on toxicity and toxicokinetics, including metabolism of botanicals and botanical preparations, should be conducted using internationally agreed protocols.

The Tiered Approach

Depending on the available knowledge, a two-level tiered approach for the safety evaluation of the botanical (preparation) is proposed. On "level A" of a risk assessment, a decision is derived based on all available data if there is (1) a safety concern, (2) no safety concern, or (3) a need for additional data. In the latter case, the requirement for further testing is specified on the subsequent level ("level B").

Level A: Presumption of Safety Approach

For "level A", the decision of "no safety concern" may be derived on the principle of a "presumption of safety". According to the guidance, a "presumption of safety" could be applied when available data would allow for the conclusion that exposure to known levels of the botanical (preparation) has occurred in large population groups for many years without reported adverse effects, and that thus no additional data are judged necessary for the safety evaluation. A "presumption of safety" decision requires that not only use levels, but also chemotypes of the botanicals and the chemical composition of the botanical preparations, are in line with historically used ones, and that intakes due to the intended levels of use are within the range of intake levels derived from the European Member States' average diets. This approach relies mainly on the objective of not significantly increasing exposures beyond the levels linked to the safe history of use. The "presumption of safety" approach has been applied in the BfR risk assessment of synephrine containing *Citrus aurantium* extracts in weight loss products (BfR 2012a). Based on exposure estimates, BfR concluded that synephrine intake from *Citrus aurantium* extracts in weight loss products should be at maximum 6.7 mg of synephrine per day (this corresponds to the intake via conventional foods with maximum synephrine levels for average consumers). This level guarantees that the total intake from traditional foods and weight loss products does not, for a vast majority of consumers, exceed the value of 25.7 mg of synephrine per day (corresponds to the intake of synephrine via conventional foods with maximum synephrine levels for frequent consumers).

Evaluations on "level A" may focus on specific compounds of concern. For a botanical (preparation) with a potential to contain toxic, addictive, psychotropic, or other substances that may be of concern, the "presumption of safety" approach can only be applied if there is convincing evidence that these undesirable substances are either absent or significantly reduced or inactivated during processing. In these cases, a "presumption of safety" of the botanical (preparation) is only justified when the overall exposure to the substances of concern is not too high compared to existing health-based guidance values such as the acceptable/tolerable daily intake (ADI/TDI). A consideration of exposure to the substance of concern in relation to the Threshold of Toxicological Concern (TTC) values may also be helpful (EFSA 2016e). When the botanical (preparation) contains substances that are both genotoxic and carcinogenic, the "Margin of Exposure" (MOE) approach (EFSA 2005c) is recommended. Advice is also given with respect to read-across between two different preparations of a botanical or between different botanicals for which equivalence of composition data and consumption patterns regarding the substances of concern are a precondition.

Level B

For those botanicals or botanical preparations for which a "presumption of safety" was not justified on "level A", e.g. the anticipated intake is significantly higher than the estimated historical intake level, or the historical intake level cannot be assessed, the evaluation proceeds on "level B" by deciding which additional studies are needed. The EFSA guidance (EFSA 2009c) recommends deducing the study requirements on "level B" from the "Guidance on submissions for food additive evaluations by the Scientific Committee on Food (SCF 2001)".[15] The toxicological data asked for primarily focus on results from toxicokinetics, genotoxicity, and subchronic toxicity studies. Depending on their outcome, further studies may be required, e.g. on reproductive toxicity, developmental toxicity, neurotoxicity, immunotoxicity, and chronic toxicity/carcinogenicity.

The two-level tiered approach was tested with a selected number of botanicals and the results of this investigation were published in a report illustrating how to apply the guidance document (EFSA 2009f).

[15]This guidance was replaced in 2012 by the "Guidance for submission for food additive evaluations by the EFSA Panel on Food Additives and Nutrient Sources added to Food (ANS)" (EFSA 2012d).

An additional tool for risk evaluation of botanicals is the "Compendium of botanicals reported to contain naturally occurring substances of possible concern for human health when used in food and food supplements" (EFSA 2012f; Silano et al. 2011). In this document, EFSA listed more than 1200 plant genus, species, and varieties; however, there is no final judgment as to whether botanicals listed in the Compendium are safe or not safe for food use.

4.19.4 Regulations and Legal Background

The use and the occurrence of plants and plant components in food are subject to different national and international regulatory frameworks. These frameworks are generally based on the outcome of the scientific risk assessments of competent national or international authorities such as the Joint FAO/WHO Expert Committee on Food Additives (JECFA) or the EFSA.

Global Framework

In 1963, FAO and WHO established "*The Codex Alimentarius Commission*" which provides a global framework. It develops international food standards, guidelines, and codes of practice for consumer protection that are also relevant for botanical components in food. Specifications and safety evaluations, including health-based guidance values, for food additives of botanical origin are published by the Joint FAO/WHO Expert Committee on Food Additives (JECFA). JECFA's publications include the Evaluation of Certain Food Additives and Contaminants reports as part of its WHO Technical Report Series, and the WHO Food Additives Series [e.g. Curcumin (E 100): JECFA 2004, 2006; EFSA 2014g]. On the basis of these monographs, the Codex Committee on Food Additives may endorse permitted maximum levels for individual food additives. Another Codex panel, namely the Codex Committee on Contaminants in Foods, establishes permitted maximum levels or guidelines levels for contaminants, including naturally occurring toxicants in food and feed. It is stated in the Codex Alimentarius standards that grains and pulses shall be free from the following seeds in amounts which may represent a hazard to human health: Crotalaria (*Crotalaria* spp.), Corn cockle (*Agrostemma* githago L.), Castor bean (*Ricinus communis* L.), Jimson weed (*Datura* ssp.) (CODEX STAN 153 1985; CODEX STAN 171 1989a; CODEX STAN 172 1989b; CODEX STAN 199 1995a; CODEX STAN 201 1995b).

EU Legislation

The use of botanicals and botanical preparations in food is regulated in Europe by national and EU legislation. The EU legislation will be the main focus of the following paragraphs.

In general, botanicals and botanical preparations occurring in food fall under the basic regulations of the General Food Law (Regulation (EC) No 178/2002/EC). Article 14 of the General Food Law protects against unsafe food and attributes responsibility for the safety of the products on the market to the food business operators (EU 2002b).

It is known that conventional plant-derived food such as certain fruits, vegetables, herbs, and spices may contain a variety of naturally occurring substances of concern. Examples of these substances include alpha-solanine, lupine alkaloids, furocoumarins, cyanogenic glycosides, (EFSA 2016c) propenyl alkoxy benzenes, capsaicin, coumarin, and thujone. The intake of these substances from a normal diversified diet generally only amounts to low levels that are not known to cause any health damage in humans. However, national authorities communicate nutritional recommendations to the public as a preventive action to avoid potential risks of high or excessive individual exposure to a certain foodstuff. One example is that they give advice to consume Cassia cinnamon, containing relatively high levels of the hepatotoxic coumarin, only in moderate amounts (BfR 2012b). Also, they warn about excessive intake of extremely hot chilli sauces with high concentrations of capsaicin, the combination of which may cause irritation of mucous membranes, nausea, vomiting, and hypertension (BfR 2011b). Specific regulations exist for some of these natural food ingredients, but they refer only to flavouring purposes [Regulation (EC) No 1334/2008 (EU 2008c)]. Thus, an increase in exposure is avoided by not allowing the addition of the isolated substances to food (e.g. capsaicin, coumarin, estragol, and thujone) and restricting their concentration in compound food containing flavourings or flavouring food ingredients (coumarin, estragol, and thujone).

Plant-Derived Food Additives

Food additives of plant origin are used for different functional purposes such as colouring [e.g. paprika extract (E 160c)], emulsifying [e.g. lecithins (E 322)], thickening [e.g. guar gum (E 412)], locust bean gum [E 410)], or sweetening [e.g. steviol glycosides (E 960)] food (EFSA 2014f). They underlie the same legislation as chemically synthesised food additives and have to undergo an approval procedure [Regulation (EC) No 1331/2008], need labelling as an additive [Directive 2000/13/EC (EU 2000b)], and have to meet the purity criteria set out in Commission Regulation (EU) No 231/2012. At present (November 2019), uncertainties exist regarding the differentiation between natural food colours as food additives [e.g. beetroot red (E 162), an extract of beetroots] and colouring foods (e.g. concentrated beetroot juice). Both examples are considered to be "characteristic" natural food ingredients. These uncertainties are complicating exposure assessment of food additive colours. A criterion to determine whether or not a food colour is an additive is that of "selective extraction of the pigments". This is based on a definition associated with the manufacturing of a colour considered as a food additive [Annex I of Regulation (EC) No 1333/2008 (EU 2008c)].

Botanicals and Botanical Preparations in Food Supplements

The use of botanicals and botanical preparations for food supplements has not yet been harmonised within the EU but is regulated in the Member States by differing national rules. Partly "positive lists" of safe botanicals and/or "negative lists" of botanicals, which are banned or restricted for food use, have been established. In some EU Member States, plant-based food supplements are subject to a notification procedure before introduction to the market (Bast et al. 2002). The EU Directive 2002/46/EC (EU 2002a) gives general provisions for food supplements and specifically addresses vitamins, minerals, and "other substances with a nutritional or physiological effect". It is generally understood that the latter include botanicals and botanical extracts along with substances such as amino acids, enzymes, pre- and probiotics, and essential fatty acids (Silano et al. 2011).

If a harmful effect on health is suspected for a botanical or botanical preparation, a procedure based on Article 8 of the food fortification legislation [Regulation (EC 1925/2006 (EU 2006b)] is initiated by the European Commission. This may result in a placement of the botanical (preparation) in Annex III of this regulation. On the basis of a risk assessment by EFSA, it can be banned (Annex III, Part A), restricted in use (Annex III, Part B), or in case of uncertainties be re-evaluated on the basis of additional safety data (Annex III, Part C). Only *Ephedra* herb and its preparations originating from *Ephedra* species (Annex III, Part A) and Yohimbe bark and its preparations originating from Yohimbe [*Pausinystalia yohimbe* (K. Schum) Pierre ex Beille] (Annex III, Part C) have been regulated by this means on the basis of existing EFSA opinions (EFSA 2013c, d).

Nutrition and health claims regarding botanical food supplements are regulated by Regulation (EC) 1924/2006 (EU 2006a), which does not foresee an assessment of the safety of the product carrying the claim.

Botanical Contaminants

As far as botanical contaminants are concerned, Article 2 of the Council Regulation (EEC) No 315/93 stipulates that, where necessary, maximum tolerances for specific contaminants shall be established for the protection of public health. Thus, Commission Regulation (EC) No. 1881/2006 (EU 2006c) lays down maximum levels for certain contaminants in foodstuffs. For natural toxicants at present (1/2018), only tropane alkaloids and erucic acid besides various mycotoxins including ergot alkaloids are addressed.

4.19.5 Future Perspectives

Exposures of consumers to plant-derived substances are increasing since plant-based preparations are widely present in different categories of food commodities with expanding market volume. The need to have better chemical and toxicological characterisation along with a harmonised systematic approach of scientific risk assessment of botanicals and botanical preparations is growing. The establishment of the EFSA guidance document for the safety assessment of botanicals and botanical preparations (EFSA 2009c) was an important step to achieve similar standards in safety evaluations performed by the national competent authorities of Member States (EFSA 2009c). A challenge for exposure assessment is the generation of total exposure estimates for a substance/group of substances of concern occurring in a diversity of plants relevant for a variety of different foods. The establishment of suitable models is a future need. Reliable estimates are especially needed for exposure of the general population to naturally occurring substances of concern via traditional food. This will enable the interpretation of health effects reported from epidemiological studies and also serve as a comparator in the presumption of safety approach.

Using the EFSA guidelines, the approach according to Article 8 of Regulation (EC) 1925/2006 (EU 2006b) for botanicals and their preparations in food supplements may be used more frequently to extend the lists in Annex III of this regulation. This is especially for botanicals or botanical preparations used in products, where the borderline between food supplements (not subjected to an authorisation procedure in the EU) and medicinal products (subjected to an authorisation procedure in the EU) is difficult to define, and the need for a harmonised legislation based on validated safety evaluations becomes evident. This approach should guarantee that intakes of toxicologically relevant botanical components are within safe limits.

In addition, activities are expected on the definition of colouring foodstuffs. A catalogue of clear criteria will possibly be established about how to distinguish colouring foods, which are requiring neither safety evaluation nor approval, from food additive colours which need an authorisation. For consumer safety, unambiguous definitions of colouring foods and their applications are considered necessary to ensure that their uses do not lead to toxicologically unacceptable higher exposures (higher than would result from normal dietary intake) for colouring components, accompanying substances, and impurities of these sources.

Further, recent EFSA risk assessments of botanical contaminants, namely of opium alkaloids in poppy seeds as well as of pyrrolizidine alkaloids or of THC in food and/or feed (EFSA 2011c, d, 2015g) may result in legal consequences. A decision by the European Commission may be made in these cases if there is a need for legal measures in the frame of Council Regulation (EEC) No. 315/93 laying down community procedures for contaminants in food to protect public health. Other measures may also be taken, such as the development of codes of practice. This has been specially recommended in a discussion paper by the Codex Committee on Contaminants in Foods to prevent and reduce pyrrolizidine alkaloid contamination of food products (FAO/WHO 2001; FAO 2011).

The EU Member States and professional stakeholder organisations are recommended to monitor and report on occurrence data for relevant botanical contaminants. On this basis, exposure assessments will be updated and allow for further decisions in risk management.

4.20 Nutrients

Birgit Niemann

Nutrients are elements or compounds needed by humans for normal growth, development, and health maintenance. One characteristic aspect of nutrients is their daily intake by the ingestion of food. In food, five major classes of nutrients have been found: the substances providing energy and body building materials like carbohydrates, proteins, and lipids, and the substances called micronutrients covering the more complex vitamin compounds and the elementary minerals. The micronutrients are free of energy but needed for various metabolic processes in small amounts. Beyond that, food components like different types of fibres that are indigestible for humans but feeding their intestinal microbiota belong to the nutrients, as does water.

The main characteristic aspect of micronutrients is the inability of the human body to synthesize them from appropriate precursor substances, and this makes their daily intake essential for human health. In contrast to non-essential chemical substances derived from food, there is a level of intake below which a risk of deficiency conditions or suboptimal functioning arises. Therefore, most micronutrients are subject to homeostatic regulation through adaptation of absorptive, excretory, or metabolic processes, and most of them can be stored in the human body. These mechanisms provide a measure of protection against occasional exposures below and above the usual intakes from balanced diets. The aspect of deficiency has been addressed by nutritional expert bodies as there are the formerly Scientific Committee of Foods (SCF) and the EFSA in the EU, and the Food and Nutrition Board of the Institute of Medicine (FNB/IOM) in the USA for establishing different dietary reference values (DRV's). The most important DRV's for this purpose are the Estimated Average Requirement (EAR) and the Recommended Dietary Allowance (RDA). The Estimated Average Requirement (EAR) represents the daily intake value that is estimated to meet the nutritional requirements in half of the healthy individuals in a given population or subpopulation. This average requirement is the basis of deriving the RDA, which is an estimate of the intake level that meets the nutrient requirements of nearly all (97–98%) healthy individuals in the respective population. The RDA can be calculated by adding two standard deviations (SD) of the EAR to the EAR value, if the individual requirements for a specific nutrient in the population underlie a normal distribution with a known variability. In the case where the individual requirements for a nutrient do not underlie a normal distribution and the data arc insufficient to calculate twice the SD, a 10% coefficient of variation (CV) can be added to the EAR (Fig. 4.26). The exposure assessment methods, which for nutrients are usually referred to as nutritional intake assessment, serve as an instrument to measure the distribution of the nutrient intake in a given population of healthy people needed to derive the dietary reference values.

However, like other food-derived substances, micronutrients may have adverse effects if consumed at levels significantly above those obtained from endogenous nutrients in traditional foods over longer time periods. This would be possible in the case of the regular consumption of food supplements and/or of nutrient-fortified food products. In the European Union, the marketing of these food categories was harmonised by the Directive 2002/46(EC) related to food supplements and the Regulation 1925/2006/EC (EU 2006b) related to the addition of vitamins, minerals, and certain other substances to foods.

4.20.1 Exposure Assessment in the Framework of the Directive 2002/46/EC and the Regulation 1925/2006/EC

The European legislation related to food supplements and fortified foods specify both the kind of nutrients and the specific nutrient compounds that can be added to food supplements and fortified food products placed on the European market. Both regulatory rules take into account the safety aspects of micronutrients by calling for minimum and maximum amounts of micronutrients that can be added to specific products. The minimum amounts are regulated in the context of food labelling laws, which prescribe a nutrient content of at least 15% of its labelling reference value per 100 g food if the nutrient should be advertised. The need for maximum amounts is addressed in the regulations mentioned above, which declare in the recital 13 resp. 14 that

> excessive intake of vitamins and minerals may result in adverse effects and therefore necessitate the setting of maximum safe levels for them in food supplements,

resp.

> when they are added to foods.

This was specified in the recital 14 resp. 15 of the rules which explain:

> When maximum levels are set, therefore, account should be taken of the upper safe levels of the vitamins and minerals, as established by scientific risk assessment based on generally acceptable scientific data, and of intakes of those nutrients from the normal diet resp. from other foods.

Furthermore, the nutrient reference intake values should be taken into account in the setting of maximum safe levels.

These statements, implemented in the article 5 resp. 6 of Article 5 of the Directive 2002/46/EC (EU 2002a) related to food supplements, and Article 6 of the Regulation 1925/2006/EC (EU 2006b) on the addition of vitamins, minerals, and certain other substances to foods clearly indicate that no special exposure assessment methods are prescribed for nutrients by the regulatory rules. These methods are rather embedded in a complex framework of a scientific risk assessment and management process that has to be performed using the best available knowledge. The risk assessment model for micronutrients implies the derivation of an additional dietary reference value, the so-called nutrient-specific tolerable Upper intake Level (UL), which reflects the maximum daily intake level at which no risk of adverse health effects is expected for almost all individuals in the general population, including sensitive individuals, when the nutrient is consumed over long periods of time.

Nutrient-specific UL's were developed by the same authorized expert bodies, as were the other dietary reference values mentioned above. In contrast to them, UL's are not recommended levels of intake. They are estimates of the highest level of intake (from all sources) judged to be unlikely to pose a risk of adverse health effects to humans (IOM 1998; EFSA 2006a). For the development of a UL, three toxicological values commonly used in assessing the risk of exposures to chemical substances are adapted for assessing nutrients: the no-observed-adverse-effect level (NOAEL), the lowest-observed-adverse-effect level (LOAEL), and uncertainty factors (U's). In contrast to chemicals, the adverse endpoints, as well as the NOAEL's or LOAEL's, are mostly being identified from human studies. The data from animals and other experimental studies are only used to support the human data. Furthermore, instead of generalized uncertainty factors for intra- and interspecies variability, nutrient-specific uncertainty factors considering the recommended dietary allowance (RDA) of the micronutrient of interest have to be used. In principle, this modified concept should generate an UL value above the RDA value for the specific nutrient considering its adequate intake (see Fig. 4.27).

UL values were derived for 16 micronutrients and for the element boron by the SCF/EFSA in the EU (EFSA 2006a), and for 26 micronutrients, the non-essential metabolite choline, plant-derived carotenoids, and for the elements boron, nickel, and vanadium by the FNB/IOM in the US (IOM 1998, 1999, 2000a, b, 2001). For some of the micronutrients (vitamin K, B1, B2, B12, pantothenic acid, biotin, and potassium), a UL could not be derived because no adverse effects could be identified inside the range of the observed intake levels in humans from all sources (including pharmaceuticals), or the scientific data for determining the dose–effect relationships were insufficient at the time.

The European Commission was requested to propose maximum amounts (MAs) for the addition of vitamins and minerals to supplements and foods by 19th January 2009 according to the regulations. Therefore, different models have been developed for the micronutrients by national experts or stakeholder organisations to translate the UL's into safe MAs for this purpose (Dufour et al. 2010; Flynn et al. 2003; Hirvonen et al. 2007; Kloosterman et al. 2007; Rasmussen et al. 2006;

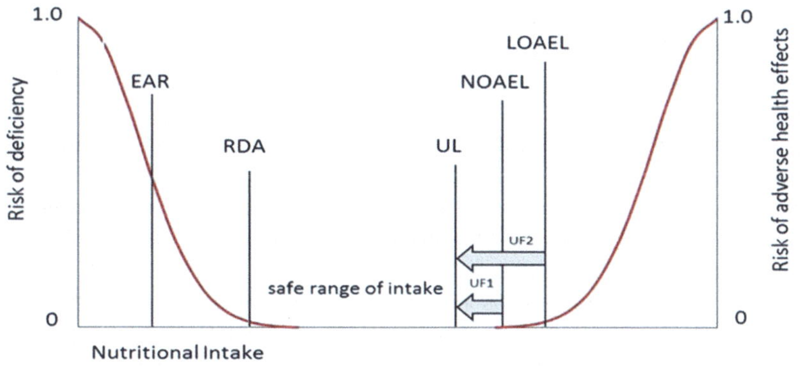

Fig. 4.27 Derivation of the tolerable Upper intake Level (UL) for a micronutrient, using toxicological threshold values combined with uncertainty factors (UF's), representing different situations of unsafety. NOAEL - no-observed-adverse-effect Level; LOAEL - lowest-observed-adverse-effect Level (modified to ILSI)

Richardson 2007). Based on the criterion that the total intake of a nutrient from all sources should not exceed the UL, all models at first calculate a safe surplus amount by subtracting the basic nutrient intake at a high intake percentile (usually the 95th or the 97.5th percentile) in a given population or subpopulation from its respective UL. The "basic nutrient intake" refers in this regard to the nutrient intake from traditional food products with only their endogenous nutrient contents, without intake from food supplements and fortified foods. However, food products with reconstituted vitamin contents lost by conservation processes and food products like margarines (containing the added vitamins A and D in amounts which ensure their nutritional equivalence to butter) are also counted as traditional food products.

In the following steps, each model describes its own way to distribute the safe surplus amount of the nutrient between supplements and other food products that are suitable for fortification from a technological view. The models frequently differ in the following points:

- Either both supplements and food products or only one of these categories were considered.
- Micronutrients were classified in different risk categories related to the magnitude of the safe range of intake or they were not classified.
- Different subpopulations were used as a reference for the calculation of safe MAs that can be added to food products.
- Different proportions of food products were considered as qualified for fortification.
- The energy intake from fortified foods, either assessed or nutrient specific measured, was considered or not considered.
- The MAs were calculated related to energy portions, or to the weight of the food (usually standardised to 100 g or 100 ml).

In the years before 2009, the available models were discussed and tested by experts from different EU member states to develop a common framework for setting safe MAs for vitamins and minerals according to the regulatory rules 2002/46/EC (EU 2002a) and the (EC) No 1925/2006 (EU 2006b). The discussion started with a model developed for fortified foods by the International Life Science Institute (ILSI) Europe (Flynn et al. 2003), and a model developed for nutritional supplements by the European Responsible Nutrition Alliance (ERNA) (Richardson 2007). Both models claimed the whole safe surplus amount of the nutrient of interest, either for supplements or for fortified food products, arguing that the intake from the respective other food category is integrated with the intake at the P95 from the chosen population used in the specific model. However, different modifications of the ILSI-Model that consider the intake from food supplements more directly have been developed by some EU member states (Domke et al. 2004, 2006; Kloosterman et al. 2007; Rasmussen et al. 2006), and during the discussion the demand for considering the mutual dependency of the MA's for both food categories in a common model was established. Therefore, a modified ILSI-Model was presented by A. Flynn at an ILSI-Workshop on this topic held from 16 to 18 April 2008 in Gubbio, Italy (ILSI report 2009). In the so-called "Gubbio-Model" (Box 4.1), the main diverging point of the discussion has been resolved. Therefore, this model achieved a high degree of acceptance among the participating experts. However, no common proposal for setting MA's could be established in the EU until now due to the complex nature of the issue and the divergent views that were expressed by the Member States and the interested stakeholders. However, different EU member states have their own national MA's for micronutrients in food supplements that were established before or after the Directive 2002/46/EC (EU 2002a) had been passed.

Box 4.1 The "Gubbio"-Model for the setting of safe maximum amounts for the adding of vitamins and minerals to food supplements and fortified foods, as presented by A. Flynn (ILSI report 2009)

1. The total intake of a given nutrient from fortified foods should not exceed the UL:

$$UL - (CI_{95} + SI)$$

UL	the tolerable Upper intake Level for the nutrient
CI_{95}	the 95th percentile of the habitual current intake distribution of the nutrient from non-fortified foods
SI	a realistic high intake scenario for that nutrient from dietary supplements

(continued)

Box 4.1 (continued)

2. The maximum amount that can be set for the nutrient in fortified foods (MAFF) expressed per 100 kcal of fortified food is given by:

$$MA = \frac{UL - (CI_{95} + SI) \times 100}{(EI_{50} \times PFFn)}$$

EI_{50} the total energy intake at the 50th percentile of the habitual daily energy intake distribution for the population/relevant age group

PFFn the proportion of EI_{50} that comes from foods fortified with the nutrient in high consumers (P_{95}) of those foods

This expression defines a fixed relationship for any given nutrient between the MAs that can be set for fortified foods and dietary supplements such that the margin between its intake from the base diet and its upper safe level will not be exceeded.

3. An example, demonstrated in the figure below, was given for the setting of MAs for fortified foods and dietary supplements with Vitamin D using intake data of Irish adults and PFFn-values of 5% (solid line) and 10% (dotted line):

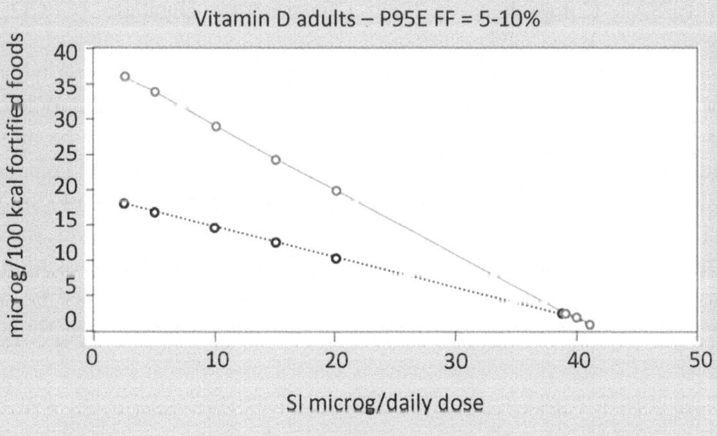

There are three points about the exposure assessment methods used to calculate MA's for micronutrients in the different models: First, they should assess the intake distribution for the micronutrient of interest in the chosen population. Second, they should model a variety of anticipated intake scenarios using a range of hypothetical added amounts of the nutrient in a variety of carrier foods for identifying suitable fortification conditions, and third, they should test the safety of the final conditions for supplements and fortified foods in all parts of the population, whereby the term “safety” means an anticipated intake distribution showing that only a negligible part of the population may have a daily intake of the micronutrient under consideration above its UL if all product categories (basic diet, fortified foods, and food supplements) were considered.

The exposure assessments used in all steps of a specific model are usually based on the same food intake data, which were linked with actual measured nutrient content data to assess the intake distribution from the basic diets, and with hypothetical nutrient contents to model the different anticipated intake scenarios, which can be analysed according to different criteria. The quality of the results depends to a great extent on the quality of the food intake data. The food intake data used should represent the whole population of a country and allow for their stratification according to gender, age groups, and other social parameters. Furthermore, separate assessments related to the user of food supplements as well as the user of specific foods or suitable aggregated food groups should be possible. Therefore, the use of food intake data generated by validated survey methods providing data on the level of individual persons and specific food products is strongly recommended. This includes food intake data gathered by retrospective survey methods like the 24-h recall combined with food frequency questionnaires and diet history interviews, as well as by prospective methods like weighed food records or duplicate diet methods. Generally, food surveys initiated by governmental authorities in the framework of national health reports mostly fulfil these criteria.

In the EU, 32 national food surveys from 22 different member states are summarized in EFSA’s Comprehensive Food Consumption Database (CFCD; EFSA 2015a).

The second important factor is the quality of the food composition table that is used. Ideally, the food composition table should contain compositional data for all nutrients of interest on the level of unique food products that can be aggregated according to the specific purposes of the intake assessment under consideration. Furthermore, the table should reflect all main food products currently sold in the food market, including food supplements and fortified foods. However, the latter is one of the conditions most difficult to fulfil because of the rapid innovation rate of these product groups in the market. Therefore, additional data related to the composition of food supplements and fortified foods wherever they can found could be used.

4.20.2 Exposure Assessment According to Legislations Related to Novel Foods or Food Ingredients, and to Genetically Modified Food and Feed

The first premarketing approval procedure for food products was established in the European Union with the Regulation (EC) No 258/97 related to novel foods and novel food ingredients. Originally, this regulation covered both novel foods or food ingredients, and foods produced from genetically modified organisms (GMO). Later, the regulation (EC) 1829/2003 specific for food and feed produced from GMO was adopted in the EU in 2003, and the regulation (EC) No 258/97 and was replaced by the Regulation (EU) 2015/2283 in January 2018. However, both categories of food products require an authorisation or notification (the latter only applies to novel foods, which are used traditionally for nutrition outside the EU) before they can be placed on the European market. The key feature of the authorisation procedure is a comprehensive risk assessment of the specific food product under the scope of the respective regulation, which is described in detail elsewhere (Pöting et al. 2014).

Novel foods and food ingredients are defined as a type of food that does not have a significant history of consumption in the European Union prior to 15 May 1997, the date of entry into force of the regulation (EC) No 258/97. The guidance for preparing of the application dossier and the risk assessment of a novel food or food ingredient was given in the Commission Recommendation 97/618/EC. An updated guidance document was published by EFSA in the context of the renewed Novel Food Regulation (EU) 2015/2283 (EFSA 2016f). All guidance documents recommend that the risk assessment of novel food products should be performed on a case-by-case approach considering the composition of the specific novel food product, possible effects of the production process, the previous exposure to the sources of the novel food, and its anticipated consumption. Furthermore, microbiological, nutritional, and toxicological information substantiating the safe use of the novel food product under the proposed conditions has to be presented in the dossier.

The assessment of the anticipated intake constitutes an essential part of the risk assessment because the safety of a novel food product will be assessed under the proposed conditions of use. Therefore, the applicant should provide specific information about the type of use (whole food or food ingredient), the food categories in which an ingredient is proposed to be used, and the maximum amounts of the ingredient in products as consumed. Based on this information, anticipated daily intakes of the novel food (per kg of BW and in absolute amounts) should be estimated in appropriate exposure scenarios providing mean and high (at least 95th percentile) anticipated daily intakes for different target population groups (e.g. children, pregnant and lactating women, the elderly, or individuals with specific health risks).

The intakes should be estimated using actual food consumption data.

In this context, the summary statistics of the CFCD can be used as a quick screening tool to assess chronic and acute exposure to a variety of foods in the European member states, to identify the subpopulation with the highest intake of all food products of interest. All foods in the surveys from the contributing member states are recoded by the unified hierarchical food classification system FoodEx2, developed by EFSA (EFSA 2011f, h, i), enabling comparability of the data. The output data contain mean and median intakes for each food (including the standard deviations) as well as low and high percentiles for the total population of each country, as well as only for consumers. All data are stratified by age groups (infants, toddlers, other children, adolescents, adults, elderly, and very elderly) and expressed in both units grams/day and grams/kg BW per day (EFSA 2011f). Data for the cumulative intake of a novel ingredient from a variety of foods can be generated by the Food Additive Intake Model (FAIM), developed by EFSA (2013f) to assess the intake of food additives. The FAIM algorithm, which is also based on summary statistics of the Comprehensive Database, calculates the high-level intakes from combined food categories by summarising the highest 95th Consumers-only intake from a specific food in a given (sub)population to the average intakes from all other foods intended to contain the novel ingredient of the whole (sub)population (Tennant 2016).

All screening estimates generated by the EFSA tools may be sufficient for the risk assessment if the high intake estimates are well below the established safe upper intake levels for the novel food product under consideration. If not, additional intake calculations based on individual data from national food consumption surveys may be performed to create more refined estimates.

Beyond the anticipated intake of the novel food product according to the intended conditions, any other potential source of intake should be taken into account. This could be the intake from the background diet in case the novel ingredient is a purified or concentrated natural food constituent. In such cases, the extent of the additional intake of the constituent resulting from its intended use as a novel food ingredient, in relation to existing dietary intake, should be assessed.

Specific examples for the assessment of the anticipated intake of novel foods or food ingredients can be found in the non-confidential parts of current novel food applications provided on the website of the Advisory Committee on Novel Foods and Processes (ACNFP) of the British national competent food assessment body, the Food Safety Agency (FSA) (https://acnfp.food.gov.uk/assess/fullapplics).

Foods and food ingredients containing or consisting of GMOs, as well as foods and food ingredients produced from but not containing GMOs, underlie an obligatory authorisation procedure according to the Regulation (EC) 1829/2003. With this regulatory rule, a single risk assessment is introduced, and a single authorisation is granted for a GMO and its possible uses including both food and feed products. However, further aspects of GMO's, e.g. the cultivation, the labelling, and the traceability, are regulated by further laws. The strategy for the risk assessment has been outlined in the Guidance for risk assessment of food and feed from genetically modified plants by EFSA (2011j). The strategy is based on the concept of substantial equivalence, which makes use of the idea that the conventional crop plant with a history of safe use for consumers and/or animals can serve as a comparator when assessing the safety and nutritional value of the modified plant and derived food or feed. The assessment is also performed on a case-by-case approach starting with a comprehensive molecular characterisation of the genetic modification followed by a comparative analysis of compositional, phenotypic, and agronomic characteristics of the GM plant in relation to its comparator grown under the same environmental conditions. Identified differences are then subject to further evaluation with regard to their potential impacts on human and animal health.

As already mentioned for novel foods, the estimation of the anticipated intake is also an essential element in the risk assessment of food and feed derived from GMO. It is assumed that GMO-derived food products are intended to replace the same food products produced from the conventional crop plant, which already exist on the market. Therefore, the intake assessment is concentrated on the newly expressed proteins, which are not existent in the conventional counterparts of the foods derived from GMO. Furthermore, identified changes in endogenous constituents resulting from the genetic modification, probably caused by disrupting specific metabolic pathways, should also be considered.

The exposure assessment for the newly expressed proteins should be based on representative consumption data for the conventional counterparts of the respective foods derived from GMO. These consumption amounts can be combined with the concentrations of the proteins under consideration stated in the GMO application. These data are usually given as μg new protein per g whole protein in the edible parts of the plant. As these data are related to the unprocessed crop plant, the influences of processing, storage, and expected treatment of the food and feed in question should be taken into account.

The prime opportunity to calculate the intake amounts for the newly expressed proteins from GMO-derived food products is the use of predefined excel spreadsheets consisting typical food categories from maize, soy beans, rapeseed, potatoes, rice, and sugar beets provided by EFSA (https://www.efsa.europa.eu/de/applications/gmo/tools). These tools are based on the summary statistics of the CEFDB and designed to perform the exposure assessment in applications for authorisation of genetically modified plants. The acute or the chronic consumption of all people as well as of only the consumer in the predefined age groups in the European population can be expressed as grams per day (g/day) and grams per kg of BW per (g/kg of BW per day).

In case a GM plant contains new constituents other than the newly expressed proteins or has a modified nutritional profile, a more refined assessment of consumer exposure relevant to European consumption patterns would be required. The focus of such an exposure assessment should be long term in the habitual diet of the general population and subsets of consumers particularly at risk. In this regard, other types of consumption data, such as the FAO food balance sheets or literature, can help in identifying the countries or geographical areas with the highest consumption of the respective foods. In situations where information at the level of individual consumers is not be available for the country identified, a country with a comparable consumption pattern from the CECFD can be used to assess the anticipated intake.

4.21 Pesticide Residues

Christian Sieke

4.21.1 Introduction

Pesticides are defined here as any substance or mixture of substances used as plant protection products to destroy, suppress, or alter the life cycle of any pest that could cause a problem of food security. They include bactericides, baits, fungicides, growth regulators, herbicides, insecticides, lures, rodenticides, and repellents.

The European Commission (COM) inventory on pesticides intended for application in agriculture reports more than 1300 active principles (substances). Some of them are not authorized within the European Union or are considered obsolete; however, they are still present as stockpiles in some Third Countries from which food is imported (EU 2016).

Pesticides can be also used in commercial, domestic, and urban environments not strictly related to food production. Within this context, the European Union frames their uses other than for crop protection within the biocides legislation (Regulation no. 528/2012; EU 2012): the active principle(s) could be the same, but the intended use and the commercial formulation make the difference, also in terms of risk assessment and authorisation.

This means that an aggregated exposure to the same active principle (i.e. the neonicotinoid imidacloprid as insecticide) could result from both pesticide residue in food and from the biocide used to fight mosquitoes, and baits in the indoor environment via dermal absorption, inhalation, and dust ingestion. Because approaches to consider both regulatory frameworks are still under development, the assessment must be done by standard model calculations, if needed, as described in Sect. 3.3.

Among all agents covered by regulatory authorisation systems, the field of pesticides is one of the longest established and most harmonised frameworks. Before marketing plant protection products, their toxicological properties and their safety for workers, consumers, and the environment have to be assessed. In most countries, this assessment is performed by national or supranational governmental bodies under consideration of harmonised criteria defined, for example in OECD Guidelines and Guidance documents.

The procedure for assessing residues of pesticides used as active agents in plant protection products follows common approaches used in many parts of the world (e.g. Australia, Canada, the European Union, or the USA). This procedure includes data requirements and endpoints that need to be addressed in the assessment with corresponding studies to be submitted by applicants according to harmonised Guidelines. Due to the high level of international harmonisation under the OECD framework, key studies are normally used globally to apply for authorisation of pesticides. In principle, the whole food chain has to be considered for a consumer risk assessment. Starting with the direct application to the crop and the metabolic behaviour of active substances in the plants themselves, the nature and magnitude of the residue in foods and feeds must be known. Also, secondary phenomena like the degradation during processing of foods and the carry-over into foods of animal origin are important necessary information. Therefore, the following key aspects have to be addressed in the assessment of the consumer exposure and for the dietary risk assessment of pesticides. In Fig. 4.28, aspects covered by regulatory frameworks are presented.

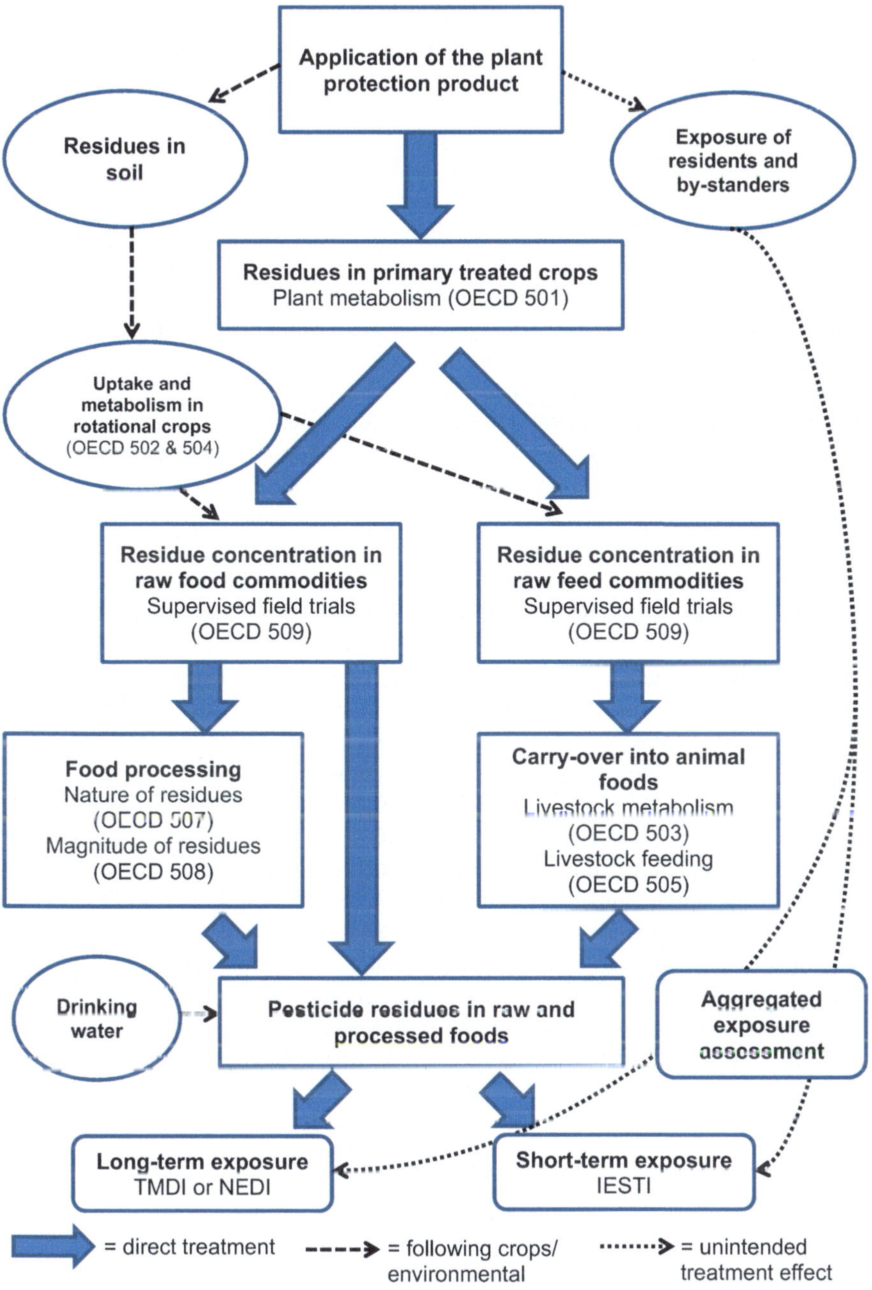

Fig. 4.28 Overview of pesticide assessment principles for consumers

4.21.2 *Residue Behaviour Assessment Principles for the Authorisation of Plant Protection Products and Pesticides*

To understand the fate of pesticides after application, their uptake into plants, their translocation, and finally their degradation/biotransformation have to be investigated according to the OECD technical guideline TG 501 (OECD 2007a). In general, metabolism studies in plants have to be provided for at least one member of each crop group to be treated (e.g. cereal grains, fruits, or root crops). When three different crop groups were investigated and a comparable metabolic pattern has been identified, the similarity of the residue profile in all plants can be assumed. Typically, ^{14}C-radiolabelled compounds are applied to allow identification of metabolites at the lowest detection limits possible. The general aim in assessing these studies is to derive the residue definition for enforcement (or monitoring) purposes and the residue definition for dietary risk assessment. These definitions should address the following criteria:

Residue definition for enforcement purposes: The main principle for deriving a residue definition for enforcement purposes is to identify a characteristic marker for the pesticidal active substance, which is suitable to establish legal limits. Food and feed commodities generally must comply with legal limits (maximum residue levels or limits, "MRLs") before being put on the market. The selected marker should be easy to analyse by enforcement laboratories (e.g. by multi-residue methods), preferably by selecting the parent molecule or by inclusion of single metabolites. It is also important that other sources of the marker compound besides pesticidal use are avoided.

The residue definition (RD) for dietary risk assessment purposes includes all metabolites/degradation products of the pesticide posing a significant contribution to the dietary risk for consumers. In contrast to the residue definition for enforcement purposes, other compounds that are difficult to analyse (conjugates or stereospecific enantiomers, etc.) have to be taken into account to ensure consumer protection. When additional sources exist for compounds in the residue definition for dietary risk assessment purposes, the total exposure has to be considered for the exposure estimation. In case of deviating residue definitions for enforcement and risk assessment purposes, a conversion factor (CF) is calculated for each crop:

$$\text{Conversion Factor(CF)} = \frac{\text{Concentration according to RD Risk Assessment}\left(\frac{\text{mg}}{\text{kg}}\right)}{\text{Concentration according to RD Enforcement}\left(\frac{\text{mg}}{\text{kg}}\right)} \qquad (4.2)$$

Example:
Simple residue definition:
Enforcement and dietary risk assessment for plant commodities: Examplethrin
Complex residue definition:
Enforcement for plant commodities: Examplestrobin
Dietary risk assessment for plant commodities: Sum of examplestrobin and its metabolites containing the 6-chloropyridinyl moiety, expressed as examplestrobin
Conversion factor:
Pome fruit: 1.3

Metabolism and magnitude of residues in rotational crops (OECD TG 502 and TG 504 (OECD 2007b, d)): Plant protection products are often applied to annual crops, which are completely removed from the field after harvests like cereal grains or potatoes. Some pesticides or their decomposition products show high persistence in soil, potentially making them available for uptake via roots into crops grown later on the same field. The nature of the residues in these crops may deviate significantly from metabolic patterns found in the primary treated plants due to uptake of compounds exclusively formed in soil. Therefore, studies with ^{14}C-labelled compounds have to be conducted involving bare soil treatment addressing actual seasonal rates of the pesticide. After intervals representing usual cropping patterns (up to 1 year), characteristic model crops are planted and grown to commercial harvest. The metabolic profile found in these crops is also taken into consideration for defining the residue definitions for enforcement and dietary risk assessment purposes. In parallel, rotational crop field studies representing common agricultural practice are conducted on representative crops to estimate residue concentrations in harvested products.

Livestock animal metabolism and feeding studies (OECD TG 503 and 505 (OECD 2007c, e)): Comparable to plant metabolic studies, information on the composition of residues in foods of animal origin is required when potential feed items contain residues either after direct treatment or by uptake from soil as a rotational crop. Normally, the ^{14}C-radiolabelled active substance or its relevant metabolites present in feed are administered to livestock animals (e.g. lactating goats and/or laying hens). The transfer of residues into edible tissues, eggs, and milk is determined as well as the metabolic profile in the various matrices. As for plant matrices, separate residue definitions for enforcement and dietary risk assessment purposes are derived for animal commodities. In addition, feeding studies using non-radiolabelled substances are conducted for a period of at least 28 days to estimate the concentrations of residues in tissues, milk, and eggs.

Supervised field trials (OECD TG 509 (OECD 2007h)): A set of supervised field trials conducted with active substances under actual field conditions is required to estimate the magnitude of residues in food and feed commodities. These trials have to reflect the most critical use of the plant protection product, probably leading to the highest residue concentrations in harvested commodities likely to occur according to the authorized use. The most critical use is normally represented by the highest amount of the pesticide applied in combination with the shortest timing interval before harvest. From each of these individual trials, one residue concentration value measured in a representative field sample is selected. The ranking of all results provides the basis for the estimation of the supervised trial median residue (STMR) and the highest residue (HR) used for the consumer dietary risk assessment.

Example:
Concentrations of compound A in tomato fruits from a series of field trials:
0.01, 0.03, 0.03, 0.05, **0.06**, 0.07, 0.07, 0.10, **0.12** mg/kg
STMR: 0.06 mg/kg
HR: 0.12 mg/kg

Based on the same data set, MRLs according to a harmonised OECD procedure (OECD 2011) are calculated. Since estimations of the consumer exposure are normally conducted prospectively to the authorisation of plant protection products, supervised field trials often provide the only source of information on residue concentrations of the pesticide. Occurrence data from food monitoring or human biomonitoring programmes are unavailable at this stage.

Nature and magnitude of residues after processing (OECD TG 507 (OECD 2007f)) and TG 508 (OECD 2007g)): As a harmonised basis for global trade, MRLs are usually set on the raw agricultural commodity ("RAC") like wheat grain or rapeseeds. However, most foods are consumed in processed form, which may have an impact on the composition of the residue as well as on concentrations found in processed products. Studies investigating changes in the composition of residues utilise hydrolytic conditions representative of typical processing procedures (pasteurisation, cooking/baking, and sterilisation). Degradation products formed under these conditions are also considered when establishing residue definitions.

These investigations are complemented by studies using field treated raw agricultural commodities for simulating industrial processes like juicing or brewing. The observed transfer of residues into processed commodities is expressed by the "processing factor" (PF). It is used in dietary exposure estimates and to decide on the compliance with MRLs (set for the RAC) by enforcement laboratories, when residues are found in market samples of processed foods. PFs may be either based on the residue definitions for enforcement or dietary risk assessment purposes, as required.

$$\text{Processing Factor(CF)} = \frac{\text{Concentration in processed products}\left(\frac{\text{mg}}{\text{kg}}\right)}{\text{Concentration in raw agricultural commodity}\left(\frac{\text{mg}}{\text{kg}}\right)} \quad (4.3)$$

Data sources for processing factors are compilations, e.g. by the German Federal Institute for Risk Assessment (Scholz et al. 2017), or by EFSA (2018a) or the Joint Meeting on Pesticide Residues (JMPR, FAO 2016).

4.21.3 Principles of Consumer Exposure Assessment for Pesticide Residues

Before authorisation of plant protection products and the establishment of MRLs for active substances, their consumer safety must be assessed based on a dietary risk assessment (besides other areas of concern such as worker safety and the impact on the environment). Most regulatory systems address both the long- and the short-term dietary exposure to assess related chronic or acute health risks.

For both scenarios, toxicological reference values are established for the pesticidal active substances and, if necessary, for their relevant metabolites. The long-term dietary exposure is compared to the Acceptable Daily Intake value (ADI), representing the maximum daily amount of one agent (or related agents like its metabolites), which individuals can be orally exposed to for their whole lifetime without adverse effects.

In addition, for agents potentially inducing adverse health effects even after single exposure events, the Acute Reference Dose (ARfD) was introduced (Solecki et al. 2005). The ARfD represents the maximum amount of one agent individuals may be orally exposed to within 24 h without adverse health effects. It is compared to the short-term dietary exposure. Deriving an ARfD only becomes necessary for active substances potentially inducing acute toxicity even after a single exposure.

Long-Term Dietary Exposure

In most countries, the estimation of the long-term dietary exposure follows the concept of the "Theoretical Maximum Daily Intake" (TMDI), published by WHO in 1989 (WHO 1989), which was refined into the "International" or "National Estimated Daily Intake" (IEDI or NEDI, respectively) in 1997 (Global Environment Monitoring System - Food Contamination Monitoring and Assessment Programme (GEMS/Food), (WHO 1997a)). The long-term dietary exposure is estimated by multiplying the average daily amount of each food commodity consumed by their respective average residue concentration. By adding the contribution of residues via all foods consumed within one day, the total daily exposure is calculated, normally expressed on a body weight basis:

$$\text{TMDI or IEDI }[\text{mg/kg bw}] = \sum_{\text{Food 1}}^{n} \frac{\text{Residue Concentration (Food } n) \times \text{Portion size (Food } n)}{\text{Body weight}} \tag{4.4}$$

While the TMDI relies on MRLs as a residue concentration for each of the commodities, making it a very conservative approach for estimating the average long-term exposure on a daily basis, the IEDI/NEDI use STMR instead as a more realistic basis. If appropriate, residue concentrations may be adjusted by processing factors (e.g. for fruit juices).

The procedure to express portion sizes and body weights for the long-term dietary exposure estimate depends on the regulatory framework. Within Codex Alimentarius, the long-term dietary exposure estimation of pesticide residues is currently based on data from the WHO Global Environment Monitoring System (GEMS), taking into account the "Principles and methods for the risk assessment of chemicals in food" (WHO/FAO 2009, 2012). By using the information on global trade of food commodities, a cluster analysis was conducted to identify 17 clusters of individual countries with comparable consumption patterns. For each of these clusters, the total amounts of imported/exported foods were derived. When expressed on a daily per capita basis, a conservative estimate of the portion sizes for each commodity becomes possible. A template for the IEDI model can be obtained from the GEMS website (http://www.who.int/foodsafety/areas_work/chemical-risks/gems-food/en/).

In contrast to the GEMS approach, national or supra-national regulatory systems mostly rely on consumption data from food surveys (see Sect. 4.6) on the general population or on vulnerable subpopulations (e.g. children). In general, average portion sizes used for long-term dietary exposure estimates are based on the arithmetic mean of the consumed amounts reported for each food commodity. These mean values also include days without reported consumption events (zero consumption days), since not all foods are consumed on a daily basis. However, the exact methods used to determine portion sizes differ between national models. In early long-term exposure models like the German Diet for a 4–6 year old girl, the portion size was directly expressed as the mean amount of food reported within a consumption survey for all individuals in the target population. These portion sizes were afterwards correlated to the mean body weight. Recent models tend towards a more sophisticated consideration of the intra-individual variability and on food frequencies on an individual level to avoid an underestimation of the long-term portion sizes. Portion sizes are normally expressed as the amount of food consumed per kg of body weight, calculated on an individual level for each person. In addition, the impact of food consumption frequencies for the long-term exposure may be estimated by calculating the mean for each individual (observed individual mean = OIM) before the overall means for each food commodity are derived. Also, parametric modelling of the average portion sizes can be conducted (e.g. LogNormal-Normal-Modelling). A comparison of available methodologies for adjusting consumption data to long-term dietary exposure scenarios was conducted in an External Scientific Report for EFSA (van Klaveren et al. 2012).

Due to the nature of an authorisation system, a prospective dietary risk assessment becomes necessary before the plant protection products are placed on the market. In contrast to other regulatory frameworks like those for contaminants, total diet studies (TDS) are rarely used for the estimation of the consumer exposure to pesticides since residues of active substances in foods can only be found for compounds already in trade. Most exposure models in the field of pesticide residues rely on types of surveys reporting consumption values on a per-day basis (24 h-recalls, food protocols, etc.), allowing estimation of both the maximum daily and the average daily portion sizes. For the long-term exposure, dietary histories and food frequency questionnaires may also be used.

Short-Term Dietary Exposure

In 1997, it became apparent that some pesticides may potentially cause consumer health risks even after exposure to a single high dose (e.g. organophosphates or carbamates). For the estimation of such a short-term dietary exposure, the Geneva Conference (WHO 1997b) developed the "International Estimate of Short-Term Intake" (IESTI) concept, which introduced three different exposure scenarios.

In all scenarios, the large portion size ("LP") for the food commodity to be evaluated was defined to represent the 97.5th percentile of the portion size for each food commodity. Since no "zero consumption days" are taken into account for the LP, this approach is considered to provide a conservative deterministic estimate for a single daily portion size. However, the percentiles of the underlying population for which the LP relates differ, depending on the consumption frequency of the respective food. For foods that are consumed daily (e.g. milk), the LP is close to the true 97.5th percentile of the whole population whereas for rarely eaten foods, the LP represents a much higher percentile due to the low frequency of consumption events. A general definition for the level of protection to be considered on a population basis has yet to be established.

Due to the use of the IESTI primarily in the authorisation of plant protection products, the STMR or the HR derived from supervised field trials (see above) is used as occurrence data. As for the long-term exposure, the IESTI is always expressed on a body weight basis for comparison with the ARfD. In the current IESTI-equations, the body weight is normally expressed as the arithmetic mean body weight of the target population. A change to individual consumption/body weight ratios in the IESTI-equations has been proposed, but is not yet accepted in regulatory frameworks (see "Revisions of the IESTI-Concept" below).

The IESTI-concept consists of four equations describing different scenarios for short-term dietary exposure called "Cases":

The Cases 1 and 3 apply to all foods of small size, defined either as raw agricultural products of low weight (<25 g per piece) like strawberries, probably originating from one single field (Case 1), or as products that are mixed/blended on large scale or industrially processed (Case 3, e.g. cereal grains or juices). The equations used for these cases involve simple multiplication of the LP, either with the HR (Case 1) representing a scenario of a single high residue concentration or with the STMR (Case 3), providing a robust estimate of the typical residue for mixed commodities.

Cases 2a and 2b also assume that the respective foods may originate from one single field, but only apply to commodities with weights of at least 25 g per piece ("unit-weight"). The implication of the unit-weight for the short-term dietary exposure estimation becomes apparent when the number of individual units representing the LP and the variability of residue concentrations between these units are considered. Investigations have shown (Hamilton et al. 2004) that single units may contain much higher residue concentrations than measured in composite analytical samples used in supervised field trials, or for MRL compliance testing by governmental enforcement laboratories. These composite samples, usually consisting of 12–24 homogenized individual units, provide a robust estimate for the average residue concentration of the lot under investigation. For small commodities like berry fruits, a sufficiently large number of individual fruits are consumed with the LP to level out the variability of residue concentrations between the units. However, since consumers eat within one day only a very limited number of units with at least 25 g per piece (e.g. 1–2 apples, oranges or mandarins) or even only a part of very large commodities like watermelons, the variability in residue concentrations has to be considered to avoid an serious underestimation of the short-term exposure.

Within the IESTI-concept, this variability is addressed by the use of the "variability factor", defined as the ratio of the 97.5th percentile of residue concentrations on individual units divided by the arithmetic mean of all units. This factor is multiplied with the residue concentration found in the composite sample to extrapolate to a single unit with above average residues. When the LP is bigger than the average unit-weight of a commodity (more than one piece is eaten), it is assumed that only the first unit fully eaten contains higher residue concentrations and is multiplied with the variability factor (Case 2a). The rest of the LP is expected to contain residue concentrations as measured for the composite sample. For large-sized commodities with unit-weights higher than the LP (only a part of one piece is eaten), the whole portion may contain above average residue concentrations and is therefore multiplied completely with the variability factor (Case 2b).

The variability of residues was investigated in multiple studies involving analysis of numerous field treated individual units and calculating the variability factors observed under natural conditions. It became apparent that the average variability factor was approximately 3 for all commodities, with a slightly lower factor of 2.8 for representative samples collected from one treated field and a slightly higher factor of 3.6 for samples of unknown origin collected in the market (EFSA 2005b; BfR 2010). Within regulatory frameworks, two different sets of default variability factors are currently used. In the Codex Alimentarius framework, the JMPR uses a default variability factor of 3 for all commodities since 2003 (FAO 2003b). In the European Union, the more conservative factors of 5 and 7 are used.

For details on the equations, please refer to the corresponding WHO Consultation report (WHO 1997b) and the progress report on the IESTI-equation (FAO 1999).

Revision of the IESTI-Concept

In 2015 an International Workshop was held in Geneva to discuss possible revisions of the IESTI-equations (EFSA 2015i). It became apparent over the years of usage that various deficiencies exist which should be addressed by revisiting each input parameter of the equations.

The first aspect discussed was the discrepancy between the HR/STMR used as an input parameter for the residue concentration in the IESTI-equations and the MRLs, which are always set at a higher level. As a consequence, foods on the market complying with MRLs may result in an exceedance of the ARfD when assessed with the IESTI-equations, indicating a possible acute intake concern. To avoid such a discrepancy, a general replacement of the HR/STMR with the MRL was suggested. For commodities that are proven to be mixed/blended on a large scale, the STMR is maintained to avoid an unrealistic overestimation of the residue concentration. To address deviating RDs for enforcement and dietary risk assessment, the use of conversion factors was implemented (see "Plant metabolism" above).

The second aspect discussed was related to the LP. Currently, the 97.5th percentile of the portion size expressed as g per day is calculated and afterwards correlated to the mean body weight of all individuals in the food survey used. For a better linkage between the amount of food consumed and the body weight, it was suggested to express the LP as the 97.5th percentile of all portion sizes to body weight ratios calculated on an individual level. For IESTI Case 2a, which also takes into account the mean unit-weight of the commodity, a general removal of the unit-weight from the equations was proposed to unify Case 2a and Case 2b.

The third aspect dealt with the usage of different default variability factors in the Codex Alimentarius framework and the European Union. A general consensus was reached that a default variability factor of 3 sufficiently describes the range of residue concentrations on single units within a lot.

For details on the new IESTI-equations, please refer to the workshop report published by EFSA.

The acceptance of the proposed changes in regulatory frameworks is still pending. The impact of the modifications on the level of protection for consumers as well as the implications for the authorisation of plant protection products and for MRL setting were investigated (Breysse et al. 2018; Richter et al. 2018; van der Velde-Koerts et al. 2018a, b) but no regulatory changes have been made yet.

However, independent to the proposed changes, further challenges for the dietary consumer risk assessment still have to be solved. The revised IESTI-equations still lack the possibility to address the exposure of one active substance arising from multiple food items within one day. Other aspects are intermittent exposure events, where individuals are exposed to high amounts of one agent for a limited time of their life, which requires new concepts for the assessment, both on the exposure and on the toxicological side.

Drinking Water and Aggregate Risk Assessments

In principle, the consumer exposure is not limited to the oral exposure route based on residues in food commodities of plant and animal origin following the application of the pesticide as plant protection product. Further uses of the active substance in other regulatory frameworks (e.g. as biocidal products or veterinary drugs) also need to be considered. For such compounds, efforts to harmonize the setting of MRLs as well as the methodology to estimate the consumer exposure are ongoing.

Another source for consumer exposure is the presence of active substances in the environment. Principally, drinking water, air, soil, or house dust should be taken into account for the total exposure estimation. For most pesticidal active substances, the environmental contribution as background to the total exposure is insignificant. For example, drinking water limit values are set at such low levels that residues in foods following direct treatment result in exposure levels orders of magnitude higher. However, special attention is required for pesticides with natural sources like copper compounds, whose background concentrations in nearly all foods already utilize a large percentage of the tolerable daily intake. Another group of compounds present in the environment are persistent organic pollutants, which are internationally banned according to the Stockholm Convention (Vanden Bilcke 2002) but still present in the food chain.

Finally, consumers may also be exposed via direct spraying as residents, or as spray drift for bystanders from nearby treated areas. This exposure pathway is normally addressed within the worker safety assessment. For example, EFSA published the "Guidance on the assessment of exposure of operators, workers, residents, and bystanders in risk assessment for plant protection products" (EFSA 2014d), which also included models for the exposure estimation for residents and bystanders. Methodologies for an aggregate exposure assessment for consumers, taking into account all possible exposure pathways, are still under development. In the EU, the ACROPOLIS Projects (Aggregate and Cumulative Risk of Pesticides: an on-line integrated Strategy) were initiated, aiming for the combination of oral, dermal, and inhalative exposure to pesticide residues. However, ACROPOLIS is not yet implemented into the EU regulatory framework for the authorisation of plant protection products.

Multiple Residues, Cumulative Risk Assessment, and Probabilistic Methods

One of the biggest challenges in the dietary risk assessment of pesticide residues is the evaluation of multiple residues and their combined impact on human health. It became apparent that foods on the market often contain more than one pesticide and that multiple compounds may all induce the same toxicological effect. Starting in 2002, the US Environmental Protection Agency (US-EPA) initiated a program for the cumulative risk assessment of pesticides belonging to the group of organophosphates, carbamates, triazines, chloroacetanilides, and pyrethroids. In 2005, the EU also implemented the consideration of cumulative effects from multiple residues into the regulatory framework for establishing MRLs according to regulation (EU) No. 396/2005 (EU 2005b). However, before assessment of cumulative risks on a routine basis is introduced, a harmonised methodology has to be developed which is still an ongoing process.

The sources of exposure to various pesticidal active substances and their metabolites are numerous. Starting with the usual variation in food items consumed by each individual within one day and the specific residue pattern in each of these, a broad range of possible combinations for the co-occurrence of pesticide residues already exists. In addition, plant protection products like herbicides, fungicides, and insecticides are frequently used for the cultivation of each crop. Since the different pests arise independently, the simultaneous use of all kinds of plant protection products in the same field is common. Many of these products already contain up to four individual active substances for efficacy or resistance management reasons, and result per se in a multiple residue situation. Another source for multiple residues are mixing and blending effects, either for large-scale traded commodities like cereal grains or oilseed, and also for fruits and vegetables in packing stations by sorting according to trade classes.

For addressing the mixture toxicity of multiple compounds, US-EPA started with an approach to group pesticides according to their chemical classes. EFSA suggests a different concept (EFSA 2013a) by defining cumulative assessment groups (CAG) based on their toxicological effect, irrespective of chemical similarity (e.g. neurotoxicity). The German BfR (Stein et al. 2014) suggests assuming dose addition of the individual toxicological potentials for each compound until more detailed data on the modes of action on a cellular level are available. In a tiered approach, the "hazard index" (HI) is calculated based on the overall toxicity, target organ-specific toxicity, or finally on a direct mode of action comparison. When the HI exceeds a value of 1, a dietary risk arising from the present mixture cannot be excluded.

All the concepts share the general approach to group chemicals with similar properties (either chemical or toxicological) and normalize the individual toxicological potentials against a selected reference compound. However, the method for normalisation differs strongly and is normally selected on a case-specific basis depending on the question raised and the available data. While lower tier approaches assume simple addition of effects on target organ/system level, higher tier assessments include considerations of specific modes of action and mechanistic studies to derive relative potency factors for each of the compounds within one CAG. The methodology to address mixture toxicity is generally still under development. Multiple ongoing research projects are investigating mechanistic effects on a cellular level to provide a better qualitative and quantitative hazard characterisation for complex mixtures.

For the dietary exposure assessment, the regulatory deterministic approaches presented above for the long- and short-term dietary exposure were primarily designed to assess the risk arising from single compounds. In principle, the IEDI/NEDI concept for the long-term dietary exposure could also address multiple compounds for an estimate of the average exposure; however, the variation in the co-occurrence of residues cannot be described by only using mean concentrations as input parameter. For the short-term dietary exposure estimation, the IESTI-equation can be used for the assessment of multiple compounds found in one specific sample. But the whole range of food items consumed within one day and the plethora of possible residue combinations surpasses the IESTI-Concept focused on the consumption of one food commodity only.

Therefore, current approaches on assessing the cumulative exposure against multiple active substances rely on probabilistic methodologies. The major advantage compared to deterministic methodology is the consideration of all available data, including the full range of values and variability for each parameter. While the IEDI/NEDI and IESTI approaches select only a single representative value both for the consumption and for the residue concentration, probabilistic methods are capable of considering the complete daily consumption of all foods for each individual in combination with large occurrence data sets obtained from annual food monitoring programs. Especially for the estimation of the cumulative dietary exposure, the consideration of all available data allows correlation of occurrence frequencies for multiple compounds already on the sample level, making them capable of assessing both the long- and the short-term dietary exposure with the same tool.

Probabilistic models for pesticide residues are available in most regulatory systems. In the USA, the first probabilistic methods and guidance to assess the cumulative dietary risk were established in 2002 (US-EPA 2002). In Europe, the Dutch National Institute for Public Health and the Environment (RIVM 2015) developed the Monte-Carlo Risk Assessment tool (MCRA, see also Sect. 4.10) (van der Voet et al. 2015), which is compliant to EFSA's Guidance on Probabilistic Modelling for Estimating the Dietary Exposure to Pesticide Residues (EFSA 2012c).

Joint research between EFSA and RIVM initiated by the European Commission is ongoing to adapt MCRA for the assessment of cumulative risks of pesticide residues, both by dietary exposure and by aggregate exposure considering all exposure pathways including inhalation and dermal exposure. However, a commonly harmonised methodology for the assessment of mixtures has not yet been finalized (Kienzler et al. 2016), and still demands further research both for the hazard characterisation and for the exposure estimation.

4.22 Nanoparticles in Food

Francesco Cubadda

4.22.1 Naturally Occurring Particles of Biological Origin Versus Environmental Particles

Nanosized particles of natural or anthropogenic origin are unavoidably present at background concentrations in the human diet. Nanoparticles originating from environmental sources are exogenous constituents adding to the large assortment of naturally-occurring nanosized entities found in food items, which include a number of endogenous components such as proteins, starch granules, casein nanoparticles, and fibres (Yeh 2017). Some of these endogenous nanoparticulate structures appear only partially digested in the gastric environment. An example is ferritin, which is approximately 12 nm in diameter with the protein shell, and contains an iron oxide core of 6–8 nm diameter (itself made up of 2–3 nm subunits). Nanoparticulate ferritin may be endocytosed and utilised by gut epithelial cells as a source of dietary iron, highlighting that uptake of such an endogenous nanosized entity may be associated with a beneficial role (i.e. nutrient capture) (Powell et al. 2010). Apart from food, there are endogenous nanoparticles physiologically produced from ions in the human gastrointestinal tract. For example, the precipitation of calcium and phosphate creates porous amorphous magnesium-substituted calcium phosphate particles, typically 75–150 nm in diameter, that may be absorbed as such (Powell et al. 2015).

Food-borne particles of environmental origin are, as mentioned above, exogenous constituents that do not have a biological origin. Most environmental nanoparticles consist of solid and potentially persistent (“hard”) particulate that may be transferred in the food chain up to humans. In areas impacted by substantial contamination sources, it is conceivable that the background level of environmental particles significantly exceeds the levels normally found in the plant and animal tissues used for human consumption, even though these processes and exposure routes have not been characterised in detail, and neither has the potential risk, if any. Food processing may contribute additional particles with small sizes, in the micro- to the nano-range, due to atmospheric deposition, release from metallic machinery, and other sources. Again, this is a poorly investigated area in exposure science and the potential relevance to human health is unknown.

4.22.2 Man-Made Particles

The interest about nanoparticles in food and related dietary exposure, with a view to characterising their impact on human health and any potential associated risks, is a result of the recent development of nanotechnologies. Nanotechnologies consist of the manipulation and control of matter in the nanoscale, i.e. approximately 1–100 nm, in order to make use of unique size- and structure-dependent properties and phenomena distinct from those associated with larger sizes of the same materials. Engineered nanomaterials are predicted to find three main areas of applications in the food sector, namely agricultural production (e.g. nano-formulated agrochemicals and animal feeds/drugs), food processing (e.g. nanosized ingredients, nutritional supplements, functional foods, and food additives), and food contact materials (Rossi et al. 2014). In each of these areas, making materials smaller can certainly generate novel and useful properties, but actual use requires that risks for human health and the environment are excluded.[16] Potential hazards that are of concern for human health arise from size-dependent biological effects of nanomaterials, in particular the toxicokinetic behaviour, the capacity to cross biological barriers, and particle–cell interactions. Risk assessment of nanotechnology applications in the food and feed sector is based on the identification and characterisation of nano-specific hazards (i.e. hazards that may be associated to size-related properties of nanomaterials) and the estimation of potential human exposure.

It has to be pointed out that size-dependent effects of relevance to human health are not rigidly related to specific size thresholds of particles. They normally occur even when the particles constituting the nanomaterial have a length above 100 nm, in a manner that is strictly material dependent (i.e. dependent on its physicochemical properties). Therefore, for whatever material containing small particles and presenting properties that are characteristic of the nanoscale, e.g. related to the large specific surface area or a toxicokinetic behaviour differing from larger particles, a nano-specific risk assessment is required. At the same time, it is important to recognize that not all nanomaterials actually have new hazard properties compared to larger sized counterparts or, when they exist, to the corresponding ionic or molecular forms (i.e. the soluble counterparts) and a case-by-case assessment is necessary (SCENIHR 2010).

A key point is that, in addition to the size and chemical composition of the material, a number of other properties may be associated with the adverse health effects, such as morphology (including aspect ratio), crystalline phase, surface properties, solubility, and others. If this is the case, these properties have to be taken into account in the risk assessment of a nanomaterial. This circumstance

[16]In the EU, nanotechnology applications in the food and feed sector need to undergo an approval procedure before being placed on the market. This is achieved by incorporating nano-specific provisions in existing legislation. For instance, food consisting of engineered nanomaterials (including vitamins, minerals, or other substances that contain or consist of engineered nanomaterials) are considered novel food and are regulated according to the relevant provisions (Regulation EU 2015/2283). Authorisation is based on a scientific evaluation by the European Food Safety Authority (EFSA), which requires the adoption of state-of-the-art approaches - both in terms of testing methods and measurement methodologies - to demonstrate safety. The same applies to, e.g. nano-sized food additives, additives used in food contact materials, feed additives, and nanopesticides. The existing guidance of the EFSA Scientific Committee on the risk assessment of the application of nanoscience and nanotechnologies in the food and feed chain (EFSA 2011k) is being updated and the new guidance is expected to become available in 2018.

highlights the need for a thorough physicochemical characterisation of the material for risk assessment purposes (EFSA 2011k).

Finally, it has to be noted that nanoparticles, with their reactive surfaces, are known to bind various moieties and, if absorbed in the gastrointestinal tract, may transport potentially harmful substances to various organs and into cells. Whereas adsorption of contaminants on particles may be an important route in environmental processes (especially in the transport of contaminants in environmental compartments) (Plathe et al. 2013), the relevance of such a mechanism to human health, especially if it may result in significant systemic exposure to other toxicants, has to be investigated further.

4.22.3 Dietary Exposure Assessment of Man-Made Nanoparticles

In general, the principles of the dietary exposure assessment of manufactured nanomaterials are the same as in the exposure assessment of conventional chemicals (EFSA 2011k). On the basis of the available consumption data, the anticipated average and high intakes in the different groups of the general population have to be estimated. Probabilistic methods may be useful to determine the ranges of plausible values rather than point estimates. If possible, particular subgroups of the population with an expected high exposure should be identified, and this should be considered in the risk assessment. Information on the consumption of food supplements is generally limited, both in terms of amounts and frequency, and thus exposure assessment of nano-formulated supplements may represent a particular challenge.

An essential difference in the risk assessment of nanoparticles compared to all other chemicals is that, in principle, the dietary exposure is not straightforwardly assessed by using a single metric, that is the concentration in food (i.e. occurrence data), which is then combined with consumption data. Adequate characterisation of the particles actually present in the diet (i.e. in the different food items) may require an assessment of other relevant physicochemical properties (see above). If these properties remain undetermined, there is no clear relationship with the structure/properties of the nanomaterial used in the toxicity studies for hazard identification/characterisation and a number of assumptions have to be made for risk characterisation (EFSA 2011k). For instance, titanium dioxide particles may be produced as either anatase or rutile, and evidence exists that the two crystal forms exert different toxicological effects and present a different toxicokinetic behaviour (Heringa et al. 2016). Therefore, the crystal form is an important property to thoroughly characterise dietary TiO_2 particles and to possibly refine the exposure assessment.

In summary, the physicochemical characterisation of nanomaterials is demanding and the analytical determination of nanoparticles in food is an even more challenging issue (Linsinger et al. 2013). For each particle of a given chemical composition, along with the mass concentration (that remains the most convenient metric for exposure assessment), the number concentration of particles and their size, and size distribution should be determined. Accurate identification of the chemical composition and measurement of particle size is not trivial, and requires a considerable effort and the use of complex analytical methods and state-of-the-art techniques, including electron microscopy combined with energy-dispersive X-ray spectroscopy (EDX) (Dudkiewicz et al. 2011; De Temmerman et al. 2012; Rasmussen et al. 2013; Lim et al. 2015; Verleysen et al. 2015; Contado et al. 2016; Yang et al. 2016) and, for inorganic nanoparticles, elemental (atomic) mass spectrometry-based techniques, e.g. field-flow fractionation-ICP-MS (Loeschner et al. 2013; Peters et al. 2014; Aureli et al. 2015; Barahona et al. 2016; Contado et al. 2016), and single-particle ICP-MS (Peters et al. 2014; Verleysen et al. 2015). Characterisation of other properties of the particles in complex matrixes such as food may be even more challenging. The difficulty in obtaining reliable occurrence data and adequate characterisation of the particles has to be appreciated since it increases the uncertainty in the exposure assessment.

For some material types, particles with the same chemical composition that are not man-made, but result from geogenic, biogenic or anthropogenic origins, may be present concurrently in the food item. It is difficult to distinguish the intentionally added nanomaterial from the background levels of such environmental particles. However, in most cases, it is expected that such background levels are negligible compared to the usage levels applied for a specific technological purpose.

4.22.4 Different Scenarios for Human Exposure to Man-Made Nanoparticles

Anticipated exposure scenarios from the proposed uses of manufactured nanomaterials can be outlined and are the first step in a tiered approach for assessing dietary exposure. Table 4.18 provides a summary of current and projected applications of nanotechnology in the food production chain that may lead to direct or indirect consumer exposure (EFSA 2011k; Rossi et al. 2014).

In general, where there is a direct use of a nanomaterial in food items (e.g. as ingredient or food additive), the type and quantity added are known or can be estimated from usage levels. In this case, it is possible to outline an exposure scenario in a rather straightforward way. Actual occurrence data can be obtained by analytical measurements for refined exposure assessment.

For nanotechnology applications where convincing evidence is provided that demonstrates, by appropriate analytical methods, that the manufactured nanomaterial is completely degraded/solubilised to a non-nanoform, i.e. where there is no persistence of the nanoparticles in preparations/formulations as marketed or there is complete transformation to a non-nanoform in the food matrix (or feed matrix, if feed additives are dealt with), then the exposure can be anticipated to be essentially zero. However, this scenario leads to exposure to degradation products (ions or molecules) that have to be assessed in terms of consumer safety.

Table 4.18 Summary of current and projected applications of nanotechnology in the agri-food sector that may result in direct or indirect consumer exposure via food

Nanotechnology	Application	Potential consumer exposure
Agricultural production		
Nanosized fertilisers	Improved delivery of fertilisers in the field	Indirect (carry-over *via* plant food)
Nano-pesticides	Nano-emulsions and encapsulates for increased efficacy or triggered (local) release	Indirect (carry-over of residues *via* plant food or food of animal origin)
Nano-veterinary medicines	Nanodelivery systems for better efficacy and better control of application/dose	Indirect (carry-over of residues *via* food of animal origin)
Nanosized feed additives	Nanosized additives for feed applications, e.g. as nutrient sources	Indirect (carry-over *via* food of animal origin)
Food processing		
Nanosized ingredients	Nanosized nutrient sources for enhanced absorption and bioavailability, processed nano-structured or nano-textured food products, nanoemulsions	Direct
Nano-supplements and nutraceuticals	Nanosized nutrient sources and nano-carrier systems for delivery of nutrients (nano-encapsulates)	Direct
Nanosized food additives	Nanosized additives for various technological purposes	Direct
FCMs (food storage and handling)		
Plastic nano-additives and nanocomposites	Improving strength of materials, durability, barrier properties, incorporating nanoparticles into a polymer to form a composite	Indirect (following migration)
Nano-coatings and "active" FCMs	Incorporating nanomaterials onto the packaging surface or using them for intentional release into the packaged food	Indirect (depending on migration and/or residual levels in food)
Surface biocides	Coating surfaces of storage containers, equipment for food processing, handling and preparation	Indirect (depending on migration and residual levels in food)

If identification of the nanomaterial in the food matrix is not possible or in case there is indication of an alteration of the nanomaterial that cannot be ascertained, no conclusion can be reached about the exposure apart from what was initially added to the food, with the understanding that the nature of the nanomaterial to which exposure takes place is unknown or altered (EFSA 2011k). It should, therefore, be assumed that all of the added nanomaterial is present, ingested and absorbed as the originally employed nanosized particles.

Indirect exposure to manufactured nanomaterials takes place when there is particle migration from food contact materials or carry-over of the nanomaterial from the feed via animals to food (e.g. due to use of nano-feed additives), or from a nano-pesticide or a nano-fertilizer to crops. In the case of indirect exposure, the type and amount of particles in food items have to be characterised by means of analytical determination.

Nanotechnology-derived food packaging materials are the largest category of current nanotechnology applications in the food sector. Relatively low levels of nanoparticles are sufficient to change the properties of plastic food contact materials (e.g. make them stronger in terms of mechanical and thermal performance, or less permeable to gases) without significant changes in density, transparency, and processing characteristics. In the EU, most manufactured nanomaterials have been authorized as plastic additives under the condition of absence of migration. However, effective entrapment of nanoparticles in polymer nanocomposites and other food contact materials has to checked, e.g. using food simulants and analytical techniques for nanoparticle determination with detection limits according to the state of the art. If migration occurs, consumers will consequently be exposed to the particles leaching into food.

For some nanoparticles, e.g. those having a certain degree of solubility, human gastrointestinal digestion may substantially alter their physicochemical properties and even lead to the loss of the nanoparticulate nature (EFSA 2011k). When evidence is provided that convincingly demonstrates, by appropriate analytical methods, that a nanomaterial quickly and completely dissolves/degrades in the gastrointestinal tract, no systemic absorption of the nanoparticles takes place (but this may happen for the resulting non-nano degradation products). Exposure to a nanomaterial can occur if the nanomaterial does not quickly dissolve. In vitro studies simulating the digestive tract conditions can be useful to assess the extent and rate of dissolution/degradation. Even though systemic exposure is absent, potential local effects on the upper gastrointestinal tract have to be considered (EFSA 2011k).

Apart from materials that may degrade/dissolve, human digestion appears to play an important role in determining the form in which the particles come into contact with the intestinal epithelium and become absorbed. Particles can undergo complex interactions in foods. In view of these interactions, pristine particles may be reversibly coated with a variety of biomolecules, depending on the composition of the food matrix, and form temporary agglomerates or (although more unlikely) permanent aggregates. The agglomerates may be reduced/destroyed as a result of the gastrointestinal process, therefore leading to a higher number of small-sized particles compared to the original food matrix, where larger secondary particles (agglomerates) may predominate (Peters et al. 2012). Therefore, as a general rule, the exposure estimates should take into consideration the evidence on the transformation of the particles, especially their agglomeration/disagglomeration, in digestive tract conditions (if known).

Another issue to be considered is that some nanomaterials are specifically designed to achieve a higher oral bioavailability (e.g. nutrient sources). Especially in these cases where the ADME profile is considerably altered, it is essential that hazard data are relevant to the specific nanosized agent being considered since the same external exposure (estimated by exposure assessment) may translate in a significantly higher internal exposure and thus risk.

4.22.5 Assessment of Dietary Exposure to Particulate Food Additives

Although the applications of nanotechnology in the food sector are still nascent, there are a number of materials that have been in use for a long time that are composed, mostly or partially, of nanosized particles. This is the case of synthetic amorphous silica (SAS), a food additive used mainly as anti-caking agent (E551), which has been produced since the last century with methods that efficiently yield SiO_2 nanoparticles ("nanosilica"), some of which may subsequently agglomerate or aggregate depending on the production process (Aureli et al. 2012; De Temmerman et al. 2012; Rasmussen et al. 2013; Barahona et al. 2016; Contado et al. 2016; Yang et al. 2016). Another example is titanium dioxide, a white colouring agent (E171), which is composed of small particles (around 200 nm) and always contains a nanosized fraction (Chen et al. 2013; Weir et al. 2012; Peters et al. 2014; Yang et al. 2014). Metallic silver (E174) is much less widely used compared to the previous ones, but it has also been demonstrated to contain a nano-fraction (Verleysen et al. 2015). In general, all bulk materials resulting from the application of production processes that are aimed at reducing the average diameter of a material's particles (e.g. micronisation) contain a fraction of very small particles, including some smaller than 100 nm. The existing attempts to effectively estimate consumer exposure to dietary nanoparticulate entities have been directed towards some food additives that fall in this category.

Investigations on intestinal biopsy samples from Western human subjects showed that macrophages at the base of human gut associated lymphoid tissue (GALT) become loaded early in life with dark granular pigment deposits that consist of particles of titanium dioxide (size 100–200 nm), aluminosilicates (size <100–400 nm) and silicates (of variable morphology and without aluminum, size 100–700 nm) (Powell et al. 1996). Food additives, along with pharmaceuticals, health supplements, and toothpaste, are thought to be the sources of this particulate material resistant to gastrointestinal degradation that accumulates mainly in the Peyer patches of the terminal ileum (Powell et al. 2010). Overall, the intake of dietary inorganic microparticles in the UK has been estimated to be about 40 mg/person/day (~35 mg for the silicates and ~5 mg for titanium dioxide), which has been calculated to equate to a daily exposure of 10^{12}–10^{14} particles/person (Powell et al. 2010).

More recently, several authors studied TiO_2 oral intake using different approaches and targeting either dietary or aggregated exposure, e.g. intake from diet and toothpaste (Rompelberg et al. 2016) or diet and medicines (Bachler et al. 2015). Exposure estimates ranged from 0.1 to 1.7 mg/kg of BW per day in the general adult population, but higher values were found for children, i.e. 1.0–5.5 mg/kg of BW per day (Weir et al. 2012; Bachler et al. 2015; EFSA 2016d; Rompelberg et al. 2016). This is why, apart from higher intake per body weight of children relative to adults, titanium dioxide as a food additive (E171) is commonly used in chewing gums, candies, chocolates, and sweets, i.e. in items largely consumed by children (Weir et al. 2012; Peters et al. 2014). Interestingly in this respect, it has been found that the vast majority of the TiO_2 particles in sugar-coated chewing gum are easily released from the coating and swallowed by gum chewers (Chen et al. 2013).

EFSA performed exposure assessment to food-grade TiO_2 under different assumptions, i.e. comparing a brand-loyal with a non-brand-loyal consumer scenario (EFSA 2016d). In the brand-loyal scenario, it is assumed that the population is exposed over a long period of time to the substance present at the maximum reported use/analytical level for one food category, and to a mean reported use/analytical level for the remaining food categories. The exposure estimates ranged, at the mean, from 0.4 mg/kg of BW per day for infants and the elderly, to 8.8 mg/kg of BW per day for children. At the 95th percentile (P95), the exposure estimates ranged from 1.1 mg/kg of BW per day for the elderly to 30.2 mg/kg of BW per day for children. In the non-brand-loyal scenario, it is assumed that the population is exposed over a long period of time to the substance present at the mean-reported use/analytical levels in all relevant food categories. The exposure estimates ranged, at the mean, from 0.2 mg/kg of BW per day for infants and the elderly, to 5.5 mg/kg of BW per day for children. P95 exposure estimates ranged from 0.5 mg/kg of BW per day for the elderly, to 14.8 mg/kg of BW per day for children. Since brand loyalty to a specific food category was not identified, the non-brand-loyal scenario covering the general population was considered to be the more appropriate and realistic scenario for risk characterisation (EFSA 2016d).

Some authors specifically investigated the percentage of particles in food-grade TiO_2 with a size <100 nm. The exercise has some limitations since, depending on the technique used, smaller particles (e.g. <20–50 nm) may remain undetected (Peters et al. 2014). Overall, the percentage of nanosized particles ranged from 5–10% (Peters et al. 2014), to 17–35% (Yang et al. 2014), to 36% (in this case only particles within a size range up to 220 nm were detected) (Weir et al. 2012), to 28–44% (Chen et al. 2013). In terms of mass, these percentages correspond to a low proportion of the bulk material (<3.2% and often ~1%) (EFSA 2016a), which translates into dietary intake estimates of nanosized TiO_2 about 2 orders of magnitude lower compared to the bulk material. In particular, Rompelberg et al. (2016) estimated the mean oral intake of TiO_2 nanoparticles from food, toothpaste, and supplements for the Dutch population to range from 0.19 μg/kg of BW per day in elderly, to 0.55 μg/kg of BW per day for 7–69-year-old people, to 2.16 μg/kg of BW per day in young children. P95 values were 0.74, 1.61, and 4.16 μg/kg of BW per day, respectively. The products contributing the most to the TiO_2 intake were found to be toothpaste (in young children only), candy, coffee creamer, fine bakery wares, and sauces.

As far as SAS is concerned, Dekkers et al. (2011) estimated an average-to-worst-case dietary intake for the Dutch population. Food products were selected based on publicly available information about the presence of food additive E551, and included ready-to-eat meals, soups and sauces, coffee creamers, seasonings, and supplements. To calculate the silica content, the total silicon concentration was determined in these products by inductively coupled plasma atomic emission spectrometry (ICP-AES). Dietary intake estimates for the whole population were obtained by combining occurrence data with consumption data from the third Dutch Food Consumption Survey, along with the expert judgment. The estimated average daily dietary intake was 9.4 mg SAS/kg of BW per day (70 kg adult). It was experimentally demonstrated that part of the SAS particles (including aggregates and agglomerates) had an external diameter smaller than 200 nm (measured by HDC-ICPMS, which indicated external sizes ranging from 30 to 200 nm) and the corresponding estimated intake was 1.8 mg/kg of BW per day (adult 70 kg). These nanoparticles were found in food products, but were also detected when these products were introduced into an in vitro digestion model simulating the physico-chemical conditions of the gastrointestinal tract (Dekkers et al. 2011; Peters et al. 2012).

Finally, it has to be highlighted that for both SAS and TiO_2, a relatively low uptake from the gastrointestinal tract has been found compared with the total administered dose expressed as a mass in animal studies. However, in view of the small dimensions of the particles, a low mass dose still consists of a high number of particles. Therefore, even when a nanomaterial is absorbed to a limited extent, if it is poorly eliminated from tissues, it is possible that it accumulates and persists in organs. This circumstance, which seems to the rule for poorly soluble particles, brings along the potential for long-term adverse effects. In order to deal with low level repeated exposure via foodstuffs and the possibility of persistence and thus accumulation in tissues, an alternative risk assessment approach based on the exposure of internal organs was proposed for non-degradable particles (van Kesteren et al. 2015; Heringa et al. 2016). In such an approach, the estimated nanoparticle concentration in the target organ in humans was compared, by means of toxicokinetic modelling, with the measured or estimated concentrations in the same target organ in animal toxicity studies in which adverse effects were found. For both SAS and TiO_2 it was found that more studies are warranted since the assessments carried out with this approach suggested that a health risk cannot be excluded (van Kesteren et al. 2015; Heringa et al. 2016).

References

Abraham K, Wöhrlin F, Lindtner O, Heinemeyer G (2010) Toxicology and risk assessment of coumarin: focus on human data. Mol Nutr Food Res, Special Issue: Special: Novel approaches for risk assessment of phytochemicals in food 54(2):170–306. https://doi.org/10.1002/mnfr.200900281

Abraham K, Gürtler R, Berg K, Heinemeyer G, Lampen A, Appel KE (2011a) Toxicology and risk assessment of 5-Hydroxymethylfurfural in food. Mol Nutr Food Res 55:667–678. https://doi.org/10.1002/mnfr.201000564

Abraham K, Andres S, Palavinskas R, Berg K, Appel KE, Lampen A (2011b) Toxicology and risk assessment of acrolein in food. Mol Nutr Food Res 55:1277–1290. https://doi.org/10.1002/mnfr.201100481

Adolf T, Schneider R, Eberhardt W, Hartmann S, Herweg A, Heseker H, Hünchen K, Kübler W, Matlaske B, Moch KJ, Rosenbauer J (1995) Ergebnisse der Nationalen Verzehrsstudie (1985–1988) über die Lebensmittel- und Nährstoffaufnahme in der Bundesrepublik

Deutschland. VERA-Schriftenreihe (Hgg: Kübler W, Anders HJ, Heeschen W), Wiss Fachverlag Dr Fleck 1995 XI

Agogo GO (2016) A zero-augmented generalized gamma regression calibration to adjust for covariate measurement error: a case of an episodically consumed dietary intake. Biom J 59:94–109. https://www.ncbi.nlm.nih.gov/pubmed/27704599

Arab L, Wittler M, Schettler G (1987) European food composition tables in translation. Springer, Berlin, p 155

Arab L, Wesseling-Perry K, Jardack P, Henry J, Winter A (2010) Eight self-administered 24-hour dietary recalls using the Internet are feasible in African Americans and Whites: the energetics study. J Am Diet Assoc 110(6):857–864

Arcella D, Soggiu ME, Leclercq C (2003) Probabilistic modelling of human exposure to intense sweeteners in Italian teenagers: validation and sensitivity analysis of a probabilistic model including indicators of market share and brand loyalty. Food Addit Contam 20(Suppl 1):S73–S86

Arcella D, Le Donne C, Piccinelli R, Leclercq C (2004) Dietary estimated intake of intense sweeteners by Italian teenagers. Present levels and projections derived from the INRAN-RM-2001 food survey. Food Chem Toxicol 42(4):677–685

Aureli F, D'Amato M, De Berardis B, Raggi A, Turco AC, Cubadda F (2012) Investigating agglomeration and dissolution of silica nanoparticles in aqueous suspensions by dynamic reaction cell inductively coupled plasma-mass spectroscopy in time resolved mode. J Anal At Spectrom 27:1540–1548

Aureli F, D'Amato M, Raggi A, Cubadda F (2015) Quantitative characterization of silica nanoparticles by asymmetric flow field flow fractionation coupled with online multiangle light scattering and ICP-MS/MS detection. J Anal At Spectrom 30:1266–1273

Bachler G, von Goetz N, Hungerbuhler K (2015) Using physiologically based pharmacokinetic (PBPK) modeling for dietary risk assessment of titanium dioxide (TiO_2) nanoparticles. Nanotoxicology 9:373–380

Banasiak U, Heseker H, Sieke C, Sommerfeld C, Vohmann C (2005) Abschätzung der Aufnahme von Pflanzenschutzmittel-Rückständen in der Nahrung mti neuen Verzehrsmengen für Kinder. Bundesgesundheitsbl – Gesundheitsforsch – Gesundheitsschutz 48:84–98

Barahona F, Ojea-Jimenez I, Geiss O, Gilliland D, Barrero-Moreno J (2016) Multimethod approach for the detection and characterisation of food-grade synthetic amorphous silica nanoparticles. J Chromatogr A 1432:92–100

Bast A, Chandler RF, Choy PC, Delmulle LM, Gruenwald J, Halkes SB, Keller K, Koeman JH, Peters P, Przyrembel H, de Ree EM, Renwick AG, Vermeer ITM (2002) Botanical health products, positioning and requirements for effective and safe use. Environ Toxicol Pharmacol 12:195–211

Bemrah N, Leblanc JC, Volatier JL (2008) Assessment of dietary exposure in the French population to 13 selected food colours, preservatives, antioxidants, stabilizers, emulsifiers and sweeteners. Food Addit Contam Part B Surveill 1(1):2–14

Bernard A, Broeckaert F, De Poorter G, De Cock A, Hermans C, Saegerman C, Houins G (2002) The Belgian PCB/dioxin incident: analysis of the food chain contamination and health risk evaluation. Environ Res 88:1–18

BfR (2004) BfR recommends provisional daily upper intake level and a guidance value for morphine in poppy seeds. BfR health assessment no. 012/2006, 27 December 2005. http://www.bfr.bund.de/cm/349/bfr_recommends_provisional_daily_upper_intake_level_and_a_guidance_value_for_morphine_in_poppy_seeds.pdf. Accessed 8 Mar 2018

BfR (2006) BfR entwickelt neues Verzehrsmodell für Kinder. Information Nr. 016/2005 des BfR vom 2, Mai 2005. http://www.bfr.bund.de/cm/343/bfr_entwickelt_neues_verzehrsmodell_fuer_kinder.pdf. Accessed 8 Mar 2018

BfR (2007) Salad mix contaminated with groundsel containing pyrrolizidine alkaloids. BfR opinion no. 028/2007, 10 January 2007. http://www.bfr.bund.de/cm/349/salad_mix_contaminated_with_groundsel_containing_pyrrolizidine_alkaloids.pdf. Accessed 8 Mar 2018

BfR (2010) Variability factors for the acute dietary risk assessment of pesticides. BfR information no. 014/2010. http://www.bfr.bund.de/cm/349/variability_factors_for_the_acute_dietary_risk_assessment_of_pesticides.pdf. Accessed 27 Oct 2019

BfR (2011a) Chemical analysis and toxicity of pyrrolizidine alkaloids and assessment of the health risks posed by their occurrence in honey. BfR opinion no. 038/2011, 11 August 2011. http://www.bfr.bund.de/cm/349/chemical-analysis-and-toxicity-of-pyrrolizidine-alkaloids-and-assessment-of-the-health-risks-posed-by-their-occurence-in-honey.pdf. Accessed 27 Oct 2019

BfR (2011b) Too Hot Isn't Healthy – Foods with very high capsaicin concentrations can damage health. BfR opinion no. 053/2011 of 18 October 2011. http://www.bfr.bund.de/cm/349/too-hot-isnt-healthy-foods-with-very-high-capsaicin-concentrations-can-damage-health.pdf. Accessed 27 Oct 2019

BfR (2012a) Gesundheitliche Bewertung von synephrin- und koffeinhaltigen Sportlerprodukten und Schlankheitsmitteln. Stellungnahme Nr. 004/2013 des BfR vom 16, November 2012. http://www.bfr.bund.de/cm/343/gesundheitliche-bewertung-von-synephrin-und-koffeinhaltigen-sportlerprodukten-und-schlankheitsmitteln.pdf. Accessed 27 Oct 2019

BfR (2012b) New insights into coumarin contained in cinnamon. BfR opinion no. 036/2012, 27 September 2012. http://www.bfr.bund.de/cm/349/new-insights-into-coumarin-contained-in-cinnamon.pdf. Accessed 27 Oct 2019

BfR (2013) Pyrrolizidine alkaloids in herbal teas and teas. BfR opinion no. 018/2013 of 5 July 2013. http://www.bfr.bund.de/cm/349/pyrrolizidine-alkaloids-in-herbal-teas-and-teas.pdf. Accessed 27 Oct 2019

BfR (2016a) Pyrrolizidine alkaloids: levels in foods should continue to be kept as low as possible. BfR opinion no. 030/2016, 28 September 2016. http://www.bfr.bund.de/cm/349/pyrrolizidine-alkaloids-levels-in-foods-should-continue-to-be-kept-as-low-as-possible.pdf. Accessed 27 Oct 2019

BfR (2016b) Report including a decision tree: for combining data from TDS and Food Monitoring programs in risk assessment, TDS Exposure, deliverable D.7.5, To be requested at: http://www.tds-exposure.eu/project-objectives-0/contact-us/. Accessed 27 Oct 2019

BfR (2017) Risikobewertung des Alkaloidvorkommens in Lupinensamen. Stellungnahme 003/2017 des BfR, 27 March 2017. http://www.bfr.bund.de/cm/343/risikobewertung-des-alkaloidvorkommens-in-lupinensamen.pdf. Accessed 27 Oct 2019

BgVV (German Federal Institute for Consumer Health Protection and Veterinary Medicine) (1997) Einsatz von Hanf in Lebensmitteln kann gesundheitlich problematisch sein. BgVV Pressedienst (Press Release) 026/97, 22.10.1997. http://www.bfr.bund.de/de/presseinformation/1997/26/einsatz_von_hanf_in_lebensmitteln_kann_gesundheitlich_problematisch_sein-829.html. Accessed 27 Oct 2019

BgVV (German Federal Institute for Consumer Health Protection and Veterinary Medicine) (2000) BgVV empfiehlt Richtwerte für THC (tetrahydrocannabinol) in hanfhaltigen Lebensmitteln. BgVV Pressedienst (Press Release) 07/2000, 16.03.2000. http://www.bfr.bund.de/de/presseinformation/2000/07/bgvv_empfiehlt_richtwerte_fuer_thc__tetrahydrocannabinol__in_hanfhaltigen_lebensmitteln-884.html. Accessed 27 Oct 2019

Bingham SA (2002) Biomarkers in nutritional epidemiology. Public Health Nutr 5:821–827

Biro G, Hulshof KF, Ovesen L, Amorim Cruz JA (2002) Selection of methodology to assess food intake. Eur J Clin Nutr 56(Suppl 2):S25–S32

Blume K, Lindtner O, Heinemeyer G, Schneider K, Schwarz M (2010) Aufnahme von Umweltkontaminanten über Lebensmittel – Ergebnisse des Forschungsprojektes LExUKon. Informationsbroschüre des Bundesinstitutes für Risikobewertung. ISBN 3-938163-70-4, ISSN 1614-5089 (Online) http://www.bfr.bund.de/cm/350/aufnahme_von_umweltkontaminanten_ueber_lebensmittel.pdf

Boobis A, Cerniglia C, Chicoine A, Fattori V, Lipp M, Reuss R, Verger P, Tritscher A (2017) Characterizing chronic and acute health risks of residues of veterinary drugs in food: latest methodological developments by the joint FAO/WHO expert committee on food additives. Crit Rev Toxicol 47(10):885–899. https://doi.org/10.1080/10408444.2017.1340259

Boon PE, Svensson K, Moussavian S, van der Voet H, Petersen A, Ruprich J, Debegnach F, de Boer WJ, van Donkersgoed G, Brera C, van Klaveren JD, Busk L (2009) Probabilistic acute dietary exposure assessments to captan and tolylfluanid using several European food consumption and pesticide concentration databases. Food Chem Toxicol 47:2890–2898. https://doi.org/10.1016/j.fct.2009.01.040

Boon PE, van Donkersgoed G, Christodoulou D, Crépet A, D'Addezio L, Desvignes V, Ericsson B-G, Galimberti F, Ioannou-Kakouri E, Jensen BH, Rehurkova I, Rety J, Ruprich J, Sand S, Stephenson C, Strömberg A, Turrini A, van der Voet H, Ziegler P, Hamey P, van Klaveren JD (2015) Cumulative dietary exposure to a selected group of pesticides of the triazole group in different European countries according to the EFSA guidance on probabilistic modelling. Food Chem Toxicol 79:13–31. https://doi.org/10.1016/j.fct.2014.08.004

Boon PE, te Biesebeek JD, van Donkersgoed G (2017) Dietary exposure to lead in the Netherlands. RIVM Letter report 2016-0206. National Institute for Public Health and the Environment (RIVM), Bilthoven. www.rivm.nl

Boushey CJ, Kerr DA, Wright J, Lutes KD, Ebert DS, Delp EJ (2009) Use of technology in childrens dietary assessment. Eur J Clin Nutr 63(Suppl 1):S50–S57

Boushey CJ, Spoden M, Zhu FM, Delp EJ, Kerr DA (2016) New mobile methods for dietary assessment: review of image-assisted and image-based dietary assessment methods. Proc Nutr Soc:1–12

Box GEP, Cox DR (1964) An analysis of transformations. J R Stat Soc Ser B Methodol 26 (2):211–252

Breysse N, Vial G, Pattingre L, Ossendorp BC, Mahieu K, Reich H, Rietveld A, Sieke C, van der Velde-Koerts T, Sarda X (2018) Impact of a proposed revision of the IESTI equation on the acute risk assessment conducted when setting maximum residue levels (MRLs) in the European Union (EU): a case study. J Environ Sci Health B 53(6):352–365. https://doi.org/10.1080/03601234.2018.1439809

Burke BS (1947) The dietary history as a tool in research. J Am Diet Assoc 23:1041–1046

BVL (2018) Berichte zur Lebensmittelsicherheit (bis) 2017 – Monitoring. Gemeinsamer Bericht des Bundes und der Länder. Reihe: BVL-Reporte. Berlin, DE: Bundesamt für Verbraucherschutz und Lebensmittelsicherheit (BVL). Verfügbar unter https://www.bvl.bund.de/DE/01_Lebensmittel/01_Aufgaben/02_AmtlicheLebensmitteluebenwachung/04_Monitoring/lm_monitoring_node.html

Codex Alimentarius Commission (CAC) (2018) Procedural manual, 26th edition, Secretariat of the Joint FAO/WHO Food Standards Programme, FAO, Rome. http://www.fao.org/3/i8608en/I8608EN.pdf. Accessed 27 Oct 2019

Caldas ED, Tressou J, Boon PE (2006) Dietary exposure of Brazilian consumers to dithiocarbamate pesticides – a probabilistic approach. Food Chem Toxicol 44:1562–1571. https://doi.org/10.1016/j.fct.2006.04.014

Carroll RJ, Midthune D, Subar AF, Shumakovich M, Freedman LS, Thompson FE, Kipnis V (2012) Taking advantage of the strengths of 2 different dietary assessment instruments to improve intake estimates for nutritional epidemiology. Am J Epidemiol 175(4):340–347

CDC (2016a) National Health and Nutrition Examination Survey (NHANES). http://www.cdc.gov/nchs/nhanes.htm. Accessed 22 Nov 2019

CDC (2016b) National Health and Nutrition Examination Survey (NHANES). Dietary Web Tutorial – Module, 18. http://www.cdc.gov/nchs/tutorials/Dietary/Advanced/ModelUsualIntake/index.htm. Accessed 22 Nov 2019

CDC (Centers for Disease Control and Prevention) (2017) 2003–2004 National Health and Nutrition Examination Survey (NHANES). Post-examination questionnaires. Food frequency (formerly called Food Propensity): mailed to respondent, Atlanta. https://wwwn.cdc.gov/nchs/nhanes/continuousnhanes/questionnaires.aspx?BeginYear=2003. Accessed 28 April 2017

Chaisson CF, Petersen B (1996) Harmonization of strategies for utilizing available data on a pan-European basis. Food Addit Contam 13(4):461–465

Chen XX, Cheng B, Yang YX, Cao A, Liu JH, Du LJ, Liu Y, Zhao Y, Wang H (2013) Characterization and preliminary toxicity assay of nano-titanium dioxide additive in sugar-coated chewing gum. Small 9:1765–1774

Chinnock A (2007) Validation of a diet history questionnaire for use with Costa Rican adults. Public Health Nutr 11(1):65–75

CODEX STAN 171 (1989a) Codex standard for certain pilses. www.codexalimentarius.net/download/standards/56/CXS_171e.pdf www.codexalimentarius.net/download/standards/51/CXS_153e.pdf. Accessed 22 Nov 2019

CODEX STAN 172 (1989b) Codex standard for sorghum grains. www.codexalimentarius.net/download/standards/57/CXS_172e.pdf www.codexalimentarius.net/download/standards/51/CXS_153e.pdf. Accessed 22 Nov 2019

CODEX STAN 199 (1995a) Codex standard for wheat and durum wheat. www.codexalimentarius.net/download/standards/62/CXS_199e.pdf www.codexalimentarius.net/download/standards/51/CXS_153e.pdf. Accessed 22 Nov 2019

CODEX STAN 201 (1995b) Codex standard for oats. www.codexalimentarius.net/download/standards/64/CXS_201e.pdf www.codexalimentarius.net/download/standards/51/CXS_153e.pdf. Accessed 22 Nov 2019

CODEX STAN 153 (1985) Codex standard for maize (CORN). www.codexalimentarius.net/download/standards/51/CXS_153e.pdf. Accessed 22 Nov 2019

Connolly A, Hearty A, Nugent A, McKevitt A, Boylan E, Flynn A, Gibney MJ (2010) Pattern of intake of food additives associated with hyperactivity in Irish children and teenagers. Food Addit Contam Part A Chem Anal Control Expo Risk Assess 27(4):447–456

Contado C, Mejia J, Lozano Garcia O, Piret JP, Dumortier E, Toussaint O, Lucas S (2016) Physicochemical and toxicological evaluation of silica nanoparticles suitable for food and consumer products collected by following the EC recommendation. Anal Bioanal Chem 408:271–286

Conway JM, Ingwersen LA, Vinyard BT, Moshfegh AJ (2003) Effectiveness of the US Department of Agriculture 5-step multiple-pass method in assessing food intake in obese and nonobese women. Am J Clin Nutr 77(5):1171–1178

COT (Committee on Toxicity of Chemicals in Food, Consumer, Products and the Environment) (2008) Statement on pyrrolizidine alkaloids in food. https://cot.food.gov.uk/sites/default/files/cot/cotstatementpa200806.pdf. Accesed 22 Nov 2019

Craft BD, Chiodini A, Garst J, Granvogl M (2013) Fatty acid esters of monochloropropanediol (MCPD) and glycidol in refined edible oils. Food Addit Contam Part A Chem Anal Control Expo Risk Assess 30(1):46–51

Crews C, Castle L (2007) A review of the occurrence, formation and analysis of furan in heat-processed foods. Trends Food Sci Technol 18:365–372

Crispim SP, Geelen A, Siebelink E, Huybrechts I, Lillegaard ITL, Margaritis I, Rehurkova I, Slimani N, Ocke MC, de Boer E, van't Veer P, de Vries JHM, Consortium E (2012) Design aspects of 24 h recall assessments may affect the estimates of protein and potassium intake in dietary surveys. Public Health Nutr 15(7):1196–1200

Crispim SP, Nicolas G, Casagrande C, Knaze V, Illner AK, Huybrechts I, Slimani N (2014) Quality assurance of the international computerised 24 h dietary recall method (EPIC-Soft). Br J Nutr 111(3):506–515

Currie LA (1997) Detection: international update, and some emerging dilemmas involving calibration, the blank, and multiple detection decisions. Chemom Intell Lab Syst 37(1):151–181

de Boer WJ, van der Voet H, Bokkers BGH, Bakker MI, Boon PE (2009) Comparison of two models for the estimation of usual intake addressing zero consumptions and non-normality. Food Addit Contam Part A Chem Anal Control Expo Risk Assess 26:1433–1449. https://doi.org/10.1080/02652030903161606

de Henauw S, Brants HA, Becker W, Kaic-Rak A, Ruprich J, Sekula W, Mensink GB, Koenig JS (2002) Operationalization of food consumption surveys in Europe: recommendations from the European Food Consumption Survey Methods (EFCOSUM) project. Eur J Clin Nutr 56(Suppl 2):S75–S88

De Temmerman PJ, Van Doren E, Verleysen E, Van der Stede Y, Francisco MA, Mast J (2012) Quantitative characterization of agglomerates and aggregates of pyrogenic and precipitated amorphous silica nanomaterials by transmission electron microscopy. J Nanobiotechnol 10:24

Dehne LI, Klemm C, Henseler G, Hermann-Kunz E (1999) The German food code and nutrient data base (BLSII.2). Eur J Epidemiol 15:355–359

Dekkers ALM, Slob W (2012) Gaussian Quadrature is an efficient method for the back-transformation in estimating the usual intake distribution when assessing dietary exposure. Food Chem Toxicol 50(10):3853–3861

Dekkers S, Krystek P, Peters RJB, Lankveld DPK, Bokkers BGH, Van Hoeven-Arentzen PH, Bouwmeester H, Oomen AG (2011) Presence and risks of nanosilica in food products. Nanotoxicology 5:393–405

Dekkers ALM, Verkaik-Kloosterman J, van Rossum CTM, Ocké MC (2014) SPADE, a new statistical program to estimate habitual dietary intake from multiple food sources and dietary supplements. J Nutr 144(12):2083–2091

Diogo JS, Silva LS, Pena A, Lino CM (2013) Risk assessment of additives through soft drinks and nectars consumption on Portuguese population: a 2010 survey. Food Chem Toxicol 62:548–553

Diouf F, Berg K, Ptok S, Lindtner O, Heinemeyer G, Heseker H (2014) German database on the occurrence of food additives: application for intake estimation of five food colours for toddlers and children. Food Addit Contam Part A Chem Anal Control Expo Risk Assess 31(2):197–206

Dixit S, Purshottam SK, Khanna SK, Das M (2011) Usage pattern of synthetic food colours in different states of India and exposure assessment through commodities preferentially consumed by children. Food Addit Contam Part A Chem Anal Control Expo Risk Assess 28(8):996–1005

Dodd KW, Guenther PM, Freedman LS, Subar AF, Kipnis V, Midthune D, Tooze JA, Krebs-Smith SM (2006) Statistical methods for estimating usual intake of nutrients and foods: a review of the theory. J Am Diet Assoc 106(10):1640–1650

Doell DL, Folmer DE, Lee HS, Butts KM, Carberry SE (2016) Exposure estimate for FD&C colour additives for the US population. Food Addit Contam Part A Chem Anal Control Expo Risk Assess 33(5):782–797

Domke A, Großklaus R, Niemann B, Przyrembel H, Richter K, Schmidt E, Weißenborn A, Wörner B, Ziegenhagen R (2004) Use of vitamins in food. Toxicological and nutritional aspects. Part I. BfR-Science 04/2005, Berlin

Domke A, Großklaus R, Niemann B, Przyrembel H, Richter K, Schmidt E, Weißenborn A, Wörner B, Ziegenhagen R (2006) Use of minerals in food. Toxicological and nutritional aspects. Part II. BfR-Science 01/2006, Berlin

Douglass JS, Barraj LM, Tennant DR, Long WR, Chaisson CF (1997) Evaluation of the budget method for screening food additive intakes. Food Addit Contam 14:791–802. https://doi.org/10.1080/02652039709374590

Dudkiewicz A, Tiede K, Loeschner K, Soegaard Jensen LH, Jensen E, Wierzbicki R, Boxall A (2011) Characterization of nanomaterials in food by electron microscopy. Trends Anal Chem 30:28–43

Dufour A, Wetzler S, Touvier M, Lioret S, Gioda J, Lafay L, Dubuisson C, Calamassi-Tran G, Kalonji E, Margaritis I, Volatier JL (2010) Comparison of different maximum safe levels in fortified foods and supplements using a probabilistic risk assessment approach. Br J Nutr 104:1848–1857

Dusemund B, Soffers AEMF, Rietjens IMCM (2012) Plant-derived contaminants in food. In: D Schrenk (ed), Chemical contaminants and residues in food. Woodhead Publishing series in food science, technology and nutrition: No. 235, Woodhead Publishing Limited, Oxford, pp 394–417

Dutch National Institute for Public Health and the Environment (RIVM) (2015) MCRA 8.1 reference manual. https://mcra.rivm.nl. Retrieved 8 Feb 2018

Dwyer J, Picciano MF, Raiten DJ, Basiotis PP, Bender MM, Bindewald BK, Carriquiry AL, Courtney AK, Crane NT, Dodd KW, Egan K, Ellwood KC, Gebhardt SE, Guthrie JF, Harnly JM, Holden JM, Johnson C, Krebs-Smith SM, Kuznesof PM, Lang CE, McDowell M, Moshfegh A, Pehrsson PR, Radimer K, Subar AF, Swanson CA, Wolf WR (2003) Collection

of food and dietary supplement intake data: what we eat in America-NHANES. J Nutr 133 (2):590S–600S
Efron B, Tibshirani R (1993) An introduction to the bootstrap. Chapman & Hall, New York. https://www.crcpress.com/An-Introduction-to-the-Bootstrap/Efron-Tibshirani/p/book/9780412042317. Accessed 8 Mar 2018
EFSA (2004) Discussion paper on "Botanicals and Botanical Preparations widely used as food supplements and related products: coherent and comprehensive risk assessment and consumer information approaches, Brussels, 25 June 2004. EFSA/SC/26 Final. https://www.efsa.europa.eu/sites/default/files/assets/af040930-ax2.pdf. Accessed 27 Oct 2019
EFSA (2005a) Opinion of the scientific committee on a request from EFSA related to exposure assessments. EFSA J 249:1–26. http://onlinelibrary.wiley.com/doi/10.2903/j.efsa.2005.249/pdf. Accessed 27 Oct 2019
EFSA (2005b) Opinion of the scientific committee on a request from EFSA related to a harmonised approach for risk assessment of substances which are both genotoxic and carcinogenic. EFSA J 282:1–31 http://onlinelibrary.wiley.com/doi/10.2903/j.efsa.2005.282/abstract. Accessed 27 Oct 2019
EFSA (2005c) Opinion of the PPR panel related to the appropriate variability factor(s) to be used for acute dietary exposure assessment of pesticide residues in fruit and vegetables. EFSA J 3,3: 177. http://onlinelibrary.wiley.com/doi/10.2903/j.efsa.2005.177/abstract. Accessed 27 Oct 2019
EFSA (2006a) Tolerable upper intake levels for vitamins and minerals. Scientific committee on food scientific panel on dietetic products, nutrition and allergies. European Food Safety Authority 2006 ISBN: 92-9199-014-0. http://www.efsa.europa.eu/sites/default/files/efsa_rep/blobserver_assets/ndatolerableuil.pdf. Accessed 27 Oct 2019
EFSA (2006b) Opinion of the scientific committee related to uncertainties in dietary exposure assessment. EFSA J 438:1–54. http://onlinelibrary.wiley.com/doi/10.2903/j.efsa.2007.438/abstract. Accessed 27 Oct 2019
EFSA (2007) EFSA calculation model pesticide residues intake model PRIMo revision 2. www.efsa.europa.eu/applications/pesticides/tools. Accessed 27 Oct 2019
EFSA (2008a) Opinion of the scientific panel on plant protection products and their residues to evaluate the suitability of existing methodologies and, if appropriate, the identification of new approaches to assess cumulative and synergistic risks from pesticides to human health with a view to set MRLs for those pesticides in the frame of Regulation (EC) 396/2005). EFSA J 6 (5):705. http://onlinelibrary.wiley.com/doi/10.2903/j.efsa.2008.705/abstract. Accessed 27 Oct 2019
EFSA (2008b) Opinion of the scientific panel on contaminants in the food chain on a request from the European Commission on marine biotoxins in shellfish – okadaic acid and analogues. EFSA J 589:1–62. http://onlinelibrary.wiley.com/doi/10.2903/j.efsa.2008.589/full. Accessed 27 Oct 2019
EFSA (2008c) Guidance document for the use of the Concise European Food Consumption database in exposure assessment. http://www.efsa.europa.eu/sites/default/files/efsa_rep/blobserver_assets/datexfooddbguidance.pdf. Accessed 27 Oct 2019
EFSA (2008d) Statement of EFSA on the risks for public health due to the presence of dioxins in pork from Ireland. EFSA J 911:1–15. http://onlinelibrary.wiley.com/doi/10.2903/j.efsa.2008.911/full. Accessed 27 Oct 2019
EFSA (2009a) Scientific opinion of the panel on contaminants in the food chain on a request from the European Commission on cadmium in food. EFSA J 980:1–139. http://www.efsa.europa.eu/en/efsajournal/pub/980. Accessed 27 Oct 2019
EFSA (2009b) Uranium in foodstuffs, in particular mineral water. EFSA J 1018:1–59. http://onlinelibrary.wiley.com/doi/10.2903/j.efsa.2009.1018/full. Accessed 27 Oct 2019
EFSA (2009c) Safety assessment of botanicals and botanical preparations intended for use as ingredients in food supplements. EFSA J 1249:1–19. http://www.efsa.europa.eu/en/efsajournal/pub/1249.htm. Accessed 27 Oct 2019

EFSA (2009d) Guidance of the scientific committee on a request from EFSA on the use of the benchmark dose approach in risk assessment. EFSA J 1150:1–72. http://onlinelibrary.wiley.com/doi/10.2903/j.efsa.2009.1150/full. Accessed 27 Oct 2019

EFSA (2009e) General principles for the collection of national food consumption data in the view of a pan-European dietary survey. Document 1435. EFSA J 7(12):1–51. http://onlinelibrary.wiley.com/doi/10.2903/j.efsa.2009.1435/abstract. Accessed 27 Oct 2019

EFSA (2009f) Advice on the EFSA guidance document for the safety assessment of botanicals and botanical preparations intended for use as food supplements, based on real case studies. EFSA J 7(9):280. http://www.efsa.europa.eu/en/supporting/doc/280r.pdf Accessed 27 Oct 2019

EFSA (2010a) Scientific opinion on the reevaluation of lutein (E 161b) as a food additive on request of the European Commission. EFSA J 8(7):1678. http://onlinelibrary.wiley.com/doi/10.2903/j.efsa.2010.1678/full. Accessed 27 Oct 2019

EFSA (2010b) Scientific opinion on lead in food. EFSA J 8:1570, 1147 pp. http://onlinelibrary.wiley.com/doi/10.2903/j.efsa.2010.1570/full. Accessed 27 Oct 2019

EFSA (2010c) Management of left-censored data in dietary exposure assessment of chemical substances. EFSA J 8(3):1557, 96 pp. http://onlinelibrary.wiley.com/doi/10.2903/j.efsa.2010.1557/full. Accessed 27 Oct 2019

EFSA (2010d) Standard sample description for food and feed. EFSA J 8(1):1457, 54 pp. https://www.efsa.europa.eu/de/efsajournal/pub/1457. Accessed 27 Oct 2019

EFSA (2010e) Guidance on the data required for the risk assessment of flavourings to be used in or on foods. EFSA J 8(6):1623. http://onlinelibrary.wiley.com/doi/10.2903/j.efsa.2010.1623/epdf. Accessed 27 Oct 2019

EFSA (2010f). Guidance on human health risk-benefit assessment of food. EFSA J 8:1673. http://onlinelibrary.wiley.com/doi/10.2903/j.efsa.2010.1673/abstract. Accessed 8 Mar 2018

EFSA (2011a) Use of the EFSA comprehensive european food consumption database in exposure assessment. EFSA J 9(3):2097. http://www.efsa.europa.eu/en/food-consumption/comprehensive-database. Accessed 27 Oct 2019

EFSA (2011b) Scientific opinion on the re-evaluation of lutein preparations other than lutein with high concentrations of total saponified carotenoids at levels of at least 80%. EFSA J 9(5):2144. http://www.efsa.europa.eu/de/efsajournal/doc/2144.pdf. Accessed 27 Oct 2019

EFSA (2011c) Scientific opinion on the risks for public health related to the presence of opium alkaloids in poppy seeds. EFSA J 9(11):2405. http://onlinelibrary.wiley.com/doi/10.2903/j.efsa.2011.2405/pdf. Accessed 27 Oct 2019

EFSA (2011d) Scientific opinion on pyrrolizidine alkaloids in food and feed. EFSA J 9(11):2406. http://www.efsa.europa.eu/de/efsajournal/doc/2406.pdf. Accessed 27 Oct 2019

EFSA (2011e) Update on furan levels in food from monitoring years 2004–2010 and exposure assessment. EFSA J 9(9):2347. https://efsa.onlinelibrary.wiley.com/doi/10.2903/j.efsa.2011.2347

EFSA (2011f) Report on the development of a Food Classification and Description System for exposure assessment and guidance on its implementation and use. EFSA J 9(12): 2489. 84 pp. http://onlinelibrary.wiley.com/doi/10.2903/j.efsa.2011.2489/abstract. Accessed 27 Oct 2019

EFSA (2011g) Overview of the procedures currently used at EFSA for the assessment of dietary exposure to different chemical substances. EFSA J 9(12):2490. https://www.efsa.europa.eu/de/efsajournal/pub/2490. Accessed 27 Oct 2019

EFSA (2011h) Technical report – The food classification and description system FoodEx 2 (draft-revision 1). Supporting Publications 2011: 215 [438pp.]. http://onlinelibrary.wiley.com/doi/10.2903/sp.efsa.2011.EN-215/abstract. Accessed 27 Oct 2019

EFSA (2011i) Evaluation of the FoodEx 2 the food classification system applied to the development of the EFSA comprehensive European food consumption database. EFSA J 9(3):1970. http://onlinelibrary.wiley.com/doi/10.2903/j.efsa.2011.1970/full. Accessed 27 Oct 2019

EFSA (2011j) Guidance for risk assessment of food and feed from genetically modified plants. EFSA J 9(5):2150. http://onlinelibrary.wiley.com/doi/10.2903/j.efsa.2011.2150/abstract. Accessed 27 Oct 2019

EFSA (2011k) Scientific opinion of the scientific committee on Guidance on the risk assessment of the application of nanoscience and nanotechnologies in the food and feed chain. EFSA J 9(5):2140. http://onlinelibrary.wiley.com/doi/10.2903/j.efsa.2011.2140/full. Accessed 27 Oct 2019

EFSA (2011l) Scientific opinion on the risk to public health related to the presence of high levels of dioxins and dioxin-like PCBs in liver from sheep and deer. EFSA panel on contaminants in the food chain (CONTAM panel). EFSA J 9(7):2297. doi: https://doi.org/10.2903/j.efsa.2011.2297. http://onlinelibrary.wiley.com/doi/10.2903/j.efsa.2011.2297/full. Accessed 27 Oct 2019

EFSA (2012a) Cadmium dietary exposure in the European population. EFSA J 10(1):2551, 37 pp. http://onlinelibrary.wiley.com/doi/10.2903/j.efsa.2012.2551/full. Accessed 27 Oct 2019

EFSA (2012b) Scientific opinion on ergot alkaloids in food and feed. EFSA J 10(7):2798. http://www.efsa.europa.eu/en/efsajournal/doc/2798.pdf. Accessed 27 Oct 2019

EFSA (2012c) Guidance on the use of probabilistic methodology for modelling dietary exposure to pesticide residues. EFSA J 10(10):2839, 95 pp. http://onlinelibrary.wiley.com/doi/10.2903/j.efsa.2012.2839/full. Accessed 27 Oct 2019

EFSA (2012d) Guidance for submission for food additive evaluations. EFSA J 10(7):2760. http://onlinelibrary.wiley.com/doi/10.2903/j.efsa.2012.2760/epdf. Accessed 27 Oct 2019

EFSA (2012e) Food additives intake model (FAIM) Template – Version 1.0 – December 2012. http://www.efsa.europa.eu/sites/default/files/assets/faimtemplateinstructions.pdf. Accessed 27 Oct 2019

EFSA (2012f) Compendium of botanicals reported to contain naturally occurring substances of possible concern for human health when used in food and food supplements. EFSA J 10 (5):2663. http://onlinelibrary.wiley.com/doi/10.2903/j.efsa.2012.2663/full. Accessed 27 Oct 2019

EFSA (2012g) EFSA's approach to identifying emerging risks in food and feed: taking stock and looking forward. EFSA J 10:s1015. http://onlinelibrary.wiley.com/doi/10.2903/j.efsa.2012.s1015/abstract. Accessed 27 Oct 2019

EFSA (2013a) Scientific opinion on the identification of pesticides to be included in cumulative assessment groups on the basis of their toxicological profile. EFSA J 11(7): n/a–n/a. http://onlinelibrary.wiley.com/doi/10.2903/j.efsa.2013.3293/full. Accessed 27 Oct 2019

EFSA (2013b) Guidance on methodological principles and scientific methods to be taken into account when establishing Reference Points for Action (RPAs) for non-allowed pharmacologically active substances present in food of animal origin. EFSA J 11(4):3195 [24 pp.]. http://onlinelibrary.wiley.com/doi/10.2903/j.efsa.2013.3195/abstract. Accessed 27 Oct 2019

EFSA (2013c) Scientific opinion on safety evaluation of Ephedra species in food. EFSA J 11 (11):3467. http://onlinelibrary.wiley.com/doi/10.2903/j.efsa.2013.3467/epdf. Accessed 27 Oct 2019

EFSA (2013d) Scientific opinion on the evaluation of the safety in use of Yohimbe (Pausinystalia yohimbe (K. Schum.) Pierre ex Beille). EFSA J 11(7):3302. http://onlinelibrary.wiley.com/doi/10.2903/j.efsa.2013.3302/epdf. Accessed 27 Oct 2019

EFSA (2013e) Guidance of EFSA – standard sample description ver.2.0. EFSA J 11(10):3424. http://onlinelibrary.wiley.com/doi/10.2903/j.efsa.2013.3424/full. Accessed 27 Oct 2019

EFSA (2013f) Food Additives Intake Model (FAIM). http://www.efsa.europa.eu/en/applications/foodingredients/tools. Accessed 8 Mar 2018

EFSA (2013g) Scientific opinion on tropane alkaloids in food and feed. EFSA J 11(10):3386. http://onlinelibrary.wiley.com/doi/10.2903/j.efsa.2013.3386/epdf. Accessed 27 Oct 2019

EFSA (2013h) European Food Safety Authority. Refined exposure assessment of ethyl lauroyl arginate based on revised proposed uses as a food additive. EFSA J 11(6):3294, 15 pp. http://onlinelibrary.wiley.com/doi/10.2903/j.efsa.2013.3294/full. Accessed 27 Oct 2019

EFSA (2013i) European Food Safety Authority. Refined exposure assessment for amaranth (E 123). EFSA J 11(10):3442. 18 pp. http://onlinelibrary.wiley.com/doi/10.2903/j.efsa.2013.3442/full. Accessed 27 Oct 2019

EFSA (2014a) Statement on a conceptual framework for the risk assessment of certain food additives re-evaluated under Commission Regulation (EU) No 257/2010. EFSA J 12(6):3697. http://onlinelibrary.wiley.com/doi/10.2903/j.efsa.2014.3697/abstract. Accessed 27 Oct 2019
EFSA (2014b) Re-evaluation of Indigo Carmine (E 132) as a food additive. EFSA J 12(7):3768, 51 pp. https://www.efsa.europa.eu/de/efsajournal/pub/3768. Accessed 27 Oct 2019
EFSA (2014c) Guidance on the assessment of exposure of operators, workers, residents and bystanders in risk assessment for plant protection products. EFSA J 12(10):3874–n/a. https://www.efsa.europa.eu/de/efsajournal/pub/3874. Accessed 27 Oct 2019
EFSA (2014d) Dietary exposure to inorganic arsenic in the European population. EFSA J 12 (3):3597–3665. Fab/pop. http://onlinelibrary.wiley.com/doi/10.2903/j.efsa.2014.3597/full. Accessed 27 Oct 2019
EFSA (2014e) Scientific Opinion on the risks to public health related to the presence of perchlorate in food, in particular fruits and vegetables. EFSA J 12(10):3869. http://onlinelibrary.wiley.com/doi/10.2903/j.efsa.2014.3869/full. Accessed 27 Oct 2019
EFSA (2014f) Scientific opinion on the revised exposure assessment of steviol glycosides (E 960) for the proposed uses as a food additive. EFSA J 12(5):3639, 23 pp. http://onlinelibrary.wiley.com/doi/10.2903/j.efsa.2014.3639/full. Accessed 27 Oct 2019
EFSA (2014g) Refined exposure assessment for curcumin (E 100). EFSA J 12(10):3876, 43 pp. http://onlinelibrary.wiley.com/doi/10.2903/j.efsa.2014.3876/full. Accessed 8 Mar 2018
EFSA (2015a) Use of EFSA comprehensive European food consumption database for estimating dietary exposure to genetically modified foods. EFSA J 13(2):4034, 11 pp. https://www.efsa.europa.eu/de/efsajournal/pub/4034. Accessed 27 Oct 2019
EFSA (2015b) The 2013 European Union report on pesticide residues in food. EFSA J 13(3):4038, 169 pp. http://onlinelibrary.wiley.com/doi/10.2903/j.efsa.2015.4038/full. Accessed 27 Oct 2019
EFSA (2015c) Shaping the future of food safety, together: proceedings of the 2nd EFSA scientific conference. Milan, Italy, 14–16 October 2015. EFSA J 13(10):s1310, 137 pp. https://www.efsa.europa.eu/de/events/event/151014. Accessed 27 Oct 2019
EFSA (2015d) Scientific opinion on acrylamide in food. EFSA J 13(6):4104, 321 pp. http://onlinelibrary.wiley.com/doi/10.2903/j.efsa.2015.4104/abstract. Accessed 27 Oct 2019
EFSA (2015e) Scientific opinion on the re-evaluation of octyl gallate (E 311) as a food additive. EFSA J 13(10):4248, 39 pp. http://onlinelibrary.wiley.com/doi/10.2903/j.efsa.2015.4248/references. Accessed 27 Oct 2019
EFSA (2015f) The food classification and description system FoodEx2 (revision 2): EFSA supporting publication 12(5):EN-804, 90 pp. http://onlinelibrary.wiley.com/doi/10.2903/sp.efsa.2015.EN-804/abstract. Accessed 27 Oct 2019
EFSA (2015g) Scientific opinion on the risks for human health related to the presence of tetrahydrocannabinol (THC) in milk and other food of animal origin. EFSA J 13(6):4141. http://onlinelibrary.wiley.com/doi/10.2903/j.efsa.2015.4141/epdf. Accessed 27 Oct 2019
EFSA (2015h) Refined exposure assessment for Quinoline Yellow (E 104). EFSA J 13(3):4070, 33 pp. http://onlinelibrary.wiley.com/doi/10.2903/j.efsa.2015.4070/abstract. Accessed 27 Oct 2019
EFSA (2015i) EFSA scientific workshop, co-sponsored by FAO and WHO, Revisiting the International Estimate of Short-Term Intake (IESTI-equations) used to estimate the acute exposure to pesticide residues via food, 8/9 September 2015, Geneva, Switzerland. https://efsa.onlinelibrary.wiley.com/doi/pdf/10.2903/sp.efsa.2015.EN-907. Accessed 27 Oct 2019
EFSA (2015j) Statement on the benefits of fish/seafood consumption compared to the risks of methylmercury in fish/seafood. EFSA J 13(1):3982 [36 pp.]. http://onlinelibrary.wiley.com/doi/10.2903/j.efsa.2015.3982/full. Accessed 27 Oct 2019
EFSA (2015k) European food safety authority. Refined exposure assessment for Azorubine/Carmoisine (E 122). EFSA J 13(3):4072, 35 pp. http://onlinelibrary.wiley.com/doi/10.2903/j.efsa.2015.4072/full. Accessed 27 Oct 2019
EFSA (2015l) European food safety authority. Refined exposure assessment for Ponceau 4R (E 124). EFSA J 13(4):4073, 34 pp. http://onlinelibrary.wiley.com/doi/10.2903/j.efsa.2015.4073/full. Accessed 27 Oct 2019

EFSA (2016a) Scientific opinion on the risks for human health related to the presence of 3- and 2-monochloropropanediol (MCPD) and their fatty acid esters, and glycidyl fatty acid esters in food. EFSA J 14(5):4426, 159 pp. http://onlinelibrary.wiley.com/doi/10.2903/j.efsa.2016.4426/full. Accessed 27 Oct 2019

EFSA (2016b) The 2014 European Union report on pesticide residues in food. EFSA J 14(10):4611, 139 pp. http://onlinelibrary.wiley.com/doi/10.2903/j.efsa.2016.4611/full. Accessed 27 Oct 2019

EFSA (2016c) Acute health risks related to the presence of cyanogenic glycosides in raw apricot kernels and products derived from raw apricot kernels. EFSA J 14(4):4424, 47 pp. http://onlinelibrary.wiley.com/doi/10.2903/j.efsa.2016.4424/full. Accessed 27 Oct 2019

EFSA (2016d) Re-evaluation of titanium dioxide (E 171) as a food additive. EFSA J 14(9):4545, 83 pp. http://onlinelibrary.wiley.com/doi/10.2903/j.efsa.2016.4545/references. Accessed 27 Oct 2019

EFSA (2016e) Review of the threshold of toxicological concern (TTC) approach and development of new TTC decision tree. EFSA J 13(3):1006E. http://onlinelibrary.wiley.com/doi/10.2903/sp.efsa.2016.EN-1006/abstract. Accessed 27 Oct 2019

EFSA (2016f) Guidance on the preparation and presentation of an application for authorisation of a novel food in the context of Regulation (EU) 2015/2283. EFSA J 14(11):4594, 24 pp. http://onlinelibrary.wiley.com/doi/10.2903/j.efsa.2016.4594/full. Accessed 27 Oct 2019

EFSA (2016g) Dietary exposure assessment to pyrrolizidine alkaloids in the European population. EFSA J 14(8):4572. http://onlinelibrary.wiley.com/doi/10.2903/j.efsa.2016.4572/full

EFSA (2016h) Statement on the refined exposure assessment of tertiary-butyl hydroquinone (E 319). EFSA J 14(1):4363, 26 pp. http://onlinelibrary.wiley.com/doi/10.2903/j.efsa.2016.4363/full. Accessed 27 Oct 2019

EFSA (2016i) Update: the use of the benchmark dose approach in risk assessment. EFSA J 15 (1):4658, 41 pp. http://onlinelibrary.wiley.com/doi/10.2903/j.efsa.2017.4658/full. Accessed 27 Oct 2019

EFSA (2017a) Re-evaluation of fatty acids (E 570) as a food additive. EFSA J 15(5):4785, 48 pp. http://onlinelibrary.wiley.com/doi/10.2903/j.efsa.2017.4785/abstract. Accessed 27 Oct 2019

EFSA (2017b) Re-evaluation of glycerol (E 422) as a food additive. EFSA J 15(3):4720, 64 pp. http://onlinelibrary.wiley.com/doi/10.2903/j.efsa.2017.4720/full

EFSA (2017c) Guidance on the risk assessment of substances present in food intended for infants below 16 weeks of age. EFSA J 15(5):4849, 58 pp. http://onlinelibrary.wiley.com/doi/10.2903/j.efsa.2017.4849/full. Accessed 27 Oct 2019

EFSA (2017d) Re-evaluation of sorbitan monostearate (E 491), sorbitan tristearate (E 492), sorbitan monolaurate (E 493), sorbitan monooleate (E 494) and sorbitan monopalmitate (E 495) when used as food additives. EFSA J 15(5):4788, 56 pp. http://onlinelibrary.wiley.com/doi/10.2903/j.efsa.2017.4788/full. Accessed 27 Oct 2019

EFSA (2018a) Database of processing techniques and processing factors compatible with the EFSA food classification and description system FoodEx 2 Objective 3: European database of processing factors for pesticides in food", 26.11.2018, https://doi.org/10.2903/sp.efsa.2018.EN-1510. https://www.efsa.europa.eu/en/supporting/pub/en-1510

EFSA (2018b) Update of the risk assessment on 3-monochloropropane diol and its fatty acid esters. EFSA J 16(1):5083. https://efsa.onlinelibrary.wiley.com/doi/10.2903/j.efsa.2018.5083

EFSA/FAO/WHO (2011) Towards a harmonised Total Diet Study approach: a guidance document. EFSA J 9(11):2450. http://onlinelibrary.wiley.com/doi/10.2903/j.efsa.2011.2450/full Accessed 27 Oct 2019

Eisinger-Watzl M, Strassburg A, Ramunke J, Krems C, Heuer T, Hoffmann I (2015) Comparison of two dietary assessment methods by food consumption: results of the German National Nutrition Survey II. Eur J Nutr 54(3):343–354

Elegbede C, Papadopoulos A, Kolbaum AE, Turrini A, Mistura L, Lindtner O, Sirot V (2017) TDS exposure project: how and when to consider seasonalityin a total diet study? Food Chem Toxicol 105:119–126. https://doi.org/10.1016/j.fct.2017.03.045

El-Shaarawi AH, Esterby SR (1992) Replacement of censored observations by a constant: an evaluation. Water Resour Res 26:835–844

Estrada-Montoya MC, González-Córdova AF, Torrescano G, Camou JP, Vallejo-Cordoba B (2008) Screening and confirmatory determination of clenbuterol residues in bovine meat marketed in the northwest of Mexico. Ciencia y Tecnología Alimentaria 6:130

ETHZ (lead), NIPH/SZU, EVIRA, MATIS, INSA, RIVM, C.R.A NUT, UGR, URV, MSSSI/AECOSAN, NUID UCD, UGENT, EuroFIR AISBL, ANSES (2015) TDS exposure deliverable report D6.3-on usability testing of FoodCASE-Risk. To be requested at: http://www.tds-exposure.eu/project-objectives-0/contact-us/. Accessed 27 Oct 2019

EU (1998) Report on methodologies for the monitoring of food additive intake across the European Union (Final report submitted by the Task Co-ordinator 16 January 1998), Reports of a working Group on Scientific Co-operation on questions relating to food. Task 4.2. SCOOP/INT/REPORT/2. European Commission Directorate General III Industry, Brussels. http://eur-lex.europa.eu/legal-content/EN/TXT/HTML/?uri=CELEX:52001DC0542&from=EN. Accessed 27 Oct 2019

EU (2000a) White paper on food safety Brussels, 12 January 2000 COM (1999) 719 final. Accessible https://europa.eu/rapid/press-release_IP-00-20_en.htm. Accessed 27 Oct 2019

EU (2000b) Directive 2000/13/EC of the European Parliament and of the council of 20 March 2000 on the approximation of the laws of the Member States relating to the labelling, presentation and advertising of foodstuffs. Off J Europ Union (EC) 2000 No. L 109/29. http://eur-lex.europa.eu/legal-content/EN/TXT/HTML/?uri=CELEX:32000L0013&from=EN. Accessed 27 Oct 2019

EU (2001) Report from the commission on dietary food additive Intake in the European Union. COM (2001) 542 final, Brussels, 01.10.2001. http://eur-lex.europa.eu/legal-content/EN/TXT/PDF/?uri=CELEX:52001DC0542&from=mt

EU (2002a) Directive 2002/46/EC of the European Parliament and of the Council of 10 June 2002 on the approximation of the laws of the Member States relating to food supplements. Off J Europ Union (EC) 2002 No. L 183/51. http://eur-lex.europa.eu/LexUriServ/LexUriServ.do?uri=OJ:L:2002:183:0051:0057:EN:PDF. Accessed 27 Oct 2019

EU (2002b) Regulation (EC) No 178/2002 of the European Parliament and of the Council of 28 January 2002 laying down the general principles and requirements of food law, establishing the European Food Safety Authority and laying down procedures in matters of food safety. Off J Europ Union (EC) 2002 No. L 31/1. http://eur-lex.europa.eu/LexUriServ/LexUriServ.do?uri=OJ:L:2002:031:0001:0024:en:PDF. Accessed 27 Oct 2019

EU (2004) Regulation (EC) No 882/2004 of the Euroepan Parliament and of the Council of 29 April 2004on official controls performed to ensure the verification of compliance with feed and food law,animal health and animal welfare rules. https://www.fsai.ie/uploadedFiles/Legislation/Food_Legisation_Links/Official_Control_Of_Foodstuffs/15-Consol_Reg882_2004_01Jan07.pdf. Accessed 27 Oct 2019

EU (2005a) REGULATION (EC) NO 396/2005 on maximum residue levels of pesticides in or on food and feed of plant and animal origin and amending Council Directive 91/414/EEC. Off J Europ Union L 70. http://eur-lex.europa.eu/legal-content/EN/TXT/PDF/?uri=CELEX:32005R0396&from=EN. Accessed 27 Oct 2019

EU (2005b) Regulation (EC) No 396/2005 of the European Parliament and of the Council of 23 February 2005 on maximum residue levels of pesticides in or on food and feed of plant and animal origin and amending Council Directive 91/414/EECText with EEA relevance. http://eur-lex.europa.eu/legal-content/EN/ALL/?uri=CELEX:32005R0396. Accessed 27 Oct 2019

EU (2006a) Regulation (EC) No 1924/2006. Corrigendum to Regulation (EC) No 1924/2006 of the European Parliament and of the Council of 20 December 2006 on nutrition and health claims made on foods. Off J Europ Union (EC) L 12/3. http://eur-lex.europa.eu/legal-content/EN/TXT/HTML/?uri=CELEX:32006R1924R(01)&from=EN. Accessed 27 Oct 2019

EU (2006b) Regulation (EC) No 1925/2006 of the European Parliament and of the Council of 20 December 2006 on the addition of vitamins and minerals and of certain other substances to foods. Off J Europ Union (EC) L 404/26. http://eur-lex.europa.eu/legal-content/EN/TXT/HTML/?uri=CELEX:32006R1925&from=EN. Accessed 27 Oct 2019

EU (2006c) Regulation (EC) No 1881/2006 of 19 December 2006 setting maximum levels for certain contaminants in foodstuffs 20.12.2006. Off J Europ Union (EC) 2006 No. L 364/5. http://eur-lex.europa.eu/legal-content/EN/TXT/PDF/?uri=CELEX:32006R1881&from=EN

EU (2008a) Regulation (EC) No 1331/2008 of the European Parliament and of the Council of 16 December 2008 establishing a common authorisation procedure for food additives, food enzymes and food flavourings. Off J Europ Union (EC) 2008 No. L 354/1. http://eur-lex.europa.eu/LexUriServ/LexUriServ.do?uri=OJ:L:2008:354:0001:0006:EN:PDF. Accessed 27 Oct 2019

EU (2008b) Regulation (EC) No 1332/2008 of the European Parliament and of the Council of 16 December 2008 on food enzymes and amending Council Directive 83/417/EEC, Council Regulation (EC) No 1493/1999, Directive 2000/13/EC, Council Directive 2001/112/EC and Regulation (EC) No 258/97. Off J Europ Union (EC) 2008 No. L 354/7. http://eur-lex.europa.eu/LexUriServ/LexUriServ.do?uri=OJ:L:2008:354:0007:0015:en:PDF. Accessed 8 Mar 2018

EU (2008c) Regulation (EC) No 1333/2008 of the European Parliament and of the Council of 16 December 2008 on food additives. Off J Europ Union (EC) 2008 No. L 354. https://eur-lex.europa.eu/legal-content/EN/ALL/?uri=CELEX%3A32008R1333. Accessed 8 Mar 2018

EU (2008d) Regulation (EC) No 1334/2008 of the European Parliament and of the Council of 16 December 2008 on flavourings and certain food ingredients with flavouring properties for use in and on foods and amending Council Regulation (EEC) No 1601/91, Regulations (EC) No 2232/96 and (EC) No 110/2008 and Directive 2000/13/EC. Off J Europ Union (EC) 2008 No. L 354/34. http://eur-lex.europa.eu/LexUriServ/LexUriServ.do?uri=OJ:L:2008:354:0034:0050:en:PDF. Accessed 27 Oct 2019

EU (2009) Directive 2009/32/EC of the European Parliament and of the Council of 23 April 2009 on the approximation of the laws of the Member States on extraction solvents used in the production of foodstuffs and food ingredients (Recast). http://eur-lex.europa.eu/legal-content/EN/ALL/?uri=CELEX%3A32009L0032. Accessed 27 Oct 2019

EU (2010) Commission Regulation (EU) No 257/2010 of 25 March 2010 setting up a programme for the re-evaluation of approved food additives in accordance with Regulation (EC) No 1333/2008 of the European Parliament and of the Council on food additives. http://eur-lex.europa.eu/legal-content/EN/TXT/HTML/?uri=CELEX:32010R0257&from=EN. Accessed 27 Oct 2019

EU (2011a) Commission Regulation (EU) No 1129/2011 of 11 November 2011 amending Annex II to Regulation (EC) No 1333/2008 of the European Parliament and of the Council by establishing a Union list of food additives. http://eur-lex.europa.eu/legal-content/EN/TXT/HTML/?uri=CELEX:32011R1129&from=EN. Accessed 27 Oct 2019

EU (2011b) Regulation (EU) No 1169/2011 of the European Parliament and of the Council of 25 October 2011 on the provision of food information to consumers, amending Regulations (EC) No 1924/2006 and (EC) No 1925/2006 of the European Parliament and of the Council, and repealing Commission Directive 87/250/EEC, Council Directive 90/496/EEC, Commission Directive 1999/10/EC, Directive 2000/13/EC of the European Parliament and of the Council, Commission Directives 2002/67/EC and 2008/5/EC and Commission Regulation (EC) No 608/2004. http://eur-lex.europa.eu/legal-content/EN/TXT/HTML/?uri=CELEX:32011R1169&from=en. Accessed 27 Oct 2019

EU (2011c) Commission Regulation (EU) No 234/2011 of 10 March 2011 implementing Regulation (EC) No 1331/2008 of the European Parliament and of the Council establishing a common authorisation procedure for food additives, food enzymes and food flavourings. Off J Europ Union (EC) 2011 No. L 64/15. http://eur-lex.europa.eu/legal-content/EN/TXT/PDF/?uri=CELEX:32011R0234&from=EN. Accessed 27 Oct 2019

EU (2012) Regulation (EU) No 528/2012 of the European Parliament and of the Council of 22 May 2012 concerning the making available on the market and use of biocidal products. https://eur-lex.europa.eu/legal-content/EN/TXT/HTML/?uri=CELEX:32012R0528&from=DE

EU (2015a) Summary report of the standing committee on plants, animals, food and feed held in Brussels on 11 FEBRUARY 2015 (Section Toxicological Safety of the Food Chain). https://ec.europa.eu/food/sites/food/files/safety/docs/reg-com_toxic_20150211_sum.pdf. Accessed 27 Oct 2019

EU (2015b) Ad hoc study in preparation of the development of a common methodology for gathering of information by the Member States on the consumption and use of food additives and flavourings in the European Union Framework Contract for evaluation and evaluation related services – Lot 3: Food Chain. Final Report Submitted by: Food Chain Evaluation Consortium (FCEC), Civic Consulting – Agra CEAS Consulting – Van Dijk Management Consultants – Arcadia International, Project Leader: Arcadia International. http://www.arcadia-international.net/news.php. Accessed 27 Oct 2019

EU (2016) EU pesticides database http://ec.europa.eu/food/plant/pesticides/eu-pesticides-database. Accessed 27 Oct 2019

EU (2017) RASFF 2016 Preliminary annual report-Food and Feed Safety. https://ec.europa.eu/food/sites/food/files/safety/docs/rasff_annual_report_2016.pdf. Accessed 27 Oct 2019

Expo Net (2015) 375 million vegetarians worldwide. All the reasons for a green lifestyle. http://www.expo2015.org/magazine/en/lifestyle/375-million-vegetarians-worldwide.html. Accessed 27 Oct 2019

FAO (1999) Progress on acute dietary intake estimation – International Estimate of Short Term Intake (IESTI). FAO Plant Protection and Protection Paper 153 (Pesticide residues in food 1999. Report of the Joint Meeting of the FAO Panel of Experts on Pesticide Residues in Food and the Environment and the WHO Core Assessment Group on Pesticide Residues, Rome, Italy). http://apps.who.int/iris/handle/10665/42314. Accessed on 8 Mar 2018

FAO (2001) Food and Agriculture Organization of the United Nations. Guidelines for the preparation of working papers on intake of food additives for the Joint FAO/WHO Expert Committee on Food Additives. Geneva, January 2001. http://www.who.int/foodsafety/chem/jecfa/en/intake_guidelines.pdf?ua=1

FAO (2003a) Trade reforms and food security: conceptualizing the linkages Rome (I), 2003. http://www.fao.org/3/a-y4671e.pdf. Accessed 27 Oct 2019

FAO (2003b) Pesticide residues in food – report of the 2003 joint FAO/WHO meeting. Rome, FAO

FAO (2011) Discussion paper on pyrrolizidine alkaloids. Joint FAO/WHO food standards programme, CODEX Committee on Contaminants in Foods, 5th Session, The Hague, The Netherlands, 21–25 March 2011. http://www.fao.org/tempref/codex/Meetings/CCCF/CCCF5/cf05_INF.pdf. Accessed 27 Oct 2019

FAO (2014) Guidelines for simple evaluation of dietary exposure to food additives. CAC/GL 3-1989. Adopted 1989. Revision 2014. http://www.fao.org/fao-who-codexalimentarius/standards/list-of-standards/en/. Accessed 27 Oct 2019

FAO (2016) Submission and evaluation of pesticide residues data for the estimation of maximum residue levels in food and feed. Third edition. FAO plant production and protection paper 225. Food and Agricultural Organization of the United Nations, Rome. http://www.fao.org/fileadmin/templates/agphome/documents/Pests_Pesticides/JMPR/Manual/FAO_manual_3rd_edition_Final.pdf. Accessed 27 Oct 2019

FAO & OIE (2009) Guide to good farming practices. Rome (I), 2009. ISBN 978-92-5-006145-0. http://www.oie.int/fileadmin/Home/eng/Current_Scientific_Issues/docs/pdf/eng_guide.pdf. Accessed 27 Oct 2019

FAO/WHO (2001) Guidelines for the preparation of working papers on intake of food additives for the Joint FAO/WHO Expert Committee on Food Additives. World Health Organization, Geneva. Available online www.who.int/foodsafety/chem/jecfa/en/intake_guidelines.pdf

FAO/WHO (2016) JECFA/83/SC Joint FAO/WHO expert committee on food additives eighty-third meeting, Rome, 8–17 November 2016 summary and conclusions. 23. November 2016. http://www.fao.org/3/a-bq821e.pdf. Accessed 27 Oct 2019

Finkelstein MM, Verma DK (2001) Exposure estimation in the presence of nondetectable values: another look. Am Ind Hyg Assoc J 62:195–198

Flynn A, Moreiras O, Stehle P, Fletcher RJ, Müller DJG, Rolland V (2003) Vitamins and minerals: a model for safe addition to foods. Eur J Nutr 42:118–130

Fontcuberta M, Calderon J, Villalbí JR, Centrich F, Portaña S, Espelt A, Duran J, Nebot M (2011) Total and inorganic arsenic in marketed food and associated health risks for the Catalan (Spain) population. J Agric Food Chem 59(18):10013–10022

Freedman LS, Schatzkin A, Midthune D, Kipnis V (2011) Dealing with dietary measurement error in nutritional cohort studies. J Natl Cancer Inst 103(14):1086–1092

Freese J, Feller S, Harttig U, Kleiser C, Linseisen J, Fischer B, Leitzmann MF, Six-Merker J, Michels KB, Nimptsch K, Steinbrecher A, Pischon T, Heuer T, Hoffmann I, Jacobs G, Boeing H, Nöthlings U (2014) Development and evaluation of a short 24-h food list as part of a blended dietary assessment strategy in large-scale cohort studies. Eur J Clin Nutr 68 (3):324–329

Frenzel F, Buhrke T, Wenzel I, Andrack J, Hielscher J, Lampen A (2017) Use of in silico models for prioritization of heat-induced food contaminants in mutagenicity and carcinogenicity testing. Arch Toxicol. Jan 16. https://doi.org/10.1007/s00204-016-1924-3. [Epub ahead of print]. https://www.ncbi.nlm.nih.gov/pubmed/28091709

Gajda-Wyrębek J, Kuźma K, Świtka A, Jarecka J, Beresińska M, Postupolski J (2017) Exposure of Polish children to Southampton food colours. Food Addit Contam Part A Chem Anal Control Expo Risk Assess 34(1).1–9

Garcia-Segovia P, González-Carrascosa R, Martinez-Monzó J, Ngo J, Serra-Majem L (2011) New technologies applied to food frequency questionnaires, a current prospective. Nutr Hosp 26:803–806

German Hunters Association (2017). https://de.wikipedia.org/wiki/Deutscher_Jagdverband. Accessed 9 Mar 2018

Gersovitz M, Madden JP, Smiciklas-Wright H (1978) Validity of the 24-hr. dietary recall and seven-day record for group comparisons. J Am Diet Assoc 73(1):48–55

Gibney MJ (1999) Dietary intake methods for estimating food additive intake. Regul Toxicol Pharmacol 30:S31–S33

Gibney MJ, Lambe J (1996) Estimation of food additive intake: methodology overview. Food Addit Contam 13(4):405–410

Gilsenan MB, Lambe J, Gibney MJ (2002) Irish National Food Ingredient Database: application for assessing patterns of additive usage in foods. Food Addit Contam 19(12):1105–1115

Gilsenan MB, Thompson RL, Lambe J, Gibney MJ (2003a) Validation analysis of probabilistic models of dietary exposure to food additives. Food Addit Contam 20(Suppl 1):S61–S72

Gilsenan MB, Lambe J, Gibney MJ (2003b) Assessment of food intake input distributions for use in probabilistic exposure assessments of food additives. Food Addit Contam 20(11):1023–1033

Glass DC, Gray CN (2001) Estimating mean exposures from censored data: exposure to benzene in the Australian petroleum industry. Ann Occup Hyg 45:275–282

Goedhart PW, van der Voet H, Knüppel S, Dekkers ALM, Dodd KW, Boeing H, van Klaveren JD (2012) A comparison by simulation of different methods to estimate the usual intake distribution for episodically consumed foods. Supporting Publications 2012:EN-299. [65 pp.]. https://doi.org/10.2903/sp.efsa.2012.EN-299

Gottschald M, Knüppel S, Boeing H, Buijsse B (2016) The influence of adjustment for energy misreporting on relations of cake and cookie intake with cardiometabolic disease risk factors. Eur J Clin Nutr 70(11):1318–1324

Gürtler R (2010) Safety of food additives from a German and European point of view. Bundesgesundheitsbl – Gesundheitsforsch – Gesundheitsschutz 53(6):554–560

Gürtler R (2014) Risk assessment of food additives. In: Reichl FX, Schwenk M (eds) Regulatory toxicology. Springer, Heidelberg, pp 803–812

Ha MS, Ha SD, Choi SH, Bae DH (2013a) Assessment of Korean consumer exposure to sodium saccharin, aspartame and stevioside. Food Addit Contam Part A Chem Anal Control Expo Risk Assess 30(7):1238–1247

Ha MS, Ha SD, Choi SH, Bae DH (2013b) Exposure assessment of synthetic colours approved in Korea. Food Addit Contam Part A Chem Anal Control Expo Risk Assess 30(4):643–653

Haddad EH, Tanzman JS (2003) What do vegetariansin the United States eat? Am J Clin Nutr 78 (3):626S–632S

Haftenberger M, Heuer T, Heidemann C, Kube F, Krems C, Mensink GBM (2010) Relative validation of a food frequency questionnaire for national health and nutrition monitoring. Nutr J 9:36. https://doi.org/10.1186/1475-2891-9-36

Hamilton D, Ambrus A, Dieterle R, Felsot A, Harris C, Petersen B, Racke K, Wong SS, Gonzalez R, Tanaka K, Earl M, Roberts G, Bhula R (2004) Pesticide residues in food—acute dietary exposure. Pest Manag Sci 60(4):311–339

Hansen SC (1966) Acceptable daily intake of food additives and ceiling on levels of use. Food Cosmet Toxicol 4(4):427–432

Hansen SC (1979) Conditions for use of food additives based on a budget for an acceptable daily intake. J Food Prot 42(5):429–432

Hansen SC (1990) Toxicological evaluation of food additives. Toxicological evaluation over estimation of potential high intake to permitted levels of use of food additives and vice versa. Regul Toxicol Pharmacol 11(1):3–7

Harada M (1995) Minamata disease: methylmercury poisoning in Japan caused by environmental pollution. Crit Rev Toxicol 25(1):1–24

Hartmann BM, Bell S, Vasquez-Caicedo AL, Goetz A, Brombach C (2006) The German nutrient database – present developments, potentials and prospects. Ernaehrungs-Umschau 53:124–129

Harttig U, Haubrock J, Knüppel S, Boeing H (2011) The MSM program: web-based statistics package for estimating usual dietary intake using the Multiple Source Method. Eur J Clin Nutr 65(Suppl 1):S87–S91

Haubrock J, Nöthlings U, Volatier JL, Dekkers A, Ocké M, Harttig U, Illner AK, Knüppel S, Andersen LF, Boeing H, on behalf of the European Food Consumption Validation Consortium (2011) Estimating usual food intake distributions by using the multiple source method in the EPIC-Potsdam Calibration Study. J Nutr 141(5):914–920

Health Canada (2011) Synephrine, Octopamine and Caffeine Health Risk Assessment (HRA) Report, Robin Marles, File Number: 172091. Date review approved 2011-05-16. http://www.novelingredient.com/wp-content/uploads/2017/03/Health-Canada-May11.pdf. Accessed 8 Mar 2018

Heinemeyer G, Heiland A, Sommerfeld C, Springer A, Hausdörfer S, Treutz M, Lindtner O, Rüdiger T (2012) Phthalat-Belastung der Bevölkerung in Deutschland: Expositionsrelevante Quellen, Aufnahmepfade und Toxikokinetik am Beispiel von DEHP und DINP. Band I: Exposition durch Verzehr von Lebensmitteln und Anwendung von Verbraucherprodukten. http://www.uba.de/uba-info-medien/4326.html. Accessed 5 Jul 2018

Heinemeyer G, Sommerfeld C, Springer A, Heiland A, Lindtner O, Greiner M, Heuer T, Krems C, Conrad A (2013) Estimation of dietary intake of bis(2-ethylhexyl) phthalate (DEHP) by consumption of food in the German population. Int J Hyg Environ Health 216:472–480

Helsel DR (1990) Less than obvious-statistical treatment of data below the detection limit. Environ Sci Technol 24:1766–1774

Helsel DR (2004) Nondetects and data analysis. Wiley, New York, NY, p 268

Heringa MB, Geraets L, van Eijkeren JC, Vandebriel RJ, de Jong WH, Oomen AG (2016) Risk assessment of titanium dioxide nanoparticles via oral exposure, including toxicokinetic considerations. Nanotoxicology 10:1515–1525

Heuer T, Krems C, Moon K, Brombach C, Hoffmann I (2015) Food consumption of adults in Germany: results of the German National Nutrition Survey II based on diet history interviews. Br J Nutr 113(10):1603–1614

Hewett P, Ganser GH (2007) A comparison of several methods for analyzing censored data. Ann Occup Hyg 51:611–632

Hirvonen T, Sinkko H, Valsta L, Hannila ML, Pietinen P (2007) Development of a model for optimal food fortification: vitamin D among adults in Finland. Eur J Nutr 46(5):264–270

Hoffmann K, Boeing H, Dufour A, Volatier JL, Telman J, Virtanen M, Becker W, De Henauw S (2002) Estimating the distribution of usual dietary intake by short-term measurements. Eur J Clin Nutr 56:S53–S62
Hornung RW, Reed L (1990) Estimation of average concentration in the presence of nondetectable values. Appl Occup Environ Hyg 5:46–51
Howlett J (1996) ILSI Europe workshop on food additive intake: scientific assessment of the regulatory requirements in Europe 29–30 March 1995, Brussels summary report. Food Addit Contam 13(4):385–395
Husain A, Sawaya W, Al-Omair A, Al-Zenki S, Al-Amiri H, Ahmed N, Al-Sinan M (2006) Estimates of dietary exposure of children to artificial food colours in Kuwait. Food Addit Contam 23(3):245–251
ILSI (2009) Micronutrient landscape of Europe: comparison of intakes and methodologies with particular regard to higher consumption – summary report of a workshop held in April 2008 in Gubbio, Italy. ILSI report series. http://ilsi.org/mexico/wp-content/uploads/sites/29/2016/09/Micronutrient-Landscape-of-Europe.pdf. Accessed 8 Mar 2018
Infozentrum Schokolade (2016). https://schokoinfo.de/kakaowirtschaft-rohkakao-weltmarkt-kakao-schokoinfo.htmlAccessed 8 April 2017
IOM (1998) Dietary reference intakes: a risk assessment model for establishing upper intake levels for nutrients. National Academy Press, Washington, DC. https://www.nap.edu/catalog/6432/dietary-reference-intakes-a-risk-assessment-model-for-establishing-upper. Accessed 27 Oct 2019
IOM (1999) Dietary reference intakes for calcium, phosphorus, magnesium, vitamin D and fluoride. National Academy Press, Washington, DC. https://www.ncbi.nlm.nih.gov/books/NBK109825/. Accessed 27 Oct 2019
IOM (2000a) Dietary reference intakes for thiamin, riboflavin, niacin, vitamin B6, folate, vitamin B12, pantothenic acid, biotin and choline. National Academy Press, Washington, DC. https://www.ncbi.nlm.nih.gov/books/NBK114310/. Accessed 27 Oct 2019
IOM (2000b) Dietary reference intakes for vitamin C, vitamin E, selenium and carotenoids. National Academy Press, Washington, DC. https://www.nap.edu/catalog/9810/dietary-reference-intakes-for-vitamin-c-vitamin-e-selenium-and-carotenoids. Accessed 27 Oct 2019
IOM (2001) Dietary reference intakes for vitamin A, vitamin K, arsenic, boron, chromium, copper, iodine, iron, manganese, molybdenum, nickel, silicon, vanadium and zinc. National Academy Press, Washington, DC. https://www.ncbi.nlm.nih.gov/books/NBK222310/. Accessed 27 Oct 2019
Ireland JD, Møller A (2010) Review of international food classification and description. J Food Compos Anal 13:529–538
Jahn SD, Bullock WH, Ignacio JS (2015) A strategy for assessing and managing occupational exposures, 4th edn. American Industrial Hygiene Association, Fairview, VA, p 569
JECFA (2004) Evaluation of certain food additives and contaminants. Sixty-first report of the Joint FAO/WHO Expert Committee on Food Additives. WHO Technical Report Series 922. Geneva. http://apps.who.int/iris/bitstream/10665/42849/1/WHO_TRS_922.pdf. Accessed 27 Oct 2019
JECFA (2006) Combined compendium of food additive specifications – all specifications monographs from the 1st to the 65th meeting. FAO JECFA Monographs Series, No. 1 Volume 1–3, 2006. http://www.fao.org/3/a-a0691e.pdf. Accessed 27 Oct 2019
JECFA (2016) Safety evaluation of certain food additives and contaminants. Supplement 1: Non-dioxin-like polychlorinated biphenyls. Prepared by the eighteenth meeting of the Joint FAO/WHO Expert Committee on Food Additives (JECFA) WHO Food Additives Series: 71-S1. World Health organisation (WHO), Geneva. www.who.int/foodsafety/publications/jecfa-reports/en/
JECFA (2017) Evaluation of certain contaminants (eighty-third report of the Joint FAO/WHO Expert Committee on Food Additives). WHO Technical Report no. 1002: World Health organisation (WHO) and Food and Agriculture Organization of the United Nations (FAO),

Geneva, Rome. http://apps.who.int/iris/bitstream/10665/254893/1/9789241210027-eng.pdf. Accessed 27 Oct 2019

Juan WY, Yamini S, Britten P (2015) 38th National nutrient databank conference. Food intake patterns of self-identified vegetarians among the U.S. population, 2007–2010. Procedia Food Science 4:86–93

Kaplan EL, Meier P (1958) Non parametric estimation from incomplete observations. J Am Stat Assoc 53:457–481

Keith LH, Crummett W, Deegan J, Libby RA, Taylor JK, Wentler G (1983) Principles of environmental analysis. Anal Chem 55(14):2210–2218

Kennedy MC, Glass CR, Bokkers BGH, Hart ADM, Hamey P, Kruisselbrink JW, de Boer WJ, van der Voet H, Garthwaite D, van Klaveren JD (2015a) A European modeland case studies for aggregate exposure assessment of pesticides. Food Chem Toxicol 79:32–44. https://doi.org/10.1016/j.fct.2014.09.009

Kennedy MC, van der Voet H, Roelofs VJ, Roelofs W, Glass CR, de Boer WJ, Kruisselbrink JW, Hart ADM (2015b) New approaches to uncertainty analysis for use in aggregate and cumulative risk assessment of pesticides. Food Chem Toxicol 79:54–64. https://doi.org/10.1016/j.fct.2015.02.008

Keogh RH, White IR (2011) Allowing for never and episodic consumers when correcting for error in food record measurements of dietary intake. Biostatistics 12(4):624–636

Kienzler A, Bopp S, van der Linden S, Berggren E, Worth A (2016) Regulatory assessment of chemical mixtures: requirements, current approaches and future perspectives. Regul Toxicol Pharmacol 80:321–334

Kipnis V, Subar AF, Midthune D, Freedman LS, Ballard-Barbash R, Troiano RP, Bingham S, Schoeller DA, Schatzkin A, Carroll RJ (2003) Structure of dietary measurement error: results of the OPEN biomarker study. Am J Epidemiol 158(1):14–21

Kipnis V, Midthune D, Buckman DW, Dodd KW, Guenther PM, Krebs-Smith SM, Subar AF, Tooze JA, Carroll RJ, Freedman LS (2009) Modeling data with excess zeros and measurement error: application to evaluating relationships between episodically consumed foods and health outcomes. Biometrics 65(4):1003–1010

Klesges RC, Klesges LM, Brown G, Frank GC (1987) Validation of the 24-hour dietary recall in preschool children. J Am Diet Assoc 87(10):1383–1385

Kloosterman J, Fransen HP, de Stoppelaar J, Verhagen H, Rompelberg C (2007) Safe addition of vitamins and minerals to foods: setting maximum levels for fortification in the Netherlands. Eur J Nutr 46:220–229

Knüppel S, Clemens M, Conrad J, Gastell S, Michels KB, Leitzmann M, Krist L, Pischon T, Krause G, Ahrens W, Ebert N, Jöckel KH, Kluttig A, Obi N, Kaaks R, Lieb W, Schipf S, Brenner H, Heuer T, Harttig U, Linseisen J, Nöthlings U, Boeing H (2019) Design and characterization of dietary assessment in the German National Cohort. Eur J Clin Nutr 73:1480–1491. https://doi.org/10.1038/s41430-018-0383-8

Koleva II, van Beek TA, Soffers AE, Dusemund B, Rietjens IM (2012) Alkaloids in the human food chain – natural occurrence and possible adverse effects. Mol Nutr Food Res 56:30–52

Krems C, Bauch A, Götz A, Heuer T, Hild A, Möseneder J, Brombach C (2006) Methoden der Nationalen Verzehrsstudie II. Ernährungs-Umschau 53(6):44–50

Kroes R, Müller D, Lambe J, Löwik MRH, van Klaveren J, Kleiner J, Massey R, Mayer S, Urieta I, Verger P, Visconti A (2002) Assessment of intake from the diet. Food Chem Toxicol 40:327385. https://doi.org/10.1016/S0278-6915(01)00113-2

Kroke A, Klipstein-Grobusch K, Voss S, Moseneder J, Thielecke F, Noack R, Boeing H (1999) Validation of a self-administered food-frequency questionnaire administered in the European Prospective Investigation into Cancer and Nutrition (EPIC) Study: comparison of energy, protein, and macronutrient intakes estimated with the doubly labeled water, urinary nitrogen, and repeated 24-h dietary recall methods. Am J Clin Nutr 70(4):439–447

Langlais R (1996) Additive usage levels. Food Addit Contam 13(4):443–452

Laureano G, Torman V, Crispim S, Dekkers A, Camey S (2016) Comparison of the ISU, NCI, MSM, and SPADE methods for estimating usual intake: a simulation study of nutrients consumed daily. Nutrients 8(3):166

Lawrie CA, Rees NM (1996) The approach adopted in the UK for the estimation of the intake of food additives. Food Addit Contam 13(4):411–416

Leclercq C, Berardi D, Sorbillo MR, Lambe J (1999) Intake of saccharin, aspartame, acesulfame K and cyclamate in Italian teenagers: present levels and projections. Food Addit Contam 16 (3):99–109

Leclercq C, Arcella D, Turrini A (2000) Estimates of the theoretical maximum daily intake of erythorbic acid, gallates, butylated hydroxyanisole (BHA) and butylated hydroxytoluene (BHT) in Italy: a stepwise approach. Food Chem Toxicol 38(12):1075–1084

Leclercq C, Arcella D, Le Donne C, Piccinelli R, Sette S, Soggiu ME (2003) Stochastic modelling of human exposure to food chemicals and nutrients within the Montecarlo project: an exploration of the influence of brand loyalty and market share on intake estimates of intense sweeteners from sugar-free soft drinks. Toxicol Lett 140–141:443–457

Lee BQ, Khor SM (2015) 3-Chloropropane-1,2-diol (3-MCPD) in Soy Sauce: a review on the formation, reduction, and detection of this potential carcinogen. Compr Rev Food Sci Food Saf 14:48–66

LGL Bayern (Bayerisches Landesamt für Gesundheit und Lebensmittelsicherheit) (2016) Mutterkorn und Mutterkornalaloide in Getreide und Mehl. www.lgl.bayern.de/lebenmittel/chemie/schimmelpilze/mutterkornalkaloide/

Lim JH, Sisco P, Mudalige TK, Sanchez-Pomales G, Howard PC, Linder SW (2015) Detection and characterization of SiO2 and TiO2 nanostructures in dietary supplements. J Agric Food Chem 63:3144–3152

Lindtner O, Ehlscheid N, Berg K, Blume K, Dusemund B, Ehlers A, Niemann B, Rüdiger T, Heinemeyer G, Greiner M, Hallerbach B, Thömmes O, Thier S (2008) Event-related survey of high consumers of energy drinks. Federal Institute for Risk Assessment, ISSN 1614–3795 (Print) 1614–3841 (Online)

Lindtner O, Berg K, Blume K, Fiddicke U, Heinemeyer G (2013) The German approach to estimating dietary exposure using food monitoring data. In: Moy GG, Vannoort RW (eds) Total diet studies, Chapter 53, Springer, New York, NY, pp 521–530. https://doi.org/10.1007/978-1-4419-7689-5_53

Lineback DR, Coughlin JR, Stadler RH (2012) Acrylamide in foods: a review of the science and future considerations. Annu Rev Food Sci Technol 3:15–35

Linsinger TPJ, Chaudhry Q, Dehalu V, Delahaut P, Dudkiewicz A, Grombe R, von der Kammer F, Larsen EH, Legros S, Loeschner K, Peters R, Ramsch R, Roebben G, Tiede K, Weigel S (2013) Validation of methods for the detection and quantification of engineered nanoparticles in food. Food Chem 138:1959–1966

Lioy PJ, Smith KR (2013) Discussion of exposure science in the 21st century: a vision and a strategy. Environ Health Perspect 121:405–409

Liu B, Young H, Crowe FL, Benson VS, Spencer EA, Key TJ, Appleby PN, Beral V (2011) Development and evaluation of the Oxford WebQ, a low-cost, web-based method for assessment of previous 24 h dietary intakes in large-scale prospective studies. Public Health Nutr 14 (11):1998–2005

Livingstone MB, Robson PJ (2000) Measurement of dietary intake in children. Proc Nutr Soc 59 (2):279–293

Livingstone MB, Robson PJ, Wallace JM (2004) Issues in dietary intake assessment of children and adolescents. Br J Nutr 92(Suppl 2):S213–S222

Loeschner K, Navratilova J, Købler C, Mølhave K, Wagner S, von der Kammer F, Larsen EH (2013) Detection and characterization of silver nanoparticles in chicken meat by asymmetric flow field flow fractionation with detection by conventional or single particle ICP-MS. Anal Bioanal Chem 405:8185–8195

Loizzo A, Gatti GL, Macri A, Moretti G, Ortolani E, Palazzesi S (1984) Italian baby food containing diethylstilboestrol: three years later. Lancet 1(8384):1014–1015

Löwik MR (1996) Possible use of food consumption surveys to estimate exposure to additives. Food Addit Contam 13(4):427–441

Mancini FR, Paul D, Gauvreau J, Volatier JL, Vin K, Hulin M (2015) Dietary exposure to benzoates (E210-E213), parabens (E214-E219), nitrites (E249-E250), nitrates (E251-E252), BHA (E320), BHT (E321) and aspartame (E951) in children less than 3 years old in France. Food Addit Contam Part A Chem Anal Control Expo Risk Assess 32(3):293–306

Martyn DM, Nugent AP, McNulty BA, O'Reilly E, Tlustos C, Walton J, Flynn A, Gibney MJ (2016) Dietary intake of four artificial sweeteners by Irish pre-school children. Food Addit Contam Part A Chem Anal Control Expo Risk Assess 33(4):592–602

Mensink GBM, Thamm M, Haas K (1999) Die Ernährung in Deutschland 1998. Gesundheitswesen 61(Sonderheft 2):S200–S206

Mensink GBM, Haftenberger M, Thamm M (2001) Validity of DISHES 98, a computerised dietary history interview: energy and macronutrient intake. Eur J Clin Nutr 55(6):409–417

Mensink GBM, Bauch A, Vohmann C, Stahl A, Six J, Kohler S, Fischer J, Heseker H (2007) EsKiMo – Das Ernährungsmodul im Kinder- und Jugendgesundheitssurvey (KiGGS). Bundesgesundheitsbl – Gesundheitsforsch – Gesundheitsschutz 50:902–908

Merten C, Ferrari P, Bakker M, Boss A, Hearty A, Leclercq C, Lindtner O, Tlustos C, Verger P, Volatier JL, Arcella D (2011) Methodological characteristics of the national dietary surveys carried out in the European Union as included in the European Food Safety Authority (EFSA). Food Addit Contam Part A Chem Anal Control Expo Risk Assess 28(8):975–995

Mischek D, Krapfenbauer-Cermak C (2012) Exposure assessment of food preservatives (sulphites, benzoic and sorbic acid) in Austria. Food Addit Contam Part A Chem Anal Control Expo Risk Assess 29(3):371–382

Møller A, Unwin ID, Ireland J, Roe MA, Becker W, Colombani P (2008) The EuroFIR Thesauri 2008. EuroFIR Technical Report D1.8.22, Danish Food Information, Denmark

Moro S, Chipman JK, Wegener JW, Hamberger C, Dekant W, Mally A (2012) Furan in heat-treated foods: formation, exposure, toxicity, and aspects of risk assessment. Mol Nutr Food Res 56 (8):1197–1211

Moy GG, Vannoort RW (2013) Total diet studies. Springer Science+Business Media, New York. https://doi.org/10.1007/978-1-4419-7689-5

MRI (2008) Nationale Verzehrs Studie II, Ergebnisbericht, Teil 2. Die bundesweite Befragung zur Ernährung von Jugendlichen und Erwachsenen. MRI, Karlsruhe

MRI (2010) Bundeslebensmittelschlüssel (BLS): Version 3.01 (German Nutrient Database: Version 3.01) MRI, Karlsruhe. http://www.blsdb.de (11.08.2011)

MRI (2012) NEMONIT MRI, Karlsruhe. https://www.mri.bund.de/de/institute/ernaehrungsverhalten/forschungsprojekte/nemonit. Accessed 8 Mar 2018

MRI (2015) Nationale Verzehrsstudie II. Ergänzungsband zum Ergebnisbericht, Teil 1. Ausgewählte Ergebnisse nach Schichtindex. MRI, Karlsruhe. http://www.was-esse-ich.de/uploads/media/NVS_II_Abschlussbericht_Teil_1_mit_Ergaenzungsbericht.pdf. Accessed 27 Oct 2019

Mulder PPJ, Sánchez PL, These A, Preiss-Weigert A, Castellari M (2015) Occurrence of Pyrrolizidine alkaloids in food. EFSA Supporting publications EN-859:1–114

Mulder PPJ, de Nijs M, Castellari M et al. (2016) Occurrence of tropane alkaloids in food. EFSA Supporting publications 13:EN-1140

Mulhausen J, Damiano J (1998) A strategy for assessing and managing occupational exposures, 2nd edn. American Industrial Hygiene Association, Fairview, VA, p 349

NCI (2017) Paper-based DHQ II & C-DHQ II Forms. DHQ II: Past year, without portion size, Bethesda. https://epi.grants.cancer.gov/dhq2/forms/. Accessed 28 April 2017

NIPH/SZU (lead), EVIRA, BfR, MATIS, INSA (2015) TDS Exposure deliverable D9.5 – Report on the key elements of realised national To be requested at http://www.tds-exposure.eu/project-objectives-0/contact-us/. Accessed 8 Mar 2018

Noethlings U, Hoffmann K, Bergmann MM, Boeing H (2003) Portion size adds limited information on variance in food intake of participants in the EPIC-Potsdam study. J Nutr 133(2):510–515

Nordqvist C (2013) Energy drinks health hazards for adolescents. http://www.medicalnewstoday.com/articles/255925.php

NTP/NIEHS (2004) Bitter orange (*Citrus aurantium* var. amara) extracts and constituents (±)-p-synephrine [CAS No. 94-07-5] and (±)-p-octopamine [CAS No. 104-14-3]. Review of toxicological literature. National Toxicology Program (NTP), National Institute of Environmental Health Sciences (NIEHS), National Institutes of Health, U.S. Department of Health and Human Services, Research Triangle Park, North Carolina 2004, VIII + 73 p. https://ntp.niehs.nih.gov/ntp/htdocs/chem_background/exsumpdf/bitterorange_508.pdf. Accessed 8 Mar 2018

Nusser SM, Carriquiry AL, Dodd KW, Fuller WA (1996) A semiparametric transformation approach to estimating usual daily intake distributions. J Am Stat Assoc 91:1440–1449

Nusser SM, Fuller WA, Guenther PM (1997) Estimating usual dietary intake distributions: adjusting for measurement error and nonnormality in 24-hour food intake data. In: Lyberg L, Biemer P, Collins M, DeLeeuw E, Dippo C, Schwartz N, Trewin D (eds) Survey measurement and process quality. Wiley, New York, pp 689–709

Nutriscan (1992) An evaluation of the methodologies for the estimation of intake of food additives and contaminants in the European Community. Nutriscan, Dublin

O'Sullivan AJ, Pigat S, O'Mahony C, Gibney MJ, McKevitt AI (2016) Probabilistic modelling to assess exposure to three artificial sweeteners of young Irish patients aged 1-3 years with PKU and CMPA. Food Addit Contam Part A Chem Anal Control Expo Risk Assess 33 (11):1660–1671

OECD (2007a) Test No. 501: Metabolism in crops. OECD publishing. http://www.oecd-ilibrary.org/environment/test-no-501-metabolism-in-crops_9789264061835-en. Accessed 8 Mar 2018

OECD (2007b) Test No. 502: Metabolism in rotational crops. OECD publishing. http://www.oecd-ilibrary.org/content/book/9789264061859-en. Accessed 8 Mar 2018

OECD (2007c) Test No. 503: Metabolism in livestock. OECD publishing. http://www.oecd-ilibrary.org/content/book/9789264061873-en. Accessed 8 Mar 2018

OECD (2007d) Test No. 504: Residues in rotational crops (limited field studies). OECD publishing. http://www.oecd-ilibrary.org/content/book/9789264013384-en. Accessed 8 Mar 2018

OECD (2007e) Test No. 505: Residues in livestock. OECD publishing. http://www.oecd-ilibrary.org/content/book/9789264061903-en. Accessed 8 Mar 2018

OECD (2007f) Test No. 507: Nature of the pesticide residues in processed commodities - high temperature hydrolysis. OECD publishing. http://www.oecd-ilibrary.org/content/book/9789264067431-en. Accessed 8 Mar 2018

OECD (2007g) Test No. 508: Magnitude of the pesticide residues in processed commodities. OECD publishing. http://www.oecd-ilibrary.org/content/book/9789264067622-en. Accessed 8 Mar 2018

OECD (2007h) Test No. 509: Crop field trial. OECD publishing. http://www.oecd-ilibrary.org/content/book/9789264076457-en. Accessed 8 Mar 2018

OECD (2011) OECD MRL Calculator user guide: 69. http://www.oecd.org/env/mrl-calculator-users-guide-and-white-paper-9789264221567-en.htm. Accessed 8 Mar 2018

OIE (2017) Bovine spongiform encephalopathy (BSE). http://www.oie.int/fileadmin/Home/eng/Media_Center/docs/pdf/Disease_cards/BSE-EN.pdf. Accessed 8 Feb 2018

Oldring PKT, O'Mahony C, Dixon J, Vints M, Mehegan J, Dequatre C, Castle L (2014) Development of a new modelling tool (FACET) to assess exposure to chemical migrants from food packaging. Food Addit Contam Part A Chem Anal Control Expo Risk Assess 31:444–465. https://doi.org/10.1080/19440049.2013.862348

Orlich MJ, Jaceldo-Siegl K, Sabaté J, Fan J, Singh PN, Fraser GE (2014) Patterns of food consumption among vegetarians and non-vegetarians. Br J Nutr 112(10):1644–1653

Palazzesi S, Brambilla G, Macrì A, Loizzo A (1995) Relay activity of 17-beta oestradiol and diethylstilbestrol in a mouse-rat system. Food Addit Contam 12(6):751–757

Parr MK, Blokland MH, Liebetrau F, Schmidt AH, Meijer T, Stanic M, Kwiatkowska D, Waraksa E, Sterk SS (2017) Distinction of clenbuterol intake from drug or contaminated food of animal origin in a controlled administration trial – the potential of enantiomeric separation for doping control analysis. Food Addit Contam Part A Chem Anal Control Expo Risk Assess 34:525–535

Penttilä PL (1996) Estimation of food additive intake. Nordic approach. Food Addit Contam 13 (4):421–426

Peters R, Kramer E, Oomen AG, Rivera ZE, Oegema G, Tromp PC, Fokkink R, Rietveld A, Marvin HJ, Weigel S, Peijnenburg AA, Bouwmeester H (2012) Presence of nanosized silica during in vitro digestion of foods containing silica as a food additive. ACS Nano 6:2441–2451

Peters RJ, van Bemmel G, Herrera-Rivera Z, Helsper HP, Marvin HJ, Weigel S, Tromp PC, Oomen AG, Rietveld AG, Bouwmeester H (2014) Characterization of titanium dioxide nanoparticles in food products: analytical methods to define nanoparticles. J Agric Food Chem 62:6285–6293

Petersen B, Petersen J, Barraj LM, Muenz LR, Harrison SL (1994) An alternative approach to dietary exposure assessment. Risk Anal 14(6):913–916. https://doi.org/10.1111/j.1539-6924.1994.tb00060.x

PHE, FSA (2014) Results of the National Diet and Nutrition Survey (NDNS) rolling programme for 2008 and 2009 to 2011 and 2012. National Diet and Nutrition Survey. Food and drink diary. https://www.gov.uk/government/statistics/national-diet-and-nutrition-survey-results-from-years-1-to-4-combined-of-the-rolling-programme-for-2008-and-2009-to-2011-and-2012. Accessed 8 Mar 2018

Pietraforte D, Brambilla G, Camerini S, Scorza G, Peri L, Loizzo A, Crescenzi M, Minetti M (2008) Formation of an adduct by clenbuterol, a beta-adrenoceptor agonist drug, and serum albumin in human saliva at the acidic pH of the stomach: evidence for an aryl radical-based process. Free Radic Biol Med 45:124–135

Plathe KL, von der Kammer F, Hassellov M, Moore JN, Murayama M, Hofmann T, Hochella MF Jr (2013) The role of nanominerals and mineral nanoparticles in the transport of toxic trace metals: field-flow fractionation and analytical TEM analyses after nanoparticle isolation and density separation. Geochim Cosmochim Acta 102:213–226

Pöting A, Schauzu M, Niemann B and Schumann A (2014) Risk assessment of novel food and genetically modified food and feed. Regul Toxicol 827–848

Powell JJ, Ainley CC, Harvey RS, Mason IM, Kendall MD, Sankey EA, Dhillon AP, Thompson RP (1996) Characterisation of inorganic microparticles in pigment cells of human gut associated lymphoid tissue. Gut 38:390–395

Powell JJ, Faria N, Thomas-McKay E, Pele LC (2010) Origin and fate of dietary nanoparticles and microparticles in the gastrointestinal tract. J Autoimmun 34:J226–J233

Powell JJ, Thomas-McKay E, Thoree V, Robertson J, Hewitt RE, Skepper JN, Brown A, Hernandez-Garrido JC, Midgley PA, Gomez-Morilla I, Grime GW, Kirkby KJ, Mabbott NA, Donaldson DS, Williams IR, Rios D, Girardin SE, Haas CT, Bruggraber SF, Laman JD, Tanriver Y, Lombardi G, Lechler R, Thompson RP, Pele LC (2015) An endogenous nanomineral chaperones luminal antigen and peptidoglycan to intestinal immune cells. Nat Nanotechnol 10:361–369

Public Health England (2016) National Diet and Nutrition Survey: results from years 1–4 (combined) of the rolling programme (2008)/2009–2011/12) executive summary. https://www.gov.uk/government/statistics/national-diet-and-nutrition-survey-results-from-years-1-to-4-combined-of-the-rolling-programme-for-2008-and-2009-to-2011-and-2012. Accessed 8 Mar 2018

Rasmussen SE, Andersen NL, Dragsted LO, Larsen JC (2006) A safe strategy for addition of vitamins and minerals to foods. Eur J Nutr 45:123–135

Rasmussen K, Mech A, Mast J, De Temmerman P-J, Waegeneers N, Van Steen F, Pizzolon JC, De Temmerman L, Van Doren E, Jensen KA, Birkedal R, Levin M, Nielsen SH, Koponen IK, Clausen PA, Kembouche Y, Thieriet N, Spalla O, Giuot C, Rousset D, Witschger O, Bau S, Bianchi B, Shivachev B, Gilliland D, Pianella F, Ceccone G, Cotogno G, Rauscher H, Gibson N, Stamm H (2013) Synthetic amorphous silicon dioxide (NM-200, NM-201,

NM-202, NM-203, NM-204): characterisation and physico-chemical properties. In: JRC repository: NM-series of representative manufactured nanomaterials: European Union, pp 200

Richardson DP (2007) Risk management of vitamins and minerals: a risk categorisation model for the setting of maximum levels in food supplements and fortified foods. Food Sci Technol Bull Funct Foods 4:51–66

Richter A, Sieke C, Reich H, Ossendorp BC, Breysse N, Lutze J, Mahieu K, Margerison S, Rietveld A, Sarda X, Vial G, van der Velde-Koerts T (2018) Setting the stage for the review 5 exposure to substances by use of consumer products of the international estimate of short-term intake (IESTI) equation. J Environ Sci Health B 53(6):343–351. https://doi.org/10.1080/03601234.2018.1439807

Roe M, Finglas PM Lindroos AK, Castanheira I, Giertlová A, Westenbrink S, and Beernaert H (2013) Final list of certified FCDB compiler organizations and future plans for continuation of the certification scheme. EuroFIR Technical Report 1.7

Rompelberg C, Heringa MB, van Donkersgoed G, Drijvers J, Roos A, Westenbrink S, Peters R, van Bemmel G, Brand W, Oomen AG (2016) Oral intake of added titanium dioxide and its nanofraction from food products, food supplements and toothpaste by the Dutch population. Nanotoxicology 10:1404–1414

Rossi M, Cubadda F, Dini L, Terranova ML, Aureli F, Sorbo A, Passeri D (2014) Scientific basis of nanotechnology, implications for the food sector and future trends. Trends Food Sci Technol 40:127–148

Rothausen BW, Matthiessen J, Hoppe C, Brockhoff PB, Andersen LF, Tetens I (2012) Differences in Danish childrens diet quality on weekdays v. weekend days. Public Health Nutr 15 (9):1653–1660

Ruprich J, Rehurkova I, Boon PE, Svensson K, Moussavian S, Van der Voet H, Bosgra S, van Klaveren JD, Busk L (2009) Probabilistic modelling of exposure doses and implications for health risk characterization: glycoalkaloids from potatoes. Food Chem Toxicol 47:2899–2905. https://doi.org/10.1016/j.fct.2009.03.008

Saveyn, H., & Eder, P (2014) End-of-waste criteria for biodegradable waste subjected to biological treatment (compost & digestate): technical proposals. European Commission, Joint Research Centre – Institute for Prospective Technological Studies of Seville (E), EUR Number: 26425 EN

SCENIHR (2010) Scientific basis for the definition of the term "nanomaterial. http://ec.europa.eu/health/scientific_committees/emerging/docs/scenihr_o_032.pdf. Accessed 27 Oct 2019

SCF (1998) Opinion of the Scientific Committee of Food on the applicability of the ADI (acceptable daily intake) for food additives to infants. 17 Sept 1998. https://ec.europa.eu/food/sites/food/files/safety/docs/sci-com_scf_out13_en.pdf. Accessed 27 Oct 2019

SCF (2001) Guidance on submissions for food additive evaluations by the Scientific Committee on Food. 12 July 2001. https://ec.europa.eu/food/sites/food/files/safety/docs/sci-com_scf_out98_en.pdf. Accessed 27 Oct 2019

Schmoyer RL, Beauchamp JJ, Brandt CC, Hoffman FO (1996) Difficulties with the Log-normal model in mean estimation and testing. Environ Ecol Stat 3:81–97

Schneider K, Schwarz MA, Lindtner O, Blume K, Heinemeyer G (2014) Lead exposure from food: the German LExUKon project. Food Addit Contam Part A Chem Anal Control Expo Risk Assess 31(6):1052–1063

Scholz R, Herrmann M, Michalski BJ (2017) Compilation of processing factors and evaluation of quality controlled data of food processing studies. J Verbr Lebensm 12(1):3–14

Schroeter A, Sommerfeld G, Klein H, Hübner D (1999) Warenkorb für das Lebensmittelmonitoring in der Bundesrepublik Deutschland. Bundesgesundheitsbl Gesundheitsforsch – Gesundheitsschutz 42:77–84

Schug TT, Janesick A, Blumberg B, Heindel JJ (2011) Endocrine disrupting chemicals and disease susceptibility. J Steroid Biochem Mol Biol 127(3–5):204–215

Schwarz MA, Lindtner O, Blume K, Heinemeyer G, Schneider K (2014) Cadmium exposure from food: the German LExUKon project. Food Addit Contam Part A Chem Anal Control Expo Risk Assess 31(6):1038–1051. https://doi.org/10.1080/19440049.2014.905711

Scotter MJ (2011) Methods for the determination of European Union-permitted added natural colours in foods: a review. Food Addit Contam Part A Chem Anal Control Expo Risk Assess 28(5):527–596

Sexton K, Olden K, Johnson BL (1993) Environmental justice: the central role of research in establishing a credible scientific basis for informed decision making. Toxicol Ind Health 9:685–727

She N (1997) Analyzing censored water quality data using a non-parametric approach. J Am Water Resour Assoc 33:615–624

Sieke C, Lindtner O, Banasiak U (2008) Nationales Monitoring – Abschätzung der Verbraucherexposition Teil 1. Deutsche Lebensmittel-Rundschau, 104. Jahrgang, Heft 6: 271–279

Silano V, Coppens P, Larranaga-Guetaria A, Minghetti P, Roth-Ehrang R (2011) Regulations applicable to plant food supplements and related products in the European Union. Food Funct 2:710

Six BL, Schap TE, Zhu FM, Mariappan A, Bosch M, Delp EJ, Ebert DS, Kerr DA, Boushey CJ (2010) Evidence-based development of a mobile telephone food record. J Am Diet Assoc 110 (1):74–79

Sjöberg A, Hulthen L (2004) Assessment of habitual meal pattern and intake of foods, energy and nutrients in Swedish adolescent girls: comparison of diet historywith 7-day record. Eur J Clin Nutr 58(8):1181–1189

Slimani N, Deharveng G, Charrondiere RU, van Kappel AL, Ocke MC, Lagiou A, van Liere M, Agudo A, Pala V, Brandstetter B, Andren C, Stripp C, van Staveren WA, Riboli E (1999) Structure of the standardized computerized 24-h diet recall interview used as reference method in the 22 centers participating in the EPIC project. Comput Methods Prog Biomed 58 (3):251–266

Slimani N, Ferrari P, Ocke M, Welch A, Boeing H, Liere M, Pala V, Amiano P, Lagiou A, Mattisson I, Stripp C, Engeset D, Charrondiere R, Buzzard M, Staveren W, Riboli E (2000) Standardization of the 24-hour diet recall calibration method used in the european prospective investigation into cancer and nutrition (EPIC): general concepts and preliminary results. Eur J Clin Nutr 54(12):900–917

Slimani N, Freisling H, Illner A-K, Huybrechts I (2015) Methods to determine dietary intake. In: Lovegrove JA, Hodson L, Sharma S, Lanham-New SA (eds) Nutrition research methodologies. Wiley, Sussex, pp 48–70

Sobo EJ, Rock CL, Neuhouser ML, Maciel TL, Neumark-Sztainer D (2000) Caretaker-child interaction during childrens 24-hour dietary recalls: who contributes what to the recall record? J Am Diet Assoc 100(4):428–433

Solecki R, Davies L, Dellarco V, Dewhurst I, van Raaij M, Tritscher A (2005) Guidance on setting of acute reference dose (ARfD) for pesticides. Food Chem Toxicol 43(11):1569–1593

Soubra L, Sarkis D, Hilan C, Verger P (2007) Dietary exposure of children and teenagers to benzoates, sulphites, butylhydroxyanisol (BHA) and butylhydroxytoluen (BHT) in Beirut (Lebanon). Regul Toxicol Pharmacol 47(1):68–77

Souverein OW, Dekkers ALM, Geelen A, Haubrock J, de Vries JH, Ocké MC, Harttig U, Boeing H, van't Veer P (2011) Comparing four methods to estimate usual intake distributions. Eur J Clin Nutr 65:S92–S101

Speijers G, Bottex B, Dusemund D, Lugasi A, Toth J, Amberg-Müller J, Galli C, Silano V, Rietjens IMCM (2010) Safety assessment of botanicals and botanical preparations used as food supplements; emerging issues when testing a proposed tiered approach. Mol Nutr Food Res 54:175–185

Sprong C, Bakker M, Niekerk M, Vennemann F (2015) Exposure assessment of the food additive titanium dioxide (E 171) with use levels provided by the industry. R IVM Letter report 2015-0195. National Instititute for Public Health and the Environment (RIVM), Bilthoven. www.rivm.nl. Accessed 8 Mar 2018

Statistisches Bundesamt (2012) Geburten in Deutschland, Ausgabe 2012

Stein B, Michalski B, Martin S, Pfeil R, Ritz V, Solecki R (2014) Human health risk assessment from combined exposure in the framework of plant protection products and biocidal products. J Verbr Lebensm 9(4):367–376

Straßburg A (2010) Ernährungserhebungen Methoden und Instrumente. Ernahrungs-Umschau 57 (8):422–430

Straßburg A, Eisinger-Watzl M, Krems C, Roth A, Hoffmann I (2019) Comparison of food consumption and nutrient intake assessed with three dietary assessment methods: results of the German National Nutrition Survey II. Eur J Nutr 58(1):193–210. https://doi.org/10.1007/s00394-017-1583-z

Subar AF, Dodd KW, Guenther PM, Kipnis V, Midthune D, McDowell M, Tooze JA, Freedman LS, Krebs-Smith SM (2006) The food propensity questionnaire: concept, development, and validation for use as a covariate in a model to estimate usual food intake. J Am Diet Assoc 106 (10):1556–1563

Subar AF, Kirkpatrick SI, Mittl B, Zimmerman TP, Thompson FE, Bingley C, Willis G, Islam NG, Baranowski T, McNutt S, Potischman N (2012) The automated self-administered 24-hour dietary recall (ASA24): a resource for researchers, clinicians, and educators from the National Cancer Institute. J Acad Nutr Diet 112(8):1134–1137

Sy MM, Feinberg M, Verger P, Tangui B, Clémençon S, Crépet A (2013) New approach for the assessment of cluster diets. Food Chem Toxicol 52:180–187. https://doi.org/10.1016/j.fct.2012.11.005

Tareke E, Ryberg P, Karlsson P, Eriksson S, Törnqvist M (2002) Analysis of acrylamide: a carcinogen formed in heated foodstuffs. J Agric Food Chem 50:4998–5006

Tennant DR (2008) Screening potential intakes of colour additives used in non-alcoholic beverages. Food Chem Toxicol 46(6):1985–1993

Tennant DR (2016) Comprehensive European dietary exposure model (CEDEM) for food additives. Food Addit Contam Part A Chem Anal Control Expo Risk Assess 33(5):772–781

Thompson FE, Byers T (1994) Dietary assessment resource manual. J Nutr 124(11):S2245–S2317

Thompson FE, Subar AF (2013) Dietary assessment methodology. In: Coulston AM, Boushey CJ, Ferruzzi MG (eds) Nutrition in the prevention and treatment of disease, 3rd edn. Academic Press, San Diego, CA, pp 3–46

Thompson FE, Kirkpatrick SI, Subar AF, Reedy J, Schap TE, Wilson MM, Krebs-Smith SM (2015) The National Cancer Institute's Dietary Assessment Primer: a resource for diet research. J Acad Nutr Diet 115(12):1986–1995

Tokudome Y, Imaeda N, Nagaya T, Ikeda M, Fujiwara N, Sato J, Kuriki K, Kikuchi S, Maki S, Tokudome S (2002) Daily, weekly, seasonal, within- and between-individual variation in nutrient intake according to four season consecutive 7 day weighed diet records in Japanese female dietitians. J Epidemiol 12(2):85–92

Tooze JA, Midthune D, Dodd KW, Freedman LS, Krebs-Smith SM, Subar AM, Guenther PM, Carroll RJ, Kipnis V (2006) A new statistical method for estimating the usual intake of episodically consumed foods with application to their distribution. J Am Diet Assoc 106:1575–1587. https://doi.org/10.1016/j.jada.2006.07.003

Tooze JA, Kipnis V, Buckman DW, Carroll RJ, Freedman LS, Guenther PM, Krebs-Smith SM, Subar AF, Dodd KW (2010) A mixed-effects model approach for estimating the distribution of usual intake of nutrients: the NCI method. Stat Med 29(27):2857–2868

Touvier M, Kesse-Guyot E, Mejean C, Pollet C, Malon A, Castetbon K, Hercberg S (2011) Comparison between an interactive web-based self-administered 24 h dietary record and an interview by a dietitian for large-scale epidemiological studies. Br J Nutr 105(7):1055–1064

Tressou J (2006) Nonparametric modeling of the left censorship of analytical data in food risk assessment. J Am Stat Assoc 101(476):1377–1386

Tressou J, Leblanc JCH, Feinberg M, Bertail P (2004) Statistical methodology to evaluate food exposure to a contaminant and influence of sanitary limits: application to Ochratoxin A. Regul Toxicol Pharmacol 40:252–263

UNECE (2012) Risk management in regulatory frameworks: towards a better management of risks United Nations. New York & Geneva, 2012. http://www.unece.org/fileadmin/DAM/trade/Publications/WP6_ECE_TRADE_390.pdf. Accessed 8 Mar 2018

USDA (United States Department of Agriculture) (2017) 5-step multiple-pass approach of the Automated Multiple-Pass-Method (AMPM). https://www.ars.usda.gov/northeast-area/beltsville-md-bhnrc/beltsville-human-nutrition-research-center/food-surveys-research-group/docs/ampm-features/. Accessed 28 April 2017

USDA-ARS (US-Department of Agriculture) (2016) AMPM features. http://www.ars.usda.gov/Services/docs.htm?docid=7711. August 2016

US-EPA (2000) Assigning values to non-detected/non-quantified pesticide residues in human health food exposure assessments. Office of Pesticide Programs, U.S. Environmental Protection Agency, Washington, DC, 33pp. https://archive.epa.gov/pesticides/trac/web/pdf/trac3b012.pdf. Accessed 8 Mar 2018

US-EPA (2002) Guidance on Cumulative Risk Assessment of Pesticide Chemicals that have a common mechanism of toxicity. U.S. Environmental Protection Agency (US-EPA). Washington, DC, 20460, USA. https://www.epa.gov/sites/production/files/2015-07/documents/guidance_on_common_mechanism.pdf. Accessed 8 Mar 2018

Valsta L, Ocke M, Lindtner O (2017) Towards a harmonised food consumption survey methodology for exposure assessment. Food Safety Assessment of Pesticide Residues. www.worldscientific.com. Accessed 8 Mar 2018

van Boekel M, Fogliano V, Pellegrini N, Stanton C, Scholz G, Lalljie S et al (2010) A review on the beneficial aspects of food processing. Mol Nutr Food Res 54:1215–1247

van der Velde-Koerts T, Margerison S, Breysse N, Lutze J, Mahieu K, Reich H, Rietveld A, Sarda X, Sieke C, Vial G, Ossendorp BC (2018a) Impact of proposed changes in IESTI equations for short-term dietary exposure to pesticides from Australian and Codex perspective. J Environ Sci Health B 53(6):366–379. https://doi.org/10.1080/03601234.2018.1439812

van der Velde-Koerts T, Breysse N, Pattingre L, Hamey PY, Lutze J, Mahieu K, Margerison S, Ossendorp BC, Reich H, Rietveld A, Sarda X, Vial G, Sieke C (2018b) Effect of individual parameter changes on the outcome of the estimated short-term dietary exposure to pesticides. J Environ Sci Health B 53(6):380–393. https://doi.org/10.1080/03601234.2018.1439814

van der Voet H, Kruisselbrink J, de Boer WJ, Boon PE (2014) Model-Then-Add. Usual intake modelling of multimodal intake distributions. RIVM Letter report 090133001/2014. National Institute for Public Health and the Environment (RIVM), Bilthoven. http://rivm.openrepository.com/rivm/bitstream/10029/314361/3/090133001.pdf. Accessed 8 Mar 2018

van der Voet H, de Boer WJ, Kruisselbrink JW, Goedhart PW, van der Heijden GWAM, Kennedy MC, Boon PE, van Klaveren JD (2015) The MCRA model for probabilistic single-compound and cumulative risk assessment of pesticides. Food Chem Toxicol 79:5–12. https://doi.org/10.1016/j.fct.2014.10.014

van Kesteren PCE, Cubadda F, Bouwmeester H, van Eijkeren JCH, Dekkers S, de Jong WH, Oomen AG (2015) Novel insights into the risk assessment of the nanomaterial synthetic amorphous silica, additive E551, in food. Nanotoxicology 9:442–452

van Klaveren JD, Goedhart P, Wapperom D, van der Voet, H et al. (2012) A European tool for usual intake distribution estimation in relation to data collection by EFSA – External Scientific Report. QEFSA Supporting Publications 2012:EN-300:42. https://doi.org/10.2903/sp.efsa.2012.EN-300. http://onlinelibrary.wiley.com/doi/10.2903/sp.efsa.2012.EN-300/abstract. Accessed 9 Mar 2018

van Klaveren JD, Kennedy MC, Moretto A, Verbeke W, van der Voet H, Boon PE (2015) The ACROPOLIS project: Its aims, achievements, and way forward. Food Chem Toxicol 79:1–4. https://doi.org/10.1016/j.fct.2015.03.006

van Liere MJ, Lucas F, Clavel F, Slimani N, Villeminot S (1997) Relative validity and reproducibility of a French dietary history questionnaire. Int J Epidemiol 26:S128–S136

Van Loco J, Vandevijvere S, Cimenci O, Vinkx C, Goscinny S (2015) Dietary exposure of the Belgian adult population to 70 food additives with numerical ADI. Food Control 54:86–94 van Staveren WA, Ocké MC, de Vries JHM (2012) Estimation of dietary intake. In: Erdman JW, Macdonald IA, Zeisel SH (eds) Present knowledge in nutrition, 10th edn. Wiley-Blackwell, Ames, pp 1012–1026

Vanden Bilcke C (2002) The Stockholm convention on persistent organic pollutants. Rev Eur Commun Int Environ Law 11(3):328–342

Vapnek J, Spreij M (2005) Perspectives and guidelines on food legislation, with a new model food law. FAO legislative study 87. Food and Agriculture Organization of the United Nation, Rome (I) 2005 ISBN 92-5-105431-2. New York, Geneva 2012

Verger P (1996) One example of utilization of the French approach. Food Addit Contam 13 (4):417–419

Verger P, Garnier-Sagne I, Leblanc JC (1999) Identification of risk groups for intake of food chemicals. Regul Toxicol Pharmacol 30(2):S103–S108

Verkaik-Kloosterman J, Dodd KW, Dekkers AL, van't Veer P, Ocké MC (2011) A three-part, mixed-effects model to estimate the habitual total vitamin D intake distribution from food and dietary supplements in Dutch young children. J Nutr 141:2055–2063. https://doi.org/10.3945/jn.111.142398

Verleysen E, Van Doren E, Waegeneers N, De Temmerman PJ, Abi Daoud Francisco M, Mast J (2015) TEM and SP-ICP-MS analysis of the release of silver nanoparticles from decoration of pastry. J Agric Food Chem 63:3570–3578

Verly E Jr, Fisberg RM, Marchioni DML (2012) Is the food frequency consumption essential as covariate to estimate usual intake of episodically consumed foods? Eur J Clin Nutr 66 (11):1254–1258

Verly E Jr, Oliveira DCRS, Fisberg RM, Marchioni DML (2016) Performance of statistical methods to correct food intake distribution: comparison between observed and estimated usual intake. Br J Nutr 116(05):897–903

Vin K, Connolly A, McCaffrey T, McKevitt A, O'Mahony C, Prieto M, Tennant D, Hearty A, Volatier JL (2013) Estimation of the dietary intake of 13 priority additives in France, Italy, the UK and Ireland as part of the FACET project. Food Addit Contam Part A Chem Anal Control Expo Risk Assess 30(12):2050–2080

Von Tungeln LS, Walker NJ, Olson GR, Mendoza MC, Felton RP, Thorn BT et al (2017) Low dose assessment of the carcinogenicity of furan in male F344/N Nctr rats in a 2-year gavage study. Food Chem Toxicol 99:170–181

Wagstaffe PJ (1996) The assessment of food additive usage and consumption: the Commission perspective. Food Addit Contam 13(4):397–403

Weir A, Westerhoff P, Fabricius L, Hristovski K, von Goetz N (2012) Titanium dioxide nanoparticles in food and personal care products. Environ Sci Technol 46:2242–2250

WHO (1957) General principles governing the use of food additives. First report of the Joint FAO/WHO Expert Committee on Food Additives. FAO Nutrition Meetings Report Series, No. 15, 1957. WHO Technical Report Series, No. 129. http://whqlibdoc.who.int/trs/WHO_TRS_129.pdf. Accessed 27 Oct 2019

WHO (1958) Procedures for the testing of intentional food additives to establish their safety for use (Second report of the Joint FAO/WHO Expert Committee on Food Additives). FAO Nutrition Meetings Report Series, No. 17, 1958. WHO Technical Report Series, No. 144, 1958. http://whqlibdoc.who.int/trs/WHO_TRS_144.pdf. Accessed 27 Oct 2019

WHO (1978) Evaluation of certain food additives. Twenty-first report of the Joint FAO/WHO Expert Committee on Food Additives. WHO Technical Report Series, No. 617. World Health Organization, Geneva. http://whqlibdoc.who.int/trs/WHO_TRS_617.pdf. Accessed 27 Oct 2019

WHO (1985) Guidelines for the study of dietary intakes of chemical contaminants. Global Environmental Monitoring System (GEMS). World Health Organization, Geneva. WHO Offset Publication No. 87. http://apps.who.int/iris/handle/10665/39255. Accessed 27 Oct 2019

WHO (1987) Principles for the safety assessment of food additives and contaminants in food. International Programme on Chemical Safety. Environmental Health Criteria 70. World Health

Organization, Geneva. http://www.inchem.org/documents/ehc/ehc/ehc70.htm. Accessed 27 Oct 2019

WHO (1989) Guidelines for predicting dietary intake of pesticides residues GEMS/Food (E, S, F), World Health Organization, Geneva. http://www.who.int/foodsafety/publications/chem/en/pesticide_en.pdf. Accessed 27 Oct 2019

WHO (1995) Reliable evaluation of low-level contamination of food. Report on a workshop in the frame of GEMS/Food-EURO 26–27 May 2005, Kulmbach, Germany. World Health Organization Regional Office for Europe, Geneva, 8 pp. http://toolbox.foodcomp.info/References/LOD/GEMS-Food-EURO%20%20-%20%20Reliable%20Evaluation%20of%20Low-Level%20Contamination%20of%20Food.pdf. Accessed 27 Oct 2019

WHO (1997a) Global Environment Monitoring System – Food Contamination Monitoring and Assessment Programme (GEMS/Food) Guidelines for predicting dietary intake of pesticide residues (revised). WHO/FSF/FOS/97.7: 41. www.who.int/iris/handle/10665/63787. Accessed 27 Oct 2019

WHO (1997b) Food consumption and exposure assessment of chemicals – Report of a FAO/WHO Consultation, Geneva, Switzerland 10–14 February 1997 (WHO/FSF/FOS/97.5). http://apps.who.int/iris/handle/10665/63988. Accessed 27 Oct 2019

WHO (2000) Human exposure assessment. Environmental Health Criteria 214, pp. 163. http://www.inchem.org/documents/ehc/ehc/ehc214.htm. Accessed 27 Oct 2019

WHO (2008) Dietary exposure assessment of chemicals in food. Report of a Joint FAO/WHO Consultation, 2–6 May 2005, Annapolis, Maryland, USA. WHO, Geneva. http://apps.who.int/iris/bitstream/10665/44027/1/9789241597470_eng.pdf. Accessed 22 Nov 2019

WHO (2012) GEMS Food consumption database. http://www.who.int/nutrition/landscape_analysis/nlis_gem_food/en/. Accessed 22 Nov 2019

WHO (2016) Joint FAO/WHO Expert Committee on Food Additives. Eighty fourth meeting, Food Additives. Rome, 6 to 15 June 2017. List of Substances scheduled for evaluation and request for data. Published 12 September 2016. http://www.who.int/foodsafety/JECFA84_Call_for_data.pdf?ua=1. Accessed 22 Nov 2019

WHO & FAO (2006) Updating the principles and methods of risk assessment: MRLs for pesticides and veterinary drugs. In: Pesticide residues in food 2006; Report of the Joint Meeting of the FAO Panel of Experts on Pesticide Residues in Food and the Environment and the WHO Core Assessment Group on Pesticide Residues, Rome, Italy, 3–12 October 2006. FAO Plant Production and Protection Paper 187. http://www.fao.org/fileadmin/templates/agphome/documents/Pests_Pesticides/JMPR/JMPRrepor2006.pdf. Accessed 22 Nov 2019

WHO & FAO (2009) Principles and methods for the risk assessment of chemicals in food. Environmental Health Criteria; 240. World Health Organization, Geneva. http://www.who.int/foodsafety/publications/chemical-food/en/. Accessed 22 Nov 2019

WHO & FAO (2012) Joint FAO/WHO expert meeting on dietary exposure assessment methodologies for residues of veterinary drugs: ISBN 978 92 4 156449 6 (NLM classification: WA 701). http://www.fao.org/fileadmin/user_upload/agns/pdf/jecfa/Dietary_Exposure_AssessmentMethodologies_for_Residues_of_Veterinary_Drugs.pdf. Accessed 22 Nov 2019

WHO/FAO & CAC (2016) Procedural manual, 25th ed. Rome (I) 2016. http://www.fao.org/fao-who-codexalimentarius/sh-proxy/en/?lnk=1&url=https%253A%252F%252Fworkspace.fao.org%252Fsites%252Fcodex%252FShared%20Documents%252FPublications%252FProcedural%20Manual%252FManual_25%252FManual_25e.pdf. Accessed 22 Nov 2019

Wientzek A, Vigl M, Steindorf K, Brühmann B, Bergmann MM, Harttig U, Katzke V, Kaaks R, Boeing H (2014) The improved physical activity index for measuring physical activity in EPIC Germany. PLoS One 9(3):e92005

Willett WC (1998) Food-frequency method. In: Willett WC (ed) Nutritional epidemiology, 2nd edn. Oxford University Press, New York, pp 74–94

Willett WC (2013) Nutritional epidemiology. Oxford University Press, Oxford

Yan H, Xu D, Meng H, Shi L, Li L (2014) Food poisoning by clenbuterol in China. Qual Assur Saf Crops Foods 7:27–35

Yang Y, Doudrick K, Bi X, Hristovski K, Herckes P, Wersterhoff P, Kaegi R (2014) Characterization of food-grade titanium dioxide: the presence of nanosized particles. Environ Sci Technol 48:6391–6400
Yang Y, Faust JJ, Schoepf J, Hristovski K, Capco DG, Herckes P, Westerhoff P (2016) Survey of food-grade silica dioxide nanomaterial occurrence, characterization, human gut impacts and fate across its lifecycle. Sci Total Environ 565:902–912
Yeh A I (2017) Natural and processed food nanostructures. In: Nanotechnologies in Food, Edition 2, Editors: Qasim Chaudhry, Laurence Castle, Richard Watkins. RSC, pp. 81–96
Zhang S, Midthune D, Guenther PM, Krebs-Smith SM, Kipnis V, Dodd KW, Buckman DW, Tooze JA, Freedman L, Carroll RJ (2011) A new multivariate measurementerror model with zero-inflated dietary data, and its application to dietary assessment. Ann Appl Stat 5(2B):1456–1487

Chapter 5
Exposure to Substances by Use of Consumer Products

Annette Bitsch, Annegret Blume, Christiaan Delmaar, Stefan Hahn, Astrid Heiland, Gerhard Heinemeyer, Stefanie Klenow, Wolfgang Koch, Friederike Neisel, Ralph Pirow, Thomas Rüdiger, Yasmin Sommer, and Michal Wiecko

5.1 Introduction

Gerhard Heinemeyer

This chapter focuses on consumer product-related exposure issues. It covers all non-food products that can release chemical substances and consumers may come into contact with. They include, for example, products used inside and around the home for personal hygiene, home maintenance, do it yourself, automotive care, hobby and craft, leisure, playing, home office, pet care, home pesticides, furniture,

A. Bitsch · S. Hahn · W. Koch
Department of Chemical Safety and Toxicology, Fraunhofer Institute for Toxicology and Experimental Medicine (ITEM), Hannover, Germany
e-mail: annette.bitsch@item.fraunhofer.de; stefan.hahn@item.fraunhofer.de; koch.wolfgang@item.fraunhofer.de

A. Blume · R. Pirow
Department of Chemicals and Product Safety, German Federal Institute for Risk Assessment, Berlin, Germany
e-mail: annegret.blume@bfr.bund.de; ralph.pirow@bfr.bund.de

C. Delmaar
Institute for Public Health and the Environment (RIVM), Centre for Safety of Substances and Products (VSP), Bilthoven, The Netherlands
e-mail: christiaan.delmaar@rivm.nl

A. Heiland · S. Klenow · F. Neisel · T. Rüdiger · Y. Sommer · M. Wiecko
Department of Exposure, German Federal Institute for Risk Assessment, Berlin, Germany
e-mail: astrid.heiland@bfr.bund.de; stefanie.klenow@bfr.bund.de; friederike.neisel@bfr.bund.de; thomas.ruediger@bfr.bund.de; yasmin.sommer@bfr.bund.de; michal.wiecko@bfr.bund.de

G. Heinemeyer (✉)
Department of Exposure, German Federal Institute for Risk Assessment (Retired), Berlin, Germany
e-mail: gerhardheinemeyer@o2mail.de

G. Heinemeyer et al. (eds.), *The Practice of Consumer Exposure Assessment*, https://doi.org/10.1007/978-3-319-96148-4_5

clothing, and yard and garden maintenance.[1] They are regulated by a number of different regulations (see the following subchapters). Consumer exposures to these products may be direct or indirect. As pointed out in Sect. 3.2.7, they can be characterised by the release of the agent from the source, the transport from the source to the target person (pathway of exposure), and the kind of contact (route of exposure). The release is determined by the chemical's concentration in an item (consumer product), its properties (e.g. volatility), how it is bonded in the matrix, and what stressors (e.g. temperature, and mechanical activity by the user) are present. These conditions of exposure can be described in general terms in the exposure scenario and by subsequent iterations in a tiered approach. This chapter gives an overview about exposure assessment with regard to consumer products, separated into mixtures and articles. It also focuses on specific issues of exposure assessment with respect to (European) chemicals and cosmetic regulation. Further, this chapter addresses particular issues for exposure estimation of nanoparticles and sprays and aerosols, as well as secondary exposures from house dust.

5.2 Definition and Characteristics of Consumer Products

Astrid Heiland

People come into contact with countless consumer products during their lifetime. These differ in terms of their shape, size, material, composition, functionality, etc. (Fig. 5.1). Some of them can be spent (used up) during their use, e.g. cosmetics, while others are durable and are only altered due to wear and washing, e.g. textiles.

There is no generally accepted definition of a "consumer product". Moreover there are differences in the meaning of consumer product and related terms like article, commodity, mixture, preparation, chemical product, etc. In principle, a "consumer product" can be a substance on its own, a mixture, an article, or any combination which is purchased from retail outlets by the general population. It is defined and categorised for different purposes such as to clarify the scope in the various regulations, e.g. Canada Consumer Product Safety Act (CCPSA, Section 2), United States Consumer Product Safety Act, and German Product Safety Act (Article 2).

Because different terms are used in a similar way, or else, the same term can have different meanings depending on the context, it is important to check the background in which a particular terminology is used.

---

[1]Household Products Database offered by the U.S. National Library of Medicine. This database will be closed by NLM december 2019 https://householdproducts.nlm.nih.gov KEMI: https://www.kemi.se/en/guidance-for/consumers and https://www.kemi.se/global/broschyrer/chemicals-in-products.pdf

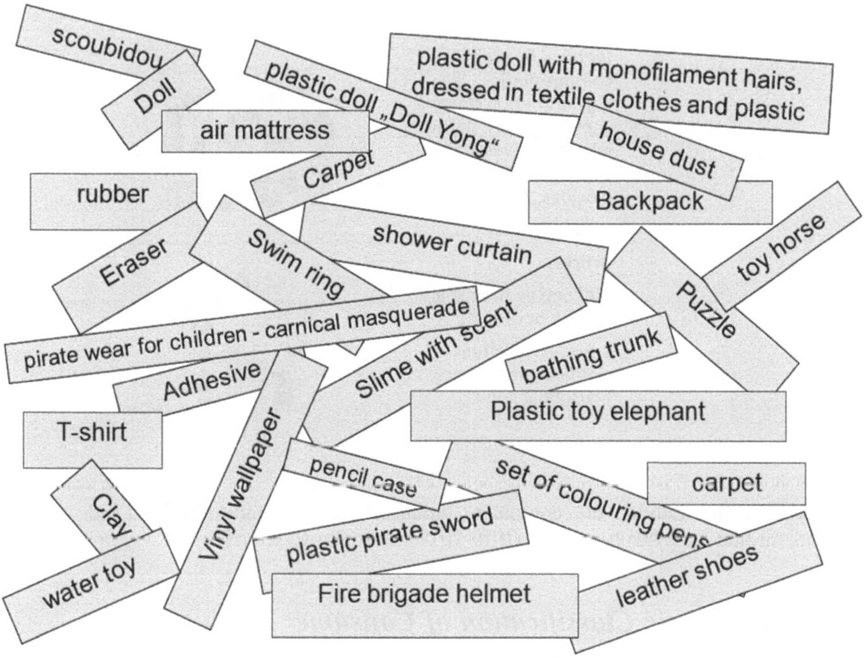

Fig. 5.1 Variety of consumer products (taken from Heinemeyer et al. 2012a) (© Bundesinstitut für Risikobewertung (BfR) & Umweltbundesamt (UBA) 2018. All Rights Reserved)

Consumer products are a subset of the broader categories "mixture" and "articles" (Fig. 5.2). Mixtures are characterised by their chemical composition, while for articles, the shape, surface or design are decisive for the function. This distinction is important both for the development of the exposure scenarios and for the legal obligations of the stakeholders. For most products, the assignment as a mixture or an article is unambiguous. In some cases, this decision whether shape, surface or design are more relevant for the function of the object than the chemical composition especially for assembled products might be difficult. The ECHA Guidance on requirements for substances in articles (ECHA 2017c) provides advice on determining whether an object is a substance/mixture or an article or consists of a substance/mixture *and* an article. The application of a decision tree is illustrated by using several examples. As a result, a wax crayon is to be regarded as a mixture, a postcard as an article, a printer cartridge as a combination of an article and a substance/mixture, a thermometer as an article, and the liquid within as an integral part of it.

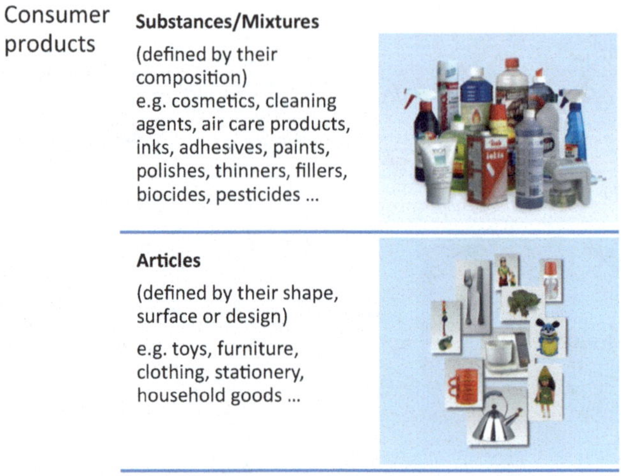

Fig. 5.2 Characterisation of consumer products as substance/mixtures and articles (© Bundesinstitut für Risikobewertung (BfR) 2018. All Rights Reserved)

5.2.1 Systematic Classification of Consumer Products

A systematic characterisation (classification) of consumer products is a key element in exposure assessment. The characteristics of a product determine the chemical release and therefore the level of exposure to consumers and the environment. The latter can lead to indirect exposure via air, water, soil, or food. Grouping is necessary to reduce the large number of individual products.

Classification systems are essential for information management and information exchange. They were developed within the different sectors, e.g. regulation, enforcement, and exposure science. They are implemented into (product) databases (e.g. IUCLID, SPIN) and exposure estimation tools (e.g. ConsExpo, ECETOC TRA).

Classification schemes follow specific product characteristics like functionality, use pattern, application form and area, material, chemical composition, etc. They can in principle be hierarchically or relationally structured or as a combination (Sect. 6.6.2).

Category Systems in Regulations

Category (classification) systems are developed to fit specific purposes, e.g. to define the scope and to support risk assessment in the context of the regulations. For example, the Cosmetic Regulation (see Sect. 5.5) classifies (in the preamble to Annex II to VI) cosmetics with regard to the affected body part (as hair, skin, lip, face, nail, eye, or oral product and as a product applied on mucous membranes) and to the application form (rinse-off and leave-on product). The Biocidal Regulation (Annex V) distinguishes 22 product types that are related to the four main groups (“disinfectants”, “preservatives”, “pest control”, and “other” biocidal products). The Toy Directive defines toys as “functional toy”, “aquatic toy”, “activity toy”, “chemical toy”, “olfactory board game”, “cosmetic kit”, and “gustative game” (Article 3).

A use descriptor system was introduced with the implementation of the REACH legislation which is described in the ECHA Guidance on Information Requirements and Chemical Safety Assessment, Chapter R.12: Use description (ECHA 2015). This system currently contains seven use descriptor categories to describe a use of a substance in each stage of its life cycle: “life cycle stage”, “sector of use”, “product category”, “process category”, “environmental release category”, “article category”, and “technical function”. For each descriptor, lists with standardised entries are linked. Only a few of the descriptors are suitable for consumer use, i.e. the “sector of use” and the “process category” indicate types of industry where the processes take place. Substances and mixtures for consumer applications are described mainly by their product category, e.g. “adhesives, sealants”, “air care products”, “coating and paints, thinners, paint removers”, “finger paints”, “inks and toners”, “washing and cleaning products”, etc. The article categories are mainly material-based, e.g. leather, paper, wood, plastic, rubber, “fabric, textiles, and apparel”, “stone, plasters, cement, glass, and ceramic”. These were refined by subcategories in 2015 taking the development of the OECD category system (OECD 2017) into account. As a result, main drivers of exposure were added: (1) large surface, (2) direct and intense skin contact, (3) articles intended for children, and (4) contact with food. In addition, categories exist for complex articles, i.e. vehicles, and electrical/electronic articles. Categories describing technical functions on the basis of the US-EPA proposal for OECD harmonised categories were also adopted for global industries to support their efforts to reach compliance for uses and to provide exposure information for fulfilling their regulatory obligations.

A European-wide project has been initiated[2] involving poison centres, ECHA, and the Commission following the amendment of Annex VIII to the CLP Regulation from 22 March 2017. The purpose is to harmonise the data relating to emergency health response and preventive measures across all Member States. This implies the creation of a database (managed by ECHA) for all products placed on the EU market (by 2020 for consumer products, by 2021 for professional products, and by 2024 for industrial products). Each product will be identified with a “Unique Formula Identifier” (UFI), and for each UFI in the database, information will be provided about the product’s classifications and its components, with range of concentrations according to their hazards. This also includes the development of a harmonised European Product Categorisation System (EuPCS). For this purpose, a study was carried out in which existing product categorisation systems were analysed. A hierarchical system based on the German TDI categorisation system of agents (TDI-CSA/TKS[3]) with additional categories and cross-references to REACH

[2]http://ec.europa.eu/growth/sectors/chemicals/poison-centres/

[3]TKS = TDI-Kategorie-System: https//:klinitox.de/3.0.html

product categories was proposed (EU 2017). Together with stakeholders from Member States, industry associations, and the Commission, the EuPCS was developed and published in 2018.[4] This system is primarily used to categorise mixtures for which there is an intended use (for further formulation or as an end product) and which are classified according to their health and physical effects. This will gradually replace the various national product category systems for notified products by a single harmonised system in the EU.

The US-EPA, Health Canada, and Environment Canada have jointly developed harmonised categories for consumer and commercial products to facilitate the information exchange between the two countries. These comprise 32 harmonised consumer and commercial product categories and some unique categories to accommodate differences between Canada and the USA. The classification systems of the US-EPA and Health Canada have been compared with the REACH use descriptor system by the OECD (OECD 2012).

Global Developments for Product and Article Use Categories

To extend this work, the OECD Task Force on Exposure Assessment has compiled existing category systems from various sources, e.g. from the US-EPA's Chemical Data Reporting, ECHA's Use Descriptor Lists, categories of the Substances in Products in Nordic Countries (SPIN) database, the "Consumer Specialty Products Association (CSPA)" (now "Household and Commercial Products Association (HCPA))", Japan's use categorisation codes, and information from the Kirk-Othmer Encyclopedia of Chemical Technology, and Ullmann's Encyclopedia of Industrial Chemistry (OECD 2017). The consolidated lists contain 107 functional groups, 91 use categories, and 69 article categories. Definitions and examples are implemented for each category at various levels of granularity. Functional use categories indicate the specific function of the used chemical in a process and/or product, while respective product and article use categories are defined by their end-use application. Here the term "product" refers to mixtures/preparations only.

Exposure scenarios for consumers are usually related to chemical product or article categories. Thus, the product use categories were also reviewed with regard to existing exposure assessments models and scenarios. As a result, 19 product use categories with several subcategories (a total of 91) are proposed and listed in the report with their defined scope. One category, for example, is "adhesives and sealants", which contains seven subcategories: single-component glues and adhesives, two-component glues and adhesives, one-component caulks, two-component caulks, fillers and putties, solder, and hot-melt adhesives. Seven article use categories were merged with nine material-based article categories and extended by five

[4] https://poisoncentres.echa.europa.eu/eu-product-categorisation-system

complex article categories and "other", resulting in a total of 69 subcategories. Further, the OECD concluded that the application to exposure assessment shall, in lieu of more specific categories, be supported by specific user-defined scenarios.

Product Category Systems Used by Poison Centres

A specific exposure situation is given in the case of a poisoning. Poison centres are established to provide expert treatment advice and assistance over the telephone in case of exposures to poisonous or hazardous substances. Therefore, information must be readily available about household products, pharmaceuticals, pesticides, plants, etc. Such data are usually stored in databases.

In Germany, a product database was established in the late 1960s as a collaboration of industry, the poison centres, and the German Federal Health Office (BGA). The purpose of this database was to provide the German poison centres with product formulation data. Later, the German EVA project (1994) was initiated, and it developed a thesaurus consisting of three levels, differentiating between three major groups, chemicals (household) products, cosmetics, and pesticides. Later, the more complex TDI-CSA/TKS was developed for the categorisation of the agents which is still in use in German poison centres.

Attempts to compare product classification schemes between poisons centres were part of the CEFIC-funded DeNaMiC project "*De*scription of the *Na*ture of Accidental *Mi*suse of *C*hemicals and Chemical Products" (2006–2009) that demonstrated that there was acceptable comparability between some parts of individual classification schemes but not between others (Wyke et al. 2009).

According to Annex VIII of the CLP Regulation (see above) from 2020 onwards, manufacturers of hazard mixtures must provide standardised information to poison centres by using the European Product Categorisation System (EuPCS).

Product Classification in Exposure Assessment

Fact sheets developed, for example, by RIVM or US-EPA provide information about exposure to chemicals from consumer products. They support the preparation of quantitative exposure estimates by providing standard exposure scenarios for a number of product categories. In order to reduce the large number of consumer products down to a limited number of categories, the assumption is that the type(s) of exposure within each category is very similar.

RIVM published several fact sheets for cosmetics (Bremmer et al. 2006a), children's toys (Bremmer and van Veen 2002), cleaning products (Meesters et al. 2018), pest control products (Bremmer et al. 2006b), paint products (Bremmer and van Engelen 2007), and do-it-yourself (DIY) products (ter Burg et al. 2007[5]). The latter classified the DIY products into 26 product categories - 10 subcategories are related to "glues" and 1 for "sealant cartridge". The glues and sealants are characterised in terms of several properties, e.g. purpose, utility form, surface size, and container.

Similar to the fact sheets published by authorities, "Specific Consumer Exposure Determinants (SCEDs)" were developed by industry associations to describe typical habits and practices of consumer products.[6] The intent of the SCEDs is to cover the direct exposure for the main product categories in an industry association's product portfolio. The following SCEDs have been published: motor fuels and lubricants by Concawe (2017); washing and cleaning products by A.I.S.E. (2017), mainly based on the results of the HERA-project (2005); and adhesives and sealants by FEICA (2018).

Further, the product categories and their corresponding defaults are implemented in exposure calculation tools (Sect. 5.10).

National Product Categories for Monitoring

In Germany, new technical feasibilities and structures are currently developed as part of the "data structure monitoring" project (AVV DatA) to improve the data communication for the national control and monitoring system.[7] In the framework of this project, a directory of catalogues (ADV Catalogues) mainly for food is already established and is described in detail in Sect. 4.4. It contains also cosmetics, personal care products, household chemicals, and other commodities like toys and apparel. The products are characterised by their function and in some cases additionally by their material (e.g. made of ceramic, glass, metal, plastic, etc.). Because the directory is focused on items with intended oral and/or skin contact, it contains, for instance, only one type of glue that is used for crafts and model kits.

Conclusion

All category systems are created to be "fit for their specific purpose", and they differ in general regarding their characterisation criteria. Nearly all of them are hierarchically structured with a different number of levels. Thus, it is difficult to transfer information from one system into the other or to gather and combine information from diverse systems, e.g. for exposure assessment purposes.

This issue was managed in the food sector by the introduction of a relational system (LanguaL) with several facets for different types of information in which the

[5]An update is in progress and will be published in 2020.

[6]https://echa.europa.eu/csr-es-roadmap/use-maps/use-maps-library

[7]https://www.bvl.bund.de/EN/01_Food/_01_tasks/02_OfficialFoodControl/11_Datamanagement/Datamanagement_node.html

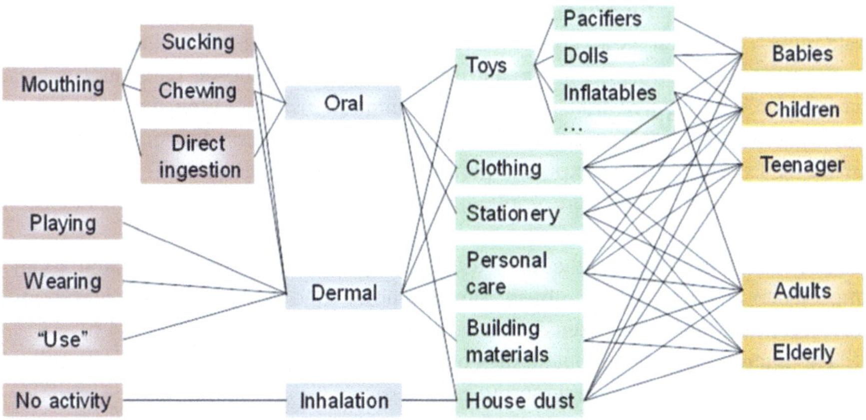

Fig. 5.3 The various combinations of human activity patterns when using consumer products. (© Bundesinstitut für Risikobewertung (BfR) 2018. All Rights Reserved)

information is structured hierarchically (Sect. 4.4). This system is more flexible than a purely hierarchical one because it allows the adaption to new facets if new information is relevant for classification. Development of an analogous system for consumer products is a real challenge but could be beneficial (Sect. 6.6.2).

5.2.2 Activity Pattern Data: Use of Consumer Products

People differ in age, gender, lifestyle, interests, their habits, and practices regarding consumer products (Fig. 5.3). These activities can lead to oral, dermal, and/or inhalation exposures irrespective of whether or not they are the result of an intended use (e.g. "use for" or "recommended for") by the manufacturer of a consumer product. The type of use (intended use, unintended use, reasonably foreseeable use, or misuse) is of primary importance in defining the scope of the exposure assessment but does not *per se* determine the exposures in the population.

The REACH Regulation defines "use" as follows:

> Use: means any processing, formulation, consumption, storage, keeping, treatment, filling into containers, transfer from one container to another, mixing, production of an article or any other utilisation.

Patterns of product use by consumers are essential information in consumer exposure scenarios. Thus, the frequency, duration, amount of products used, and location are parameters of interest. Fact sheets for consumer exposure determinants developed by authorities (e.g. RIVM and US-EPA) or industry associations (e.g. Concawe, A.I.S.E., and FEICA) provide default values for parameters largely based on expert judgement. Ideally, the patterns of product use can be derived from (representative) empirical studies. However, surveys that characterise the habits and behaviour of persons using chemical products are rare.

The default value for a parameter can be determined by characterising the "central tendency" or chosen as a "conservative estimate". Thus, the data sources and the assumptions, as well as the scope and limitations for the chosen parameter values, have to be documented transparently and comprehensibly. Often the defaults are implemented in exposure calculation tools. In this process, the origins and the designated purpose of the exposure parameters frequently get lost. This can lead to errors if these parameters are applied to products for which they have not been collected.

The effort to improve the database for consumer activities is related to the importance attached to the subject. For example, food is essential for everybody; therefore, time- and cost-intensive surveys have been carried out (Sect. 4.6).

The interest in carrying out studies on consumer behaviour is largely determined by the nature of exposure. Triggering factors are:

- The use of products with hazardous substances, e.g. biocides
- The special behaviour of children as a particularly vulnerable group, e.g. mouthing (see below)
- The intended contact of products with the (mucosa) skin, e.g. by cosmetics or textiles

Conversely, for products and their uses that do not meet the above criteria, only limited exposure information is available. This conclusion is in line with the outcome of a project that was carried out by FoBiG in cooperation and on behalf of the BfR. The aim of this project was to compile and evaluate primary data about consumer use patterns in terms of mixtures regulated under REACH (Heiland et al. 2016). An inventory of parameter values on the amount, duration, frequency, and location of consumer product uses was established. Overall, about 40 relevant studies were identified that allowed extraction of parameter values, resulting in 822 data sets of which 43% related to use frequency, 26% to product amount, and 23% to use duration. The location of consumer product use is documented in relatively few data sets (9%). Looking at the product types, most data sets (57%) addressed cleaning activities followed by the use of air care products (10%) and paints (8%). To evaluate the quality of parameter values, a scoring system based on features was developed to characterise empirical studies (e.g. study design, number of subjects, and level of detail). The majority of data sets yielded a score of more than 60 (out of 120) for almost all parameters, but only a small fraction (about 10%) received a score of 90 or higher.

Usually, more data are available for the frequency of use than for the product amount and application duration because the latter two are especially difficult to collect. Therefore, in a pilot study, several methods of collecting data have been checked to develop hypotheses and instruments to perform more comprehensive and representative systematic survey about use habits in the German population (Schneider et al. 2019).

There are some product types where direct contact is intended, e.g. dermal contact through cosmetics like creams and shower gels, inhalation through air care products and perfumes, and oral contact through personal care items like a toothbrush and toothpaste, or oral contact especially for baby care and little children items like pacifiers and teething rings. Non-intended contact is of great importance but is less documented in the literature. Only for the so-called "mouthing" behaviour of babies and little children, characterised by putting things or part of the body (fingers) into the mouth, do some studies exist. The frequency (Tulve 2002; Xue 2007) and duration (Groot 1998; Juberg 2001; Smith and Norris 2003) as well as the identification of the mouthed objects are important for mouthing exposure assessment. Chemicals could be contaminants on the object surface or leached-out from within an object over time. For the latter, the mouthing duration is of particular interest.

In conclusion, the published knowledge base of product use is weak, and further consumer behaviour studies are needed.

5.3 Exposure from Mixtures/Preparations

Astrid Heiland

Mixtures/preparations[8] contain two or more substances. Normally other substances (flavours, stabilizers, colouring agents, etc.) are added to the active substance(s). Thus, people using consumer products are mainly in contact with mixtures that are characterised by their chemical composition.

The REACH Regulation, Article 3 defines a substance and mixture as follows:

Substance: means a chemical element and its compounds in the natural state or obtained by any manufacturing process, including any additive necessary to preserve its stability and any impurity deriving from the process used, but excluding any solvent which may be separated without affecting the stability of the substance or changing its composition.
Mixture: means a mixture or solution composed of two or more substances.

The purpose of the mixture determines the regulation under which the substances are controlled (e.g. cosmetics, biocides, plant protection products, pharmaceuticals, etc.). The characteristics (composition, application form, physical form, etc.) and the way in which a mixture is used by a person determine the exposure.

Typical mixtures for consumers are household products (e.g. cleaning and washing agents, impregnation sprays, polishes, and air care products), DIY products (e.g. glues, sealants, and paints), and cosmetic/personal care products (Sect. 5.5). These products can be used, e.g. by wiping, polishing, brushing, and spraying (Sect. 5.9). Most of them are liquids. A mixture can be applied in the pure form or

[8]With the entry into force of the CLP Regulation in 2008 and with the aim of harmonisation in terms of the regulation, the term "preparation" was replaced in the REACH Regulation by the generic term "mixture". It was also introduced in further regulations, e.g. for biocides and cosmetics, but others still use "preparation", e.g. the pharmaceutical legislation.

can be mixed (e.g. the dilution of cleaning agents with water or the colouring of paints). They can be emitted to air and thus lead predominately to exposure by inhalation and dermal contact.

As a rule, the exposure of a single substance or a substance class in a mixture is considered. The properties of a substance in a mixture might differ from those of the pure substance. However, mixtures are often very complex, so that exposure estimates are usually based on the substance properties (e.g. volatility). For higher tier calculations, information about the product properties should be considered in the models, and experimental data are needed.

5.3.1 Exposure Scenario

The aim of an exposure assessment determines the exposure scenario. For building the exposure scenario, parameters related to the consumer product(s) and their considered substance(s), the type of application, and the affected group of consumers have to be defined. The purpose and effort of the exposure calculation as well as the availability of data determine the complexity of the scenario (Heinemeyer 2004).

The use of a mixture can cover several steps, such as the mixing and loading phase, the actual use, and the cleaning of the equipment after the use. The exposure can be calculated for the application itself, for the post application, or as an overall exposure for the user(s) as well as for the bystander(s). The use can be described by only one broad exposure scenario or by several single ones. Broad exposure scenarios allow to cover several products in the same category (e.g. cleaning or painting agents) if the parameter values are sufficiently conservative. This applies equally to different population groups, consumer activities, or contributing activities for the same use. Normally the product amount, the use duration, and frequency differ in such a group, e.g. oven cleaner is used less frequently than a dishwashing detergent. For individual exposure estimates, broad scenarios can be broken down to various exposure scenarios.

There are several other options to refine an exposure estimate (tiered approach), e.g. the model (simple → complex) or the parameters in an algorithm (conservative defaults → measured data, generic → specific, single points → distributions). There is no consistent course of action for describing what a tiered approach should be, and there is no unique meaning of what a tier 1, tier 2, or other tier of an exposure assessment is. Usually the options cannot be chosen independently, e.g. a more complex model requires more parameters. However, there is consensus that higher tier exposure assessments are very data intensive.

Therefore, many aspects have to be considered and decided during the exposure scenario building. A clear description of the scope and the aim of the assessment is a precondition for choosing the appropriate model and the appropriate parameter values to estimate the exposure.

5.3.2 Exposure Models

With the aim of simplifying the reality, a model describes the concept and mathematical representation of the exposure. Models can therefore reflect different steps in the exposure pathway, e.g. the emission of substances during or after application, transport via media like air or dust, that leads to inhalation, dermal and/or oral exposure (see also Sect. 3.3).

A collection of several models is recorded in the user guides of various exposure calculation tools (Sect. 5.10). For example, the user guide for the Consumer Exposure Model (CEM) tool of the US-EPA describes 21 models for mixtures and articles (US-EPA 2019): six emission models, three inhalation models, five ingestion models, and seven dermal models. The ECETOC TRA consumer tool (ECETOC 2012), with its focus on consumer mixtures and articles regulated under REACH, offers one model for each product/article category and exposure route. ConsExpo (Delmaar and Schuur 2017) provides inhalation models to estimate the exposure to vapour (which includes three emission sub-models), to spray, and to the emission from solid materials, furthermore five dermal and two oral models (with three sub-models for direct ingestion).

The available models for mixtures with focus on substance emission and external exposure by inhalation, dermal, and/or oral contact are compiled below. The exposure by spraying leads to a special exposure situation and is described in Sect. 5.9.

Inhalation Exposure

Inhalation exposure may occur directly during a use, e.g. during painting or afterwards (indirect exposure), if the substance is volatile and reaches the breathing zone of the consumer. Among other factors, the exposure is determined by the concentration of the substance in the inhaled air and the duration of exposure.

$$D_{\text{inh}} = \frac{F_{\text{resp}} * c_{\text{air}} * \text{IH}_{\text{air}} * t}{\text{BW}} n \tag{5.1}$$

D_{inh} Inhalatory dose
F_{rep} Respirable fraction of inhaled substance
c_{air} Concentration of substance in the room air
IH_{air} Ventilation rate of person
t Exposure duration
n Mean number of events per day
BW Body weight

The concentration of the considered substance in the room varies with time and place (distance from the source). It increases due to the release from the source/product and simultaneously decreases due to the normal air exchange rate in the room and is limited by the saturated vapour pressure. Local variations with regard to the concentration in a room or building can be simulated by introducing different zones (near-field-far-field model, two-zone model of indoor environment; (US-EPA 2019)).

The simplest approach to calculate the concentration of a substance in the air is the instantaneous release model. Here, the complete quantity of the substance in the product is released into the room air at once. This can be expressed without including the ventilation in the following equation:

$$c_{\mathrm{air}} = \frac{m_{\mathrm{prod}} * \mathrm{wf}}{V_{\mathrm{room}}} \tag{5.2}$$

m_{prod} Product amount
wf Weight fraction of substance in product
V_{room} Room volume

This can be refined by introducing the ventilation into the model, with an exponential decline over time:

$$c_{\mathrm{air}} = \frac{m_{\mathrm{prod}} * \mathrm{wf}}{V_{\mathrm{room}}} * \exp^{-qt} \tag{5.3}$$

q Ventilation rate
t Exposure duration

Higher tier models take the release rate of a substance into account. One approach is to use a constant rate that is defined by the time needed to release a certain amount of the substance from the product into the air. Normally the constant rate is not known, but it can be estimated by the division of the maximal air concentration by a reasonable estimate of the time for a complete evaporation. If no reasonable estimation of the emission duration is possible, the exposure time can be taken as a surrogate, provided that the exposure time is shorter than the emission duration. This results in higher exposure estimates but is still less conservative than the instantaneous release model.

More common is to model the evaporation in a detailed way as well. This requires many additional parameters, e.g. application time/temperature, vapour pressure, molecular weight, mass transfer coefficient, and surface area of the product.

The main drivers for evaporation are the pressure difference between of the partial pressure of the substance in the room air and the vapour pressure, the release area, and the mass transfer coefficient that reflects how fast the evaporated substance is shifted from the product surface to the room air. There are differences between whether the product is applied with a constant rate on a surface (increasing release area) or all at once (constant release area).

For solid materials (and articles), the emission is also determined by the diffusion of the substance from the bulk material into the boundary layer and then into the indoor air. This requires several parameters, such as the diffusion coefficient, the product thickness and density, and the product/air partition coefficient.

Dermal Exposure

Dermal exposure can occur during or after product use if skin contact is intended (e.g. cosmetics, Sect. 5.5) or is unintended but likely (e.g. cleaning agents).

A simple approach is to assume that all substances come into direct contact with the skin (instant application model):

$$D_{\text{derm}} = \frac{m_{\text{prod}} * \text{wf}}{\text{BW}} n \tag{5.4}$$

D_{derm} Dermal dose
m_{prod} Product amount
wf Weight fraction of substance in product
n Mean number of events per day
BW Body weight

The exposure can likewise be expressed as dermal load L_{der} (amount of substance on skin area A_{skin} per event):

$$L_{\text{derm}} = \frac{m_{\text{prod}} * \text{wf}}{A_{\text{skin}}} \tag{5.5}$$

Another model is based on the assumption that the layer thickness of the product on the skin is 0.01 cm and that the proportion of the substance in this layer constitutes the external exposure. This model is modified by introducing a dermal transfer factor that takes into account that only a fraction of the substance is transferred to the skin. But this information is rarely available. A similar approach is to use the diffusion coefficient to model the flux of the substance through the product to the skin according to Fick's law.

If details on how the skin is exposed to the product are available and the skin exposure increases (e.g. during painting by brushing or rolling), then the "constant rate" model can be applied that calculates exposure over a period:

$$D_{\text{derm}} = \frac{R * t * \text{wf}}{\text{BW}} \tag{5.6}$$

Respectively

$$L_{\text{derm}} = \frac{R * t * \text{wf}}{A_{\text{skin}}} \tag{5.7}$$

R Rate at which the product is applied to the skin
t Loading time

Skin contact with a substance can also occur after application (secondary exposure), e.g. after impregnation. With the "rubbing off" model, the dislodgeable product amount of a treated article which comes into skin contact can be estimated.

Oral Exposure

Mixtures may come into oral contact because from residues of dishwashing detergents on dinnerware and utensils, exposure originating from dermal contact through hand-mouth contact, etc. This can lead to ingestion which can be expressed in a way similar to the ingestion via food:

$$D_{\text{oral}} = \frac{m_{\text{prod}} * \text{wf}}{\text{BW}} n \tag{5.8}$$

D_{oral} Oral dose
m_{prod} Product amount
wf Weight fraction of substance in product
n Mean number of events per day
BW Body weight

5.4 Exposure from Articles

Ralph Pirow

5.4.1 Characteristics and Definition of Articles

Articles are a type or subcategory of consumer products (ECHA 2016) and different definitions exist for this subcategory. Articles can be defined from the producer's perspective (production process and raw/starting materials) and the consumer's perspective (use aspect/activity pattern). Qualifying features are the intrinsic properties of an article (composition, shape, surface, and design), the article function, and the place of use. Article 3(3) of the REACH Regulation (EC) No. 1907/2006 (EU 2006) provides that an:

> article means an object which during production is given a special shape, surface or design which determines its function to a greater degree than does its chemical composition.

A more elaborate definition is given in ECHA's guidance for substances in articles (ECHA 2017c):

> An article is generally understood to be an object composed of one or more substances or mixtures given a specific shape, surface or design. It may be produced from natural materials, such as wood or wool, or from synthetic ones, such as polyvinyl chloride (PVC). Objects may be simple, like a paper sheet, but can also be very complex, like a laptop computer, consisting of many articles. Most of the commonly used objects in private households and industries are or incorporate articles, e.g. furniture, clothes, vehicles, books, toys, kitchen equipment and electronic equipment. Buildings are not considered to be articles, so long as they remain fixed to the land on which they stand.

General and Particular Regulatory Context

Articles may be subject to different European legislations with respect to substances they contain. For example, toys are subject to the particular chemical safety requirements of the Toy Safety Directive 2009/48/EC (EU 2009a) and also to a few Annex XVII restrictions under REACH (see Sect. 5.6). In addition, the Toy Safety Directive contains generic restrictions for hazardous substances with a harmonised CMR classification under the CLP regulation.

For articles other than toys, a product-specific legislation does not exit. Instead, there are some Annex XVII restrictions under REACH for certain hazardous chemicals in articles, or parts of those, for which a higher probability of exposure exists due to the specific route of exposure and the frequency and duration of use. An example is the restriction of certain azo dyes in "textile and leather articles that may come into direct and prolonged contact with the human skin or oral cavity" (entry 43 of REACH Annex XVII). Certain polycyclic aromatic hydrocarbons (PAHs) are restricted in rubber or plastic components of articles

> that come into direct as well as prolonged or short-term repetitive contact with the human skin or the oral cavity, under normal or reasonably foreseeable conditions of use

(entry 50 of REACH Annex XVII). Other examples for restrictions under REACH are:

- Lead and cadmium in jewellery articles
- Nickel in post assemblies which are inserted into pierced ears and other pierced parts of the human body and in articles intended to come into direct and prolonged contact with the skin
- Chromium (VI) in leather articles coming into contact with the skin
- Nonylphenol ethoxylates in textile articles
- Polybrominated biphenyls, tris (2,3-dibromopropyl) phosphate, and tris (aziridinyl)phosphine oxide in textile articles
- Tributyltin, dibutyltin, and dioctyltin in articles
- Dimethyl fumarate in articles

Recently, the European Commission published Regulation (EU) 2018/1513 (EU 2018) to restrict certain CMR 1A and 1B substances in apparel, footwear, and other textile consumer products by adding the new entry 72 to REACH Annex XVII.

Articles treated with biocidal products are subject to the Biocidal Products Regulation No 528/2012 (EU 2012). This legislation distinguishes different biocidal product types. Treated articles such as textiles are covered by product-type 9 (preservatives to protect fibre, leather, rubber, and polymerised materials from microbiological deterioration) and product-type 18 (insecticides, acaricides, and products to control other arthropods).

Articles not specifically regulated otherwise are subject to the general safety requirements of the General Product Safety Directive 2001/95/EC (EU 2001).

Dermal Exposure Assessment Approaches for Articles

The exposure to substances released from articles may occur by direct dermal contact with the article. In case of semi-volatile organic compounds (SVOCs), which are emitted from articles into indoor air, a dermal exposure may also result from the air-to-skin transfer as a consequence of the partitioning of SVOCs between the gas phase and the skin (Weschler and Nazaroff 2012, 2014; Gong et al. 2014; Weschler et al. 2015). The focus of this section is on the dermal exposure resulting from direct contact with articles.

The dermal exposure ($\mathrm{Exp}_{\mathrm{ext.derm}}$) to substances released from articles in direct contact with the skin can be calculated from the contact area (A_{contact}), the dermal load (L_{derm}), the use frequency (f_{use}), and the body weight (BW) (Sect. 3.2.3, Table 5.1):

$$\mathrm{Exp}_{\mathrm{ext.derm}} = \frac{A_{\mathrm{contact}} * L_{\mathrm{derm}}}{\mathrm{BW}} * f_{\mathrm{use}} \tag{5.9}$$

The dermal load L_{derm} (μg/cm^2) is the amount of substance loaded onto the skin per unit area. An adjustment can be made if the exposed skin area deviates from the contact surface area of the article (Delmaar et al. 2005). Estimates for L_{derm} are obtained by considering the release of the substance from the article's contact surface. This can be done in migration experiments in which the release into a liquid medium (e.g. sweat simulant or other solvents) or the transfer to surrogate skin (e.g. cotton wipe) is measured. There are different ways of calculating L_{derm}. The simplest way is to multiply the migration rate per unit area (J_{migr}) with the duration of contact (t_{contact}):

$$L_{\mathrm{derm}} = J_{\mathrm{migr}} * t_{\mathrm{contact}} \tag{5.10}$$

The guidance document for consumer exposure assessment under REACH (ECHA 2016) contains a calculation of L_{derm} based on the article's weight (m_{article}),

Table 5.1 Parameter and parameter values used by dermal exposure assessment approaches for articles

Parameter	Unit	Description
$\mathrm{Exp}_{\mathrm{ext.derm}}$	$\mu g\ (kg\ BW)^{-1}\ day^{-1}$	External dermal exposure
A_{contact}	cm^2	Contact area
L_{derm}	$\mu g\ cm^{-2}$	Amount of substance loaded onto skin per unit skin area
f_{use}	day^{-1}	Frequency of use (= contact events per day)
BW	kg	Body weight
t_{contact}	s	Duration of contact
m_{article}	g	Article weight
ρ_{A}	$g\ cm^{-2}$	Area density of the article (e.g. textile fabric)
c_0	$\mu g\ g^{-1}$	Initial content of the substance in the article
C_0	$\mu g\ cm^{-3}$	Initial concentration of the substance in the article
C	$\mu g\ cm^{-3}$	Substance concentration in the article at a given x and time point
x	cm	Coordinate along which the diffusion proceeds
Δx	cm	Thickness of the contact layer
D	$cm^2\ s^{-1}$	Diffusion coefficient of substance in article matrix
J_{migr}	$\mu g\ cm^{-2}\ s^{-1}$	Migration rate per unit area (= flux)
f_{migr}	$\mu g\ \mu g^{-1}\ s^{-1}$	Relative migration rate (= fraction released per unit time)
F_{migr}	$\mu g\ \mu g^{-1}$	Migration fraction (= fraction released per wearing event)

the initial substance content (c_0) in the article, and the relative migration rate (f_{migr}) (= fraction of c_0 released per unit time):

$$L_{\mathrm{derm}} = \frac{m_{\mathrm{article}}}{A_{\mathrm{contact}}} * c_0 * f_{\mathrm{migr}} * t_{\mathrm{contact}} \tag{5.11}$$

For sheetlike articles such as textiles, it is more convenient to use the area density ρ_A (g/cm^2) of the fabric instead of fabric weight. For clothing textiles, it is also common to combine the relative migration rate and the contact duration into a new parameter called migration fraction (F_{migr}) that describes the fraction of the initial substance content released during a wearing event (Krätke and Platzek 2004; BfR 2012):

$$L_{\mathrm{derm}} = \rho_A * c_0 * F_{\mathrm{migr}} \tag{5.12}$$

Different theoretical models exist to estimate L_{derm}. Of practical relevance are the contact layer thickness approach, the diffusion layer approach, and the diffusion approach.

The contact layer thickness approach assumes a layer on the article's surface from which the substance present in that layer is released and completely transferred onto the skin upon direct dermal contact with the article. This approach is mentioned in the Technical Guidance Document for new and existing substances (EU 2003) and is used in the ECETOC TRA consumer tool for a first-tier or screening exposure assessment (ECETOC 2009a, 2012; ECHA 2016). The dermal load is calculated from the initial substance concentration (C_0) and the contact layer thickness (Δx):

$$L_{\mathrm{derm}} = C_0 * \Delta x \tag{5.13}$$

In the current version 3.1 of the ECETOC TRA consumer tool, the contact layer thickness for solid articles is set to an arbitrary default value of 0.001 cm (= 10 μm) (ECHA 2016). The model equation does not take into account the duration of the dermal contact.

The diffusion layer approach was proposed by Delmaar et al. (2013) to overcome limitations of the contact layer thickness model. It modifies the above approach by equating the contact layer thickness with the diffusion distance, which depends on the duration of the contact (t_{contact}) and the diffusion coefficient (D) of the substance in the article matrix:

$$L_{\mathrm{derm}} = C_0 * \sqrt{2 * D * t_{\mathrm{contact}}} \tag{5.14}$$

The diffusion approach is based on Fick's second law of diffusion (Crank 1975), which is a partial differential equation. The solution of this equation requires specification of the article's shape and of initial and boundary conditions. For slab-like articles (e.g. flooring), which have a relatively large contact area in comparison to the other article dimensions, the diffusion problem reduces to the one-dimensional case. Taking x as the space coordinate along which the diffusion within the article proceeds and C as the substance concentration at a given x position and time point, the substance flux J_{migr} (μg/cm^2/s) across the article's contact surface is given by:

$$J_{\mathrm{migr}} = D * \left.\frac{\partial C}{\partial x}\right|_{x = \text{at surface}} \tag{5.15}$$

The flux is driven by the concentration gradient that exists at the article's surface and which may change over time, so that this equation needs to be integrated over the time of contact to obtain the dermal load (L_{derm}). To apply the model in specific situations, appropriate initial and boundary conditions needs to be specified. Often, numerical methods are used to solve the specific diffusion problem.

The example in Fig. 5.4 demonstrates the prediction of a concentration profile within the article (Fig. 5.4a), the substance flux across the article's contact surface (Fig. 5.4b), and the dermal load (Fig. 5.4c) in dependence of the time of contact.

Analytical solutions of Fick's second law of diffusion exist for a few specific cases. One example is the release of the substance across the contact surface of a slab-like article in dependence on the contact duration (t_{contact}) and the diffusion coefficient (D) (Crank 1975; Piringer 1994; Clausen et al. 2016):

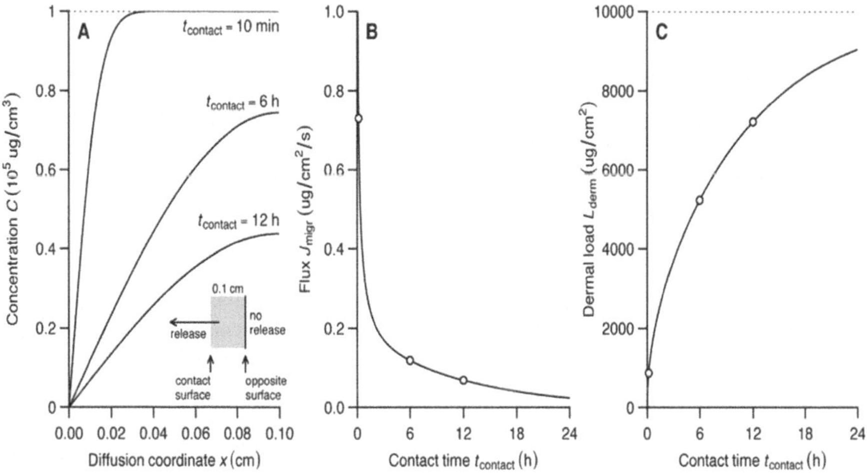

Fig. 5.4 Application of the diffusion approach to a slab-like article with a thickness of 0.1 cm, a uniform initial substance concentration (C_0) of 10^5 µg/cm^3, and a diffusion coefficient (D) of 10^{-7} cm^2/s. The substance is released at the contact surface, at which the concentration is zero; there is no release at the opposite surface. (**a**) Concentration profiles in the article at three different contact times ($t_{contact}$). (**b**) Substance flux (J_{migr}) across the article's contact surface in dependence of contact time. Circles indicate the flux at selected contact times. (**c**) Dermal load (L_{derm}) in dependence of contact time (© Bundesinstitut fur Risikobewertung (BfR) 2018. All Rights Reserved)

$$L_{derm} = C_0 * 2 * \sqrt{D * \frac{t_{contact}}{\pi}} \tag{5.16}$$

This equation describes well the substance release in the initial phase of migration experiments (Piringer 1994), when substance concentration in the extraction solvent can be assumed to be close to zero and when only a fraction of the substance content is released from the article. The predictions by this model are virtually identical to those shown in Fig. 5.4c for $t_{contact} < 6$ h. An analytical solution also exists for the diffusion from slab-like articles into a stirred solution of limited volume (Crank 1975), which is of relevance for migration modelling in the field of food contact materials (Brandsch et al. 2002).

Depending on the approach chosen, different parameter values need to be specified. The ideal situation is when experimental data are available for the dermal load (L_{derm}) or the migration rate per unit area (J_{migr}). The dermal load has been measured, for example, for bisphenol A in volunteers holding thermal printer paper (Biedermann et al. 2010). Migration experiments have been performed, for example, for dyes and insecticides in textiles (ETAD 1983, 1997; Snodgrass 1992; BfR 2012), and for phthalates and PAHs in polymeric materials (Clausen et al. 2016; Bartsch et al. 2016). If only the content or concentration of the substance in the article is known, it may be possible to use default values for the migration fraction (F_{migr}). Such default values have been proposed for dyes and textile auxiliaries in clothing (Krätke and Platzek 2004; BfR 2012). Alternatively, theoretical models enable the prediction of the dermal load based on the diffusion coefficient (D) and the duration of dermal contact (t_{contact}). Overviews about the diffusion coefficients for different substance-article matrix combinations are available (Delmaar et al. 2013; Clausen et al. 2016). Also, default values for the initial content, the contact area, the article thickness, and the exposure duration are available for certain article categories predefined in ECETOC TRA (Delmaar et al. 2013).

The dermal exposure assessment approaches described above provide estimates of the potential external dose that is available for dermal uptake. Estimates for the dermal load as obtained from diffusion models (Eqs. 5.15 and 5.16) represent upper-bound estimates. This is because the release from the article's surface is only limited by the diffusive mobility of the substance in the article matrix. This also applies to migration experiments under dynamic conditions with solvents having a higher solubility for the substance than the article matrix. Under real exposure conditions, with direct contact between the article and the skin, the substance transport from the article matrix into the skin may be limited, for example, by the low solubility in the skin surface film or by the diffusional resistance of the stratum corneum (Delmaar et al. 2013; Clausen et al. 2016). Such transport limitations can be taken into account in experimental settings with direct contact between article and the skin. Experiments with simultaneous migration and dermal absorption have been performed, for example, for insecticides and dyes in different textile fabrics (Snodgrass 1992; Heine et al. 1996; BfR 2012) and for PAHs in polymers and elastomers (Bartsch et al. 2016).

Oral Exposure Assessment Approaches for Articles

An oral exposure to substances in articles may occur when an article is placed into the mouth and the substance is released into the saliva in the oral cavity. Other scenarios are the accidental swallowing of small articles or the ingestion of material that is scraped off the article's surfaces. All these are foreseeable scenarios for small children due to their typical mouthing behaviour (see Sect. 5.2.2). An oral exposure is also possible via indirect exposure pathways. For example, substances that are transferred to the fingers during dermal contact may be taken up orally due to hand-to-mouth contact. An oral exposure may also result from the swallowing of the non-respirable fraction of inhaled airborne particles containing absorbable (semi) volatile substances emitted from articles. The focus of this section is on direct exposure pathways.

For the mouthing scenario with a substance released into the saliva, the oral exposure (Exp_{oral}) is calculated from the contact area (A_{contact}), the migration rate per unit area (J_{migr}), the contact time (t_{contact}), the use frequency (f_{use}), and the body weight (BW) (Table 5.2, Eq. 5.17):

Table 5.2 Parameters used in the oral exposure assessment approaches for articles

Parameter	Unit	Description
$\mathrm{Exp}_{\mathrm{oral}}$	$\mu g\ (kg\ BW)^{-1}\ day^{-1}$	Oral exposure
$A_{contact}$	cm^2	Contact area of the mouthed article
J_{migr}	$\mu g\ cm^{-2}\ s^{-1}$	Migration rate per unit area (= flux)
$t_{contact}$	s	Duration of contact
f_{use}	day^{-1}	Frequency of use (= events per day)
BW	kg	Body weight
$m_{article}$	g	Amount of the ingested/swallowed article material
c_0	$\mu g\ g^{-1}$	Initial content of the substance in the article

$$\mathrm{Exp}_{\mathrm{oral}} = \frac{A_{contact} * J_{migr} * t_{contact}}{\mathrm{BW}} * f_{use} \tag{5.17}$$

For the swallowing of small articles and the ingestion of scraped-off material, the oral exposure is calculated from the amount of swallowed or ingested article material ($m_{article}$) and the initial content (c_0) of the substance in the article (Eq. 5.18):

$$\mathrm{Exp}_{\mathrm{oral}} = \frac{m_{article} * c_0}{\mathrm{BW}} * f_{use} \tag{5.18}$$

The above model assumes that the substance is completely released from the article matrix into the gastrointestinal tract.

For children, standard values exist for the contact area of toys that can be mouthed and for the amount of toy material that can be swallowed. In the absence of data on the migration rate (J_{migr}), predictions can be made by using the diffusion approach (Eq. 5.16) if the diffusion coefficient is known.

Inhalation Exposure Assessment Approaches for Articles

The assessment of inhalation exposure to (semi)volatile substances emitted from articles is generally linked to indoor scenarios. The substance concentration in indoor air can be used directly as an exposure metric or as a parameter for the calculation of the inhalation exposure.

The inhalation exposure (Exp_{inh}) is calculated from the respirable fraction of the inhaled substance (F_{resp}), the substance concentration in indoor air (C_{air}), the respi ration rate of the person (V_{resp}), the exposure duration (t_{inh}), the frequency of exposure (f_{event}), and the body weight (BW) (Table 5.3, Eq. 5.19):

$$\mathrm{Exp}_{inh} = \frac{F_{resp} * C_{air} * V_{resp} * t_{inh}}{\mathrm{BW}} * f_{event} \tag{5.19}$$

The indoor air exposure concentration depends on the article's emission rate and on removal processes such as room ventilation. C_{air} is further influenced by room

Table 5.3 Parameters used for inhalation exposure assessment approaches of articles

Parameter	Unit	Description
Exp_{inh}	μg (kg BW)$^{-1}$ day^{-1}	Inhalation exposure
F_{resp}	μg μg^{-1}	Respirable fraction of the inhaled substance
C_{air}	μg m^{-3}	Concentration of the substance in indoor air
V_{resp}	m^{-3} h^{-1}	Ventilation rate of the person
t_{inh}	h	Duration of inhalation exposure
f_{event}	day^{-1}	Frequency of exposure (= events per day)
BW	kg	Body weight
R	μg h^{-1}	Emission rate per article
q	h^{-1}	Room ventilation rate (= volume fraction exchanged per unit time)
V	m^{-3}	Room volume
t	h	Time
t_{em}	h	Duration of the emission phase

volume, article size, and geometry. Experimental approaches use climate-controlled emission testing systems such as emission test chambers and cells, which enable the measurement of air concentration and the article's emission rate under indoor-related conditions and over defined time scales (Salthammer and Bahadir 2009).

Different theoretical models exist to predict the indoor air concentration in dependence of the substance emission from articles. Of practical relevance are the constant emission rate approach and the diffusion approach.

The constant emission rate approach assumes a substance emission at a constant rate for a certain duration (t_{em}), while the substance in indoor air is simultaneously removed from the air by room ventilation (HVBG 2001; Delmaar et al. 2005). The following two equations describe the concentration-time profile $C_{air}(t)$ for a substance in indoor air for the emission phase at times $t \leq t_{em}$ (Eq. 5.20) and for the post-emission phase at $t > t_{em}$ (Eq. 5.21):

$$C_{air}(t) = \frac{R}{q * V} * \left(1 - e^{-qt}\right) \tag{5.20}$$

$$C_{air}(t) = \frac{R}{q * V} * \left(1 - e^{-qt_{em}}\right) * e^{-q(t-t_{em})} \tag{5.21}$$

The model parameters are the article's emission rate (R), the room ventilation rate (q), and the room volume (V). A practical example is given below for the nickel emission from burning candles.

The diffusion approach describes the transport of the substance into indoor air by diffusion within the article matrix and by mass transfer across a boundary layer at the article-air interface (Delmaar et al. 2013; Huang et al. 2017). Apart from room ventilation, additional removal processes such as the sorption to interior surfaces and to airborne and settled particles (e.g. dust) can be taken into account (Huang et al. 2017; ICF 2016). The modelling of the emission process requires information on the diffusion coefficient of the substance in the article matrix, the material-air partition coefficient, and the mass-transfer coefficient (Delmaar et al. 2013).

Experimental emission data exist for a several substance-material combinations such as formaldehyde in wood-based materials (Salthammer et al. 2010), plasticisers in PVC (Schossler et al. 2011), fragrances in plasticised PVC (Masuck et al. 2011), VOCs in leather (Frank et al. 2012), and formamide in EVA-foamed materials (ANSES 2011). Theoretical models can be used to predict the emission rates and indoor air concentrations if information on the diffusion coefficient, the material-air partition coefficient, and the mass-transfer coefficient is available. Provide A compilation of such parameter values is available (Delmaar et al. 2013).

Practical Example 1: Dermal Exposure to Permethrin in Impregnated Clothing

Outdoor textiles such as clothing (e.g. battle-dress uniforms and forestry worker clothing), sleeping bags, and picnic blankets can be impregnated with the insecticide permethrin that acts as a contact repellent. Direct skin contact with impregnated textiles may lead to a dermal exposure of consumers if the permethrin is released from the textile fibres. Equation 5.9 can be used to predict the external dermal exposure ($\text{Exp}_{\text{ext.derm}}$) for a single wearing event ($f_{\text{use}} = 1$/day) with a dermal load (L_{derm}) of 0.625 μg/cm^2, which is obtained by multiplying the fabric-impregnation load of 125 μg/cm^2 (WHO 1997; US-EPA 2009b) with the migration fraction (F_{migr}) of 0.005 (Snodgrass 1992). Default values recommended by Krätke and Platzek (2004) were used for the exposed skin area ($A_{\text{skin}} = 10{,}000$ cm^2) and the body weight (BW = 60 kg). This yields an external dermal exposure of 104 μg/kg of BW per day. For assessing the risk for possible systemic effects, the external dermal exposure can be multiplied with the dermal absorption fraction of 0.03 (Ireland 2014) to arrive at an internal dermal exposure of 3 μg/kg of BW per day. Both exposure estimates apply to brand-new textile articles. To account for effects of wear and laundering, which would result in a reduced migration in the long term, an additional factor (default: 0.1) can be included (Krätke and Platzek 2004).

Practical Example 2: Inhalation Exposure to Nickel in Candles

Nickel compounds are used as a catalyst for the hydrogenation/hydro-treatment of base materials for candles. The nickel catalyst is removed after hydro-treatment, but residual amounts may be present in the final candle product. A maximum value for nickel in candles of 0.5 mg/kg has been proposed as specific safety requirement to be met by EU standards for candles pursuant to the General Product Safety Directive (EU 2001). Nickel is assumed to be released into indoor air during the burning of a candle. For assessing the additional lifetime cancer risk resulting from the inhalation

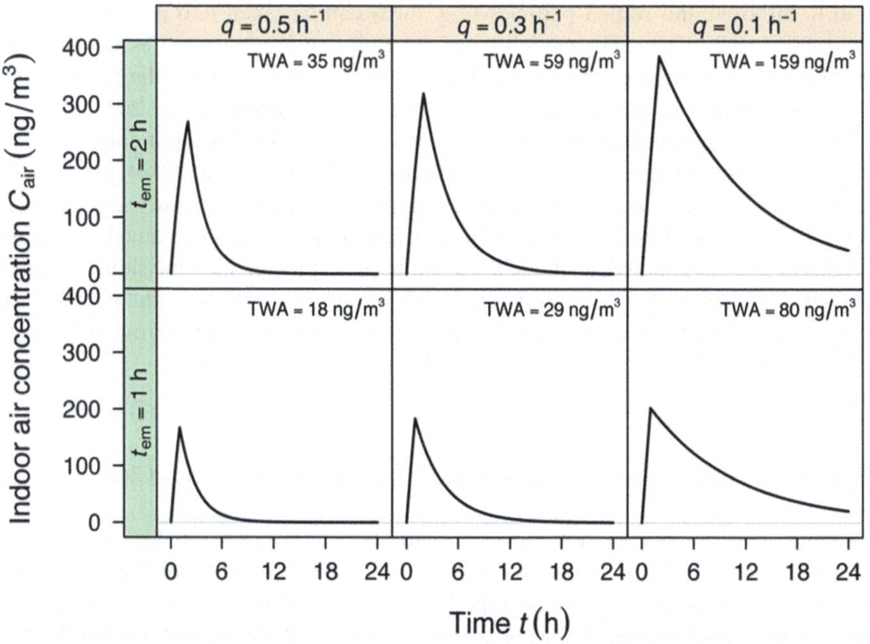

Fig. 5.5 Simulated time courses of indoor air concentration (C_{air}) of nickel following the burning of a candle for two different burning durations (t_{em}) and three different room ventilation rates (q). In addition, the time-weighted average (TWA) of the indoor air concentration is provided (© Bundesinstitut für Risikobewertung (BfR) 2018. All Rights Reserved)

exposure to nickel compounds, the indoor air concentration is an appropriate exposure metric that can be compared with health-based guidance values.

The constant emission rate approach (Eqs. 5.20 and 5.21) can be used to predict the concentration-time profile for nickel in indoor air (C_{air}) for a burning candle in a room volume (V) of 20 m^3. The candle's emission rate (R) of 4.25 µg/h was obtained by multiplying the maximum nickel content of 0.5 µg/g with the burning rate of 8.5 g wax per hour (upper-bound value). The room ventilation rate (q) was set to three different values (Fig. 5.5) and the candle burning duration (t_{em}) to two different values to cover uncertainties about the use habits.

The simulated concentration-time profiles (Fig. 5.5) provide information on peak concentrations, which can be compared with acute health-based guidance values, and time-weighted averages (TWA), which can be used for comparisons with chronic health-based guidance values. Remaining uncertainties in the exposure estimation result from the limited information about the frequency of candle use, the loading ratio (= number of candles per unit volume), and the chemical identity of the nickel compounds and from idealised assumptions about the fate of the emitted nickel species (nickel may be associated with emitted fine particulate matter which is removed faster from indoor air than predicted by the model due to deposition on interior surfaces).

5.5 Cosmetics Regulation

Annegret Blume

Cosmetic products are subject to a relatively high level of regulation compared to other consumer products. The reason for this mainly lies in the high frequency and extent of exposure of the consumers of these products. In addition to the toxicological profile of the ingredients, a detailed knowledge of the consumer exposure is indispensable for the required safety assessment under the European Cosmetics Regulation 1223/2009 (EU 2009b).

5.5.1 Definition of a Cosmetic Product

The European Cosmetics Regulation 1223/2009 (EU 2009b) defines a cosmetic product in Article 2, 1(a), as follows:

> Cosmetic product' means any substance or mixture intended to be placed in contact with the external parts of the human body (epidermis, hair system, nails, lips and external genital organs) or with the teeth and the mucous membranes of the oral cavity with a view exclusively or mainly to cleaning them, perfuming them, changing their appearance, protecting them, keeping them in good condition or correcting body odours.

In the preamble to the regulation 1223/2009, a clear distinction is made between cosmetics, medicinal products, medical devices, and biocidal products (recital 6). However, the preamble acknowledges that the assessment of whether or not a product is a cosmetic product may have to be made on a case-by-case basis, taking into account the characteristics of the product. Sometimes this decision is not easily reached, and there is an expert group on the EU level, the Borderline Working Group of the European Commission, where doubtful cases are discussed. The agreements of the Borderline Working Group discussions are published. Regulation 1223/2009 also lists a number of products that are clearly cosmetic products (recital 7).

Interestingly, the definition of a cosmetic product includes products that are intended to come into contact with the teeth or the oral cavity, but not those intended for contact with the nasal cavity. Thus, toothpaste and mouthwash are cosmetic products, while creams intended for the inside of the nose are not.

5.5.2 Exposure Scenarios

The most important document for the risk assessment of cosmetic ingredients is "The SCCS Notes of Guidance for the testing of cosmetic products and their safety evaluation". The Notes of Guidance outline possible exposure scenarios for cosmetics and how to use them for the risk assessment process. This document has been compiled and is constantly being updated by the expert committee of the European Commission, the Scientific Committee on Consumer Safety (SCCS). It is presently in its tenth revision (SCCS 2018).

Dermal Route

Human exposure to cosmetic products occurs mainly via the dermal route. The outer layer of the skin, the stratum corneum, constitutes a quite effective barrier against a lot of substances that are used as ingredients in cosmetics. Two main categories of cosmetic products are on the market. One category is the so-called "rinse-off" products that stay on the body only for a short period of time before they are washed away; examples are body wash, soap, or hair shampoo. The other product category comprises the so-called "leave-on" products that are intended to remain on the skin for a prolonged period of time; examples are many kinds of creams, such as hand or facial creams. Also included are body lotions and perfumes. These different exposure scenarios have to be taken into account for a risk assessment of cosmetic products. In fact, a number of cosmetic ingredients are restricted as only for use in rinse-off products.

The key parameter to estimate systemic exposure to cosmetics is the determination of the dermal or percutaneous absorption. The SCCS distinguishes the process into three steps: (1) penetration [entry of a substance into a particular skin layer (e.g. stratum corneum)], (2) permeation (penetration through one skin layer into another one), and (3) absorption (uptake into the vascular system, systemic availability) (SCCS 2018). Besides toxicity due to systemic availability, cosmetic ingredients can also cause local reactions like irritation, corrosion, and sensitisation. These local effects can vary between different skin areas, e.g. mucous membranes or the sensitive area around the eyes where the skin is less thick than in other body regions like the back. Therefore, the cosmetics regulation restricts the use of some ingredients in certain skin areas.

An important point to be taken into account is metabolism. What happens during metabolism, e.g. the metabolism rates and the metabolites formed, depends on the substance. Similar to phase 1 metabolism and phase 2 metabolism of xenobiotics in the liver, skin metabolism may result in the toxification or detoxification of substances, or in metabolites with a different skin penetration profile. Both phase 1 and phase 2 enzymes are active in the skin, but phase 2 enzymes are more prominent. The possible conversion of a substance to a skin sensitizer by skin metabolism is a problem that has to be kept in mind during risk assessment.

Oral Route

Oral exposure to cosmetics occurs inadvertently. However, lipstick may eventually be swallowed, and toothpaste or mouthwash may be accidentally swallowed. In its Notes of Guidance, the SCCS recommends estimates of the amounts of product that are used daily by the consumer and uses a retention factor. These estimates are directed at dermal exposure and include ones for lipstick, toothpaste, and mouthwash (SCCS 2018). In its guidance on the safety assessment of nanomaterials in cosmetics, an approach for the calculation of oral exposure is given which is similar to the estimation of dermal exposure. That is, the systemic exposure dose is calculated as an estimated daily exposure, obtained by multiplying the concentration of the ingredient in the product by the oral absorption. For toothpaste, the use of a retention factor is proposed to account for the fact that only a small amount is swallowed in everyday use.

Inhalation Route

Inhalation is a possible exposure route for some cosmetic products that contain volatile ingredients. These products include perfumes or nail products like nail polish or nail polish remover that release solvents. In addition, some cosmetics like deodorants or sun protection products are marketed in sprayable form, either as a pump spray (vapour) or as an aerosol. Powders like face powder are the third group of cosmetic products that may be inhaled.

The anatomy of the respiratory tract includes a wide lumen in the upper parts like nose, mouth, throat, and tracheobronchial area but a small lumen in the alveoli. The areas of the respiratory tract also include differences in bifurcations that impact particle deposition patterns. This results in a dependency of the inhalation of substances on the particle or droplet size. Particles/droplets with mass median aerodynamic diameter (MMAD) of >100 μm are considered as not inhalable, those with an MMAD between 100 and 10 mm can be inhaled and reach the upper parts of the respiratory system; particles/droplets of >7 μm are subject to mucociliary clearance within 24 h from the tracheobronchial compartment (MAK 2012; Phalen and Oldham 2006). This clearance can result in an uptake via the oral route. As the respiratory fraction (RF), particles/droplets with an MMAD <10 μm can reach the deeper parts of the lung, trachea, and alveoli. Because of the large surface area of the deeper lung compartments, the systemic availability of inhaled substances that reach the deeper parts of the lung is probable (Snipes 1989; Valentine and Kennedy 2008).

Propellant-driven aerosols often have a particle size <10 mm (Bremmer et al. 2006b; Eickmann 2007), while pump sprays typically produce particles in the range of 70 μm. However, droplet size depends on the architecture of the nozzle of the spray pump (Quadros and Marr 2011; Losert et al. 2015), and both propellant-driven and pump sprays can produce particles in the range of the RF.

An important point to consider is that the droplets can vary in size during their way through the air. They may become smaller due to evaporation but could also, e.g. in the case of nanomaterials, aggregate. Since these processes can influence the RF, the droplet size before settling has to be taken into account for risk assessment.

Inhalation of substances from cosmetic products can result in local toxicity such as irritation in the upper respiratory tract, local toxicity in the deep lungs, or respiratory sensitisation. Inhalation can also result in systemic toxicity due to an uptake from the alveoli. In addition, two special scenarios have to be taken into account, one being that a high dose of inert material can result in a particle overload, a situation associated with inflammation (Greim et al. 2001; Mühle und Mangelsdorf 2003). Also, fine droplets of lipids like oil or lipophilic substances can cause an acute respiratory syndrome (Vernez et al. 2006).

5.5.3 Models and Parameters

Dermal Exposure

Dermal exposure estimation can be supported by in vivo or in vitro studies. Test protocols are available from the EU (EC B.44 skin absorption, in vivo method; EC B.45 (EU 2008a, b) skin absorption, in vitro method) and OECD (2004a, b) (OECD 427, skin absorption, in vivo method; OECD 428, skin absorption, in vitro method). The SCCS and its predecessor also issued guidance documents on the execution and interpretation of these studies (SCCNFP 1999; SCCS 2010b).

Diffusion cells are used to measure dermal absorption in in vitro studies. They consist of a donor chamber and a receptor chamber, between which a membrane is positioned. The membrane may be non-viable skin to measure diffusion only, or it can be a fresh metabolically active skin to measure diffusion and skin metabolism simultaneously. The receptor chamber constitutes a fluid reservoir. The diffusion of the applied substance into and across the skin to this fluid reservoir is measured. Diffusion cells may be static or flow-through. The receptor fluid is collected at time points throughout the experiment and analysed for the test chemical and/or metabolites. The validity of the data obtained by these in vitro models strongly depends on the conditions applied. They should, as far as possible, mimic the intended use conditions with regard to dose, concentrations, and amounts (SCCS 2018). The SCCS in its Notes of Guidance lists *14 specific parameters that need to be considered when developing a dermal absorption* study. Examples include consideration of the technicalities of the model, the choice of receptor fluid (e.g. lipophilic substances will not readily migrate into a hydrophilic receptor fluid), the quality of the skin preparation, doing a mass balance analysis, and evaluating the recovery data. At the end of an experiment, the amounts of substance localised in the epidermis (but not the stratum corneum), dermis, and receptor fluid will be summarised to calculate dermal absorption.

The systemic exposure dose (SED; mg/kg of BW per day) may be calculated as dermal absorption of the test substance in μg/cm^2 (Section 4.2 in SCCS 2015, updated 2018):

$$\mathrm{SED} = \frac{\mathrm{DA}_a * 10^{-3} * \mathrm{SSA} * F}{60\ \mathrm{kg}} \tag{5.22}$$

where

DA_a (μg/cm^2) Amount of dermal absorption per area reported as amount/cm^2, resulting from an assay under in-use mimicking conditions. This term will be multiplied by 10^{-3} μg/mg.

SSA (cm^2) Skin surface area expected to be treated with the finished cosmetic product. See Table 1 in Section 4.2 of SCCS (2015, updated 2018) for SSA values per product type.

F (day^{-1}) Frequency of application of the finished product (default: $F = 1$).

BW: Standard default human body weight (60 kg).

The dermal absorption as percentage of the amount of substance applied can be calculated as follows:

$$\mathrm{SED} = A * C * \mathrm{DA}_p \tag{5.23}$$

where

A (mg/kg of BW per day) Estimated daily exposure to a cosmetic product per kg body weight, based upon the amount applied and the frequency of application (for calculated relative daily exposure levels for different cosmetic product types, Section 4.2 of the SCCS notes of Guidance (SCCS 2015), updated 2018)

C (%) Concentration of the substance under study in the finished cosmetic product on the application site

DA_p (%) Dermal absorption expressed as a percentage of the test dose assumed to be applied in real-life conditions

A number of computer-based models exist that aim to estimate exposure. Among them are models like Targeted Risk Assessment ECETOC TRA (ECETOC 2012) or Crème RIFM (Safford et al. 2015) that assume 100% dermal absorption. Other models like the SHEDS-HT (Isaacs et al. 2014), ConsExpo (Delmaar et al. 2005), or PACEM (Delmaar et al. 2015; Dudzina et al. 2015) use skin permeation coefficients. Some of these platforms can take multiple exposure scenarios into account. This can be important, e.g. in the case of the substance applied to the skin being volatile, only a part of the applied amount will penetrate the skin, while another part might be inhaled. The above models aim to enable high-throughput screening of exposures to cosmetic ingredients. However, at present, a serious limitation for these high throughput screening models is the lack of data available for the development and validation of the models.

Oral Exposure

As mentioned earlier, the oral exposure to cosmetics is marginal; however, it may have to be considered in the case of nanomaterials. The SCCS "Guidance on the safety assessment of nanomaterials in cosmetics" (SCCS 2012) suggests a calculation of oral absorption that is similar to the one for dermal absorption:

$$\mathrm{SED} = A * C * \mathrm{DA}_{\mathrm{oral}} \tag{5.24}$$

A (mg/kg of BW per day)	Estimated daily exposure to a cosmetic product per kg body weight, based upon the amount applied and the frequency of application (for calculated relative daily exposure levels for different cosmetic product types, see Table 2 in the SCCS (2015, updated 2018)).
C (%)	Concentration of the substance under study in the finished cosmetic product on the application site.
$\mathrm{DA}_{\mathrm{oral}}$ (%)	Oral absorption expressed as a percentage of the test dose assumed to be applied in real-life conditions. In some cases, a retention factor is used to account for the fact that the whole amount is not ingested (e.g. toothpaste).

Inhalation

For a risk assessment of inhalation toxicity of a given substance, the measured or modelled local consumer exposure is compared with the no-observable-adverse-effect concentration (NOAEC). If a NOAEC is not available, a route-to-route extrapolation from oral studies may be possible, but this has to be decided on a case-by-case basis (ECHA 2012). The margin of safety (MoS) is calculated on the basis of a NOAEC or NOAEL from a local or systemic toxicity study, which is compared to the SED.

There are some studies about typical exposures from deodorants or hairsprays (Hall et al. 2007, 2011; EU 1996; Loretz et al. 2006). The SCCS Notes of Guidance contain tables of the estimated daily exposure levels for different cosmetic product types (SCCS 2018). It has been estimated that 85% of a hairspray will reach the skin or hair, which leaves 15% for inhalation (Bremmer et al. 2006c). The estimates for the duration of exposure vary, e.g. while some authors use a worst case of 10–20 min, RIVM assumes 5 min (Bremmer et al. 2006a).

Exposure models that assume an even distribution of the substance in the room are mostly not suitable to estimate inhalation exposure for cosmetics. Instead, a so-called two box model has been proposed. It takes into account two different scenarios: it is assumed that during the first two minutes of exposure, a spray that is directed towards the consumer (e.g. deodorant or hairspray or perfume) produces a cloud of substance of about 1–2 m^3 around a facial or body area. During the subsequent 18 min, the cloud disperses into an air volume of about 10 m^3, a volume roughly comparable to a small bathroom (Bremmer et al. 2006c). The inhalation is calculated for each of the scenarios and summarised in the end, taking into account all variables and necessary parameters like sprayed amount, concentration of ingredient, proportion of non-propellant spray in formulation, airborne fraction, breathing rate and exhalation, daily application, and body weight to obtain the SED (Rothe et al. 2011; Steiling et al. 2014).

A number of computational tools are available for the prediction of inhalation exposure. The RIVM has developed ConsExpo (Delmaar et al. 2005; Delmaar and Bremmer 2009), which allows the modelling of consumer exposure to vapours or sprays on the basis of the inhalable fraction The model takes into account product-specific parameters like the airborne fraction, the weight fraction of the ingredient, the mass generation rate, or the aerosol size distribution. External parameters include the duration of exposure or spraying, the volume of the room, and the ventilation rate. The model embedded in the ConsExpo tool is a so-called multiple path particle deposition model, which means that the physiological differences between the various parts of the respiratory tract are taken into account. The special exposure scenario mentioned above can be modelled by the ConsExpo tool by reducing the room volumes to respective values and selecting a high room ventilation rate. Similar approaches are the BG-Spray (Eickmann 2007) and the SprayExpo (Koch et al. 2012). A discussion of the advantages and disadvantages of the various models has been provided by Eickmann et al. (2007). The SCCS has published opinions in which inhalation toxicity has been taken into account for the risk assessment of a substance, e.g. the opinion on dihydroxyacetone (DHA) (SCCS 2010a).

There are in vitro methods that use co-culture cell models of human reconstructed tissue of the respiratory tract. However, since these models do not take into account the different parts of the airways with their different properties, only supportive information can be derived from these models (Sauer et al. 2013).

Exposure scenarios as used in standard inhalation toxicity studies, e.g. OECD guideline 413 (OECD 2009), are not appropriate to estimate inhalation of cosmetics because they do not account for the droplet maturation mentioned above. Also, at present, there is no computational model that accounts for droplet maturation in a dependable way. Therefore, the direct measurement of sprays during the use of products by humans (as is the case for cosmetic products) is preferable to modelling. According to the SCCS, the assessment of the inhalation risk and the decision about which model is appropriate has to be done on a case-by-case basis (SCCS 2018).

Practical Examples

Example 1: Exposure to 2-Nitro-*p*-Phenylenediamine in Hair Dye

2-Nitro-*p*-phenylenediamine (2NPPD) is a colourant that has been used in semi-permanent and oxidative hair dyes as a colour enhancer. According to the European Cosmetics Regulation, 2NPPD is no longer allowed in hair dyes (Annex 2, no. 1319).

A surveillance authority in Germany had measured a concentration of 1.9% 2NPPD in a hair dye and asked for a risk assessment. A 1993 Scientific Committee on Cosmetics (SCC 1993) opinion had assessed a concentration of 0.3% 2NPPD as safe for the consumer with regard to acute toxicity, reproduction toxicity, and mutagenicity/carcinogenicity. Since the publication of this opinion, new data on the sensitising properties of 2NPPD were generated, including a study that showed that the risk of an occupational allergic contact dermatitis (OACD) for hair dressers increased between 1980–1993 (first study) and 1994–2003 (second study; patch-test, 78.3% vs. 58.8%; OACD 58.0 vs. 48.8%). Also, sensitisation to 2NPPD increased from 2.1% to 7.3% (Valks et al. 2005). One case study reported an anaphylactic shock after use of a hair dye that contained 2NPPD (Kuck-Koot and Lahey-De Boer 2005).

An older in vivo study showed that from using a concentration of 1.36% 2NPPD in an oxidative hair dye, 3.4 μg/cm^2 was absorbed by the human scalp (corresponding to 0.143% absorption and a scalp surface area of 650 cm^2; Wolfram and Maibach 1985). This was confirmed by a more recent study that measured percutaneous penetration in vitro using human skin in a flow-through diffusion cell. In this case, 5.4 μg/cm^2 were absorbed from a semi-permanent formulation with 0.7% 2NPPD (Yourick und Bronaugh 2000). For a penetration of 5 μg/cm^2, a scalp surface of 580 cm^2 and an average adult body weight of 60 kg (scalp surface area and average body weight according to SCCS Notes of Guidance), the estimated exposure dose was 48 μg/kg of BW.

The concentration of 2NPPD in the sample measured by the surveillance authority was 1.9%, which roughly corresponds to 3 times of the amount used in the study of Yourick und Bronaugh (2000). This would result in an exposure of about 140 μg/kg of BW. A study directly measuring the skin absorption of 1.9% 2NPPD was not available. The frequency of exposure for semi-permanent hair dyes was estimated to be once in 4–6 weeks. This toxicological endpoint was considered to be the most relevant one since new data on the sensitisation risk of 2NPPD were available, and there is no threshold for the sensitisation risk. Thus, a concentration of 1.9% 2NPPD was judged to be a serious risk for the health of the consumer, irrespective of the contribution of other toxicological endpoints.

Example 2: Exposure to Benzene in Nail Polish

Benzene is a carcinogen that causes acute myeloic leukaemia in humans (Infante et al. 1977; Hayes et al. 2001; Rinsky et al. 1987; IARC 2012). Its acute toxicity is low; chronic effects of benzene include neurotoxicity, haematotoxicity, endocrine activity, and toxic effects on the circulation, liver, kidney, and respiratory system (reviewed by Bahadar et al. 2014). The exact mechanisms of benzene toxicity are not known but involve toxic metabolites like benzene oxide or benzoquinones in addition to benzene itself (Snyder 2000, 2002). These substances form adducts with macromolecules like DNA or proteins, leading to cellular toxicity in organs and carcinogenesis (Waidyanatha et al. 1998).

Consumers are mainly exposed to benzene via contaminated air, e.g. from exhaust fumes. Breathable air is responsible for about 96–99% of the total exposure (EU RAR Benzene 2008c).

Exposure to benzene from nail polish occurs not only primarily via the inhalation route but also from ungual (through the nail itself), dermal (through the skin around the nails), or oral (from nail biting) routes. Nail products are not addressed separately in the SCCS Notes of Guidance (SCCS 2018). Therefore, probabilistic exposure estimation was used for the assessment (Ficheux et al. 2014). The key exposure parameters of this study are comparable to the "cosmetic fact sheet" of the RIVM (Bremmer et al. 2006a). A conservative approach (worst-case scenario) based on P95 values was used. According to Ficheux, exposure to nail polish is highest for women aged between 18 and 34 years, with an exposure (in mg/kg of BW per week) of 5.27 for inhalation, 8.55 ungual, 2.31 dermal, and 0.94 oral. For a measured maximum content of benzene in nail polish of 182 mg/kg, this results in an exposure of 1 for inhalation, 1.6 for the ungual, 0.4 for the dermal, and 0.17 for the oral route (in μg/kg of BW per week).

These values correspond to the external exposure. The absorption was taken into account to estimate how much of this external exposure will be bioavailable. Dermal and oral external exposure were considered to be negligible in comparison to the other routes. Data on the ungual absorption of benzene are not available. The dermal absorption is about 0.2% (Franz 1984). On the one hand, the nail plate differs structurally from the stratum corneum, and its high degree of keratinisation makes it less permeable for diffusing substances (Brown et al. 2009; Walters et al. 2012). Therefore, it can be assumed that absorption through the nail plate is less than 0.2%. On the other hand, the nail plate could be compared to a hydrogel, forming small pores which allow diffusion of small molecules. Ungular absorption is strongly dependent on the properties of the penetrating molecules and their vehicles and on a number of other factors like pH or electric charge (Murdan 2002). Thus, a conclusive estimate of ungual exposure was not possible. Since benzene is highly volatile (vapour pressure 110 hPA), it can be assumed, however, that the majority of the benzene contained in nail polish ends up in the breathable air during the hardening process of the polish after it has been painted on the nail. Thus, the ungual contribution to the total exposure to benzene was also considered negligible in this case.

From the above paragraphs, the exposure to benzene was calculated to be 1 μg/kg of BW per week. A conservative assumption of one exposure per week and a distribution of benzene not throughout the whole headspace but in a 1 m^3 volume (because of the small distance between the nails and nose) result in an amount of 60 μg benzene per m^3 breathable air or 18.48 ppb (using the conversion factors of 1 ppm = 3.247 $mg/^3$ and 1 mg/m^3 = 0.308 ppm). The absorption of benzene via the alveoli from breathable air is dependent on the concentration; it is about 50% in the ppm range (EU RAR Benzene 2008c). The conservative assumption that this value is also applicable for the ppb range results in a systemic exposure to benzene of 1.38 μg per week (using 1 m^3 = 1000 l; a respiration volume per minute for women of 4.6 l/min, and a 10 min exposure).

The use of worst-case scenarios for external as well as internal exposure in the estimation above suggests that actual exposure to benzene from nail polish is even less. The estimated daily total uptake of benzene from breathable air is between

70–75 μg/day for a non-smoker in rural areas and 814–819 μg/day for a smoker, who lives in a city and carries out work that is physically strenuous (Duarte-Davidson et al. 2001). Compared to these values, weekly uptake of benzene from nail polish of 1.38 μg is small.

5.6 Chemicals Regulation (REACH)

Michal Wiecko and Thomas Rüdiger

5.6.1 Basic Principles in the REACH Regulation

The Regulation concerning the Registration, Evaluation, Authorisation and Restriction of CHemicals (REACH) (EU 2006) is the main chemicals legislation in Europe. The Regulation was adopted to ensure a high level of protection of human health and the environment, together with an enhanced competitiveness and innovation, and to make a free movement of substances and the functioning of the EU market possible. REACH implements responsibilities for industry and authorities relevant for consumer protection and consumer exposure, including mechanisms to regulate substances used in consumer-related products or in articles. In principle, the use of chemicals in consumer products shall be "safe" under foreseeable conditions without any special requirements regarding consumer behaviour, e.g. the observation of safety instructions or the use of protective equipment. Notably, in many REACH processes, the socio-economic impact needs to be considered if the use of substances is regulated and cost-benefit-related calculations become relevant (Sect. 7.2).

The scope of the Regulation is very broad, and most chemicals that are used in the EU will at some point of their life cycle lead to responsibilities under REACH for the manufacturer, importer, and downstream user of the substance. Very few areas are completely excluded from the scope, but some elements do not apply to all uses of a substance. For example, a risk assessment regarding human health from consumer use of cosmetic products is not required under REACH. The Guidance Documents of the European Chemicals Agency or information from national helpdesks for REACH should be consulted for details on the rules applicable to the specific substance or use.

The authors of this section will focus on the basic principles of the Regulation and its challenges for authorities and industry from the exposure assessor's perspective since extensive information on the implementation of the regulation is available on the website of ECHA. Since exposure assessment is of key importance for the functioning of the Regulation, the relevance and consequence of exposure assessment will be presented for each of the major REACH processes.

Regulatory Situation in the EU Before REACH

The REACH Regulation was preceded by the Dangerous Substances Directive (EU 1967) and the Existing Substances Regulation (EU 1993). The previous system discriminated between *new* substances and *existing* substances, i.e. substances present on the EU market before 1981. In 2001 the EU Commission issued a *white paper* identifying major deficiencies of the previous system. A general lack of knowledge about the properties and the uses of substances was criticised, and the processes were found to be too slow and resource-intensive. Moreover, the allocation of responsibilities was found to be inappropriate because authorities were mainly responsible for the assessment instead of the enterprises that produce or use the substances. In addition, information regarding the exposure to chemicals was difficult to obtain. In the *white paper*, the EU Commission concluded that a future chemicals regulation shall shift main responsibility for the safety of chemicals to industry. Further, improved exposure information was found to be of key importance for risk assessment and that exposure estimates or analytical determination of exposure should be obligatory. These principles were implemented in the REACH Regulation.

Authorities and Key Players

European Commission

The European Commission adopts measures to update and complete the REACH Regulation. These measures include authorisation decisions, restrictions, and adaptations to technical progress. Also, the Commission can make decisions where a unanimous agreement in the MSC (Sect. 7.2.4) could not be reached.

European Chemicals Agency (ECHA)

ECHA plays a major role in REACH implementation and administration of REACH processes. The role and tasks of ECHA are described in detail in Sect. 7.2.

Member States and National Authorities

The Member States send representatives to the MSC (Sect. 7.2) and the EU Commission's deciding Committees. Also, the national authorities perform substance evaluations, prepare dossiers for authorisation requirements or restriction proposals, and are responsible for enforcement activities. The Member States experts perform most of the scientific work on the chemical substances in REACH.

5.6.2 Instruments of the REACH Regulation

Registration: No Data, No Market

The manufacturers or importers of substances in the EU for amounts above one ton per year are required to submit a registration for the substance to ECHA (*no data, no market*). The data requirements depend on the tonnage band (1–100 tons per year, 100–1000 tons per year, and >1000 tons per year), with higher requirements for high tonnages. Companies registering the same chemical need to form a *Substance Information Exchange Forum* (SIEF) and agree on data and cost sharing. The registration consists of a registration dossier containing information regarding potentially hazardous properties of the substance and meeting the data requirements laid down in the REACH Annexes VII–X. Information on uses of the substance existing in the supply chain of the respective registrant is also an integral part of the dossier. Thus, the registrants usually not only need to perform studies (e.g. animal tests) but also need to collect information from customers (downstream users) on the uses of the substance and for the respective products present on the market or in the registrant's supply chain. Parts of the registration data like identified uses and study summaries are publicly available on ECHA's website.

In principle, all of the registered intended uses shall be safe for consumers, workers, and the environment, and all identified hazards of the substance need to be taken into account. Regarding consumers, this means that the reasonably foreseeable use - not limited to the intended use - of a product by consumers is required to be safe, as demonstrated by exposure scenarios and a comparison of exposure estimates to toxicological reference levels (usually DNELs). This assessment is documented in a Chemical Safety Assessment (CSA) according to Article 14, which is required for substances produced or imported above 10 tons per year.

The conditions of use, the covered products or articles, the operational conditions, risk management measures, and the exposure assessment need to be compiled in exposure scenarios and documented in the Chemical Safety Report (CSR). The resulting exposure scenarios also become part of the (extended) safety data sheet to ensure supply chain information and communication of safe uses or use conditions supported by the registration and of the uses advised against. The downstream users shall compare their uses with uses supported by the exposure scenarios and either inform their supplier if the use is not supported so that it can be included in the registration or they can prepare their own CSR and submit a downstream user report to ECHA.

Evaluation: Data Gaps and Risk Assessment

Two different evaluation processes were implemented by the REACH Regulation: the dossier evaluation (DEv), focussing on individual registrations, and the substance evaluation (SEv), focussing on registered substances.

Only the dossiers' technical completeness is verified during the initial registration. The compliance of the submitted data with the REACH information requirements is checked by ECHA during the dossier evaluation process for 5% of the dossiers for each tonnage band. Although in principle all parts of the dossier are within the scope of a compliance check, to date the dossier evaluation is focused on the data regarding intrinsic properties (i.e. the hazard) rather than uses of the substance and the exposure assessment in the Chemical Safety Report. The resulting lack of reliable exposure data poses a challenge for authorities. Also, this may affect the downstream use of substances, since the uses, products, or conditions of use may be supported in the safety data sheets even though a conclusion of a safe use is not warranted.

In the substance evaluation process, the Member States can evaluate substances that were previously considered to give reasons for a *concern* and are listed in the *Community Rolling Action Plan* (CoRAP; Sect. 7.2.4), which is annually updated including proposals of the Member States. The concern triggering a substance evaluation is usually caused by a combination of hazardous properties (e.g. suspected carcinogenic properties) and potential for exposure. For the SEv, all available information for the substance, including all registration dossiers, is taken into account. As a result of the evaluation, a conclusion should be reached whether further regulatory action to address the concern is required. To this end, new exposure assessments may need to be performed by the Member States' authorities, and these can be based on the registration dossiers but may also need to go beyond by including other relevant uses, aggregated exposures, or a higher tier modelling. If the available information is not sufficient to conclude on a potential risk resulting from uses of the substance, the evaluating Member State and ECHA may issue a decision requesting further data from the registrants of the substance. Such an information request can go beyond REACH registration requirements if justified, e.g. the request could include detailed information regarding the products in the registrants' supply chain.

All Member States need to agree on the decision in MSC; otherwise, the decision is referred to the European Commission. The SEv process is limited to 12 months or, if further information is requested, to 12 months after the information request is fulfilled.

Authorisation: Driving Force for Substitution

In principle the *Authorisation* in REACH consists of consequent processes with different regulatory consequences: (1) inclusion of substances into the *candidate list*; (2) inclusion into REACH Annex XIV, and (3) granting an authorisation (if applicable).

The complete substitution of *Substances of Very High Concern* (SVHC) over time is one of the goals of the REACH Regulation. These substances are identified based on their intrinsic properties alone, according to Article 57, due to CMR or PBT/vPvB properties or due to properties posing an equivalent level of concern. The

SVHC substances are considered as candidates for inclusion into REACH Annex XIV, meaning that an Authorisation will be required to use the respective substance. The identification as an SVHC already installs certain responsibilities. Inter alia, upon request the supplier of an article needs to inform the consumer about the presence of an SVHC substance in a concentration above 0.1%. For some retailers, the *candidate list* has become a *blacklist*, and they do not accept articles intended for sale to the general public from their suppliers if they contain an SVHC.

For identification as SVHC, a member state prepares a dossier in accordance with Annex XV, which also includes available exposure information. Once on the candidate list, substances are prioritised for inclusion into REACH Annex XIV (authorisation list) based on information given in the Annex XV SVHC dossier. Priority is normally given to substances with *wide dispersive* use or if used in high volumes and if the consumer use is usually considered as *wide dispersive* use. Once the recommendation by ECHA is received, the EU Commission decides on the inclusion of a substance in Annex XIV. Subsequently, the substance can only be used if an application for authorisation has been submitted to ECHA. The applicant bears the burden of proof in the application for authorisation, and the application therefore needs to describe in detail how risks resulting from the use of the SVHC are controlled or, if this is not possible, how exposure is minimised. In any case, a detailed exposure assessment needs to be included and higher tier assessments or exposure measurements may be required for a successful application. Taking into account the opinions of the RAC regarding risks and the SEAC regarding the socio-economic impact and costs, the EU Commission can approve the authorisation and will set a review period. Some uses regulated elsewhere are not subject to REACH Authorisation.

The downside of the authorisation procedure is the fact that bringing consumer articles to the market is not considered as a *use* of the substance under REACH (while production of articles is). Thus, imported articles are not subject to authorisation requirements even if they contain substances listed in Annex XIV. Companies also criticize this since it may lead to an advantage in competition for producers of articles outside the EU.

Restriction: A Complex Task

Restriction within the context of REACH means the banning of a substance or a group of substances for particular uses or under specific conditions. A restricted substance is listed in Annex XVII of REACH together with the provisions for the restriction. In contrast to authorisation, where industry is required to demonstrate the safe use and/or the socio-economic benefit of further use of a substance, authorities of the Member States or ECHA need to demonstrate an *unacceptable risk* resulting from the uses of the substance to restrict a substance. In addition, the relevance and necessity for EU-wide action needs to be demonstrated. To this end, a detailed risk characterisation must be included in a restriction dossier showing that the consumer, worker, or the environment is exposed to the substance at levels leading to a risk. For restrictions addressing risks for consumers, this may need to include robust data on consumer products or articles, e.g. from market surveillance or product registers which is not always readily available. Experience shows that the level of detail of the information submitted is often not sufficient to reach the conclusion of *unacceptable* risk even in complete and compliant registrations. At best a tier 1 exposure assessment can usually be performed based on registration data alone due to broad product categories and limited data on articles. The exposure of consumers may also result from uses or products not covered by any registration. It may be challenging in such cases to show that the risks outweigh the socio-economic costs of a particular restriction.

After a public consultation, the European Commission will decide on the inclusion of the substance in Annex XVII, taking into consideration the opinions of RAC and SEAC according to Article 68(1). In addition, for CMR substances in consumer products and articles, the European Commission can initiate a simplified "fast track" restriction according to Article 68(2). A socio-economic assessment is not required for this procedure.

Other Processes

In addition to the legally defined processes described above, the authorities involved in REACH implementation have established other standardised procedures that complement the REACH instruments.

RMOA

The Risk Management Option Analysis (RMOA) was developed early after the coming into force of the REACH to facilitate the common decision-making and discussions between Member States. Initially the RMOA was an informal and non-public document where a member state would summarise the considerations that led to the decision to initiate a certain regulatory action rather than a different one (e.g. an authorisation versus a restriction or a regulatory action outside of REACH). The RMOA development has become a standard process that in many cases will precede other REACH regulatory action of the Member State. Also, the development of the RMOA is now officially announced, and a stakeholder consultation may be performed. The decision-making is accompanied by scientific discussions in expert groups, and RMOA conclusion documents are publicly available. Although not mandatory, an assessment of exposure and resulting potential risks is usually relevant to reach a conclusion on the most suitable regulatory action.

5.6.3 The REACH Exposure Scenario

Gerhard Heinemeyer

According to the EU-Regulation EC No. 1907/2006 (EU 2006), the exposure scenario is defined as follows:

> An exposure scenario is the set of conditions that describe how the substance is manufactured or used during its life-cycle and how the manufacturer or importer controls, or recommends downstream users to control, exposures of humans and the environment. These sets of conditions contain a description of both the risk management measures and operational conditions which the manufacturer or importer has implemented or recommends to be implemented by downstream users.

The consequence of this definition is, in terms of consumer exposure and in contrast to the "classical" definition of the exposure scenario (Sect. 3.2.1), the legal request to guarantee that a product is safe for use. Basically, consumers are not involved in the supply chain because they are out of the chain of responsibility. The legislation is not requesting for recipes or ingredient lists. However, the manufacturer or importer of chemical products must declare how the use of the products in the form being marketed is safe. The producers of chemical products have different options to characterise the safety.

Examples of a scenario characterisation under REACH include:

1. A producer may declare that the concentration in the product category is in a range which will not lead to concentrations of concern if the product is used as employed.
2. The product is safe, due to implementation of efficient risk management measures. In the view of the authors of this book, non-communicated (product-integrated) risk management measures are considered to be efficient; there are no needs to make any actions for the consumers (Bruinen de Bruin et al. 2007).

REACH exposure scenarios are normally constructed as "broad" scenarios, which means that they cover a broad range of uses. For example, a solvent may occur in a great number of different products that might be characterised by product categories, e.g. as pointed out in the ECHA Guidance R.12 (ECHA 2015). The broad scenario might be represented by products or a category of products at an aggregated level having high concentrations and respective uses to produce the highest possible exposure. If the estimate of this scenario will not exceed the critical DNEL, then any other exposures with other products having the substance as an ingredient will be lower. This result also applies for risk management measure application. Risk management measures are mostly model parameters for consumer exposures, e.g. the concentration in the product or the *controlled* release from the product.

van Engelen et al. (2007) proposed the use of so-called sentinel scenarios. A sentinel scenario is represented by the scenario that reveals the highest exposure of all uses and therefore represents a very conservative estimate. However, sentinel

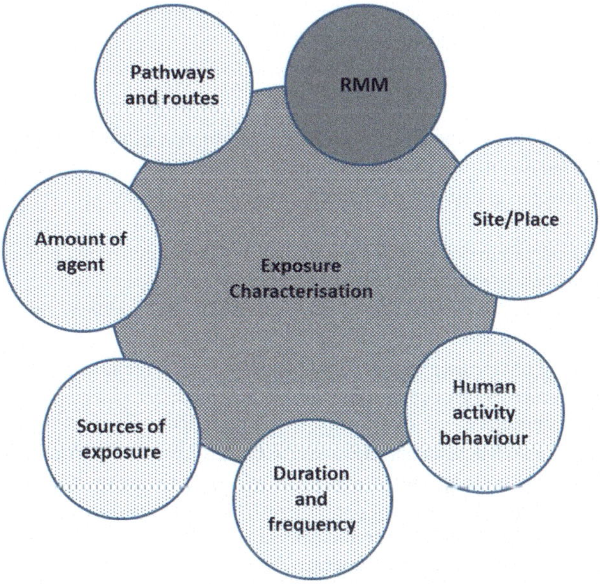

Fig. 5.6 Exposure scenario elements extended by risk management measures

scenarios do not consider possible aggregated exposure and in some (theoretical) cases lead to underestimation.

Basically, the REACH exposure scenario is not different from the classical one. REACH introduces two terms in the scenario characterisation that should be separated: (1) the operational conditions and (2) the risk management measures. These two terms need particular attention. Both can be seen as exposure parameters as explained in Sect. 3.4. The operational conditions are parameters that characterise the exposure scenario, the population and use, and the product. The risk management measures characterise those parameters that can be used to control the exposure, i.e. that keep it at an acceptable level. In the REACH scenario, the main focus is put on measures that limit the exposure at a particular level, e.g. by limitation of the concentration of the substances of concern in the products.

The elements of an exposure scenario as shown in Fig. 3.2 and discussed in Sect. 3.2.1 are extended by an additional element, becoming an integrated part of it (Fig. 5.6). Normally the REACH exposure scenario is a low tier scenario, characterised by the three exposure scenario determinants (operational conditions) that describe exposure on a generic basis: first, the exposure source, covering a wide range of any products in which the substance of concern is an ingredient; second, the aggregated amount of this substance in the products; and third, the paths of exposure.

The term *risk management measure* (RMM) is in accordance with the aim of REACH, which is to control (manage) the risk. For estimation of consumer exposure, RMMs are normal variables in an exposure model, and, from this view, they

Table 5.4 Risk management measures that are also parameters in an exposure scenario and are thus variables in the exposure model

RMM parameter	Objective	Remarks
1. Communicated risk management measures (not to be considered in exposure scenarios)		
User instruction in general	To instruct the user how to use the product, e.g. to take a particular quantity	The user will often not follow the instructions
Ventilation rate	To increase room ventilation with the objective to reduce the exposure level	Instruction: open windows during use
Duration of use	Restriction of the duration of use	Do not use longer than one hour
Place of use	To instruct the user whether or not to use the product in particular rooms or environments	This communicated RMM is often made for sprays, e.g. not be used indoor
2. Product integrated risk management measures (to be considered in exposure scenarios)		
Product composition	To avoid direct hazardous effects during use and to reduce the amount of ingredients in products, e.g. food to achieve acceptable/tolerable limit values	Providing only low amounts or concentrations of certain hazardous substances, e.g. corrosives
Packaging design	Restrict the amount used, size of portion	Packaging that dispenses only a limited amount of product at a time
Limit values	For limiting concentrations of substances in consumer products and in food to avoid critical exposures, i.e. the maximal daily intake from the entire sources of the substance (includes food and other consumer products, e.g. cosmetics, toys, etc.)	Limit values are determined by regulation and after scientific evaluation of the risks of the substance of concern

can be used in any exposure scenario and model. Some risk-reducing measures typical for consumer products are listed in Table 5.4. The tiered approach in the REACH exposure scenario can be understood by introducing risk management measures.

RMMs can be roughly distinguished into communicated and non-communicated ones. The latter are also called product-integrated RMMs (Bruinen de Bruin et al. 2007).

The communication of RMMs to consumers happens by use instructions. Normally these instructions are the recommendations for use that are printed on the labelling or on use instruction folders. It has been questioned whether the use instructions are typically followed by the users. Therefore, Bruinen de Bruin et al. (2007) recommend to resign for communicated use instructions for exposure estimations. This recommendation has been picked up in the technical guidance documents for REACH (ECHA 2012).

In this context, for example, although the amount of a product to use may be communicated by instructions to the user (but should not be accounted for exposure), it may also be restricted by the packaging, e.g. by limiting the amount available per packaged form or by a restriction by changing the physical characteristics of the product such as the viscosity of the composition. Only the latter ones might be considered in an exposure scenario to justify safety. Changing the composition to avoid certain ingredients eliminates the exposure to and risk from the avoided substance but may bring exposure and risk arising from a replacement ingredient.

Tiered Approach (CS-Reporting, Restriction, and Approval)

In view of the registrants' responsibility for many different substances, the desire to have a high throughput of data for those substances using minimal capacity is high. Having to fulfil the legal data requirements for the hazard assessment makes it logical to follow a screening approach for the exposure assessment (and thus risk assessment) side. Therefore, in the registration process under REACH, the exposure assessment under REACH follows a tiered approach. In the first tier, the exposure is assessed using simple and straightforward estimations where the assessment is based on worst-case assumptions for weight fraction, product amount, conditions of use, and residential settings. Several computerised tools have been developed to support this type of assessment. For example, the ECETOC TRA consumer tool version 3.1 (ECETOC 2012, 2014) contains a set of fixed parameters that should cover the exposures resulting from the use of products within product categories and article categories. The categories serve as sentinels for all products that fall within a category. As stated by ECETOC, the purpose is to filter out those exposure scenarios where a possible risk cannot be excluded. Implicitly, the purpose is not to acquire a realistic exposure estimate. Despite not being realistic, if considered safe, the exposure assessment determines the conditions of use throughout the supply chain. As of this writing, the ECETOC TRA consumer tool (all versions) is used in practically all registration dossiers that describe the consumer use of products (mixtures) and articles (Sect. 5.4).

There are two options possible when the exposure estimate derived with the first-tier assessment leads to the conclusion that risks are not excluded. The first option of advice is "do not use it", the second is "refine the assessment" (ECHA 2016). Here, the refinements concern the exposure assessment only. The higher tier estimates can be obtained by either refining the input parameters to better suit the exposure scenario of a specific product or activity or by moving on to more sophisticated exposure models (incl. introduction of RMM). The recommended model for higher tier consumer exposure assessments under REACH is ConsExpo 4.1 (see Sect. 5.10.5). Exposure measurements in the domain of consumer exposure are seldom performed. However, the input parameters may be based on exposure measurements. Since 2016 ConsExpo is available as a Web-based tool (see Sect. 5.10.5).

Practical experience with the first-tier assessments may indicate that exposure scenarios fail the screening and are subject to refinement. The sentinel product and associated defaults are supposed to cover a wide range of products that in fact prove difficult to group into one category. The exposure estimate in many cases may turn out to be too worst case. Some branch organisations therefore developed their own models, e.g. the A.I.S.E. REACT tool (see Sect. 5.10.13), or developed the so-called Specific Consumer Exposure Determinants (SCEDs) to be used in the ECETOC TRA consumer tool version 3.1 (see Sect. 5.10.10). This latest version includes the possibility to overwrite certain default settings, thus creating one's own default scenarios. The SCEDs contain a set of input parameters per product and article categories for a subset of products that are marketed within that branch. At of this writing, SCEDs have been prepared for several categories cleaning products, e.g. laundry products and hand dishwashing liquids, and for other products.

Higher tier models, such as ConsExpo, present a more realistic way to determine the exposure. However, in practice the ConsExpo tool is used together with the fact sheets where the users apply the default scenarios that are described. The fact may be that the model itself still describes a first-tier equation similar to ECETOC TRA consumer tool but uses more appropriate input parameters. On the other hand, other default scenarios describe the use of higher tier models that are physics-based and are "more" realistic. The use of higher tier models generally requires more input and knowledge from the user, and the results are more specific for the product of interest (ECHA 2016).

The tiered approach is generally valid for all processes accepted under REACH for registration purposes. However, first-tier exposure assessments are generally too simplistic for substances of very high concern (SVHC, this includes CMR, PBT, vPvB, and ELoC substances) that fall under the restriction, authorisation, and substance evaluation processes or those processes preceding these processes in the RMOA or SVHC identification. This is especially when the risk assessment is for a substance without a threshold (e.g. genotoxic carcinogens) and the assessment is supposed to demonstrate that the exposure of interest is as low as reasonably achievable. First-tier assessments are not to be used for this purpose. It is therefore strongly recommended by ECHA that exposure assessments for authorisation processes are at a higher tier level and specific for the use for which authorisation is applied for. It is to be noted that consumer products should not contain substances above 0.1% v/v that are CM (carcinogenic, mutagenic) or 0.3% v/v that are R (reprotoxic). Such limits do not exist for articles, unless restricted.

Short-Term and Long-Term Exposures

The starting point in the risk assessment under REACH is the long-term exposure. In the exposure scenario concerning consumer products, the long-term exposure is by default determined as daily event exposures. This conservative approach can lead to the conclusion that risks are not controlled. One of the refinement options is to adapt the exposure assessment based on frequency of use data. If the events occur on fewer than 15 days per year, the exposure may be considered as infrequent and short term. The guidance prescribes that registrants may seek alternative DNELs that better match the exposure, e.g. DNELs based on 28-d or 90-d toxicity studies. Additional

Table 5.5 Actions needed to perform exposure assessments for acute and chronic risks under REACH

Scenario	Frequency duration	DNEL derivation approach/risk characterisation	Remark
Chronic	365 d/a, 70 a, 24 h/d	Derive long-term DNEL Compare event exposure to the long-term DNEL	Conservative starting point (see step 1 below)
Chronic adjusted for shorter daily duration	365 d/a, 70 a, <<24 h/d	Adjust long-term DNEL by applying modified Haber's law For practical reasons: divide the exposure concentration by the adjustment factor and compare to chronic DNEL	The following default adjustment factors (rounded) are based on exposure duration (calculation see step 3 below) Up to 0.25 h/d = >adjustment factor 4.5 Up to 1 h/d = > adjustment factor 3 Up to 3 h/d = >adjustment factor 2 Up to 8 h/d = >adjustment factor 1.5
Infrequent	<15 d/a 24 h/d	Derive DNEL for infrequent exposure Compare event exposure to DNEL for infrequent exposure	Alternative starting point: use the result of an appropriate short-term study (e.g. 28 days) and apply appropriate AFs to derive the DNEL for short-term conditions (see step 4 below)
Infrequent adjusted for shorter daily duration	<15 d/a <<24 h/d	In addition, adjust DNEL for infrequent exposure by applying modified Haber's law For practical reasons: divide the exposure concentration by the adjustment factor and compare to DNEL for infrequent exposure	The following default adjustment factors (rounded) are based on exposure duration (calculation see step 3 below) • Up to 0.25 h/d = > adjustment factor 4.5 • Up to 1 h/d = > adjustment factor 3 • Up to 3 h/d = > adjustment factor 2 • Up to 8 h/d = > adjustment factor 1.5
Assessment for acutely toxic substances	n.a.	Derive acute DNEL Compare exposure peak with acute DNEL (convention, take 15 min average)	Complementary assessment

substance-specific data are required to ensure that internal exposure to a substance does not lead to long-term exposure.

Some exposures may only last for a short period, e.g. a few minutes, during the day. The ECHA Guidance implemented correction factors for day of exposure duration (see Table 5.5). Performing a short-term exposure assessment is not mandatory in any case but is an option provided to registrants for further refinement. It should be noted that short-term exposure in the occupational exposure assessment refers to durations shorter than 8 hours; more specifically in most cases with short-term, a 15-min time-weighted average is meant.

If a substance is classified as acutely toxic, it should be kept in mind that the generic scenario does not distinguish between short-term and long-term exposure. The generic REACH exposure scenario is a short-term scenario. Therefore, it can be taken to be compared with the acute hazard level. If needed, a tiered approach that particularly characterises the short-term event might be performed.

5.6.4 REACH Exposure Models

Christiaan Delmaar

All standard algorithms needed for exposure calculations at low tiers for exposure estimations under REACH are simple equations. They are described in detail in the ECHA Guidance document R.15 (ECHA 2016). The estimation can be done using basic arithmetic rules and by using standard calculation tools. It should be mentioned that at this level of the tiered approach, the formulas used in the model equations are similar. The selection of the correct type of product and the respective model parameter (Sect. 5.6.5) must be considered.

Dedicated software tools for exposure evaluations under the REACH legislation by using the formulas are proposed in the ECHA R.15 guidance (ECHA 2016).

A number of dedicated tools for consumer exposure assessment are available and can be used under REACH. The tools specifically developed for the use in REACH include ECETOC TRA, EGRET, and A.I.S.E. REACT (see Sect. 5.10). They represent different iteration steps in the tiered approach. An example of a more general tool that is suitable for use under REACH is ConsExpo which provides a number of tiered models to calculate an iterative assessment, including deterministic and probabilistic estimations.

Example

Case: a substance is used in a surface cleaner. For the approval of this substance, the safety of its use in a spray product is to be demonstrated.

Tier 1

ECETOC TRA (2012) specifies a default exposure assessment for sprays in the category "PC35 Washing and cleaning products" with a sub-model for "Cleaners, trigger sprays (all-purpose cleaners, sanitary products, glass cleaners)".

(continued)

The model used by ECETOC TRA is the equation for average the daily exposure:

$$\mathrm{Exp} = \frac{\mathrm{PI} * A * \mathrm{FQ} * \mathrm{DF} * \mathrm{ET} * \mathrm{IR}}{V * \mathrm{BW}} \quad (5.25)$$

and the model inputs (exposure parameters{XE "exposure:parameter"}) are given in Table 5.6. This equation assumes that the substance is completely released into the room of volume V. The resulting air concentration remains constant during the exposure event and does not diminish, for example, with ventilation or the deposition of aerosol droplets to the floor.

This (constant) air concentration is inhaled with a rate IR during the exposure duration ET. The resulting event exposure is averaged using the frequency per day (FQ) and normalised using the body weight.

With a user-specified input of the product ingredient of 0.01 (1% by weight), and using the ECETOC TRA provided defaults for all other parameters, an inhalation exposure (daily dose) of 0.47 mg/kg of BW per day is calculated.

If this exposure is above a derived DNEL for this substance, a refinement of the exposure assessment is an option in a higher iteration of the risk assessment.

Tier 2

The A.I.S.E. REACT tool provides a model "SURFACE CLEANERS (A.I.S.E. C7, PC35); Spray (neat)" that seems appropriate for the product being considered in an iteration of the exposure assessment.

The equation used in this model is:

$$D = \left[\frac{\mathrm{wf} * A_{\mathrm{spray}}}{V}\right] * \left[\frac{Q_{\mathrm{inh}} * T * \mathrm{FQ} * F_{\mathrm{resp}} * F_{\mathrm{dep}} * F_{\mathrm{abs}}}{\mathrm{BW}}\right] \quad (5.26)$$

The main difference in the model is the inclusion of fractions of respirable particles F_{resp} and F_{dep}, the fraction of the inhaled mass that deposits in the lung.

Using the same input of the concentration of the substance in the product of 0.01 (1%) and the assumption that the complete aerosol is respirable, A.I.S.E. calculates (using default values for all other parameters) that 0.012 mg/kg of BW per day is the exposure, a reduction of a factor of 40 over the first iteration using ECETOC TRA. The difference is mainly due to the default for the exposure duration T, which is reduced to 10 minutes in A.I.S.E. REACT from 4 h (the default in ECETOC TRA) and also the reduction due to the deposition fraction of inhaled mass in the lung (which is 20%, rather than the 100% implied in the ECETOC TRA) (comp. Table 5.7).

(continued)

Tier 3

Higher tier exposure can also be calculated in the ConsExpo Web tool. This tool features two models that can be used to estimate exposures from spray products: a low tier model "Spraying: instant release" and a more mechanism-based, higher tier model, the "Spray: spraying" model. The "instant release" model assumes immediate dispersion of the spray and considers only removal of the substance from the indoor environment by ventilation. The "spraying" model simulates exposure from sprays and takes into account (in a simplified manner) dispersion of the sprayed mass and removal of the substance from the room by ventilation and by deposition of the aerosol particles.

The ConsExpo fact sheet on cleaning products provides a default spray cleaner scenario. Using the settings of this default scenario with additional input on the weight fraction of the substance in the product of 1%, ConsExpo calculates an inhalation exposure of 0.054 mg/kg per day for the "Instantaneous release" model and 0.0006 mg/kg per day for the higher tier "spraying" model.

The "Instantaneous release" mode in the ConsExpo model is useful as a low tier model. It does not consider limited deposition in the respiratory tract as does A.I.S.E. REACT.

Of interest is that this iteration step will not result in a lower exposure estimate compared to step 2. This is explained by changing the model concept from model 2 to model 3. Iteration step 4 is consistent with step 3 and reveals a reduction of exposure due to having more detail.

Conclusions

In higher tier models, more detailed processes in the exposure evaluation and more product-specific input values are taken into account. Thus, the exposure calculated tends to decrease (a lot) in higher iterations (higher tier evaluations). However, many aspects of the higher tier methods are difficult to quantify or defend in a product safety assessment that should be protective (e.g. deposition in the lung, aerosol size distributions, and near-field versus far-field concentrations). This introduces significant uncertainties in the higher tier assessments. These uncertainties should be discussed and preferably quantified.

Note: The tier from the ECETOC TRA to the A.I.S.E. tool includes several iterations: (1) the first one is the enlarge the model by consideration of the physiological inhalation characteristics, and (2) the second one is an iteration of the model parameters for the exposure time and, to a minor degree, the differences in the defaults for the amount of product and room volume. These changes might balance each other in the assessment calculations.

In any case of using different tools for tiering the exposure assessment, the basic information on model equations and parameters is needed.

Table 5.6 Input parameters used in the ECETOC TRA and A.I.S.E. tool for assessment of exposure from use of a surface cleaner

Symbol	Parameter	ECETOC TRA	A.I.S.E.	Unit	Origin
PI	Product ingredient	0.01	0.01	–	User input
A	Amount of product	35	30	g	Tool default
FQ	Frequency of exposure	1	1	per day	Tool default
F	Fraction released into air	1			User input
DF	Dilution fraction	0.29			Tool default
F_{dep}	Fraction deposited in lung		0.2		Tool default
F_{resp}	Respirable fraction		1		User input
F_{abs}	Fraction absorbed		1		Tool default
ET	Exposure time	4	0.167	hours	Tool default
IR	Inhalation rate	1.37	1.08	m^3/h	Tool default
V	Room volume	20	15	m^3	Tool default
BW	Body weight	60	60	kg	Tool default

Table 5.7 Exposure estimates for "surface cleaner" using different standard tools in a tiered approach

Model	Daily average dose (mg/kg/day
ECETOC TRA	0.47
A.I.S.E. REACT	0.012
ConsExpo spray: instant release model	0.054
ConsExpo spray: spraying model	0.00056

5.6.5 Exposure Parameters in the REACH Regulation

Friederike Neisel

The exposure parameters used in consumer exposure assessments for registrations under the REACH Regulation reflect the operational conditions and risk management measures of the exposure scenarios. These are documented in the Chemical Safety Report (CSR) of the REACH registration dossier and in the (extended) safety data sheet to ensure supply chain communication about safe uses. When comparing their uses to the uses supported by the registration, the downstream users will have to rely on the communicated operational conditions and risk management measures that define the relevant exposure parameters.

The tools used in the tiered approach of consumer exposure estimation under REACH generally provide default values for the exposure parameters used in their exposure models (see Sect. 5.10).

Some exposure parameters used in the tier 1 equations of the ECHA Guidance R.15 (ECHA 2016) are presented (Tables 5.8, 5.9, 5.10, 5.11, 5.12), and the way how they can be influenced by risk management measures is described. The abbreviations in the tables are indicated as given in the ECHA Guidance; however, others

Table 5.8 REACH exposure parameters (anthropometrics)

Abbreviation				
ECHA	Also used	Parameter	Unit	Remark
BW		Body weight	kg	Default = 60 kg (for adults) or 10 kg (for children)
IH_{air}	IR	Inhalation rate of person	m^3/h	Default = 20 m^3 per day (for adults)

Table 5.9 REACH exposure parameters (product-related parameters)

Abbreviation				
ECHA	Also used	Parameter	Unit	Remark
Fc_{prod}	PI, w_f	Weight fraction of substance in product/article (before dilution)	*g/g*	The limitation of concentrations in products should be understood as a non-communicated RMM
C_{prod}		Concentration of substance in product before dilution	g/cm^3	
RHO_{prod}		Density of product before dilution	g/m^3	

Table 5.10 REACH exposure parameters (building-related parameters)

Abbreviation				
ECHA	Also used	Operational condition	Unit	Remark
V_{room}	V	Room volume	m^3	Default = 20 m^3
q	VR	Ventilation rate (room)	h^{-1}	Also RMM

also used in dedicated tools (e.g. ECETOC TRA, A.I.S.E.; see Sect. 5.10) are also mentioned.

Anthropometrics

(see Table 5.8)

In consumer exposure assessment tools used for registrations under REACH, the values for body weights are derived from the ConsExpo General Fact Sheet, most recently updated in 2014 (te Biesebeck et al. 2014). Exposure calculations are normally targeted to adults, and only exposure assessments for selected child-specific articles and mixtures are targeted to small children. However, it should be noted that according to the Guidance (ECHA 2016):

> when performing an exposure assessment to a chemical from consumer products within a certain Article Category (AC) or Product Category (PC), the possibility of a separate exposure assessment for children should be always considered, unless they are clearly not exposed.

Table 5.11 REACH exposure parameters (consumer activity pattern data)

Abbreviation				
ECHA	Also used	Operational condition	Unit	Remark
Q_{prod}	A	Amount of product/article used (before dilution)	g	The limitation of the amount of a product should be understood as a communicated risk management measure
V_{prod}		Volume of product used before dilution	cm^3	Due to Q_{prod}
V_{appl}		Volume of diluted product actually contacting the skin Volume of diluted product per event in contact with the mouth	cm^3	Due to Q_{prod} and D
D	DF	Dilution factor	–	Default = 1
N		Mean number of events per day	–	
$T_{contact}$	ET	Duration of contact per event	d	Default = 1 day
A_{skin}		Surface area of the exposed skin	cm^2	Defaults are available for different skin areas

Table 5.12 REACH exposure parameters (parameters characterising contact)

Abbreviation	Operational condition	Unit	Remark
TH_{der}	Thickness of product layer in contact with the skin or the mouth	cm	For assessment of dermal exposure and exposure from mouthing behaviour, default = 0,01 (mixture) or 0,001 (article) cm
F_{oral}	Fraction of V_{appl} that is ingested		Default = 1
Fc_{der}	Fraction of the applied product remaining on the skin		Default = 1
F_{resp}	Respirable fraction		Default = 1

While the anthropogenic parameters cannot be changed by risk management measures, the exposure of children can be prevented (or minimised) by product integrated measures like childproof packaging. Advice against child use can also be given; however, compliance should not be assumed in the quantitative exposure estimation.

Product-Related Parameters

(see Table 5.9)

These parameters refer to the product composition which is basically known by the product manufacturer. The knowledge on product-related parameters ideally derives from supply chain communication between downstream users and registrants. However, experience has shown that the assessments in the registration dossiers are often based on standard assumptions provided in the exposure assessment tools. Many of these assumptions have been collected to serve as a database in ConsExpo fact sheets for the different product groups. These fact sheets have been developed to provide background information and default exposure parameters for consumer exposure estimations with the ConsExpo tool (Delmaar and Schuur 2017, see Sect. 5.10.6) (http://www.rivm.nl/en/Topics/C/ConsExpo/Fact_sheets).

Iterations of the concentration of a substance in a product are often used for refined calculations in a tiered assessment when exposure assessments are reviewed in substance evaluations under REACH. It should be noted that such product-integrated risk management measures are a matter of communication along the supply chain.

Building-Related Parameters

(see Table 5.10)

The standard room volume used in tier 1 assessments may be modified in refined exposure estimations if data show that the use is specific to certain rooms with other dimensions such as kitchens, bathrooms, or garages. Also, for short-term local exposure assessments, the room volume may be reduced to a 1 m^3 "personal volume".

In the tools used for exposure assessments under REACH, the values for room volumes are normally derived from the ConsExpo General Fact Sheet (te Biesebeck et al. 2014). The values in this fact sheet are based on well-documented Dutch data, but other data are also reported. This is also the case for the default values for ventilation that are used in higher tier exposure assessments.

While the building characteristics themselves cannot be changed by risk management measures, the choice of a location (outdoors or different rooms indoor) or the management of ventilation devices (doors, windows, or mechanical ventilation) is up to the consumers and their habits and practices, and it can be addressed by the communication of risk management measures (see below).

Ventilation rates are used in many exposure models at higher tiers. However, the definition of a ventilation rate, particularly if it is higher than the basic default of 0.6/h given in the ConsExpo General Fact Sheet, should be understood as a communicated risk management measure that can be indicated in the CSR but should not be accounted for in the quantitative risk characterisation. This view is supported by the REACH Guidance (ECHA 2012):

> Consumer instructions cannot be expected to be highly effective, unless consumer behavioural data suggest that a sufficient degree of implementation can be assumed. Therefore consumer RMMs that depend on instructions should as a general rule only be introduced when the use of such RMMs can be shown to be effective, necessary and well adhered to by consumers.

Activity Pattern Data (Consumers)

(see Table 5.11)

The amount or volume of a product used by the consumers is a key parameter in tier 1 exposure estimations under REACH. This also applies for the dilution factor and the applied volume of a product that has been diluted by the consumer. The duration of contact per event can be related to the amount used. It is also related to the application technique which may vary among products and also among consumers. The mean number of use events per day and the frequency of uses over the year, another parameter that is important for refinement options, will vary a lot between different products and consumer populations. The surface area of the exposed skin is frequently deduced from the surfaces of potentially exposed body parts. All of these parameters are specific to the single-product categories and are influenced by cultural habits.

The amount or volume used and the dilution factor can be addressed by product integrated risk management measures, i.e. individually packaged amounts like tablets, separate sachets, or other devices for dosing and application. A packaging that, e.g. releases only small amounts or an application device can also restrict the exposed surface of the skin. In addition, all behaviour-related exposure parameters can be addressed by the advice that is communicated, and the surface area of the exposed skin can be reduced by use of personal protective equipment such as gloves, but compliance to these measures should not be assumed in the quantitative exposure assessment.

The default values for parameters related to consumer habits and practices that are used in the registration dossiers under REACH are often based on standard assumptions provided in the exposure assessment tools. Many of these assumptions are based on the ConsExpo fact sheets for different product groups, but others are based on expert judgement.

Another source for behaviour-related exposure parameters under REACH are the Specific Consumer Exposure Determinants (SCEDS). These parameter values were developed by industry's downstream sector organisations in order to feed information for refined consumer exposure estimations to ECETOC TRA 3.1 and CHESAR, but they can also be used for exposure assessments with other tools. According to the Guidance on SCEDS (DUCC and CONCAWE 2014), each value within the SCEDs has to be substantiated by reference to suitable information sources that are open access and have been published and ideally peer reviewed (the "rationale"). However, a reference to defaults contained in existing ECETOC TRA reports is also deemed sufficient.

Knowledge about the distributions of parameters related to consumer habits and practices can be obtained by different methods. These include questionnaires, observational studies, laboratory observational studies, and measured data. However, it has to be noted that few data on behaviour-related exposure parameters have been published in the open literature for most consumer product categories. In a literature study on published consumer behaviour data performed on behalf of BfR in the year 2015 (Heiland et al. 2016, see also Sect. 5.2.2), 37 studies were identified. From these, 822 data sets on the consumer exposure parameters "amount used", "duration", "frequency", and "location of use" were extracted. Fifty-seven percent of the data sets referred to washing and cleaning products. Data for other product categories were scarce. For example, only one report presented European data on "amounts used" for the product category PC9a: Coatings, paints, thinners, removers. This means that a scientific basis for behaviour-related parameters used for consumer exposure assessments under REACH is still missing for many product categories.

Parameters Characterising Contact

(see Table 5.12)

The three fractions have been documented in the Table 5.12: as they appear in the tier 1 equations published in the REACH Guidance (ECHA 2016). However, it should be noted that the use of these fractions requires a data-based justification of the specific numerical value. If the respirable fraction is used for the assessment of exposure by inhalation, the oral exposure from swallowing the non-respirable fraction may have to be considered.

The thickness of a product layer in contact with the skin or the mouth is a parameter used for tier 1 assessments of dermal or oral exposure in cases were measured data on the migration of a substance from a product to the skin or into the mouth are missing. It is assumed that all of the substance contained in the thin product layer is released. Oltmanns et al. (2015) provided a general discussion on the exposure parameters used in different tier 1 tools under REACH and discussed the background of this parameter. Its use in exposure assessments for consumer articles has been critically discussed by Delmaar et al. (2013).

5.7 Nanoparticles and Nanomaterials

Yasmin Sommer

5.7.1 What Is "Nano" About?

Above all "nano" is a question of scale. The term "nano" derives from the Greek word for "dwarf". It is a unit prefix applied in the International System of Units (SI) to express the 10^{-9} part of a unit. Some easy to visualise examples of the relative dimensions of this scale can be found at US-EPA (2017a) and ECHA (2017b). Beyond this, the understandings on terminology regularly differ depending on the regional, scientific, and legal context. Notably, the term "nano" is often used when referring to objects with at least one external dimension in the size range of 1–100 nm, or materials with inner or surface structures in this range. An example is the set of definitions in the (DIN) ISO/TS 80004-X series (ISO 2010, 2015). According to them, a "nano-object" would overarchingly also be defined as "nanomaterial". However, different conclusions for a given material might arise, especially in a regulatory context due to differing criteria (see Box 5.1 for more information). In this chapter, the term nanomaterial will be used in the sense of the ISO definition unless otherwise indicated.

Size and structure of the nanoscaled matter may lead to phenomena and properties differing distinctly from the physical and chemical properties of the same matter at conventional (bulk) scale. These changes have made application of "nanotechnologies" a major innovation driver across different product sectors. Besides "merely" size-reduced particles of substances, sophisticated materials connecting "nano" to advanced materials and biotechnology have been created. The resulting "nano-enabled" products are covered by a broad range of regulations (e.g. due to the different product sectors effected such as personal care, cosmetics, textiles, biocides, sporting equipment, cleaning products, building/construction materials, electronics, fuel, food, feed, food contact materials, and medicine).

Regulation has to ensure that the benefits of such products within their scope can be used safely. Some areas of legislation have developed guidance or best practice documents for this task (e.g. EFSA 2018; ECHA 2014). Irrespective of whether or not such a guidance or best practice document exists, exposure assessment is an essential step in the safety assessment. Following the external exposure assessment, information on toxicokinetics based on measurements or assumptions is used to establish the internal dose (the dose that reaches the systemic circulation, organs, and tissues).

Box 5.1 Examples of "nano" terminology definitions by different bodies

The contained information is not an exhaustive overview of "nano" terms and related, existing definitions. Most of the included definitions are based on publications by ISO. Boverhof (2015) provides a more comprehensive overview regarding definitions of "nanomaterial".

Nanomaterial	• According to the definitions of nanotechnology core terms by the International Organization for Standardization (ISO), a "nanomaterial" means: A material with any external dimension in the nanoscale or having internal structure or surface structure in the nanoscale. (ISO 2010) • According to the recommendation adopted by the EU Commission in 2011, a "nanomaterial" means: A natural, incidental or manufactured material containing particles, in an unbound state or as an aggregate or as an agglomerate and where, for 50% or more of the particles in the number size distribution, one or more external dimensions is in the size range 1–100 nm.

(continued)

Box 5.1 (continued)

	In specific cases and where warranted by concerns for the environment, health, safety, or competitiveness, the number size distribution threshold of 50% may be replaced by a threshold between 1% and 50%. By derogation from the above, fullerenes, graphene flakes, and single-wall carbon nanotubes with one or more external dimensions below 1 nm are to be considered as nanomaterials. If the size distribution criteria is not fulfilled and where technical feasible and in specific regulations, instructed identification may also be based on the specific surface/volume ratio. A ratio above 60 m^2/cm^3 is considered as "nanomaterial". (EU 2011) Please note: At the time of writing (August 2017), the recommendation was under revision. • According to the working definition developed for regulatory purposes by the Australian NICNAS, nanomaterials are: "*...industrial materials intentionally produced, manufactured or engineered to have unique properties or specific composition at the nanoscale, that is a size range typically between 1 nm and 100 nm, and is either a nano-object (i.e. that is confined in one, two, or three dimensions at the nanoscale) or is nano-structured (i.e. having an internal or surface structure at the nanoscale)*". The material must include 10% or more number of particles that meet the above criteria (size, unique properties, intentionally produced). In addition, the notes to the working definition given in the cited source explain that the materials are: "*distinct from accidentally produced materials*", that: "*'unique properties' refers to chemical and/or physical properties that are different because of a material's nanoscale features when compared with the same material without nanoscale features, and result in unique phenomena (e.g. increased strength, chemical reactivity or conductivity) that enable novel applications*" and that: "*aggregates and agglomerates are considered to be nanostructured substances*". (*NICNAS* 2017)
Nanotechnology (ISO)	Application of scientific knowledge to manipulate and control matter in the nanoscale in order to make use of size- and structure-dependent properties and phenomena, as distinct from those associated with individual atoms or molecules or with bulk materials. (ISO 2008)

(continued)

Box 5.1 (continued)

Nano-object (ISO)	According to the definitions by the International Organization for Standardization (ISO), a "nano-object" means: A material with one to three external dimension in the nanoscale (nanoplate, nanofibre, and nanoparticle). (ISO 2015)
Agglomerates (ISO)	According to the definitions by the International Organization for Standardization (ISO), agglomerates can be described as collections of weakly bound particles and/or aggregates for which the resulting surface area is similar to the respective sum obtained for its individual components. (ISO 2008)
Aggregates (ISO)	According to the definitions by the International Organization for Standardization (ISO), aggregates can be described as collections of strongly bonded or fused particles for which the resulting surface area may be significantly smaller than the respective sum obtained for its individual components. (ISO 2008)
Nano-objects and their agglomerates and aggregates (ISO)	This term is abbreviated NOAA. It was introduced by ISO and pulls together a specific combination of ISO terminology on nanomaterials (regarding nano-objects and nanostructured powders) which are published in two different series. According to ISO, it relates to nano-objects and their aggregates and agglomerates greater than 100 nm. (ISO 2014)

The following chapters aim to illustrate the current status of consumer exposure assessment for nanomaterials and explain the differences/particularities in comparison to conventional (bulk) chemicals. They will focus on the assessment of the external exposure of individuals occurring directly through the use/handling/manipulation of consumer products during their use phase. It should be noted that the knowledge about consumer exposure to nanomaterials is evolving and development of appropriate assessment concepts is an area of ongoing scientific discussion. Limitations of the validity should be kept in mind, especially for more sophisticated nanomaterials and future generations (e.g. with self-assembling or biosensor properties) as these have not been in the focus of scientific debate so far. Some of the knowledge presented below has only been gained on specific types of nanomaterials (e.g. nanoparticles), and further research is needed to determine its validity for the broad category of nanomaterials.

5.7.2 *Where Is the Difference in Exposure Assessment?*

Apart from an easily confusing terminology requiring some attention, at first sight it seems there is not much of a difference in the exposure assessment of nanomaterials compared to the exposure assessment of conventional (bulk) chemicals:

Consumers may become exposed via all routes during different stages of the products life cycle.

The (primary and secondary) exposure routes to consider will depend on the product of interest and regarded activities associated with that product (e.g. a nanomaterial in a sun-blocking textile will trigger an investigation of dermal exposure, incorporation into any kind of spray will also lead to assessment of exposure via inhalation, and the routes of interest for painting an item may differ from those to consider for sanding it off again).

The assessment will be based on modelling and/or measurements.

If a modelling is performed, the tool and model must be chosen to match the assessed scenario (bearing in mind the particulate nature of the assessed matter could, in practice, mean exclusion of models based on evaporation).

Use of measured data should be checked for application of an appropriate methodical approach and analytical set-up for the determination.

Assessments may follow tiered approaches, be based on probabilistic or deterministic methods, and take account of combined exposures.

On closer look, there are some specific considerations to take into account (e.g. related to the metric; see section on exposure models below). In addition, conducting an exposure assessment for a nanomaterial is a complex task, often lacking important information and accompanied by high uncertainty.

5.7.3 *Characterisation*

One important aspect to understand is that the identity and effects of a nanomaterial do not solely depend on its chemical composition. They are a function of its physical characteristics (such as shape, particle size/range, and surface area) and chemical characteristics (e.g. composition of the core, the surface, and impurities). These characteristics determine its resulting specific physico-chemical behaviour. In consequence, different forms of a nano-object and nanomaterial can exist based on a set core composition (e.g. silver) which may exhibit different specific behaviours in respect of product properties, hazardous effects, and fate/exposure.

Noteworthy is that information on the characteristics can only be obtained via measurements. For a full data set, more than one method has to be employed. Some policy documents indicate minimum elements for characterisation, e.g. ECHA (2017a). However, the scientific discussion of the required data set for comprehensive characterisation is ongoing. Especially measurement methods for the characterisation of nanomaterials present in complex matrices (e.g. cosmetics) remain challenging and are still under development.

5.7.4 *Information on Uses*

An important factor related to the uncertainty is that knowledge about the identity of nanomaterials that consumers will come into contact with is often lacking.

Many nanomaterials used for consumer products are specifically developed and designed for a defined use (e.g. to add a specific functionality like improving transparency/strength/conductivity or adding antibacterial activity). Information on the exact identity and use (product) is often confidential.

Some publicly available information on nanomaterials incorporated into consumer products exists (e.g. from market surveys, inventories, databases, and reports from national registers). A comprehensive listing of current sources would overload this chapter. Instead, readers are referred to a list recently compiled by Larsen et al. (2015b). In addition, some European countries (FR, BE, DK, NO) have nano registers, and further countries are planning to install them. Among the already active registers, France and Belgium annually publish reports on the information received. The European Union Observatory for Nanomaterials (EUON) (ECHA 2017b) went online in 2017. Its aim is to provide factual and neutral information on nanomaterials for consumers, workers, regulators, and scientists to meet the needs of these various audiences (ECHA 2017d).

In view of publicly available sources, it is important to note that they will generally not reveal the full identity of a nanomaterial and/or product. Furthermore, some compilations are solely based on unconfirmed claims or labelling, and the existence of labelling requirements differs across product regulations in Europe.

In addition, the characteristics of a nanomaterial present in a consumer product may change during its life cycle and that of the consumer product. There can be continuous transformation and degradation processes along the production process of the product, during its use phase and in contact with body fluids or ambient media after release.

5.7.5 Release

Exposure may also occur to nanomaterials intentionally or unintentionally released from consumer products, or generated by their use.

The release can result from different mechanisms (e.g. aerosol formation, migration/leaching, or abrasion). It is impacted by different aspects of the product's characteristics. The nanomaterials in the products can be dispersed in liquids or embedded in a more or less solid matrix (e.g. gels, plastics, or coatings). The matrix's firmness and structure as well as the nanomaterial's size and the strength of its binding to the matrix influence the mobility of the nanomaterials in the product. The location of nanomaterials in the matrix can differ (e.g. sitting on the surface or distributed within). The product's characteristics also determine the activities to consider for use or handling of the product and the forces acting on it (e.g. walking over a sanded floor or sanding it off). The release may change between different stages of the service life, e.g. due to ageing- and weathering-related alterations of the matrix.

The released nanomaterials may occur in free form (e.g. nanoparticles in sprays) or as part of larger matrix fragments. Following its release, the material will likely undergo additional changes such as agglomeration or chemical reactions, e.g. oxidation. Larger matrix fragments may also disintegrate into nanomaterials at a later stage.

The resulting exposure is of course influenced by additional factors relating to the product's characteristics, consumer behaviour, and incorporated/released nanomaterial. Readers are referred to Wijnhoven et al. (2009) and Larsen et al. (2015a) for more details about these additional factors.

Release assessments can be based on models or measurements. Bott et al. (2014) developed a model for the migration of nanoparticles from polyolefins based on an existing migration model for conventional polymer additives. If theoretical models are used to assess released amounts, their outcome should be checked against analytically measured data wherever feasible. One example for such a comparison is given in a scientific opinion of EFSA considering the release of titanium nitride nanoparticles from polyethylene bottles (EFSA 2008).

5.7.6 Measurement

Mackevica and Hansen Foss (2016) reviewed the data availability of release measurements. The availability of data is increasing, especially for nanomaterials based on substances with a high market prevalence. Another notable focus is on nanomaterials becoming airborne and possibly inhaled. Presumable reasons for the prevalence are that the route can potentially lead to hazardous effects without the materials having to cross barrier functions, and considerations for worker protection also focus on this route.

Consumer products often involve complex matrices. Measurement methods to detect nanomaterials in these matrices for confirmation of their presence are still under development. One example of such activities is the project NanoDefine (2017).

For interpretation of results, it should be noted that all types of analytical results, including the characterisation and the quantification of content or release, depend on the analytical set-up and methodology chosen. Notably, results obtained by usage of different protocols are often not comparable. Efforts are ongoing to overcome the lack of harmonised protocols (Packroff et al. 2016).

Consideration of realistic release conditions requires knowledge of the consumer behaviour related to the products and the product characteristics influencing the release.

5.7.7 Activity Pattern Data: Data on Consumer Behaviour

Consumer behaviour-related data such as the use frequency, event duration, or amount of product used per event are required to perform an exposure assessment.

In principle, they could differ from conventional products (e.g. being used less often or in reduced amounts per event). In practice, the availability of specific data is generally limited.

5.7.8 *Exposure Models and Tools*

Due to the limited availability of measured data, exposure assessments for consumers are regularly based on exposure models and tools. Some specific requirements of tools for assessments of nanomaterials have been addressed in research projects [e.g. in Work Package 4 of the NANEX-Project funded under the seventh framework programme - NMP-2009-1.3-2: Exposure scenarios to nanoparticles (Micheletti et al. 2010)]. One observation was that the considered models/tools generally provide results in a mass-based metric (e.g. g/m^3). However, the dose-response relationship within a hazard assessment is sometimes indicated in another metric (e.g. particle number/m^3). When the results of an exposure assessment are to be compared against the outcome of a hazard assessment, having matching metrics is helpful. The effects of agglomeration should be considered for inhalation. The development of nano-specific tools is an ongoing scientific activity. One result of such efforts is ConsExpo nano (see Sect. 5.10.7). It enables higher tier assessments of the magnitude and duration of alveolar loads from consumer spray products, with "built-in" default scenarios in a variety of dose measure metrics. Version 1.1 of ConsExpo nano also allows for calculations of alveolar loads from a known air concentration for any inhalative exposure (outside spray scenarios) and to convert external doses from animal studies into internal exposure doses. In a recent compilation of tools used for human health-related exposure assessment (Table 5.13), all other nano-specific exposure tools addressing consumer exposure are based on qualitative exposure considerations generally used for prioritisation purposes and did not provide route-specific or quantitative results.

Notably, none of the listed tools for consumers explicitly addresses exposure of the eye (connected to secondary inhalation and oral exposure via the nasolacrimal duct).

Within OECD (see Sect. 7.5), a project was initiated which compiles information on the available models and tools for assessing consumer exposure and aims to evaluate their applicability in exposure assessment.

5.7.9 *Tiered Approach*

A tiered approach for exposure assessment and risk assessment is needed since it is impossible to measure every possible scenario.

High-throughput assessments (e.g. for usage in product development phases) are often based on life cycle assessments identifying the potentially exposed population

Table 5.13 Characteristics of selected tools used for human health-related exposure assessment of nanomaterials (adapted from Sommer (2016))

Tool	nano-specific	Tier	Result	Population		Route			Dose metric	Reference (description, tool provision)[c]
				cons.	occ.	Inhalation	Dermal	Oral		
NanoRiskCat[a]	x	L	qual. Risk Management Prio (qual. Exposure category extractable)	x	–	NA (exposure addressed irrespective of a specific route)			NA	Hansen Foss et al. (2014)
Swiss Precautionary Matrix[a,b]	x	L	semi-quant. Risk Managament Prio	x	x	NA (exposure addressed irrespective of a specific route)			NA	FOPH (2017)
ConsExpo nano (V 1.1)	x	H	quant. Exposure Calculation	x	x	x	–	–	Mass; number, surface area, and volume of nanoparticles; number, surface area, and volume of aerosol particles	RIVM (2017a, b)
NanoSafer[a,b]	x	L	quant. Control Banding	–	x	x	–	–	NA	Nanosafer (2017), Jensen (2016)
Control Banding Nanotool[b]	x	L	qual. Control Banding	–	x	x	–	–	NA	Control Banding Nanotool (2015)
ANSES tool[a]	x		qual. Control Banding	–	x	x	–	–	NA	ANSES (2010, 2017)
DREAM Nano[b]	x	L	semi-quant. Exposure Estimates (in DREAM units, grouped into categories)	–	x	–	x	–	NA	Van Duuren-Stuurmann et al. (2010)
Stoffenmanager Nano[a,b]	x	L	qual. Risk prioritizing band	–	x	x	–	–	NA	Stoffenmanager Nano (2016), Van Duuren-Stuurmann et al. (2012)

Stoffenmanager[a]	–	**L**	qual. Risk prioritizing band (dermal)	–	**x**	**x**	**x**	–	Mass	Stoffenmanager (2016)
			quant. Exposure Estimate (inhalation)							
DREAM[a]	–	**L**	semi-quant. Exposure Estimates (in DREAM units, grouped into categories)	–	**x**	–	**x**	–	NA	Van-Wendel-De-Joode (2003)
ECETOC-TRA[a]	–	**L**	quant. Risk characterization ratio (exposure calculation extractable)	**x**	–	**x**	**x**	**x**	Mass	ECETOC (2016)
ConsExpo[a]	–	**H**	quant. Exposure Calculation	**x**	–	**x**	**x**	**x**	Mass	RIVM (2016, 2017a)
E-FAST[b]	–	**L**	quant. Exposure Calculation	**x**	–	**x**	**x**	**x**	Mass	US-EPA (2016)

NA not applicable, *cons.* consumer, *occ.* occupational, *L* lower tier, *H* higher tier

[a]This tool is part of a compilation and subsequent characterisation for risk assessment performed by Larsen et al. (2015a, b)

[b]This tool is part of the information submitted in an informal survey conducted by the OECD Working Party for Manufactured Nanomaterials (WPMN) to develop an inventory of available methods and models used for exposure assessment. Its analysis was published (OECD 2015)

[c]The referenced websites have been accessed in August 2017. The tool versions provided on them may differ from those originally analysed. The versions are likely to change over time

and/or estimated release potentials based on product characteristics. Such considerations are used regularly for priority setting of the cases where a quantitative exposure assessment should be performed.

A tiered approach may be employed by moving from the use of qualitative, generic information to the consideration of quantitative or specific knowledge. In terms of employed measurements, it may shift from robust, cost-efficient, and widely available methods to more sophisticated, cost-intensive, and less available methods. The approach may also be employed to counterbalance information missing for exposure assessment. At a low tier, conservative considerations will aim to ensure that unconsidered information are covered at this stage. Higher tiers will take into account more parameters and/or specific information.

5.7.10 Conclusions

In summary of the previous chapters, the following conclusion can be drawn:

Nanomaterials in the sense of the ISO definitions are employed for the production of, contained in, and potentially released from a variety of consumer products regulated by numerous different legislations.

Differences in definitions of terminology influence the respective identification as a nanomaterial. Guidance on exposure assessment is available for some of these contexts.

Assessment concepts, information on parameters effecting the exposure, and the toolbox of methodologies for assessment of consumers are evolving. Generally, the exposure assessment takes into account parameters of exposure related to the physico-chemical properties of the regarded material, the properties of the regarded product, as well as behaviour data influencing the release of nanomaterials and the consumer's contact with a nanomaterial. Qualitative considerations about the release potential currently serve the purpose of prioritisation, whereas measured data on release are used to check modelled predictions and move exposure assessments from a conservative level to consideration of more specified realistic conditions.

There are a substantial variety of nanomaterials and applications in consumer products. In addition, nanomaterials present in consumer products can undergo transformation processes. Their physico-chemical properties during their life cycle, the use phase of the consumer product, and after their release (or a larger fragment containing them) may be subjected to considerable changes.

Limitations on the publicly available information on the properties and the combination of nanomaterials and applications in consumer products affect the scenario building. The main routes of consumer exposure depend on the use conditions of the product. Dermal and oral exposures need to be considered in addition to the inhalation exposure of airborne particles. Some cases may have to take exposure via the eye and connected nasolacrimal duct into account.

The availability of measured data varies. Development of standardised protocols and the required toolbox of analytical methods is on-going. The later especially relates to complex matrices.

Models are employed to overcome limitations of resources and the lacking availability of measured data. Whenever feasible, the results should be checked against measured data. Availability of nano-specific exposure models/tools for consumers is limited. The included defaults for parameters reflecting consumer behaviour and built-in scenarios are generally based on non-nano-specific data. The use of other tools needs to be carefully checked for boundaries and applicability.

The assessment of consumer exposure from nanomaterials in consumer products requires a case-by-case decision about the data, methods, and approaches employed. Over the years, a variety of examples in different product sectors has been published. Generally, the examples examine apparently predominant materials and applications of concern. Some recently published examples (looking at more than one product type) are Rompelberg et al. (2016) and the non-food item case studies in the Danish Environmental Project No. 1636 (Larsen et al. 2015a). The consideration of the uncertainties associated with an assessment is a notable part of assessing the quality of that assessment.

For nanoparticles in food, see Sect. 4.22.

5.8 Exposure via the House Dust Path

Stefanie Klenow

5.8.1 Introduction

House dust is composed of small particles located indoors. Many substances, notably SVOCs, can be found in house dust whose intake might pose a potential health risk (Becker et al. 2004; Bonvallot et al. 2010). Phthalates, pesticides, short-chain chlorinated paraffins, PBDEs, perfluoroalkyl acids, organotins, PCBs, and PAHs are of particular interest because of their occurrence in house dust and their toxicological profiles (Bonvallot et al. 2010). Most people in Western societies spend much of their time in indoor environments such as homes, kindergartens, schools, offices, or libraries, which makes an exposure to house dust ubiquitous. House dust can, therefore, be an important exposure pathway for some chemicals (Wormuth et al. 2006). Exposure via house dust is indirect (house dust particles are carriers for non- and semi-volatile chemicals), unconscious (people are not aware of the exposure), and unintended (apart from a special form of pica behaviour).

Exposure via house dust is estimated from the detected concentrations of the target compound in house dust and the assumed rate of house dust intake.

Reference values have been derived for the amount of house dust ingested by humans per day which should be considered for exposure assessment. These values, however, exhibit an exceptionally high degree of uncertainty about the estimated intakes of substances via house dust.

5.8.2 Definition of House Dust and Characterisation of Exposure via House Dust

The term house dust (or indoor dust) is not clearly defined in a scientific context. There is general agreement that house dust is composed of particles from indoor environments. This includes all rooms used privately and professionally as well as public buildings. However, there is ambiguity in the scientific literature about which particles are specifically described by the term house dust. It often refers only to settled particles on surfaces (floor dust and deposited dust). However, the term house dust may also encompass air-suspended dust. Some authors differentiate between airborne particles and house dust. Others describe house dust as the contents of vacuum cleaner bags, which primarily contain floor dust. The question of a house dust definition is essential to describe the exposure scenarios, the evaluation of the available data, and the design of studies investigating exposure via house dust. In order to describe house dust in the context of exposure comprehensively, the following definition is proposed:

> The term "house dust" covers all particles that occur indoors and are capable as such, or as carriers of substances, to contribute to oral, inhalative, or dermal exposure. These particles can be present on all indoor surfaces, or they may have been suspended and then deposited again on (other) surfaces. The particles differ in origin, location, particle size, and composition. A distinction based on the location can be relevant for different scenarios of house dust intake.

Exposure to substances in house dust can occur by different exposure routes: oral, inhalative, and dermal. However, these routes contribute differently to the exposure. A higher relevance of the oral exposure route is discussed specifically for toddlers as they crawl on the floor and frequently lick their fingers and objects (so-called hand- and object-mouthing). These age-related activity patterns result in a higher likelihood of ingesting considerable amounts of floor dust - in particular when related to body weight. Apart from this intake scenario, little is known about how house dust is ingested by older children and adults. Air-suspended particles might be swallowed while speaking and breathing. Inhaled dust particles can be captured in mucus and subsequently transferred into the gastrointestinal tract. House dust might be ingested with food when air-suspended dust particles settle on foods or following the contact of foods with hands and surfaces to which dust particles adhere (such as dishes, table, floor). Eating finger food, licking fingers, chewing fingernails, or possibly touching the lips with fingers or objects and subsequently licking the lips may also result in oral ingestion of dust particles. Mouth contact with bottlenecks, pens, cigarettes, and other objects might also contribute to oral ingestion of dust particles. No information is available on whether, and if so, to what extent, these potential scenarios contribute to the oral intake of house dust. Inhalation route of exposure via the lower respiratory tract is probably negligible (Oomen et al. 2008). Dermal contact with house dust can result in desorption of substances adhered to dust particles, which may subsequently penetrate the skin, but this route of exposure may also be marginal.

House dust particles can originate from indoor and outdoor sources. The diversity of sources of house dust results in a complex and extremely heterogeneous matrix. This applies for the temporal and spatial variability in particle size, shape, composition, and quantity, as well as in the concentrations of the contaminants (Maertens et al. 2004; Millette et al. 2003). House dust particles may consist of soil, fibres (paper, textiles, synthetics), abrasion from furniture or the building, combustion products (e.g. from cooking), dead organic material (hair, dandruff, and excretions from humans and animals, dead insects (e.g. mites), or parts thereof), living organisms (bacteria, fungi), and pollen (Butte 2004; Mølhave et al. 2000; Webster et al. 2009; Millette et al. 2003; Turner et al. 2005). Unlike particles in the outdoor environment (soil), contaminants have the potential to persist and accumulate in house dust, as indoor environments are not subjected to the same degradation processes as outdoor environments, such as sunlight, rain and wind, fluctuations in temperature and humidity, other weather-dependent effects, and high rates of microbial degradation (Paustenbach et al. 1997; Lewis et al. 1999).

In house dust, SVOCs can be detected as a consequence of their widespread use in biocide products, cleaning products, furnishings, and cosmetics (Bonvallot et al. 2010). Spray applications of products, or abrasion or emission from solid materials, lead to deposition in house dust. The involved processes are complex and dynamic, resulting in the occurrence of SVOCs on surfaces (notably house dust) and in indoor air (Weschler and Nazaroff 2010). Therefore, house dust along with indoor air has to be considered to assess exposure to SVOCs. In addition, house dust is also a reservoir of inorganic substances.

A number of issues need to be considered when looking at house dust. House dust is not only an indicator of indoor pollution but also an exposure pathway for substances. This applies mainly to the quantity of house dust indoors (dust load) - especially on surfaces which are in direct contact with humans - the time spent in a particular room, and the activities of people during this time (Fig. 5.7). The potential level of house dust intake correlates to the indoor dust load. It is determined by location (region, building), interior decorations and furnishings, the number of people and pets (as well as their activities), and room hygiene (frequency, location, and method of cleaning). In addition to the dust load, the particle size in house dust can also be highly variable and may determine the quantity of house dust intake. Particle size can affect adhesion to the hands and hence the amount of house dust

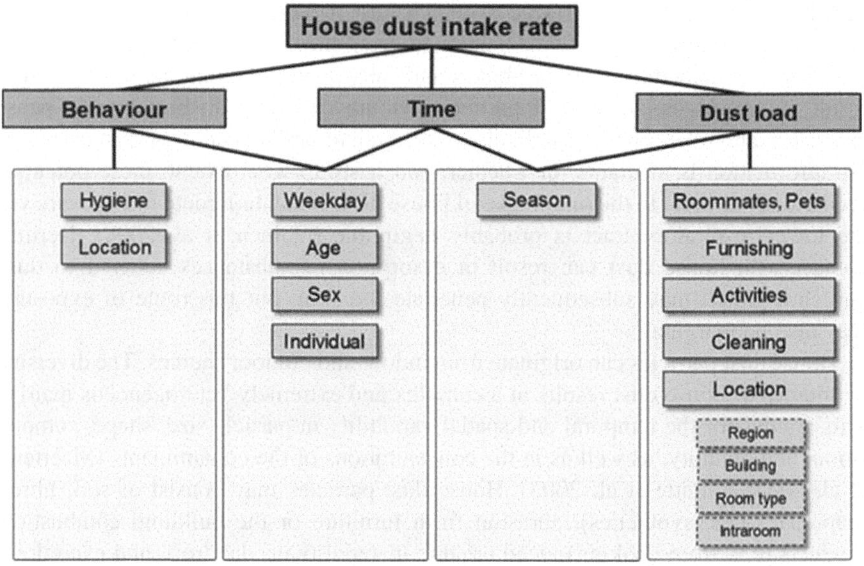

Fig. 5.7 Factors that influence the rate of house dust intake (© Bundesinstitut für Risikobewertung (BfR) 2018. All Rights Reserved)

ingested via hand mouthing. Particle size can also influence the level of exposure to substances, as these can adhere in greater quantities to smaller particles due to their relatively greater total surface area (Cao et al. 2012). Consequently, an unsuitable sampling of dust with respect to location and particle size may result in skewed exposure assessment results (Cao et al. 2012). There is at present no standard protocol for house dust sampling or particle size fractionation for an adequate exposure assessment. The variable composition of house dust particles, their potential influence on substance adhesion, and their spatial mobility affect both the substance resorption and the level of house dust intake. Time spent indoors can also be related to the level of house dust intake and is determined by season, age, day of the week, possibly gender, and individual factors. Moreover, activities (behaviour) in a room determine the potential level of house dust intake. These are influenced especially by location (such as a private or professional environment). Personal hygiene (e.g. hand-washing) plays a decisive role too. It can be assumed that there are also cultural and climate-related differences concerning the amount of dust (particle entry and indoor cleaning), residence times, and indoor activities. Because of the associated diverse and variable impact factors, a corresponding variability in house dust intake can be expected.

5.8.3 *Scenario Characterisation of House Dust Exposure*

A scenario can be characterised as follows (Heinemeyer et al. 2012a): concentrations of DEHP are measured in house dust from 600 German households with children aged 3–14 years. For exposure by oral mouthing, distributions were formulated that describe conservative estimations for house dust intake rates (Table 5.14). It is also be assumed that the substance will be completely released from the house dust and also completely absorbed in the gastrointestinal tract.

5.8.4 *Modelling House Dust Exposure*

According to the above-characterised scenario, the following model can be derived to estimate exposure via the house dust path (ExpoHD in µg/kg of BW per day):

$$\mathrm{EXPO_{HD}} = \frac{C_{\mathrm{HD}} * I_{\mathrm{HD}}}{\mathrm{BW}} * F_{\mathrm{des}} * F_{\mathrm{abs}} \tag{5.27}$$

where

C_{HD} is the concentration of the substance measured in the house dust and I_{HD} is the intake rate per day of house dust. F_{des} is a factor that comprises the desorption of the substance from the dust particles, and F_{abs} is a factor which characterises the systemic availability of the substance.

5.8.5 *Parameters for House Dust Exposure Assessment*

Concentrations of Substances in House Dust

The concentration may be influenced by the fraction of house dust sampled and chosen for measurement, as well as the analytical procedure (analytical sensitivity). Factors that affect house dust intake rates may also have an impact on the concentration of substances in house dust, such as furnishings (as a source of substances), location, and the particle sizes and particle compositions (see Sect. 5.8.1).

For quantitative estimation of house dust exposure, it is essential to characterise the analysed house dust regarding:

- Location of sampling (region, building, room type, location within the room)
- Sampling method (air sampler, wipes, vacuum cleaner bags, with key metrics)

Table 5.14 Recommendations and assumptions for oral intake of house dust to estimate exposure

Reference	Central tendency in mg/d	Upper estimate in mg/d	Age groups	Comments
US-EPA (2019)	20	60	<6 month	Estimates of dust were derived from the soil + dust values assuming 45% soil and 55% dust, rounded to one significant figure Upper estimate: upper percentile
	40	100	6 month–<1 year	
	50	100	1–<2 years	
	30	100	2–<6 years	
	40	100	1–<6 years	
	30	100	6–<12 years	
	20	60	>12 years	
Oomen et al. (2008)		100	Children	"Conservative (*upper estimate*) but realistic estimate" Expert judgement
		50	Adults	
Nordic Exposure Group (2012)	100		Children	Based on Oomen et al. (2008) and considered as being representative for the European population
	50		Adults	
ECHA (2016)		100	Children	Conservative estimate, based on Oomen et al. (2008)
Bothe (2004)	60		0–1 years	Study on soil intake, house dust intake was estimated alternatively Central tendency: arithmetic mean
	131		1–2 years	
	44		2–7 years	
	97		7–12 years	
van Holderbeke et al. (2008)	–/2.0	–/6.9	3–<6 months	Modelling deterministic/probabilistic[a] Central tendency[b]: median (deterministic) vs. arithmetic mean (probabilistic) Upper estimate: 95th percentile (probabilistic) Considering only hand mouthing
	7.0/8.5	–/31	6–<12 months	
	14/16	–/63	1–<2 years	
	12/15	–/62	2–<3 years	
	15/25	–/106	3–<6 years	
	14/18	–/63	6–<11 years	
	8.6/5.8	–/17	Adults	
Özkaynak et al. (2011)	19.80/ 8.39 (hand-*mouthing*)	73.74 (hand-*mouthing*)	3–6 years	SHEDS—probabilistic modelling (micro-activity data) Considers mouthing of hands and objects Central tendency: arithmetic mean/median Upper estimate: 95th percentile
	6.85/0.69 (object-*mouthing*)	27.23 (object-*mouthing*)		
Wilson et al. (2013)	38/36	–/140	0–6 months	Modelling (macro activities) deterministic/probabilistic Central tendency: arithmetic mean Upper estimate: 95th percentile Considers hand-*mouthing* only
	41/41	–/140	7 months–4 years	
	31/32	–/110	5–11 years	
	2.2/2.2	–/7.3	12–19 years	
	2.5/2.6	–/8.6	20–59 years	
	2.5/2.6	–/8.7	>59 years	

(continued)

Table 5.14 (continued)

Reference	Central tendency in mg/d	Upper estimate in mg/d	Age groups	Comments
Wormuth et al. (2006)	50		Toddler (up to 3 years)	Toddler: derived from Hawley (1985), Clausing et al. (1987), Stanek III and Calabrese (1995) Children ingest 16.7% and adults 1.1% of the amounts of dust ingested by toddlers (source: Hawley (1985))
	10		Children (4–10 years)	
	1		Adolescent and adults	
Hawley (1985)	50 (Apr–Oct) 100 (Nov–Dec)		2.5 years	Expert judgement based on limited quantitative data Central tendency = average intake
	3		6 years	
	0.56		Adults	

[a]Daily intake, considering intake per hour and given exposure time
[b]Based on usual assumptions and median values

- Particle size distribution of collected house dust
- Analytical method with limits of detection and quantification
- Treatment of values below LOD/LOQ (see Sect. 4.9.2)

Recommendations and Assumptions of Oral House Dust Intake to Assess Exposure

Various institutions and authors have published recommendations about house dust quantities that should be adopted for exposure assessment (Table 5.14). These values refer to oral house dust intake and were derived from studies that used varying methods. These methods include the use of tracers, biokinetic modelling, and modelling with activity patterns.

The United States Environmental Protection Agency in the 2008 version of its Exposure Factors Handbook for the first time addressed not only the ingestion of soil but also the intake of house dust (US-EPA 2008). The latest version, dating from 2017, contains recommended values for the estimation of exposure via the house dust pathway (US-EPA 2017d). Here the probable mean oral intake of house dust by children below 6 months of age as well as aged 12 years and older is given as 20 mg per day. Children aged between 6 months and <6 years are likely to take up 40 mg per day. However, if children aged 1 to <2 years is considered only, 50 mg per day should be assumed on average, whereas children aged 2 to 12 years most likely ingest 30 mg per day. For children below 6 months and those that are aged 12 years and older, 60 mg house dust per day is suggested as the upper percentile for oral intake. For all other age groups (between 6 months and 12 years), 100 mg per day should be assumed for high house dust intake. The US-EPA (2019) has extensively dealt with the limitations of individual underlying studies, such as the assumptions in tracer studies on the nonsystemic intake of tracer, or the non-consideration of other sources of exposure. Overall, the US-EPA (2019) assigns a low confidence level to the recommendations they provide.

The derived default value for house dust intake as suggested by the RIVM (100 mg per day for children and 50 mg per day for adults) is attributed as conservative but realistic. The values represent an expert judgement based on two arguments (Oomen et al. 2008; Davis and Mirick 2006). First, the assumption was based on estimates for soil ingestion by children in the Netherlands via hand mouthing. Second, the authors compiled values from the literature and interpreted those data as support for their derivation. The RIVM indicates that these default values are subject to great uncertainty.

The Nordic Exposure Group (2012) compared various recommendations for soil and house dust ingestion. Overall, the data situation is described as limited and variable. The hygienic pattern and regional differences in the behaviour of children would likely affect ingestion rates. The recommendations of the US-EPA (2009a) are considered as being most valid values. However, recommended values for house dust ingestion for adults and children from RIVM (Oomen et al. 2008) are viewed as more representative for the European population. Therefore, the Nordic Exposure Group recommended these values as default exposure factors in the context of the European regulation on the Registration, Evaluation, Authorisation and restriction of CHemicals (REACH).

According to the European Chemicals Agency (ECHA 2016), house dust may present an important path for exposure to non-volatiles that may accumulate in house dust. ECHA argues that substances in house dust are included in the conservative general default values for low tier assessments in ECETOC TRA (Sect. 5.10.10). However, ECHA recommends consideration of house dust as a path of exposure in higher tier assessments for children. For this, conservative estimates of house dust intake, such as 100 mg per day for children as proposed by the RIVM (Oomen et al. 2008), should be used.

Intake of house dust was estimated as an alternative pathway of exposure in a survey of soil ingestion by children aged up to 12 conducted in Germany that used tracer elements (Bothe 2004). The arithmetic mean of house dust intake is 82 mg per day (range in the age groups, 13–248 mg per day). In addition, an upper estimate of house dust ingestion was determined ignoring dietary intake of tracer elements. In this case, the arithmetic mean of house dust intake is 140 mg per day (range in age groups, 35–296 mg per day).

Ingestion of house dust via mouthing behaviour was modelled mechanistically in three studies (van Holderbeke et al. 2008; Özkaynak et al. 2011; Wilson et al. 2013). This approach is also referred to as activity pattern methodology, and it combines factors such as dust load, frequency of hand-to-mouth contact, transfer rate of particles from surfaces to hands, and many other factors. While the models of Wilson et al. (2013) and van Holderbeke et al. (2008) cover all age groups, Özkaynak et al. (2011) limited their estimate to children at the age of 3 to 6 years. They argue that insufficient information is available for other age groups. Moreover, Wilson et al. (2013) and van Holderbeke et al. (2008) only consider house dust ingestion via hand mouthing, while Özkaynak et al. (2011) modelled hand- and object-mouthing. This methodology resulted in average estimates of house dust ingestion that range from 2 to 41 mg per day for children and from 3 to 6 mg per day for adults. The upper estimates (95th percentile) range between 7 and 140 mg per day (children) and between 9 and 17 mg per day (adults).

Wormuth et al. (2006) derived the house dust intake estimates from Clausing et al. (1987), Hawley (1985), and Stanek and Calabrese (1995). Thus, infants and toddlers (up to 3 years) would ingest 50 mg per day, children at the age of 4–10 years ingest 10 mg per day, and teens and adults ingest 1 mg per day.

Hawley (1985) expected that children at the age of 2.5 years ingest 50 mg house dust per day in the summer and 100 mg per day in the winter. Children at the age of 6 years would ingest 3 mg house dust per day based on assumptions about how much house dust adheres on children's hands and on the skin surface in contact with the mouth. For adults, the author assumed a daily house dust ingestion of 0.56 mg per day.

The Committee on Environmental Health (AUH 1995) did not recommend any value for house dust ingestion because the studies evaluated leave many questions unanswered. The non-standardised sampling was emphasised to be a main issue. Even many years later, a recommendation of reference values for house dust intake could not be derived from available studies (Mekel et al. 2007).

Currently, the most common value for house dust ingestion to assess exposure is 100 mg per day. This (or even lower assumptions) could result in estimates covering half of the total exposure, depending on the substance under consideration (Egeghy et al. 2011; Wormuth et al. 2006).

The observation that central tendencies of house dust ingestion estimates are generally comparable may tempt exposure assessors to assume that exposure via house dust can be assessed with sufficient precision, ignoring the present lack of understanding of this path of exposure.

Body Weight

The body weight of the target population at a specific age is the model parameter which is best known and can obtained from relevant sources (see Sect. 3.4.1).

Factors

Most often F_{des} and F_{abs} are unknown and they are then set to 1 which means that the entire substance adhered to dust particles is systemically available. It should be noted that setting the factors to 1 is a very conservative approach.

Example
Taking all data and uncertainties together, Heinemeyer et al. (2012a, b) comprised the data of house dust intake rates in a probabilistic approach into three different distributions to consider the variability of the data in the apparent situation.

Scenario 1	Risktriang(0;20;100)
Scenario 2	RiskTriang{0,riskuniform(20;100);Riskuniform(100;500)}
Scenario 3	RiskCumul(0;1391;0;0;16;45;110):(0.05;0.25;0.5;0.75;0.95)
Concentration	RiskLoglogistic(0;538.46;2.5778)
Body weight	Risknormal(10.42;1.36)

The first scenario characterises a triangle with the lowest house dust intake of 0 mg/kg of BW a day, the highest intake of 100 mg per day, and the most probable intake of 20 mg per day. The second scenario indicates that the highest and the most probable intakes have ranges which are characterised by uniform distributions. The third scenario indicates the most likely percentiles and the respective numerical values of a distribution indicative for the available data. The DEHP concentrations in house dust are measured in the German Environmental Survey (GerES 2003–2006, Schulz et al. 2007), and the distribution function is derived from primary data by Heinemeyer et al. (2012a). Body weights for children aged 2 (toddlers with high probability of hand- and object-mouthing) are taken from the AUH report (1995).

Figure 5.8 shows intake estimations for DEHP by the house dust path, as obtained by using the three scenarios of house dust intake and concentrations of DEHP measured in German kindergartens.

The figure shows that estimations of scenario 1 and 3 are rather similar, while scenario 2 reveals higher estimates particularly for the central tendencies. This can be explained by the choice of the bands for the high and the most probable house dust intake rate. The example shows also the ability of probabilistic estimations to demonstrate the uncertainty of the exposure assessment result.

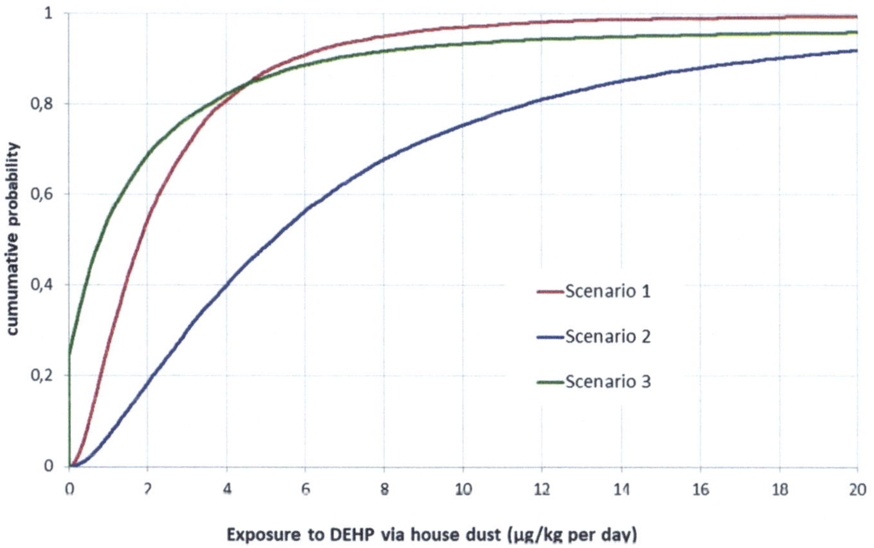

Fig. 5.8 Results from estimations of DEHP exposure from house dust intake. Data taken from Heinemeyer et al. (2012b)

5.9 Sprays, Aerosols

Annette Bitsch, Stefan Hahn, and Wolfgang Koch

5.9.1 Introduction

Spray products have many types of consumer applications, ranging from the use of household detergents in spray applications to the use of plant protection products in the private sector and to typical cosmetic products. Even more, spray products have grown in popularity during the last decades and are often preferred to other forms of products by consumers. One reason for this might be the fact that spraying allows material to be used most efficiently.

On the other hand, the regulation of chemicals and products has intensified, including the requirements for a safety assessment; the Aerosol Dispensers Directive (EU 1975), the Cosmetic Products Regulation (EU 2009b), and the REACH Regulation (EU 2006) are only few regulatory examples showing the present data requirements. With respect to this, spray products require additional considerations for potential inhalation exposure especially for consumer products.

The actual exposure to the spray product and its ingredients has to be determined. A spray product is a liquid formulation of chemical actives in solvent that is applied by a technique (packaging and nozzle design) generating spray droplets (aerosols). Detailed knowledge of the use conditions established from data on the habits and practices of consumers is necessary for the exposure assessment.

Aside from the deposition of these aerosols in the lung, the propellants and solvents often used in spray products could contribute to the health impact and have to be taken into account for the overall assessment of the inhalable chemicals in the products.

Regulatory bodies as well as several branches of industry have identified this demand and worked on exposure prediction models. These models combine data from computer simulations, knowledge of spray characteristics, individual real measured data, and use experiences from the market.

5.9.2 Definitions and Boundary Parameters

Aerosol and Spray

Throughout the following section, we use the general terms aerosol and spray. In the scientific community, an aerosol is defined as an ensemble of airborne particles of either liquid or solid state. In this sense, the airborne fraction of the spray droplets is an aerosol. In the consumer world, the term aerosol is frequently related to an aerosol can, a propellant-driven device to generate a spray (such as hairspray, deodorant spray, etc.). The scientific aerosol definition is used exclusively in the following text.

When the spray formulation contains substances of considerable vapour pressure, the related aerosol is a two-phase system consisting of the droplet phase and the vapour phase of the evaporating liquid. The inhalative exposure to both phases has to be considered separately since the uptake mechanisms via the respiratory system are quite different for particles and vapours.

If the volatile solvent has completely evaporated, the spray has matured into its secondary aerosol consisting of non-volatile material only. This process of aerosol ageing is presented in Fig. 5.9.

Air Concentration, Distribution, and Deposition in the Respiratory Tract

Inhalation of an air contaminant - here an aerosol - results in transport and distribution in the respiratory tract, deposition in designated areas of the branched tracheobronchial tree, and finally in possible pulmonary absorption that leads to an internal (or body) dose.

Thus, three different external doses should be considered for spraying: (1) the dose that is sprayed from the product can into the room, (2) the inhaled dose, and (3) the deposited dose. For respiratory uptake, the inhaled dose is the amount of

Fig. 5.9 Schematic representation of spray ageing

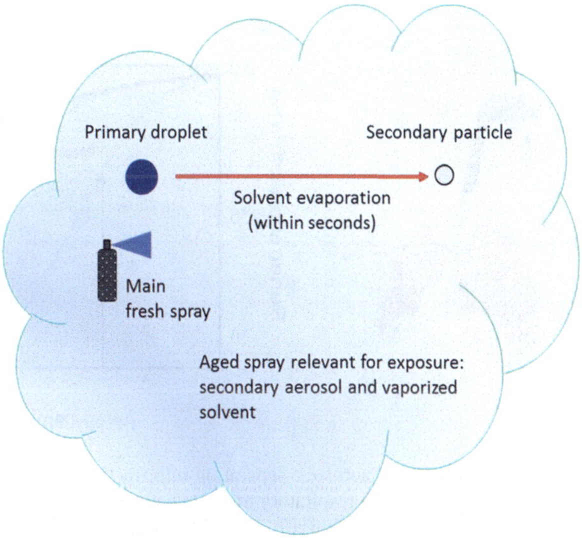

inhaled substance from which only a fraction, the so-called deposition fraction, will be available for absorption.

Particle properties such as the aerodynamic size distribution as well a breathing parameters are factors influencing the deposition fraction. The particle deposition pathway is well understood and is described by well-established and experimentally validated theoretical models (Asgharian et al. 1995; Martonen et al. 2000; Subramaniam et al. 2003; Choi and Kim 2007; Hofmann 2009). Regarding the site of deposition, a distinction between the alveolar, tracheo-bronchial, and extra-thoracic compartment is generally done, and this has been shown to be sufficient for risk assessment purposes (Fig. 5.10).

Different clearance mechanisms and lining properties in these three compartments lead to quite different residence times and bioavailabilities of deposited particulate material. The assessment of the inhalation exposure of spray products should take this fact into account. This is well fulfilled from a practical standpoint by measuring the respirable (<5 μm), the thoracic (<10 μm), and the inhalable fraction (<40 μm) of the spray-related aerosol (Fig. 5.10b). Although these conventions have been developed for workplace aerosol assessment (CEN 1993), they can be applied to assessing consumer exposure.

5.9.3 Sources of Exposure

Spray Products on the Market

As mentioned above, a multitude of spray products are available on the consumer market, and the fields of application cover household and cleaning products;

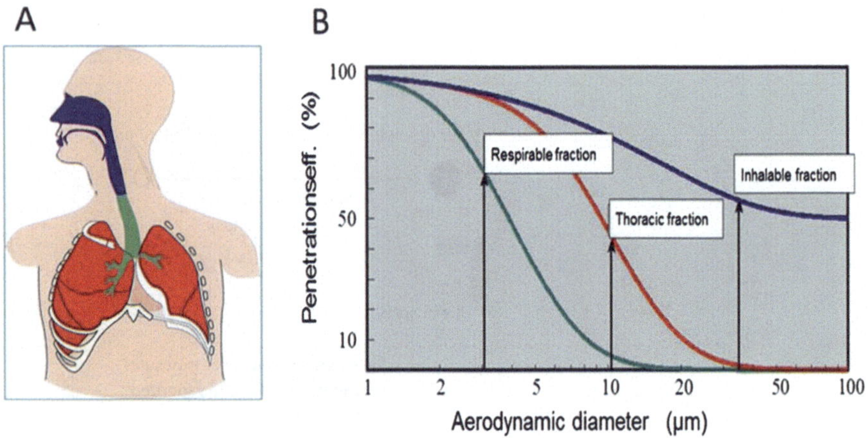

Fig. 5.10 Inhalation of aerosols: separation into fractions by particle size and depth of entry. (**a**) Presentation of the human respiratory tract indicating the entry of the different fractions given in 2b by colours. (**b**) Definition curves of health-relevant aerosol size fractions according to CEN 481 (CEN 1993)

cosmetics, biocides, and/or pesticides for the private market; healthcare products; care products for furniture, clothes, etc. The advantage of spray applications is a homogenous distribution of active chemicals on surfaces or in air, which means that less product is applied more efficiently.

All these products are available on the market and some are used on a daily basis.

Consumer spray products that gained a lot of attention in the recent years are those containing nanomaterials as active ingredients. One recent example is the use of titanium dioxide nanomaterial in sun protection sprays.

In Table 5.15, some prominent types of consumer sprays and their estimated or measured application duration per task are provided. This compilation has also been compiled by another author in a comparable manner (Steiling et al. 2014)

Special Characteristics of Spray Products

A spray product is characterised by the liquid formulation of the chemical compounds and the technology by which spray droplets are generated. The chemistry and droplet size distribution are optimised by the producers in order to meet the requirements of the intended application of the spray product. Thus, spray formulations cover a wide range of physico-chemical properties, particularly as far as the volatility of the chemical constituents is concerned. An important class of sprayed formulations contains non-volatile active substances dissolved in volatile solvents. This includes water and many organic solvents. The concentration of the non-volatile phase varies depending on the spray application. For example, it is in the low percentage regime for biocidal or fabric protection sprays; the active

Table 5.15 Use durations of prominent types of spray products

Consumer spray product	Duration(min)	Reference
Hairspray	3–4	BAMA (2008)
	11 sec release, 23 sec contact	EPA
Deodorant	1.4	Steiling et al. (2012)
Air freshener	4–5	BAMA (2008)
Spray starch	2–3	BAMA (2008)
All-purpose cleaner	Repeatedly within 24	Weerdesteijn et al. (1999)
Carpet cleaner	20–30	BAMA (2008)
Oven cleaner	10–15	BAMA (2008)
Furniture polisher	2–3	BAMA (2008)
Shoe polish spray	7.49	US-EPA (2011)
Paint	39.54	US-EPA (2011)
Spray paints for cars	42.77	US-EPA (2011)
Aerosol rust remover	18.57	US-EPA (2011)
De-icing spray	15–20	BAMA (2008)

compound makes up the main mass fraction and can even be 100% for paints and resin-based coating, one example being the liquid non-volatile silane compounds used for treatment of building surfaces.

Aerosol cans operating with a propellant gas are well known and widely used methods of spray generation. They are mainly used in consumer products as mentioned below in Table 5.15. Although spray guns, liquid spray nozzles, and fogging devices are attributed more to the professional environment for painting, surface treatment and sealing, pesticide application, etc., some of these devices may also be used by consumers. Depending on the spray technology, the mean spray droplet size covers a range from some tens of micrometres, typical for propellant sprays, up to several hundred micrometres generated by liquid spray nozzles. Accordingly, the percentage of respiratory particles, say, for example, in the thoracic fraction, is expected to vary by orders of magnitude. In any case, the maximum inhalable fraction of the active spray compound is given after complete evaporation of the solvents which takes place in time scales smaller than a few seconds for the solvents considered here and droplet sizes not larger than 100 μm. Evaporation causes the droplets to shrink to their minimum size determined by the content of the non-volatile material and thus leads to an increase in the respiratory mass fractions.

Exposed Population

Although most of these spray products are intended to be used only by adults, a thorough exposure assessment should take into account the broad range of sub-populations, including at least teenagers, who are supposed to use hairspray and deodorants as well as painting sprays, etc. In addition, children should be considered as possible bystanders staying beside and looking during the application of spray products. Sections 3.2 and 3.4 as well as 4.5 contain more information about typical behaviour and anthropometric data.

5.9.4 Typical Exposure Scenarios and Calculations

The external exposure is controlled by factors such as the properties of the spray process (mass generation rate and droplet size distribution), properties of the formulation (vapour pressure and concentrations of the active substances), and parameters characterising the exposure scenario (room size, ventilation parameter (air exchange rate) etc.).

In realistic exposure situations, the spray user is indirectly exposed to the overspray of the native spray aged for a certain period of time. In most cases due to the short evaporation time, the complete solvent evaporates before droplet sizes relevant for inhalation reach the recipient.

This status is assumed to represent the exposure condition of the consumer or professional spray user and should therefore be included when determining the exposure of non-volatile constituents of the spray released in the relevant health-related particle size regimes. Thus, there is a co-exposure to the matured secondary aerosol and the vaporised solvent. The aerosol exposure is always directly correlated with the spraying process, whereas the exposure to vaporised compounds may originate not only from the dispersed phase but also from the surfaces treated by the spray formulation. Thus, the exposure to the solvent, if toxicologically relevant, has to be performed separately (Fig. 5.10). An assessment of additive or synergistic effects is not foreseen at present. In addition, solvent evaporation from surfaces is a process not only related to sprays but occurs for every method of surface treatment with consumer products such as wiping and brushing, for example.

Characterisation of the Spray Process

An important parameter of the spray in view of inhalational exposure is the size distribution of the active material. There are principally two ways to measure the health-relevant size fractions of the spray's active compound:

Analysis of the droplet size distribution of the native spray and derivation of the secondary aged aerosol size distribution on a theoretical basis assuming complete evaporation of the solvent

Direct characterisation of the aged aerosol by means of a mass balance method

Characterisation of the Native Spray Droplet Distribution

Laser diffraction spectrometry and phase Doppler anemometry are the two methods usually employed for the measurement of the size distribution of the native spray droplets. Both methods are well described in the literature (Combellack and

Matthews 1981; Nuyttens et al. 2007) and are widely applied in industry to characterise spray devices and nozzles. The methods are nonintrusive and quickly provide results. However, they only probe a limited segment of the spray intersecting with the laser beam and are therefore sensitive to spatial variations of the spray droplet diameters. For certain types of wide-angle sprays, the determination of the integral mass fractions of specified size ranges (such as the respirable and thoracic regimes) is therefore difficult, if not impossible, because the local droplet mass flux and the size distributions have to be measured simultaneously. Furthermore, if the mass fraction in the size regime of interest is smaller than one percent, it represents the tail of the cumulative droplet distribution function on the far left of the size scale. The accuracy of quantifying this small mass fraction by the laser diffraction method is poor because of the mathematical inversion routines used to transform the measured light diffraction patterns into droplet size distributions.

Secondary Aerosol Mass Balance Method

This method addresses the measurement of released mass fractions of the non-volatile active substance in different health-relevant size fractions, e.g. the respirable and the thoracic size class as defined in the CEN standard 481 (1993). Release fractions, R_“res”, R_“thor”, are defined as the mass, m_“res”, m_“thor”, of the non-volatile material generated in the two size classes normalised to the total mass, M_“nv”, of non-volatile active material released during spraying:

$$R_{\text{res}} = \frac{m_{\text{res}}}{M_{\text{nv}}}, R_{\text{thor}} = \frac{m_{\text{thor}}}{M_{\text{nv}}} \tag{5.28}$$

The release fractions are determined via concentration measurements in a control volume (V) while simulating the mode of application and taking into account overspray formation as well as the ageing process due to solvent evaporation. A defined spray bolus is released either into the open space of the well-mixed release chamber (room disinfection, etc.) or against a surface introduced into the release chamber (surface disinfection, etc.). The analysis of the concentration-time series allows extrapolation to the concentration, c_(0,“res”),”\ \ \ “c_(0,“thor”) at the time point of spray release, and from this the calculation of the released mass can be done:

$$m_{\text{res}} = c_{O,\text{res}} * V, m_{\text{thor}} = c_{O,\text{thor}} * V \tag{5.29}$$

The release fractions determined in the way as described above deliver the source strength required as an input parameter for model-based estimation of the exposure concentration.

5.9.5 Models and Tools for Exposure Calculations

Based on the diversity of spray products and their variability in applications, as well as the complexity of the spray processes, the approaches for exposure estimations differ highly, i.e. depending if they are intended as a tier 1 approach for a rough estimation or as more precise and more sophisticated mechanistic models. An understanding of certain individual strengths and the applicability of these models is needed in a proper choice for a certain scenario. Especially for the exposures of products that fall under the regulatory scope of REACH, assessors can look at the recommendations for models to calculate consumers' exposure given in the REACH Guidance R.15 (ECHA 2016). The models noted as possible choices are very similar to those in the guidance documents for biocide exposure. A compilation of generally available exposure models is given in Sect. 7.4.

Only those models which have a dedicated focus on inhalation exposure due to spray applications are mentioned below.

Tier 1 Tools Used in a Regulatory Context

The most prominent TIER 1 tool within REACH is the ECETOC TRA for deterministic exposure estimation. In the ECETOC TRA, spraying is considered in different product categories as shown in Table 5.16. The basis for exposure calculation is however based on a rough estimation using an approach assigning the scenario to a specific product and use category; this is in agreement with its function as a tier 1 tool as described in more detail in Sect. 5.10.

The ECETOC TRA approach is refined further for special subcategories by the respective industry and/or downstream user groups, i.e. with EGRET for solvents and different other refinements for washing and cleaning products. This is also described in more detail in Sect. 5.10.

Spray applications are relevant for a large set of biocide product types. Here the exposure is calculated either with the ECETOC TRA or with the more mechanistic models used in ConsExpo or SprayExpo.

Table 5.16 ECETOC TRA product categories including one or more sprays

Descriptor	Subcategory
PC1: Adhesives, sealants	Glue from spray
PC3: Air care products	Air care, instant action (aerosol sprays)
PC9a: Coatings, paints, thinners, removers	Aerosol spray can
PC24: Lubricants, greases, and release products	Sprays
PC31: Polishes and wax blends	Polishes, spray (furniture, shoes)
PC35: Washing and cleaning products (including solvent based products)	Cleaners, trigger sprays (all-purpose cleaners, sanitary products, glass cleaners)

The focus for biocides application has shifted from spraying towards foaming, and the assumption of a much lower inhalation exposure by foaming seems plausible but has yet to be proved by further research work.

Mechanistic Models

If the spray process is characterised in terms of release fractions of the active substance as described above, exposure modelling can be carried out using the existing mass transport models based on source strength and ventilation data.

In many cases, the above release fractions are not available, and information about the spray is confined to the primary droplet spectrum and the composition of the liquid formulation. In this case, models can be applied simulating the droplet size dependent transport processes as well as the spray ageing primarily due to evaporation of material (mainly solvent) from the droplets.

Some computational tools including mechanistic exposure models have been developed in parallel for aerosol exposure and spray scenarios.

Such tools, developed to calculate the expected inhalation exposure, vary from simple to sophisticated models. The most obvious differences between these models are the number of assumed exposure zones (boxes), i.e. two-zone models vs. a single-box model.

Some prominent models that explicitly address the aerosol release of sprays in consumer products are mentioned and described in more detail in Sect. 5.10:

- RIVM ConsExpo 4.1 models
- BAuA SprayExpo 2.0 model
- BAMA/FEA Indoor Air Model
- RIFM 2-Box Indoor Air Dispersion Model

In the following chapter, a given scenario is calculated with the two more detailed spray models SprayExpo and ConsExpo.

> Example for a spray exposure scenario:
> A very simple spray scenario is chosen here to demonstrate the use of the models.
> The exposure after spraying a product containing 6% of active non-volatile substance is simulated. It is assumed that the spray is used for 30 s in a 25 m^2 room. The room ventilation is assumed to be very low or nearly zero. The droplet distribution of the released spray is characterised.
> The scenario could apply, for example, to a shoe impregnation spray but would also fit a short-term disinfection with a (non-alcoholic) disinfectant.

Calculation Using SprayExpo

Short Introduction of SprayExpo

SprayExpo is a model that has been developed specifically for spray scenarios. It calculates the airborne concentration of the respirable, thoracic, and inhalable fractions of aerosols generated during spray processes. The model and the tool are described in more detail in Sect. 7.4, and only a brief description is given as introduction here.

It is assumed for the spray scenarios that the sprayed product is composed of a non-volatile active substance dissolved in a solvent with known volatility. The model simulates the motion of released droplets, taking into account gravitational settling, turbulent mixing with the surrounding air, and droplet evaporation.

The main input parameters are the released droplet spectrum, the release rate of the spray liquid, the concentration of the active substance, the spatial and temporal pattern of the release process (surface spraying against floor, ceiling, wall; room spraying; etc.), the vapour pressure of the liquid, the size of the room, and the room ventilation rate. The path of the sprayer can be explicitly included into the model.

For surface treatment by spraying, a droplet deposition module is incorporated in the programme package. This module calculates the fraction of non-impacting droplets (overspray) that are relevant for human exposure.

Calculation of the Given Scenario

The scenario described above simulates the exposure from use of an impregnation spray. The droplet distribution of the released spray is characterised by a median primary droplet diameter of 100 μm and a geometric standard deviation of 1.8. The release rate is assumed to be 1 ml/s and the release height is 1 m. It is assumed that the spray is used for 30 s in the middle of a 25 m^2 room, and the user remains in the room for another 10 min. Figure 5.11 shows a result of a SprayExpo calculation. The two curves represent the actual exposure concentration (dark blue) and the time-averaged concentration at the receptor site (breathing zone of the user: breathing height of 1.5 m at a horizontal distance of 1 m from the end of the spray cone) of the thoracic fraction of the airborne non-volatile material. There is a steady increase in concentration at the receptor site during the release period because this is near the localised cloud of the aerosolised substance. This cloud then disperses inside the room, and the concentration becomes spatially homogeneous. There is more than a factor five between the peak concentration and the final concentration at the endpoint of the exposure. The programme also calculated the inhaled lung dose to be 79 μg on the basis of a respiratory minute volume of 10 l/min. There is a strong influence of median droplet diameter on the inhaled dose. For an aerosol size of 50 μm, the inhaled lung dose of active substance is 430 μg: for 200 μm, this is reduced to 11 μg.

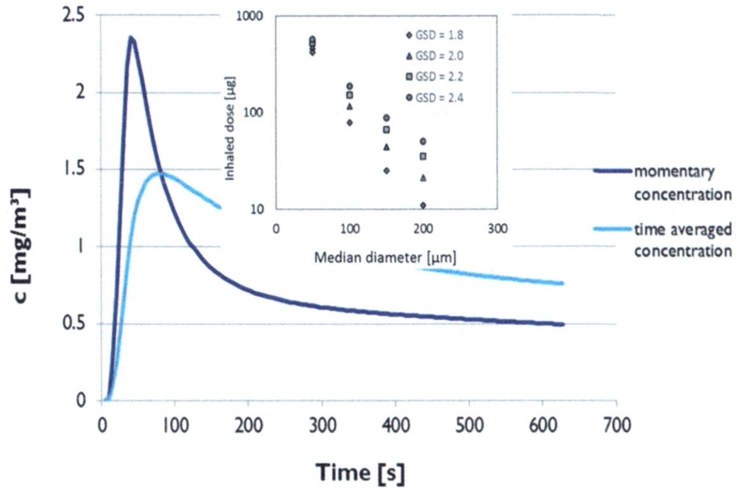

Fig. 5.11 Momentary and time-averaged concentration of the thoracic size fraction of the airborne non-volatile substance for an exposure scenario characterising shoe impregnation. The given parameters were 6% of active non-volatile substance in a spray; use duration of 30 s room size, 25 m^2; and a release rate of 1 g/s

The inhaled lung dose increases when the distribution becomes wider (see insert in Fig. 5.11).

The insert presents the inhaled dose as function of median diameter and geometric standard deviation of the spray droplet size distribution.

Instead of primary droplet distribution, the release fractions of the matured spray could be available from measurements. For example, SprayExpo can be used by assuming the non-volatile fraction to be 100% and the size distribution given by the intervals <5 µm 5–10 µm and 10–40 µm calculated from the respirable, thoracic, and inhalable release fraction.

It should be considered that SprayExpo is a short-term exposure model covering only those time scales that are typical for spraying and the release process. Long-term emissions of vapours from walls and other surfaces have to be calculated separately.

Calculation Using the Spray Module in ConsExpo

Short Introduction of the Spray Module

The spray module of the general ConsExpo computer programme (version 4.1, version 5, or ConsExpo Web version) is a deterministic mass balance model for calculating the inhalative exposure to the non-volatile fraction of a sprayed liquid remaining after instantaneous evaporation of the solvent. The balance equations include ventilation as well as the particle size-dependent settling. It is generally assumed that the released product is well mixed inside the volume at any time (well-stirred conditions). If the spray is released towards the body (hairspray, deodorant spray, etc.), the exposure can be calculated assuming a personal cloud that expands at constant rate with time. Furthermore, the oral uptake of non-respirable aerosols that deposits in the higher airways and is cleared via the gastrointestinal tract can also be taken into account.

Moreover, the spray model of ConsExpo has been validated experimentally (Delmaar and Bremmer 2009).

ConsExpo will not calculate the decreasing particle size by evaporation of the solvent after the spraying process but assumes that the particle size distribution entered in the input menu is without the solvent. it is stated in the validation report that “in view of the results of these calculations on the evaporation of solvents from a mixture, it is assumed that all of the solvent will already have evaporated from the aerosols at the instant the aerosol size distribution was recorded. This assumption will only hold for volatile solvents (at least as volatile as water) and for the small size aerosols in the distribution (diameters up to about 10 μm). These smaller aerosols, however, are precisely the ones that are of interest with respect to inhalation exposure. But it must be kept in mind that this is a crude assumption that may hold only approximately for less volatile solvents”.

Calculation of the Given Scenario

Again, the above-described scenario of an impregnation spray is simulated. The droplet distribution of the released spray is characterised by a lognormal distribution with a median primary droplet diameter of 39 μm (assuming that droplets with original 100 μm will be reduced after evaporation of the solvent) and a coefficient of variation of ln (1.8). The maximal diameter was set to 50 μm and the inhalation cut-off to 15 μm (each value is a default). It is assumed that the spray is used for 30 s in a 25 m^2 room with a room high of 3 m. The release rate is assumed to be 1 g/s with an airborne fraction of 1 and no ventilation as a worst-case. Figure 5.12 shows the graphical output of this ConsExpo calculation.

A combined evaluation for spray products containing evaporating solvents and particles (i.e. paints with their pigments) can be calculated in two separate modelling steps within ConsExpo.

The final outcomes of this scenarios show that both tools result in the same order of magnitude for the external exposure.

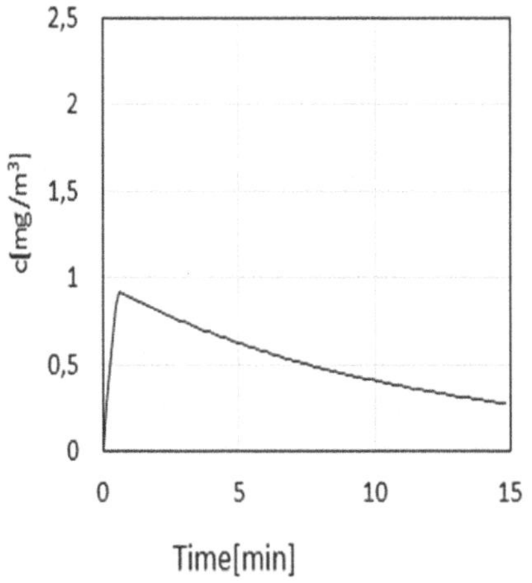

Fig. 5.12 Output graph of ConsExpo (version 4.1) for the given spray impregnation scenario, 6% of active non-volatile substance in a spray; use duration of 30 s room size, 25 m^2 with 3 m height; and a release rate of 1 g/s

5.10 Dedicated Tools for Estimation of Exposure from Consumer Products

Christiaan Delmaar and Annette Bitsch

This section provides an overview of other dedicated tools that may be useful for specific consumer applications in a tiered exposure assessment. The list is not meant to be all-inclusive but does give an overview of the most used tools.

5.10.1 List of Dedicated Exposure Modelling Tools

Many dedicated tools for consumer exposure assessment are available and can be used for exposure assessment under REACH. Some of these tools have been developed specifically for use under REACH, and some are more generally applicable.

The tools specifically developed for the use in REACH include ECETOC TRA, EGRET, and A.I.S.E. REACT. An example of a more general tool that is suitable for use under REACH is ConsExpo. Short descriptions of each of these tools are provided below.

5.10.2 CEM

The US-EPA Consumer Exposure Model (CEM) offers a set of models to estimate exposure to chemicals contained in consumer products and articles (US-EPA 2019). The products are generally consumable liquids, aerosols, or semi-solids that are used a given number of times before they are exhausted. The articles are generally solids, polymers, foams, metals, and woods. These articles are always present within indoor

Table 5.17 Overview of the models included in the Consumer Exposure Model (CEM)

Model	Description
Product Applied to a Surface Indoors Incremental Source Model	The model assumes a constant application rate over a user-specified duration of use and keeps track of the history of applied product. The emission rate is assumed to decline exponentially in time
Product Applied to a Surface Indoors Double Exponential Model	This model accounts for an initial fast release by evaporation, followed by a slow release dominated by diffusion
Product sprayed	This model assumes a small percentage of a product is aerosolised and therefore immediately available for uptake by inhalation. The remainder is assumed to contact the target surface and to later volatilise at a rate that depends on the chemical's molecular weight and vapour pressure
Product added to water	Model assumes emission at a constant rate over a duration that depends on the chemical's molecular weight and vapour pressure
Product placed in environment	Model assumes emission at a constant rate over a duration that depends on its molecular weight and vapour pressure
Article placed in environment	Model assumes emissions of semi-volatile organic compound (SVOC) additives from articles and subsequent partitioning between indoor air, airborne particles, and settled dust over time
Product applied to ground outdoors	A shallow mixing model provides algorithms and inputs to assess a number of scenarios where products such as fertilisers are applied to soil
Product swallowed	Model assumes that the product is directly ingested as part of routine use
Article mouthed	Models oral exposure using a migration rate of the chemical of interest from the article of interest
Product Applied to Skin, Fraction Absorbed Model	For articles that come into direct contact with the skin, the potential dermal dose is estimated by multiplying the thickness of the contact layer of the article by the density and weight fraction to estimate a loading on the surface of the article
Product Applied to Skin, Permeability Model	For products for which a skin permeability coefficient is known

environments for the duration of their useful life which may be several years. CEM offers a set of sub-models that simulate different products and circumstances (Table 5.17). The model provides default values for input parameters, but the defaults may be overridden by manual input. The product use categories define various kinds of products, and examples include aerosol spray paints, laundry detergent, foam-based furniture, hard-plastic toys, and motor oil.

The model evaluations are deterministic; however, analysis of variation is supported. The user can provide high, moderate, and low estimates of the input parameters and perform an analysis on both the central tendency and the higher bound of exposure.

The model covers the inhalation, ingestion, and dermal routes of exposure.

Example

CEM includes a list of substances with their physical chemical properties and an array of predefined exposure scenarios. For example, to assess exposure of a substance from a liquid cleaner, the scenario "all-purpose liquid cleaner (diluted)" can be used. This sets up a default assessment for which the exposure assessor needs to make a number of additional choices such as the exposed population to use (in particular, the age), the activity pattern, whether or not to use a near-field model option, etc. The defaults provided by CEM may be overridden.

In this scenario, CEM generates estimates for short-term dose rates and long-term (daily) doses for the inhalation and dermal routes and adds both to estimate total exposure (both acute and chronic).

CEM is now integrated in the Exposure and Fate Assessment Screening Tool (E-FAST programme, available via https://www.epa.gov/tsca-screening-tools/e-fast-exposure-and-fate-assessment-screening-tool-version-2014).

Different scenarios are expected to be applicable for exposures by spraying, i.e. product spray on surface (inhalation), fabric protector, aerosol paint, or a user-defined scenario can be used.

In addition, CEM (2.0) is available at https://www.epa.gov/tsca-screening-tools/approaches-estimate-consumer-exposure-under-tsca.

5.10.3 *Multi-Chamber Consumer Exposure Model*

The Multi-Chamber Consumer Exposure Model (MCCEM) estimates average and peak indoor air concentrations of chemicals released from products or materials in houses, flats, townhouses, or other residences (US-EPA 2017b). The model uses a mass balance approach, modelling the emission from a source (product or article), transfer through (multiple) compartments in a residence, (reversible and irreversible) sinks, and elimination by ventilation. MCCEM simulates exposure for periods ranging from 1 hour to 1 year. Seasonal or annual exposure profiles are constructed using a long-term model.

MCCEM contains data libraries with residential settings such as zone or area volumes, interzonal airflows, and whole-house exchange rates. The model also accepts manual input and can be used other indoor environments (e.g. schools and offices) if the user can supply the necessary inputs.

MCCEM estimates inhalation exposures to these chemicals, calculated as single-day doses, long-term (average daily) doses, or lifetime average daily doses. (All dose estimates are potential doses; they do not account for actual absorption into the body.)

Note:
The tool is currently not operational. It requires Windows 95 to run but will not install on more recent versions of the operating system. An updated version available as a Web tool is foreseen (see www.epa.gov/tsca-screening-tools/approaches-estimate-consumer-exposure-under-tsca).

5.10.4 US-EPA Wall Paint Exposure Model WPEM

The US-EPA Wall Paint Exposure Model (WPEM) is a special purpose model to facilitate product safety assessments for substances in interior latex and alkyd paints (US-EPA 2017c). It estimates an individual's inhalation exposure to airborne concentrations of a chemical released from latex or alkyd primer/paint, during and after the time when a building (residence, office, or standard box) is painted. The model provides both short-term and long-term exposure measures. Short-term measures include the highest instantaneous, 15-minute-average, and 8-hour-average airborne concentrations to which an individual is exposed under the conditions represented by model inputs. Long-term measures include lifetime average daily dose (LADD) and lifetime average daily concentration. The emission algorithms used in the model, and their relationship to chemical properties, are based on chamber tests specific to interior paints. At present there is no basis for applying these algorithms to other types of products. The model has no capability for Monte Carlo simulation as a means of addressing uncertainty.

Example
WPEM can be used for exposure arising specifically from paint components. The tool simulates inhalation arising from one or multiple paint jobs. For example, exposure arising from painting the walls of a bedroom using a latex paint can be simulated. The tool provides defaults for many parameters such as the application rate of the paint, ventilation rate, painted surface area, each of which may be overridden. The application of multiple layers and inclusion of primer application can be incorporated in the simulation.

The tool includes different exposed populations such as children (bystander) and adult occupants.

Lifelong exposures (average daily doses) as well as maximum 24-hour dose rates are given as output.

Note: WPEM is designed for Windows 95 and will not install under the newest versions of the Windows operating system. No information on the future developments are available at present.

5.10.5 *ADL Polymer Migration Estimation Model (AMEM)*

ADL Polymer Migration Estimation Model (AMEM) is a computer software tool that implements a number of models to simulate migration of substances from polymeric materials to gases, liquids, and solid materials in contact with a polymeric material (US-EPA 1990). The output of the model are migration rates which may be processed further in an exposure assessment tool that may use the estimated emission rates from dermal, ingestion, or inhalation exposure models. AMEM requires detailed input on partition coefficients and factors that determine mass transfer. In cases where such detailed information is lacking, the model documentation provides estimation methods. AMEM has no capabilities for sensitivity or uncertainty analysis.

5.10.6 *ConsExpo*

The RIVM Consumer Exposure modelling tool ConsExpo (Delmaar and Schuur 2017) is a Web application that offers a set of models to simulate a variety of exposure scenarios for substances in consumer products. The tool covers the inhalation, dermal, and oral routes of exposure. The models include simple, conservative low tier models as well as higher tier, physics-based models, enabling a tiered approach to exposure assessment.

Inhalation

Inhalation models include vapour and aerosol (to be used for sprays) models. Evaporation models estimate the emission of substances that evaporate from products such as cleaning agents, paints, and flooring. Different tiers of models increasingly include more physics-based aspects. The low tier models require user input on the emission of the substance. By contrast, two higher tier models simulate the physics of evaporation from liquids and solid materials, respectively:

- Tier 1: instantaneous release with zero ventilation
- Tier 1a: instantaneous release with ventilation of 0.2–1.0 h^{-1}
- Tier 2: constant rate
- Tier 3: evaporation

In every tier, iterations are possible which can modify the exposure, e.g. the ventilation rate can be changed from zero to high rates, as shown in Fig. 3.6. Also the physico-chemical characteristics of the substance can be included by consideration of its vapour pressure and molecular weight. As a further iteration, the model can be shifted from the deterministic to the probabilistic approach. This step is possible in any of the iteration steps.

The aerosol or spray models include a low tier model that assumes instant release of the sprayed material into indoor air and a more advanced higher tier model that takes into account the characteristics of the spray and generated aerosol to estimate the air concentration.

Dermal

Dermal models in ConsExpo include low tier models that formulate exposure in total amount available for dermal exposure without considering the manner and time profile of how the product is loaded onto the skin. More advanced models describe migration to the skin from products in dermal contact, transfer of residues from treated surfaces in a room to the skin, and diffusion of the substance from a viscous product applied to the skin (e.g. a cosmetic cream).

Oral

Oral models include the migration of the substance from a product that is mouthed and substances that migrate from food packaging material into consumed foods.

Fact Sheets and Default Model Input

The ConsExpo tool is supported by fact sheets that compile exposure information on a number of product groups (e.g. cosmetics, paints, cleaning products, and pest control products). The fact sheets suggest defaults for a range of products or product types. Currently, fact sheets are available for cleaning products, cosmetics, paint products, pest control products, disinfectant products, do-it-yourself products, and children's toys (Bremmer and van Veen 2002; Bremmer et al. 2006b, c; Bremmer and van Engelen 2007; Ter Burg et al. 2007; Prud'homme de Lodder et al. 2006; Meesters et al. 2018). These fact sheets will be updated in the future and extended to cover, for example, air fresheners.

Tools for Uncertainty Analysis

ConsExpo offers tools for sensitivity analysis and one-dimensional Monte Carlo simulation for uncertainty analysis. The ConsExpo tool is currently available as a Web-based application (ConsExpo Web) that is freely accessible and does not require any additional software or licence to use.

Example

As a typical example of a tiered assessment in ConsExpo, consider the emission of a solvent from a cleaning product that is applied to a large surface.

For a low tier assessment, ConsExpo features a model "exposure to vapours: instantaneous release". This model calculates exposure by assuming that the solvent is immediately released and dispersed through the room air. After the initial release, the substance is removed by ventilation. Exposure is due to inhalation of the room air and is proportional to the average air concentration of the solvent.

This first-tier evaluation requires only limited information such as the amount of product used, the concentration of the substance in the product, the room volume, and the air changeover (ventilation rate) of the room. Reasonable estimates of these parameters can usually be made.

If this first-tier fails to demonstrate safe use, a higher tier model may be used for a more refined exposure analysis. ConsExpo offers a higher tier model "exposure to vapours: evaporation" that estimates the air concentration using physics-based algorithms that describe evaporation of the solvent from a mixture. The model requires much more elaborate input than the first-tier model, e.g. physical-chemical information, product composition, and details about the way the product is applied.

Providing appropriate input values for these parameters usually requires more effort, and the provided input values will have a higher degree of uncertainty associated with them.

To account for this enhanced uncertainty in input parameters, a probabilistic evaluation may be conducted using ConsExpo, in which probability distributions accounting for the uncertainty in the precise numerical values of the parameter are used as inputs and a distribution of exposure is calculated using Monte Carlo simulation. This distribution reflects the uncertainty in the exposure estimate that results from the uncertainty in the input variables. It may be used, for example, to derive the confidence intervals of the exposure estimate.

In October 2016, ConsExpo Web was launched (RIVM 2016). ConsExpo Web contains the functionalities of ConsExpo 4.1 and the emission tool.

The update of ConsExpo Web is realised by RIVM in collaboration with the counterpart institutes ANSES (France), BfR (Germany), FOPH (Switzerland), and Health Canada. These institutes will continue to work on the further development of the instrument. In the future, ConsExpo Web can easily be expanded with new applications, such as an exposure assessment for short-term exposure. Also, the content of the fact sheets will be updated.

ConsExpo Web version 1.0.1 is released in July 2017 (RIVM 2017b). Exposure estimations made with this version of ConsExpo Web can be imported into Chesar through a ConsExpo Web export file. Chesar is the CHEmical Safety Assessment and Reporting tool of the European Chemicals Agency (ECHA). The development of this export functionality was financed by ECHA. Besides the introduction of a Chesar export functionality, the option to export graph data to excel was added and a few bugs were fixed.

5.10.7 ConsExpo Nano

In addition to ConsExpo Web RIVM developed for the exposure assessment of regular substances, a module developed specifically for the exposure assessment of nano substances (RIVM 2017a), in spray applications, is available. This module is ConsExpo nano, and it allows for detailed mechanistic estimation of the alveolar load of a nanomaterial after the use of a consumer spray containing that nanomaterial. The tool combines the ConsExpo spray model with the ICRP human deposition and clearance model (ICRP 1994; Bailey et al. 2007). The estimated alveolar load can be expressed in different exposure metrics such as mass, number of nanoparticles, and the total surface area of nanoparticles. The tool does not yet offer sensitivity analysis or probabilistic evaluations.

Example

A typical example of an application of ConsExpo nano is the exposure assessment of nano TiO_2 in sunscreen sprays. The tool requires input on the aerosol size distribution of the spray, composition of the spray (i.e. the weight fraction of the nanomaterial), and specifics of the nanomaterial used [its shape and size (distribution) and density]. For many uses, the parameter defaults from the ConsExpo fact sheet on cosmetics (Bremmer et al. 2006c) can be used.

The ConsExpo spray model can be used to model the inhalation exposure to respirable aerosol. The estimate of the inhalation exposure is used as input for the ICRP deposition and clearance model to estimate the alveolar load in the course of time. The modelling of the load takes into account that multiple exposures take place and that the load from a series of exposure events cumulates.

This load may can be expressed in various dose metrics that are deemed appropriate for risk assessment. These include, but are not limited to, the total mass of nanomaterial in the alveoli, the total surface area of nanomaterial in the alveoli, and the total volume of deposited aerosol in the alveoli.

5.10.8 SHEDS

The Stochastic Human Exposure and Dose Simulation Model for Multimedia, Multipathway Chemicals (SHEDS-Multimedia) from the US-EPA is a high tier multipathway, multi-chemical exposure model. SHEDS provides several modules, including dietary and residential exposure modules. SHEDS-Residential provides a framework for stochastic simulation of population exposures to single or multiple chemicals applied in residential homes, yards, and gardens (Isaacs et al. 2014; Özkaynak et al. 2011). SHEDS allows the estimation of population exposures for cohorts of interest incurred via inhaling contaminated air, touching contaminated surface residues, and ingesting residues from hand- or object-to-mouth activities. The model may be used for single and multi-chemical exposure assessments. It merges chemical usage, human activity data, concentrations data, and exposure factors to generate time series of exposure for simulated individuals. Sensitivity analysis and two-dimensional Monte Carlo simulation to analyse uncertainty and variability are supported.

The use of SHEDS requires the SAS software to be installed.

5.10.9 Probabilistic Aggregate Exposure Assessment Model (PACEM)

PACEM is a modelling tool that assesses the total consumer exposure from products used in combination to substances contained in these products. The basis for PACEM is realistic product usage information (used amount per product use and use frequency) in the population. This product usage information is obtained in population surveys. Currently information is available on the use of cosmetics and personal care products and household cleaning products. PACEM has been applied in a number of risk and health impact assessments (Dudzina et al. 2015; Delmaar et al. 2015; Ezendam et al. 2018). PACEM is expected to be available at (www.rivm.nl) as a publicly accessible Web tool in the course of 2019–2020.

SprayExpo is a deterministic droplet simulation model enabling the calculation of the inhaled and dermal doses users are exposed to during the release of a non-volatile substance by a spray (or fogging) technique. The typical product is composed of a non-volatile active substance dissolved in a solvent with known volatility, and the application is performed in enclosed rooms.

Although this model was originally developed for occupational exposure, SprayExpo is very applicable for consumer exposure since the modelling is not based on default settings but on consumer-relevant product and device parameters.

The model is based on a simulation of the motion of released droplets taking into account gravitational settling, turbulent mixing with the surrounding air, and droplet evaporation. Continuous spatial release patterns can be simulated in the model. The spatial distribution of the concentration is explicitly taken into account in the calculation of the inhaled dose and the dermal dose. In addition, more general model parameters, such as room dimensions and ventilation data, physico-chemical product data, and technical data, for the spraying process are taken into account. Further, the application pattern has to be specified.

The main input parameters are the released droplet spectrum, the release rate of the spray liquid, the concentration of the active substance, the spatial and temporal pattern of the release process (surface spraying against floor, ceiling, wall; room spraying, etc.), the vapour pressure of the liquid, the size of the room, and the ventilation rate. The path of the sprayer can be explicitly included into the model. There are different options to enter information about the droplet size:

1. Manual input of the percentages of droplet mass in the 7 size ranges.
2. Direct use of measured values, generated by a specific laser diffraction spectrometer.
3. Input of the parameters, i.e. the median value of the droplet diameter of the mass size distribution and geometric standard deviation, and then a lognormal distribution of the droplet mass over the droplet diameter is assumed.
4. Selection of a common application technique. Eight different techniques are currently implemented.

For surface treatment by spraying, a droplet deposition module is incorporated in the programme package. This module calculates the fraction of non-impacting droplets (overspray) that is relevant for human exposure. The stopping distance of droplets is also taken into account for the room spraying.

The model calculates the airborne concentration of the respirable, the thoracic, and the inhalable fractions, or any other meaningful size fraction of aerosols generated during spray processes. The dermal exposure includes the deposition of active substance on body surfaces by aerosol settling. The model is a short-term exposure model covering time scales typical for the release process. Long-term emissions of vapours from walls and other surfaces are not included.

The programme can be executed using MS Excel. The first level is an input form for the definition of general data such as room size, room ventilation rate, turbulent intensity, and nozzle and spray parameters, as well as relevant parameters of the spray liquid. The second level allows for the definition of the spray path and contains the results of the calculation:

1. Graphical presentation of the concentration versus time curve
2. The time-integrated inhaled and deposited dose of active substance

The model is validated by measured data specific for the use of biocides, and a sensitivity analysis has been performed (Koch et al. 2012).

The Excel spreadsheet-based model is available for downloading at the homepage of the Federal Institute for Occupational Safety and Health (BAuA) at http://www.baua.de/en/Topics-from-A-to-Z/Hazardous-Substances/SprayExpo.html.

5.10.10 ECETOC TRA Version 3.1

The ECETOC Targeted Risk Assessment (TRA) tool was developed to be used for the worker and consumer exposure estimates within the European Chemicals

Agency (ECHA) (ECETOC 2012, 2014). The intention from the beginning was to serve as a tier 1 tool specifically for safety assessment under REACH [(EC) No 1907/2006; Registration, Evaluation and Authorisation and Restriction of Chemicals]; thus, the intention was to build a screening tool giving, as a first step, a rationale for a safe use without going into too many specific details for every single product. More refined tier 1.5 or tier 2 approaches could be used if necessary.

Therefore, for low-level exposure assessments, tier 1 equations, and default model, parameters for product and article categories that are defined in the REACH documentation are implemented in the tool. Product categories or article categories were described using standard descriptors. Information on use conditions was based on knowledge about the consumer habits and practices (H&P). This tool considers each applicable phase of activity (preparation, application, post-use exposure, and cleaning/removal).

The ECETOC TRA consists of three separate models for estimating exposures to workers, consumers, and the environment. Version 3 incorporated a number of improvements in each section of the tool to increase the accuracy and utility of the tool while still keeping it as a tier 1 estimation tool.

These changes are also for the consumer exposure modelling part and should overcome some of the limitations of the older versions. It is possible with this version to better account for (limited types of) consumer H&P data when certain information is or is not available. It also allows for the calculation of exposure estimates for new subcategories within the product and article categories if the details of product use and consumer behaviour are available. Using the SpERC experiences as a reference point, ECETOC developed a format to facilitate the collection and communication of relevant H&P information for identified consumer uses (termed SCED Specific Consumer Exposure Determinants).

Additional features included for consumer exposure estimation are:

- The possibility of creating new product subcategories for which all the required parameters could be set and justified
- A refinement for inhalation exposures by (1) including an upper bound concentration (saturated vapour concentration) and (2) including a value for "standard" ventilation
- Dermal and oral transfer factors
- Calculation of combined total RCR values

In addition, some more specific models have been developed by parts of industry that address applications of special products. These models also have more specific default parameters that have been derived. Examples for this are the EGRET tool developed by the European Solvents Industry Group (ESIG) and the A.I.S.E. tool REACT for higher tier assessments for cleaning agents. More information is provided below.

Version 3 of the TRA is available in two forms: as an integrated exposure/risk assessment tool, covering worker, consumer, and environmental exposures, and as a stand-alone consumer exposure estimation tool. All elements of the TRA (workers, consumers, and the environment) are incorporated into an integrated tool with a common user interface and common data sheets. The user interface allows a user to perform assessments for a single substance for different settings (e.g. different processes, consumer articles, and environmental release classes). All data can be stored in Excel data sheets and further processed. The data sheets can store up to 80 substances and include more scenarios per substance than the user interface.

ECETOC TRA is included in the European Chemicals Agency's (ECHA) CHESAR (CHEmical Safety Assessment and Reporting) tool.

5.10.11 EGRET

The acronym EGRET stands for the European Solvents Industry Group (ESIG) Generic Exposure Scenario (GES) Risk and Exposure Tool. This tool was developed to refine the use scenarios for solvent registration under REACH (ESIG 2915). It considers the requirements when assessing exposure to solvents in consumer products. The tool is mainly based on the default assumptions and algorithms (equations) described in the ECETOC TRA (2009b) consumer module but defines additional exposure scenarios and sub-scenarios that are more specific for the particular use of solvents. EGRET covers all the consumer product categories that could contain solvents in the ESIG/ESVOC GES library. EGRET introduces adapted default values and develops additional product categories relevant to solvents (as described in Chapter R.12 of the REACH TGD), including the ability to estimate exposures from infrequent solvent uses. Thus, EGRET might be noted to be a tier 1+ model, which means that predictions are positioned between conservative tier 1 tools like ECETOC TRA (2009b) and more sophisticated higher tier models, or even measurements.

EGRET has already been used for consumer exposure and risk assessments under REACH for a number of solvents and solvent-like products, including:

- Hydrocarbon solvents
- Oxygenated solvents
- Fuels and lubricants

Typical products are categorised in groups. In addition, eight individual GES consumer workbooks were developed that cover the associated PCs:

- Coatings (consumer use)
- Cleaning
- Lubricants
- Agrochemicals
- Fuels
- Functional fluids
- De-icing
- Water treatment

EGRET does not yet include exposure scenarios for substances in articles.

It is an Excel-based tool currently available as version 2. This version enables users to control the level of detail/complexity considered appropriate for any GES (which is a compilation of the product categories and subcategories considered relevant for the particular use) (http://www.esig.org/en/regulatory-information/reach/ges-library/consumer-gessieren).

An output in XML format for CHESAR and/or the International Uniform Chemical Information Database (IUCLID) is being developed.

Some of the new functionalities include:

- The use in combination with Specific Consumer Exposure Determinants (SCEDs) which can be useful additional "information input" for consumer exposure assessments. Thus far, the exposure estimates derived from the use of SCEDs are relevant for long-term systemic endpoints, and thus a chronic long-term DNEL is needed for risk assessment.
- An assessment of additional types of exposure (other than long-term systemic) is possible.
- An updated air exchange rate for outdoor exposure scenarios.
- Frequency bands of exposure (frequent, $\geq$1/week; occasional, 1/week–1/month; infrequent, 1/month–1/6 month; and very infrequent, $<$1/6 month).

The tool has the functionality to auto-populate the control measures (e.g. product concentration for particular use) to a desired safe level (0.9 by default, or 0.5 or 0.2), and it auto-generates the supporting narrative exposure scenario using standard phrases and all relevant exposure determinants.

Although this tool is for consumer-related applications, it requires expert knowledge to use and to understand how exposures are determined and how the associated risks are evaluated.

5.10.12 *Specific Consumer Exposure Determinants (SCEDS) as Refined Input for Exposure Modelling: Predominantly for ECETOC TRA*

Specific Consumer Exposure Determinants (SCEDS) can be used to refine exposure estimates using the ECETOC TRA v.3.1 (2009a, b). SCEDS are a good source of information on exposure determinants associated with consumer product applications. They have been and are being developed by industry sectors and the downstream user groups, i.e. the solvents group developed the SCED for the consumer use of solvents mentioned earlier. Also, A.I.S.E. developed some specific SCEDS.

The A.I.S.E. SCEDs include a supporting explanation document and a fact sheet for consumer uses of each of the following product categories:

- Laundry products in the home
- Fabric conditioners in the home
- Surface cleaner in the home - non-spray application

- Liquid surface cleaner in the home - spray application
- Auto dishwashing products in the home
- Hand dishwashing liquid in the home
- Polishes and wax blends in the home - non-spray application
- Polishes and wax in the home - spray application
- Air care products in the home - non-aerosol
- Air care products in the home - aerosol

The A.I.S.E. SCEDS gives default values for contact areas, frequencies, and typical concentrations based on information about consumer's habits and practices originally developed under the HERA project and complemented with data from companies and recent A.I.S.E. consumer habits surveys.

A description of consumer uses of cleaning and maintenance products and the SCEDs fact sheets are made available from the A.I.S.E. Consumer Safety Exposure Assessment homepage (https://www.aise.eu/our-activities/product-safety-and-innovation/reach/consumer-safety-exposure-assessment.aspx).

5.10.13 REACH Exposure Assessment Consumer Tool (REACT) from A.I.S.E.

The A.I.S.E. REACT Consumer Tool (A.I.S.E. 2010) allows quantitative estimation of systemic consumer exposures to substances that are present in preparations used by consumers. The tool calculates substance exposures from inhalation, dermal, and oral routes separately. It also provides a summation of all the relevant exposure routes and is based largely on tier 1 algorithms. REACT contains several subproducts within PC3 (air freshener), PC31 (polishes and wax), and PC35 (washing and cleaning). It can be considered to be a tier 1 quantitative estimation tool containing these refined subproduct categories and the default values assigned to them.

The REACT tool can also calculate risk characterisation ratios (RCR) for the different exposure routes separately and for the overall exposure. RCRs are only calculated provided that substance-specific hazard data values (e.g. DNELs) are entered in the user input worksheet by the user. They are based on information about the identification of consumer's habits and practices originally developed under the HERA project and complemented with data from companies' data and recent A.I.S.E. consumer habits surveys (see the SCEDs information provided above).

REACT implements the British Aerosol Manufacturers' Association's (BAMA) model for air fresheners (BAMA 2008), which is applicable for both aerosol particles and gas-phase substances. The dermal exposure estimation can be refined in the REACT tool by including information on substance-specific skin permeability. The dermal RCRs that are derived should be compared to the systemic hazard value if they use permeation coefficient (Kp) or weight fraction absorbed through skin data and should not be compared with a local hazard from external doses.

DNELS can be used only in the unit mg/kg of BW, which means the tool is designated only for systemic exposures.

The tool does not cover the evaporation of volatile substances from the product. It covers the inhalation route in case of spray products; however, the vapour pressure is not a user entry and included is nearly no information about PC data.

The tool is delivered as an EXCEL version containing eight worksheets:

1	User input	User-specific entries and products selection
2	Results	Displays exposure estimates and RCRs for product categories and subcategories
3	Dermal	Displays dermal exposure estimates and RCRs for the product subcategories and shows the algorithms and parameters used for calculation
4	Inhalation	Displays inhalation exposure estimates and RCRs for the product subcategories and shows the algorithms and parameters used for calculation
5	Oral	Displays oral exposure estimates and RCRs for the product subcategories and shows the algorithms and parameters used for calculation
6	Def_AISE H&P2009a	Table of habits and practices for consumer products in Western Europe
7	Def_AISE H&P2009b	Lists additional data required to estimate adult exposure to air freshener ingredients
8	Def_TRA2009	Lists subpopulations' specific (child, adult) default parameters - same list as in ECETOC TRA (2009b)

More information is available via the user manual which provides guidance for working with the A.I.S.E. REACT Consumer Tool https://www.aise.eu/our-activities/product-safety-and-innovation/reach/consumer-safety-exposure-assessment.aspx.

5.10.14 Further Tools

Additional tools for exposure assessment are mentioned in the context of REACH for a use in tier 1 assessments. These tools focus more on worker exposures and will not be described in more detail:

- Stoffenmanager® (v 4.5)
- EMKG-tool
- Riskofderm
- MEASE
- ART

5.10.15 Further Tools Specifically for Application in Consumer Exposure

Additional tools for consumer exposure have been developed by defined user downstream groups for specific chemical industries. They can help to better characterise the sometimes very complex exposures for consumer uses of a diverse range of products, e.g. from insecticide spraying to the use of cosmetics. Some of these tools are either not publicly available or are in a phase of being further developed. Thus, this list is not a complete list of tools and very detailed information is not provided.

BAMA/FEA Indoor Air Model

This model/tool was developed by the British Aerosol Manufacturers Association (BAMA) and the European Aerosol Federation (FEA).

The BAMA Indoor Air Model is a relatively simple tool to predict the concentration of aerosol components within a room for a suitable time interval after spraying for a wide range of use conditions. It is available as an Excel sheet and intended to be used in a worst-case approach to calculate the highest concentration in air, assuming that all sprayed material is available to be inhaled. This can be performed for different use pattern (single use, repeat use, continual use, and continuous release). However, the model is based on an instantaneous, immediate distribution in the entire room. It does not consider the gravitational deposition of droplets which is of some importance for low-volatile substances. Thus, this model has some limitations, and more realistic other models should be preferred for detailed exposure assessments.

According to BAMA, experts from BRE (UK) who are actively involved in writing international standards for the measurement and control of indoor air pollutants (Rowley and Crump 2005) have independently validated this model and concluded that it is "A simple exponential decay model that assumes immediate and uniform vaporisation and dispersal of an aerosol (that) can be used to predict the concentration of aerosol components within a room after a suitable time interval after spraying". The BAMA/FEA Indoor Air Model has also been used in the A.I.S.E. REACT tool (see above). A comparison of the BAMA Indoor Air Model and the respective functionalities of ConsExpo has been performed by RIVM.

Overall, the model assumptions are simple, but the tool can be used to rapidly generate predicted "worst-case" concentrations for a wide range of use conditions for spray products. Thus, it is suited for screening of potential exposure of substances.

The tool and the user guide are available on the BAMA website (http://www.bama.co.uk) free of charge for BAMA members and with a charge for non-members.

RIFM 2-Box Indoor Air Dispersion Model

The RIFM 2-Box Air Dispersion Model is for consumer inhalation exposure assessments. It is an indoor environment air model that characterises the dispersion of a single chemical inside two connected enclosed “zones” or “boxes” (both terms are used interchangeably in the context of this model). It is also called a “room within a room” model. This model calculates peak air concentrations in the room of interest (room of use) and in the rest of the house, ideally even for air exposure from multiple sources under typical consumer use conditions. Typical room and house sizes are based on the European information and are assumed to represent conservative sizes.

The inputs to this model include chemical source parameters, zone (contact) volumes, and airflow parameters. The model outputs include a temporal profile of chemical concentrations in the two zones.

Model parameters have thus far been established for three air care product categories (aerosol air fresheners, candles/incense, and plug-in air fresheners/reed diffusers) and for personal default parameters. Calculations for the alternate 2-box scenario have been approved by RIFM, A.I.S.E., and HERA. RIFM intends to further develop the inhalation Threshold of Toxicological Concern (TTC) using this model (Api et al. 2014).

The use of this tool has been demonstrated for candles (Petry et al. 2013) and hairspray.

This tool is currently available for free for RIFM members or RIFM Database subscribers at the RIFM website (http://rifmdatabase.rifm.org/nd/Login.cfm). Non-RIFM subscribers should request access from the RIFM homepage (www.rifm.org).

CEPST

CEPST™ is an exposure tool intended to be used as a Chemical Exposure Priority Setting Tool. It has evolved from the Complex Exposure Tool© (ComET©) that the LifeLine Group™ developed for Health Canada. CEPST currently exists as a proof of concept modelling construct (http://www.thelifelinegroup.org/cepst/index.php). This is a beta version in the form of an Excel spreadsheet that unfortunately does not indicate the sources of either the population information (i.e. body weight and/or consumption data) or the scenario background information.

However, CEPST demonstrates users can set priorities relative to exposure potential for large groups of chemicals with little data and possible multiple exposure scenarios. The LifeLine Group developed an approach to do this relative to human health risks. CEPST considers multiple routes of exposure, multiple subpopulations (children, adults, elderly), and different possible durations of exposure related to typical toxicological information (acute (1 day), short-term (90 days), or long-term (365 days) exposure). Users can change the assumptions in the model and see the impacts. The full IT implementation of CEPST into a software tool is under further design and development.

References

A.I.S.E. (2010) Reach Exposure Assessment Consumer Tool, REACT. https://www.aise.eu/our-activities/product-safety-and-innovation/reach/consumer-safety-exposure-assessment.aspx. Accessed 27 Oct 2019

A.I.S.E. (2017) SCEDs – Specific Consumer Exposure determinants. A.I.S.E. supporting explanation: Version 1.1. 18 S. Brussels, BE: International Association for Soaps, Detergents and Maintenance Products Verfügbar unter https://echa.europa.eu/documents/10162/22788232/aise_sceds_explanatory_document_v1-1_oct2017_en.pdf/c2b2c440-46be-3948-cecf-4713bcdd48c7. Accessed 27 Oct 2019

ANSES (2010) Development of a specific Control Banding Tool for Nanomaterials – Scientific report of the Expert Committee (CES) on Physical Agents on Request N°2008-SA-0407 relating to Control Banding. https://www.anses.fr/en/system/files/AP2008sa0407RaEN.pdf. Accessed 27 Oct 2019

ANSES (2011) Opinion on the uses of formamide in consumer goods and health risks related to formamide in children's foam puzzle mats. ANSES Opinion, Request No 2010-SA-0302, 4 July 2011. https://www.anses.fr/en/system/files/CHIM2010sa0302Ra-2EN.pdf. Accessed 27 Oct 2019

ANSES (2017) Description of the control banding tool for nanomaterials. https://www.anses.fr/en/content/tool-control-banding-risks-associated-nanomaterials. Accessed 27 Oct 2019

Api AM, Belsito D, Magnus B, Cadby P, Calow P, Dagli ML, Dekant W, Dent M, Ellis G, Fryer AD, Fukayama M, Griem P, Hickey C, Kromidas L, Lalko JF, Liebler DC, Miyachi Y, Politano VT, Renskers K, Ritacco G, Salvito D, Schultz TW, Sipes IG, Smith B, Vitale D, Wilcox DK (2014) Criteria for the Research Institute for Fragrance Materials, Inc. (RIFM) safety evaluation process for fragrance ingredients. Food Chem Toxicol. https://doi.org/10.1016/j.fct.2014.11.014. http://lup.lub.lu.se/record/4908204. Accessed 27 Oct 2019

Asgharian B, Wood R, Schlesinger RB (1995) Empirical modeling of particle deposition in the alveolar region of the lungs: a basis for interspecies extrapolation. Fundam Appl Toxicol 27 (2):232–238

AUH (1995) Standards zur Expositionsabschätzung. Bericht des Ausschusses für Umwelthygiens. Arbeitsgemeinschaft der leitenden Medizinalbeamtinnen und – beamten der Länder. Herausgeber: Behörde für Arbeit, Gesundheit und Soziales (Hamburg) and Institut für Mathematik und Datenverarbeitung in der Medizin

Bahadar H, Mostafalou S, Abdollahi M (2014) Current understandings and perspectives on non-cancer health effects of benzene: a global concern. Toxicol Appl Pharmacol 276:83–94

Bailey MR, Ansoborlo E, Guilmette MR, Paquet F (2007) Updating the ICRP human repiratory tract model. Radiat Prot Dosim 127(1–4):31–34

Bartsch N, Heidler J, Vieth B, Hutzler C, Luch A (2016) Skin permeation of polycyclic aromatic hydrocarbons: a solvent-based in vitro approach to assess dermal exposures against benzo[a]pyrene and dibenzopyrenes. J Occup Environ Hyg 13:969–979

Becker K, Kaus S, Seifert M, Heidrich F, Rosskamp E, Schulz C, Schlüter C, Seifert B (2004) Umwelt-Survey 1998, Band V: Hausstaub. Stoffgehalte im Hausstaub aus Haushalten der Bevölkerung in Deutschland. Institut für Wasser-, Boden- und Lufthygiene des Umweltbundesamtes, Berlin, DE

BfR (2012) Introduction to the problems surrounding garment textiles. Updated BfR Opinion No. 041/2012, 6 July 2012. http://www.bfr.bund.de/cm/349/introduction-to-the-problems-surrounding-garment-textiles.pdf. Accessed 27 Oct 2019

Biedermann S, Tschudin P, Grob K (2010) Transfer of bisphenol A from thermal printer paper to the skin. Anal Bioanal Chem 398:571–576

Bonvallot N, Mandin C, Mercier F, Le Bot B, Glorennec P (2010) Health ranking of ingested semi-volatile organic compounds in house dust: an application to France. Indoor Air 20(6):458–472. https://doi.org/10.1111/j.1600-0668.2010.00667.x

Bothe M (2004) Quantifizierung der Ingestion von Boden durch Kinder. Schriftenreihe Reaktorsicherheit und Strahlenschutz. VKTA, Verein für Kernverfahrenstechnik und Analytik Rossendorf eV, Bonn

Bott J, Störmer A, Franz R (2014) A model study into the migration potential of nanoparticles from plastics nanocomposites for food contact. Food Packag Shelf Life 2(1):73–80. https://doi.org/10.1016/j.fpsl.2014.08.001. Accessed 28 Jul 2017

Boverhof DR, Bramante CM, Butala JH, Clancy SF, Lafranconi M, West J, Gordon SC (2015) Comparative assessment of nanomaterial definitions and safety evaluation considerations. Regul Toxicol Pharmacol 73(1):137–150. https://doi.org/10.1016/j.yrtph.2015.06.001. http://www.sciencedirect.com/science/article/pii/S0273230015001488. Accessed 27 Oct 2019

Brandsch J, Mercea P, Rüter M, Tosa V, Piringer O (2002) Migration modelling as a tool for quality assurance of food packaging. Food Addit Contam 19(Suppl):29–41

Bremmer HJ, van Engelen JGM (2007) Paint products fact sheet. To assess the risks for the consumer. Updated version for ConsExpo 4. National Institute for Public Health and the Environment (RIVM), RIVM report 320104008 Bilthoven, NL. https://www.rivm.nl/bibliotheek/rapporten/320104008.pdf. Accessed 27 Oct 2019

Bremmer HJ, van Veen MP (2002) Children's toys fact sheet. To assess the risks for the consumer. National Institute for Public Health and the Environment (RIVM), Bilthoven, NL. RIVM report 612810012 http://www.rivm.nl/bibliotheek/rapporten/612810012.pdf. Accessed 27 Oct 2019

Bremmer HJ, Prud'homme de Lodder LCH, van Engelen JGM (2006a) Cosmetics fact sheet. To assess the risks for the consumer. Updated version for ConsExpo 4. National Institute for Public Health and the Environment, Bilthoven, NL. https://www.rivm.nl/bibliotheek/rapporten/320104001.pdf. Accessed 6 Feb 2018

Bremmer HJ, Blom WM, Van Hoeven-Arentzen PH, Prud'homme de Lodder LCH, Van Raaij MTM, Straetmans EHFM, Van Veen MP, Van Engelen JGM (2006b) Pest control products fact sheet. To assess the risks for the consumer – Updated version for ConsExpo. 4. RIVM report 320005002. https://www.rivm.nl/publicaties/pest-control-products-fact-sheet-to-assess-risks-for-consumer-updated-version-for. Accessed 27 Oct 2019

Bremmer HJ, Blom WM, van Hoeven-Arentzen PH, Prud'homme de Lodder LCH, van Raaij MTM, Straetmans EHFM, van Veen MP, van Engelen JGM (2006c) General fact sheet – Limiting conditions and reliability, ventilation, room size, body surface area. Updated Version for ConsExpo 4. RIVM Report 320005002/2006. https://www.rivm.nl/bibliotheek/rapporten/320005002.pdf. Accessed 27 Oct 2019

British Aerosol Manufacturers' Association (BAMA) (2008) BAMA Indoor Air Model, Building Exposure Scenarios. Issue 1. https://www.bama.co.uk/product.php?product_id=11. Accessed 29 Oct 2019

Brown MB, Khengar RH, Turner RB, Forbes B, Traynor MJ, Evans CR, Jones SA (2009) Overcoming the nail barrier: a systematic investigation of ungual chemical penetration enhancement. Int J Pharm 370(1–2):61–67

Bruinen de Bruin Y, Hakkinen P, Lahaniatis M, Papameletiou D, Del Pozo C, Reina V, Van Engelen J, Heinemeyer G, Viso AC, Rodriguez C, Jantunen M (2007) Risk management measures for chemicals in consumer products: documentation, assessment, and communication across the supply chain. J Expo Sci Environ Epidemiol 17(Suppl 1):S55–S66

Butte W (2004) Synthetic Musks in House dust. Series anthropogenic compounds, vol 3X. Springer, Berlin. https://doi.org/10.1007/b14127

Cao Z-G, Yu G, Chen Y-S, Cao Q-M, Fiedler H, Deng S-B, Huang J, Wang B (2012) Particle size: a missing factor in risk assessment of human exposure to toxic chemicals in settled indoor dust. Environ Int 49:24–30. https://doi.org/10.1016/j.envint.2012.08.010

CEN (1993) EN 481: Workplace atmospheres: size fraction definitions for measurement of airborne particles. Comité Européen de Normalisation, Brussels

Choi JI, Kim CS (2007) Mathematical analysis of particle deposition in human lungs: an improved single path transport model. Inhal Toxicol 19(11):925–939

Clausen PA, Spaan S, Brouwer DH, Marquart H, le Feber M, Engel R, Geerts L, Jensen KA, Kofoed-Sørensen V, Hansen B, De Brouwere K (2016) Experimental estimation of migration and transfer of organic substances from consumer articles to cotton wipes: evaluation of underlying mechanisms. J Expo Sci Environ Epidemiol 26:104–112

Clausing P, Brunekreef BA, van Wijnen JH (1987) A method for estimating soil ingestion by children. Int Arch Occup Environ Health 59(1):73–82. https://doi.org/10.1007/bf00377681

Combellack JH, Matthews GA (1981) Droplet spectra measurements of fan and cone atomisers using a laser diffraction technique. J Aerosol Sci 12:529–540

Concawe (2017) Handbook – Specific Consumer Exposure Determinants (SCEDs) & Supporting Explanation. 46 S. Brussels, BE: Concawe Health Management Group. https://www.concawe.eu/reach/specific-consumer-exposure-determinants-sceds-documents/. Accessed 13 Feb 2019

Control Banding Nanotool (2015). http://controlbanding.net/Home.html. Accessed 05.05.15

Crank J (1975) The mathematics of diffusion, 2nd edn. Clarendon Press, Oxford

Davis S, Mirick DK (2006) Soil ingestion in children and adults in the same family. J Expo Sci Environ Epidemiol 16(1):63–75. https://doi.org/10.1038/sj.jea.7500438

Delmaar JE, Bremmer HJ (2009) The ConsExpo Spray Model: modeling and experimental validation of the inhalation exposure of consumers to aerosols from spray cans and trigger sprays. RIVM Report 320104005/2009

Delmaar JE, Schuur AG (2017) ConsExpo Web. Consumer exposure models – model documentation. RIVM Report 2017-0197. https://www.rivm.nl/bibliotheek/rapporten/2017-0197.pdf. Accessed 27 Oct 2019

Delmaar JE, van der Zee Park M, van Engelen JGM (2005) ConsExpo 4.0. Consumer exposure and uptake models. Program manual. RIVM report 320104004/2005. https://rivm.openrepository.com/rivm/bitstream/10029/7307/1/320104004.pdf. Accessed 8 Feb 2018

Delmaar JE, Bokkers BG, ter Burg W, van Engelen JG (2013) First tier modeling of consumer dermal exposure to substances in consumer articles under REACH: a quantitative evaluation of the ECETOC TRA for consumers tool. Regul Toxicol Pharmacol 65:79–86

Delmaar CJE, Bokkers B, ter Burg W, Schuur G (2015) Validation of an aggregate exposure model for substances in consumer products: a case study of diethyl phthalate in personal care products. J Expo Sci Environ Epidemiol 25(3):317–323

Duarte-Davidson R, Courage C, Rushton L, Levy L (2001) Benzene in the environment: an assessment of the potential risks to the health of the population. Occup Environ Med 58:2–13

DUCC & Concawe (2014) SCEDs. Specific Consumer Exposure Determinants. How to use the SCEDs for chemical exposure assessment under REACH – Guidance for SCEDs user. www.ducc.eu/documents/20140423_SCEDs_Template_Final.pdf. Accessed 6 Feb 2018

Dudzina T, Delmaar CJE, Biesterbos JWH, Bakker MI, Bokkers BGH, Scheepers PTJ, van Engelen JGM, Hungerbuehler K, von Goetz N (2015) The probabilistic aggregate consumer exposure model (PACEM): validation and comparison to a lower-tier assessment for the cyclic siloxane D5. Environ Int 79:8–16

ECETOC (2009a) Targeted risk assessment: user guide for the standalone consumer tool – version 2. Brussels, Belgium: European Centre for Ecotoxicology and Toxicology of Chemicals. http://www.ecetoc.org/tools/targeted-risk-assessment-tra/. Accessed 27 Oct 2019

ECETOC (2009b) Addendum to ECETOC targeted risk assessment report no. 93. Brussels, Belgium: European Centre for Ecotoxicology and Toxicology of Chemicals. http://www.ecetoc.org/wp-content/uploads/2014/08/ECETOC-TR-107-Addendum-to-ECETOC-TRA-report-93.pdf. Accessed 27 Oct 2019

ECETOC (2012) Targeted risk assessment: user guide for the standalone consumer tool – version 3. Brussels, Belgium: European Centre for Ecotoxicology and Toxicology of Chemicals. http://www.ecetoc.org/tools/targeted-risk-assessment-tra/. Accessed 27 Oct 2019

ECETOC (2014) Technical report 124 – Addendum to Technical report 114: Technical basis for the TRA V3.1. Brussels, Belgium, European Centre for Ecotoxicology and Toxicology of Chemicals. http://www.ecetoc.org/wp-content/uploads/2014/08/ECETOC_TR_124.pdf

ECETOC (2016) Targeted Risk Assessment (TRA). http://www.ecetoc.org/tools/targeted-risk-assessment-tra/. Accessed 27 Oct 2019

ECHA (2012) Guidance on information requirements and chemical safety assessment. Chapter R.8: Characterisation of dose [concentration]-response for human health. Updated version 2.1, November 2012. ECHA-2010-G-19-EN. http://echa.europa.eu/documents/10162/13632/information_requirements_r8_en.pdf. Accessed 27 Oct 2019

ECHA (2014) Human health and environmental exposure assessment and risk characterisation of nanomaterials – Best practice for REACH registrants. https://echa.europa.eu/documents/10162/5399565/best_practices_human_health_environment_nano_3rd_en.pdf; doi: 10.2823/93521. Accessed 28 Jul 2017

ECHA (2015) Guidance on information requirements and chemical safety assessment. Chapter R.12: Use description. Version 3.0 ISBN: 978-92-9247-685-4. https://echa.europa.eu/documents/10162/13632/information_requirements_r12_en.pdf. Accessed 27 Oct 2019

ECHA (2016) Guidance on information requirements and chemical safety assessment. Chapter R.15: Consumer exposure assessment. Version 3.0 ISBN: 978-92-9495-079-6. https://echa.europa.eu/documents/10162/13632/information_requirements_r15_en.pdf. 27 Oct 2019

ECHA (2017a) How to prepare registration dossiers that cover nanoforms: best practices. Version 1.0 – May 2017. https://echa.europa.eu/documents/10162/13655/how_to_register_nano_en.pdf/f8c046ec-f60b-4349-492b-e915fd9e3ca0; DOI 10.2823/128306. Accessed 8 Mar 2018

ECHA (2017b) European union observatory for nanomaterials – general information. https://euon.echa.europa.eu/de/general-information. Accessed 8 Aug 2017

ECHA (2017c) Guidance on requirements for substances in articles. Version 4.0. Helsinki, FI. doi: 10.2823/470616. https://echa.europa.eu/documents/10162/23036412/articles_en.pdf/cc2e3f93-8391-4944-88e4-efed5fb5112c. Accessed 8 Mar 2018

ECHA (2017d) ECHA Press release on the launch of the EU observatory for nanomaterials. ECHA/PR/17/11. https://echa.europa.eu/-/eu-observatory-for-nanomaterials-launched. Accessed 28 Jul 17

EFSA (2008) Scientific opinion scientific opinion of the panel on food contact materials, enzymes, flavourings and processing aids (CEF) on 21st list of substances for food contact materials. The EFSA Journal 888–890, 1–14. http://www.efsa.europa.eu/en/scdocs/doc/cef_op_ej888-890 21stlist_en,3.pdf. Accessed 8 Aug 2017

EFSA (2018) Scientific Opinion on Guidance on the risk assessment of the application of nanoscience and nanotechnologies in the food and feed chain: Part 1, human and animal health – endorsed 29. Mai 2018; https://doi.org/10.2903/j.efsa.2018.5327. https://www.efsa.europa.eu/en/efsajournal/pub/5327. Accessed 26 Feb 2019

Egeghy PP, Cohen Hubal EA, Tulve NS, Melnyk LJ, Morgan MK, Fortmann RC, Sheldon LS (2011) Review of pesticide urinary biomarker measurements from selected US-EPA children's observational exposure studies. Int J Environ Res Public Health 8(5):1727–1754. https://doi.org/10.3390/ijerph8051727

Eickmann U (2007) Methoden der Ermittlung und Bewertung chemischer Expositionen an Arbeitsplätzen. ecomed MEDIZIN. ISBN-10: 3609163909; ISBN-13: 978–3609163901

Eickmann U, Eickmann J, Tischer M (2007) Exposure to sprays – comparison of the available exposure models. Gefahrstoffe – Reinhaltung der Luft 67(7/8):305–318

ETAD (1983) Final report on extractability of dyestuffs from textiles. Project A 4007. Basel: ETAD

ETAD (1997) Extractability of dyestuffs from textiles over a normal life time of use. Project G 1033. Basel: ETAD

EU (1967) Council Directive 67/548/EEC of 27 June 1967 on the approximation of laws, regulations and administrative provisions relating to the classification, packaging and labelling of dangerous substances. http://eur-lex.europa.eu/legal-content/EN/TXT/?uri=CELEX:31967L0548. Accessed 27 Oct 2019

EU (1975) Council Directive 75/324/EEC of 20 May 1975 on the approximation of the laws of the Member States relating to aerosol dispensers. http://eur-lex.europa.eu/legal-content/EN/TXT/?uri=CELEX:31975L0324. Accessed 27 Oct 2019

EU (1993) Existing Substances Regulation (ESR). European programme for identifying risks from chemicals to human health and the environment. http://www.hse.gov.uk/chemicals/esr.htm. Accessed 27 Oct 2019

EU (1996) Technical guidance documents in support of the commission directive 93/67/EEC on risk assessment for new notified substances and the commission regulation (EC) 1488/94 On Risk Assessment For Existing Substances

EU (2001) Directive 2001/95/EC of the European Parliament and of the Council of 3 December 2001 on general product safety. https://eur-lex.europa.eu/legal-content/EN/TXT/PDF/?uri=CELEX:32001L0095

EU (2003) Technical guidance document on risk assessment in support of commission directive 93/67/EEC on risk assessment for new notified substances commission regulation (EC) No 1488/94 on Risk Assessment for existing substances. Directive 98/8/EC of the European Parliament and of the Council concerning the placing of biocidal products on the market Part I. https://echa.europa.eu/documents/10162/16960216/tgdpart1_2ed_en.pdf. Accessed 8 Mar 2018

EU (2006) Regulation (EC) No 1907/2006 of the European Parliament and of the Council of 18 December 2006 concerning the Registration, Evaluation, Authorisation and Restriction of Chemicals (REACH), establishing a European Chemicals Agency, Amending Directive 1999/45/EC and repealing Council Regulation (EEC) No 793/93 and Commission Regulation (EC) No 1488/94 as well as Council Directive 76/769/EEC and Commission Directives 91/155/EEC, 93/67/EEC, 93/105/EC and 2000/21/EC. Official Journal L 396, 30.12.2006. http://eur-lex.europa.eu/legal-content/EN/TXT/?uri=CELEX:32006R1907R(01). Accessed 27 Oct 2019

EU (2008a) EC B.44 – Skin absorption: in vivo method Council Regulation (EC) No 440/2008 of 30 May 2008 laying down test methods pursuant to Regulation (EC) No 1907/2006 of the European Parliament and of the Council on the Registration, Evaluation, Authorisation and Restriction of Chemicals (REACH). Official Journal L 142, 31/05/2008, p. 432. http://eur-lex.europa.eu/LexUriServ/LexUriServ.do?uri=OJ:L:2008:142:0001:0739:en:PDF. Accessed 8 Mar 2018

EU (2008b) EC B.45 – Skin absorption: in vitro method Council Regulation (EC) No 440/2008 of 30 May 2008 laying down test methods pursuant to Regulation (EC) No 1907/2006 of the European Parliament and of the Council on the Registration, Evaluation, Authorisation and Restriction of Chemicals (REACH). Official Journal L 142, 31/05/2008, p.438. https://publications.europa.eu/en/publication-detail/-/publication/0299d269-6576-47db-8222-a08bbd1b3fc2/language-en. Accessed 27 Oct 2019

EU (2008c) Risk Assessment Report (RAR) BENZENE, CAS No: 71-43-2, EINECS No: 200-753-7, final version of 2008. https://echa.europa.eu/documents/10162/be2a96a7-40f6-40d7-81e5-b8c3f948efc2. Accessed 27 Oct 2019

EU (2009a) Directive 2009/48/EC of the European Parliament and of the Council of 18 June 2009 on the safety of toys. http://eur-lex.europa.eu/legal-content/EN/TXT/?uri=CELEX:32009L0048. Accessed 8 Mar 2018

EU (2009b) Regulation (EC) No 1223/2009 of the European Parliament and of the Council of 30 November 2009 on cosmetic products. http://eur-lex.europa.eu/legal-content/EN/TXT/PDF/?uri=CELEX:32009R1223&from=DE. Accessed 27 Oct 2019

EU (2011) Commission Recommendation of 18 October 2011 on the definition of nanomaterial (2011/696/EU) – Official Journal of the European Union L 275/38. https://ec.europa.eu/research/industrial_technologies/pdf/policy/commission-recommendation-on-the-definition-of-nanomater-18102011_en.pdf. Accessed 8 Aug 2017

EU (2012) Regulation (EU) No 528/2012 of the European Parliament and of the Council of 22 May 2012 concerning the making available on the market and use of biocidal products. Official Journal L 167, 27.6.2012, p. 1–174. https://eur-lex.europa.eu/legal-content/EN/TXT/HTML/?uri=CELEX:32012R0528&from=en. Accessed 27 Oct 2019

EU (2017) Study on a product categorisation system for information to be submitted to Poison Centres, according to Article 45 (4) of EC Regulation No 1272/2008 (CLP Regulation). ISBN: 978-92-79-63777-3; https://doi.org/10.2873/87997. https://publications.europa.eu/en/publication-detail/-/publication/269f51ae-e6d4-11e7-9749-01aa75ed71a1/language-en. Accessed 27 Oct 2019

EU (2018) Commission Regulation (EU) 2018/1513 of 10 October 2018 amending Annex XVII to Regulation (EC) No 1907/2006 of the European Parliament and of the Council concerning the Registration, Evaluation, Authorisation and Restriction of Chemicals (REACH) as regards certain substances classified as carcinogenic, mutagenic or toxic for reproduction (CMR), category 1A or 1B. https://eur-lex.europa.eu/legal-content/EN/TXT/PDF/?uri=CELEX:32018R1513

Ezendam J, Bokkers BGH, Bilc W, Delmaar JE (2018) Skin sensitisation quantitative risk assessment (QRA) based on aggregate dermal exposure to methylisothiazolinone in personal care and household cleaning products. Food Chem Toxicol 112:242–250

FEICA (2018) FEICA guidance paper – FEICA Specific Consumer Exposure Determinants (SCEDs). Bericht: GUP-EX-H01-002, 14 S. Brussels, BE: Association of the European Adhesive & Sealant Industry. http://www.feica.eu/our-priorities/key-projects/reach.aspx. Accessed 3038 27 Oct 2019

Ficheux AS, Morisset T, Chevillotte G, Postic C, Roudot AC (2014) Probabilistic assessment of exposure to nail cosmetics in French consumers. Food Chem Toxicol 66:36–43

FOPH (2017) Precautionary Matrix Web application. https://www.bag.admin.ch/bag/en/home/themen/mensch-gesundheit/chemikalien/nanotechnologie/sicherer-umgang-mit-nanomaterialien/vorsorgeraster-nanomaterialien-webanwendung.exturl.html/aHR0cDovL3d3dy5iYWctYW53LmFkbWluLmNoLzIwMTZfdGFnbG/FiLzIwMTZfbmFub3Jhc3Rlci9wb3J0YWxfZGUucGhwP21vZD1p/bXBleCZkaXJ0PTk3JlNJRD1nYTRvOGdxZGpvN2c0N3FiOW02Z2/tmMWM4Nw==.html. Accessed 09.08.17

Frank G, Galinkina J, Schelle C (2012) Relevanz der Prüfmusterkonfektionierung und der Luftdurchflussrate für die Emissionscharakteristik von Leder in Prüfkammeruntersuchungen. Gefahrstoffe – Reinhalt Luft 72:203–210

Franz TJ (1984) Percutaneous absorption of benzene. In: MacFarland HN, Holdsworth CE, MacGregor JA, Call RW, Kane ML (eds) Advances in modern environmental toxicology. Princeton Scientific Publishers, Princeton (NJ), pp 61–70

Gong M, Zhang Y, Weschler CJ (2014) Predicting dermal absorption of gas-phase chemicals: transient model development, evaluation, and application. Indoor Air 24:292–306

Greim H, Borm P, Schins R, Donaldson K, Driscoll K, Hartwig A, Kuempel E, Oberdörster G, Speit G (2001) Toxicity of fibers and particles. Report of the workshop held in Munich, Germany. Inhal Toxicol 13:737–754

Groot ME, Lekkerkerk MC, Steenbekkers LPAB (1998) Mouthing behaviour of young children: an observational study. Landbouwuniversiteit Wageningen Wageningen, NL

Hall B, Tozer S, Safford B, Coroama M, Steiling W, Leneveu-Duchemin MC, McNamara C, Gibney M (2007) European consumer exposure to cosmetic products, a framework for conducting population exposure assessments. Food Chem Toxicol 45:2097–2108

Hall B, Steiling W, Safford B, Coroama M, Tozer S, Firmani C, McNamara C, Gibney M (2011) European consumer exposure to cosmetic products: a framework for conducting population exposure assessments part 2. Food Chem Toxicol 49:407–421

Hansen Foss S, Jensen KA, Baun A (2014) NanoRiskCat: a conceptual tool for categorization and communication of exposure potentials and hazards of nanomaterials in consumer products. J Nanopart Res 16(2):2195–2220. https://doi.org/10.1007/s11051-013-2195-z

Hawley JK (1985) Assessment of health risk from exposure to contaminated soil. Risk Anal 5 (4):289–302

Hayes RB, Yin S-N, Dosemeci M, Linet M (2001) Benzene and lymphohematopoietic malignancies in humans. Am J Ind Med 40:117–126

Heiland A, Oltmanns J, Bohlen M-L, Kaiser E, Klenow S, Schneider K (2016) Consumer behaviour – compilation and evaluation of primary data. Poster presented at the 26th Annual ISES Meeting, Utrecht, NL, October 9–13, 2016

Heine E, Thomas H, Müller BM, Höcker H (1996) Entwicklung praxisgerechter Modelle für die Exposition mit chemischen Stoffen aus Textilien). DWI Reports. Aachen: Deutsches Wollforschungsinstitut an der RWTH Aachen e.V

Heinemeyer G (2004) Expositionsszenarien. In: Reichl S (ed) Regulatorische toxikologie. Springer, Berlin, pp 230–242

Heinemeyer G, Heiland A, Sommerfeld C, Springer A, Hausdörfer S, Treutz M, Lindtner O, Rüdiger T (2012a) Phthalat-Belastung der Bevölkerung in Deutschland: Expositionsrelevante Quellen, Aufnahmepfade und Toxikokinetik am Beispiel von DEHP und DINP. Band I: Exposition durch Verzehr von Lebensmitteln und Anwendung von Verbraucherprodukten. http://www.uba.de/uba-info-medien/4326.html

Heinemeyer G, Heiland A, Sommerfeld C, Springer A, Hausdörfer S, Treutz M, Lindtner O, Rüdiger T (2012b) Phthalat-Belastung der Bevölkerung in Deutschland: Expositionsrelevante Quellen, Aufnahmepfade und Toxikokinetik am Beispiel von DEHP und DINP. Anhang zu Band I: Datenbasis zur Exposition durch Lebensmittelverzehr und Verbraucherprodukte. http://www.uba.de/uba-info-medien/4391.html

HERA (2005) Human & environmental risk assessment on ingredients of household cleaning products: guidance document methodology. http://www.heraproject.com/Library.cfm. Accessed 27 Oct 2019

Hofmann W (2009) Modelling particle deposition in human lungs: modelling concepts and comparison with experimental data. Biomarkers 14(Suppl 1):59–62

Huang L, Ernstoff A, Fantke P, Csiszar SA, Jolliet O (2017) A review of models for near-field exposure pathways of chemicals in consumer products. Sci Total Environ 574:1182–1208

HVBG (Hauptverband der gewerblichen Berufsgenossenschaften) (2001) Berechnungsverfahren und Modellbildung in der Arbeitsbereichsanalyse. BIA-Report 3/2001. Sankt Augustin: HVBG, 266 Seiten, ISBN 3-88383-588-9. http://www.dguv.de/medien/ifa/de/pub/rep/pdf/rep01/biar0301/rep3_01.pdf. Accessed 27 Oct 2019

IARC (2012) Monograph on the evaluation of the carcinogenic risk of chemicals, Vol 100F. WHO: Lyon

ICF (2016) Consumer exposure model (CEM). Draft user guide. EPA Office of Pollution Prevention and Toxics by ICF International under EPA Contract # EP-W-12-010. https://www.epa.gov/sites/production/files/2015-09/documents/cem_user_guide_beta_test.pdf. Accessed 27 Oct 2019

Infante PF, Rinsky RA, Wagoner JK, Young RJ (1977) Leukemia in benzene workers. Lancet 2 (8028):76–78

International Commission on Radiological Protection (ICRP) (1994) Human respiratory tract model for radiological protection. ICRP Publication 66. Pergamon Press, Oxford

Ireland JD (2014) Assessment report – Permethrin. http://dissemination.echa.europa.eu/Biocides/ActiveSubstances/1342-18/1342-18_Assessment_Report.pdf. Accessed 27 Oct 2019

Isaacs KK, Glen WG, Egeghy P, Goldsmith MR, Smith L, Vallero D, Brooks R, Grulke CM, Özkaynak H (2014) SHEDS-HT: an integrated probabilistic exposure model for prioritizing exposures to chemicals with near-field and dietary sources. Environ Sci Technol 48:12750–12759

ISO (2008) ISO/TS 27687: Nanotechnologies – Terminology and definitions for nano-objects – Nanoparticle, nanofibre and nanoplate. 1–14

ISO (2010) ISO TS 80004-1:2010: Nanotechnologies – Vocabulary Part 1 – Core Terms

ISO (2014) ISO/TS 12901-2:2014(en) – Nanotechnologies – Occupational risk management applied to engineered nanomaterials – Part 2: Use of the control banding approach/ https://www.iso.org/obp/ui/#iso:std:iso:ts:12901:-2:ed-1:v1:en. Accessed 10.08.2017

ISO (2015) ISO/TS 80004-2:2015: Nanotechnologies – Vocabulary – Part 2: Nano-objects. https://www.iso.org/obp/ui/#iso:std:iso:ts:80004:-2:ed-1:v1:en. Accessed 28.07.2017

Jensen KA (2016) NanoSafer v. 1.1beta. QuickQuide, National Research Centre for the Working Environment, DK. www.nanosafer.org. Accessed 9 Aug 2017

Juberg DR, Alfano K, Coughlin RJ, Thompson KM (2001) An observational study of object mouthing behavior by young children. Pediatrics 107(1):135–142. https://doi.org/10.1542/peds.107.1.135

Koch W, Behnke W, Berger-Preiß E, Kock H, Gerling S, Hahn S, Schröder K (2012) In: Bundesanstalt für Arbeitsschutz und Arbeitsmedizin (ed), Validation of an EDP-assisted model for assessing inhalation exposure and dermal exposure during spraying processes. Research Project F2137. 1st ed. BauA, Dortmund

Krätke R, Platzek T (2004) Migrationsverfahren und Modelle zur Abschätzung einer möglichen Exposition mit Textilhilfsmitteln und -farbmitteln aus Bekleidungstextilien unter Anwendungsbedingungen. Aus dem Arbeitskreis "Gesundheitliche Bewertung von Textilhilfsmitteln und -farbmitteln" der Arbeitsgruppe "Textilien" des BfR. Bundesgesundheitsbl Gesundheitsforsch Gesundheitsschutz 47:810–813

Kuck-Koot N, Lahey-De Boer AM (2005) Anaphylactic reaction following skin contact with hair dye. Nederlands Tijdschrift voor Dermatologie en Venereologie, 15, 355–358. (Evaluation based on abstract)

Larsen PB, Christensen F, Jensen KA, Brinch A, Mikkelsen SH (2015a) Exposure assessment of nanomaterials in consumer products- Environmental project No. 1636; http://www2.mst.dk/Udgiv/publications/2015/01/978-87-93283-57-2.pdf. Accessed 27 Oct 2019

Larsen PB, Christensen F, Jensen KA, Brinch A, Mikkelsen SH, Larsen Clausen AJ, Flotel FL, Koivisto AJ, Nørgaard AW (2015b) Exposure assessment of nanomaterials in consumer products – Appendix report to environmental project no. 1636; https://www2.mst.dk/Udgiv/publications/2015/01/978-87-93283-57-2_appendix.pdf. Accessed 8 Aug 2017

Lewis RG, Fortune CR, Willis RD, Camann DE, Antley JT (1999) Distribution of pesticides and polycyclic aromatic hydrocarbons in house dust as a function of particle size. Environ Health Perspect 107(9):721–726. https://doi.org/10.2307/3434657

Loretz L, Api AM, Barraj L, Burdick J, de Davis A, Dressler W, Gilberti E, Jarrett G, Mann S, Laurie Pan YH, Re T, Renskers K, Scrafford C, Vater S (2006) Exposure data for personal care products: hairspray, spray perfume, liquid foundation, shampoo, body wash, and solid antiperspirant. Food Chem Toxicol 44:2008–2018

Losert S, Hess A, Ilari G, von Goetz N, Hungerbuehler K (2015) Online characterization of nanoaerosols released by commercial spray products using SMPS–ICPMS coupling. J Nanopart Res 17:293

Mackevica A, Hansen Foss S (2016) Release of nanomaterials from solid nanocomposites and consumer exposure assessment – a forward-looking review. Nanotoxicology 10(6):641–653. https://doi.org/10.3109/17435390.2015.1132346

Maertens RM, Bailey J, White PA (2004) The mutagenic hazards of settled house dust: a review. Mutat Res 567(2/3):401–425. https://doi.org/10.1016/j.mrrev.2004.08.004

MAK Commission (2012) List of MAK and BAT values 2012: maximum concentrations and biological tolerance values at the workplace. Report 48. In: Deutsche Forschungsgemeinschaft (DFG) (Ed.), Wiley-VCH, Weinheim (German language Version: Senatskommission zur Prüfung gesundheitsschädlicher Arbeitsstoffe, Mitteilung 48, MAK- und BAT-Werte-Liste 2012). https://www.wiley.com/en-us/List+of+MAK+and+BAT+Values+2013%3A+Maximum+Concentrations+and+Biological+Tolerance+Values+at+the+Workplace-p-9783527675128. Accessed 27 Oct 2019

Martonen TB, Schroeter JD, Hwang D, Fleming JS, Conway JH (2000) Human lung morphology models for particle deposition studies. Inhal Toxicol 12(Suppl 4):109–121

Masuck I, Hutzler C, Jann O, Luch A (2011) Inhalation exposure of children to fragrances present in scented toys. Indoor Air 21:501–511

Meesters JAJ, Nijkamp MM, Schuur AG, te Biesebeek JD (2018) Cleaning products fact sheet – default parameters for estimating consumer exposure. Updated version 2018. National Institute

for Public Health and the Environment (RIVM), Bilthoven, NL. doi:10.21945/RIVM-2016-0179: https://www.rivm.nl/bibliotheek/rapporten/2016-0179.pdf
Mekel O, Mosbach-Schulz O, Schümann M, Okken P-K, Peters C, Herrmann J, Hehl O, Bubenheim M, Fehr R, Timm J (2007) Evaluation von Standards und Modellen zur probabilistischen Expositionsabschätzung. Teil 2: Empfehlungen für Expositionsfaktoren. Abschlußbericht. Umweltbundesamt, Dessau-Roßlau, DE
Micheletti C, Gerritsen R, Schmid K, Brouwer DH, Peters S, Gaelle Uand Christensen F (2010) Development of exposure scenarios for manufactured nanomaterials – NANEX-Project Report on Work Package 4 – Consumer Exposure Scenarios. http://www.nanex-project.eu/mainpages/public-documents/doc_download/89-nanexwp4final.pdf. Accessed 27 Oct 2019
Millette JR, Lioy PJ, Wietfeldt J, Hopen TJ, Gipp M, Padden T, Singsank C, Lepow J (2003) A microscopical study of the general composition of household dirt. Microscope 51(4):201–207
Mølhave L, Schneider T, Kjærgaard SK, Larsen LS, Norn S, Jørgensen O (2000) House dust in seven Danish offices. Atmos Environ 34(28):4767–4779. https://doi.org/10.1016/S1352-2310(00)00104-7
Mühle H, Mangelsdorf I (2003) Inhalation toxicity of mineral particles: critical appraisal of endpoints and study design. Toxicol Lett 11:223–228
Murdan S (2002) Drug delivery to the nail following topical application. Int J Pharm 236:1–26
NanoDefine (2017) Project. http://www.nanodefine.eu/index.php/project-overview. Accessed 9 Aug 2017
Nanosafer (2017) web application version 1.1 beta. http://www.nanosafer.org/. Accessed 9 Aug 2017
NICNAS (2017) NICNAS working definition of industrial nanomaterial. https://www.nicnas.gov.au/notify-your-chemical/data-requirements-for-new-chemical-notifications/data-requirements-for-notification-of-new-industrial-nanomaterials/nicnas-working-definition-of-industrial-nanomaterial. Accessed 30 Jun 2017
Nordic Exposure Group, norden (2012) Existing default values and recommendations for exposure assessment. A nordic exposure group project 2011, vol 2012:505. TemaNord. Nordic Council of Ministers, Copenhagen, DK. https://doi.org/10.6027/TN2012-505
Nuyttens D, Schampheleire MD, Baetens K, Sonck B (2007) The influence of operator-controlled variables on spray drift from field crop sprayers. Trans ASABE 2007(50):1129–1140
OECD 2004a) OECD Guideline for testing of chemicals – Guideline 427: Skin absorption: in vivo method. Organization for Economic Cooperation and Development, Paris, adopted 13 April 2004
OECD (2004b) OECD Guideline for testing of chemicals – Guideline 428: Skin absorption: in vitro method. Organization for Economic Cooperation and Development, Paris, adopted 13 April 2004
OECD (2009) OECD Guideline for testing of chemicals – Guideline 413: Subchronic inhalation toxicity: 90-day study. Organization for Economic Cooperation and Development, Paris, adopted 7 September 2009
OECD (2012) Crosswalk of harmonised U.S. – Canada industrial function and consumer and commercial product categories with EU chemical product and article categories. Series on testing and assessment no. 167 (trans: IOMC, Inter-Organization Programme for the Sound Management of Chemicals). OECD, Paris, FR. http://www.oecd.org/officialdocuments/publicdisplaydocumentpdf/?cote=env/jm/mono%282012%295&doclanguage=en. Accessed 27 Oct 2019
OECD (2015) Analysis of the survey on available methods and models for assessing exposure to manufactured nanomaterials, series on the safety of manufactured nanomaterials no. 56. ENV/JM/MONO(2015)20. Accessed 9 Aug 2017
OECD (2017) Internationally harmonised functional, product and article use categories. Series on testing & assessment no. 262. OECD Environment Directorate, Paris, FR. http://www.oecd.org/officialdocuments/publicdisplaydocumentpdf/?cote=env/jm/mono%282017%2914&doclanguage=en. Accessed 27 Oct 2019
Oltmanns J, Neisel F, Heinemeyer G, Kaiser E, Schneider K (2015) Consumer exposure modelling under REACH: assessing the defaults. Regul Toxicol Pharmacol 2(2):222–230

Oomen AG, Janssen PJCM, Dusseldorp A, Noorlander CW (2008) Exposure to Chemicals via House dust. RIVM, Rijksinstituut voor Volksgezondheid en Milieu, Bilthoven, NL. Report 609021064/2008. http://www.rivm.nl/bibliotheek/rapporten/609021064.pdf. Accessed 11 Feb 2018

Özkaynak H, Xue J, Zartarian VG, Glen G, Smith L (2011) Modeled estimates of soil and dust ingestion rates for children. Risk Anal 31(4):592–608. https://doi.org/10.1111/j.1539-6924.2010.01524.x

Packroff R, Völker D, Mutz, D, Bresch H, Bosse H (2016) Nanomaterials and other advanced materials: application safety and environmental compatibility, Review of the joint research strategy of the higher federal authorities. 2016. Umweltbundesamt UBA http://www.bfr.bund.de/cm/349/nanomaterials-and-other-advanced-materials-application-safety-and-environmental-compatibility-review-of-the-joint-research-strategy-of-the-higher-federal-authorities.pdf. Accessed 9 Aug 2017

Paustenbach DJ, Finley BL, Long TF (1997) The critical role of house dust in understanding the hazards posed by contaminated soils. Int J Toxicol 16(4–5):339–362. https://doi.org/10.1080/109158197227008

Petry T, Vitale D, Joachim FJ, Mascarenhas R, Schneider S, Smith B (2013) Human health assessment of scented candle emissions. Toxicologist 132(1):333

Phalen RF, Oldham MJ (2006) Aerosol dosimetry considerations. Clin Occup Environ Med 5:773–784

Piringer OG (1994) Evaluation of plastics for food packaging. Food Addit Contam 11:221–230

Prud'homme de Lodder LCH, Bremmer HJ, Pelgrom SMGJ, Van der Zee Park M, Van Engelen JGM (2006) Disinfectant products fact sheet. To assess the risks for the consumer. RIVM report 320005003. https://www.rivm.nl/documenten/disinfectant-products-fact-sheet

Quadros ME, Marr LC (2011) Silver nanoparticles and total aerosols emitted by nanotechnology-related consumer spray products. Environ Sci Technol 45:10713–10719

Rinsky RA, Smith AB, Hornung R, Filloon TG, Young RJ, Okun AH, Landrigan PJ (1987) Benzene and leukemia: an epidemiologic risk assessment. N Engl J Med 316:1044–1050

RIVM (2016) ConsExpo nano web application (V1.1). https://www.consexponano.nl/. Accessed 06.10.2016

RIVM (2017a) ConsExpo nano. https://www.consexponano.nl/. Accessed 27 Oct 2019

RIVM (2017b) Consexpo webpage. http://www.rivm.nl/en/Topics/C/ConsExpo. Accessed 27 Oct 2019

Rompelberg C, Heringa MB, van Donkersgoed G, Drijvers J, Roos A, Westenbrink S, Peters R, van Bemmel G, Brand W, Oomen AG (2016) Oral intake of added titanium dioxide and its nanofraction from food products, food supplements and toothpaste by the Dutch population. Nanotoxicology 10(10):1404–1414. https://doi.org/10.1080/17435390.2016.1222457

Rothe H, Fautz R, Gerber E, Neumann L, Rettinger K, Schuh W, Gronewold C (2011) Special aspects of cosmetic spray safety evaluations: principles oninhalation risk assessment. Toxicol Lett 205:97–104

Rowley J, Crump D (2005) Measurements of the dispersal of aerosol sprays in a room and comparison to a simple decay model. J Environ Monit 7:960–963

Safford B, Api AM, Barratt C, Comiskey D, Daly EJ, Elli G, McNamara C, O'Mahony C, Robison S, Smith B, Thomas R, Tozer S (2015) Use of an aggregate exposure model to estimate consumer exposure to fragrance ingredients in personal care and cosmetic products. Regul Toxicol Pharmacol 72:673–682

Salthammer T, Bahadir M (2009) Occurrence, dynamics and reactions of organic pollutants in the indoor environment. Clean 37:417–435

Salthammer T, Mentese S, Marutzky R (2010) Formaldehyde in the indoor environment. Chem Rev 110:2536–2572

Sauer UG, Vogel S, Hess A, Kolle SN, Ma-Hock L, van Ravenzwaay B, Landsiedel R (2013) In vivo – in vitro comparison of acute respiratory tract toxicity using human 3D airway epithelial models and human A549 and murine 3T3 monolayer cell systems. Toxicol In Vitro 27:174–190

SCC (Scientific Committee on Cosmetics) (1993) SCC 54th plenary meeting of Dec 10, 1993. In: Reports of the SCC (ninth series)), Office for Official Publications of the European Commission

SCCNFP (1999) SCCNFP/0167/99: Basic Criteria for the in vitro assessment of percutaneous absorption of cosmetic ingredients. 8th plenary meeting of 23 June 1999. https://ec.europa.eu/health/archive/ph_risk/committees/sccp/documents/out231_en.pdf. Accessed 27 Oct 2019

SCCS (2010a) Opinion on dihydroxyacetone. SCCS/1347/10, 2010. http://ec.europa.eu/health/scientific committees/consumer safety/docs/sccs o 048.pdf

SCCS (2010b) Basic Criteria for the in vitro assessment of dermal absorption of cosmetic ingredients, adopted by the SCCS during the 7th plenary meeting of 22 June 2010. SCCS/1358/10. https://ec.europa.eu/health/sites/health/files/scientific_committees/consumer_safety/docs/sccs_s_002.pdf. Accessed 27 Oct 2019

SCCS (2012) Guidance on the safety assessment of nanomaterials in cosmetics. SCCS/1484/12, 2012. https://ec.europa.eu/health/sites/health/files/scientific_committees/consumer_safety/docs/sccs_s_005.pdf. Accessed 8 Mar 2018

SCCS (2015) The SCCS Notes of Guidance for the testing of cosmetic ingredients and their safety evaluation (9th revision). http://ec.europa.eu/health/sites/health/files/scientific_committees/consumer_safety/docs/sccs_o_190.pdf

SCCS (2018) The SCCS notes of guidance for the testing of cosmetic ingredients and their safety evaluation (10th revision). SCCS/1602/18. https://ec.europa.eu/health/sites/health/files/scientific_committees/consumer_safety/docs/sccs_o_224.pdf. Accessed 27 Oct 2019

Schneider K, Recke S, Kaiser E, Götte S, Berkefeld H, Lässig J, Rüdiger T, Lindtner O, Oltmanns J (2019) Consumer behaviour survey for assessing exposure from consumer products: a feasibility study. J Expo Sci Environ Epidemiol 29(1):83–94. https://doi.org/10.1038/s41370-018-0040-2

Schossler P, Schripp T, Salthammer T, Bahadir M (2011) Beyond phthalates: gas phase concentrations and modeled gas/particle distribution of modern plasticizers. Sci Total Environ 409:4031–4038

Schulz C, Wolf U, Becker K, Conrad A, Hünken A, Lüdecke A, Müssig-Zufika M, Riedel S, Seiffert I, Seiwert M, Kolossa-Gehring M (2007). Kinder-Umwelt-Survey (KUS) im Rahmen des Kinder- und Jugendgesundheitssurveys (KiGGS). Erste Ergebnisse [German Environmental Survey for Children (GerES IV) in the German Health Interview and Examination Survey for Children and Adolescents (KiGGS). First results]. Bundesgesundheitsblatt – Gesundheitsforschung – Gesundheitsschutz 50(5/6): 889–894. doi: 10.1007/s00103-007-0252-4

Smith SA, Norris B (2003) Reducing the risk of choking hazards: mouthing behaviour of children aged 1 month to 5 years. Inj Control Saf Promot 10(3):145–154. https://doi.org/10.1076/icsp.10.3.145.14562

Snipes MB (1989) Long-term retention and clearance of particles inhaled by Mammalian species. Crit Rev Toxicol 20(3):175–211

Snodgrass HL (1992) Permethrin transfer from treated cloth to the skin surface: potential for exposure in humans. J Toxicol Environ Health 35:91–105

Snyder R (2000) Overview of the toxicology of benzene. J Toxicol Environ Health A 61:339–346

Snyder R (2002) Benzene and Leukemia. Crit Rev Toxicol 52:155–210

Sommer Y (2016) Analysis of toolkit and strategy developments for the exposure assessment of nanomaterials in consumer products; International Society of Exposure Science (ISES) 2016 Annual Meeting in Utrecht, NL, October 9–13, 2016 – Poster We-Po-27, 12. October 2016

Stanek EJ III, Calabrese EJ (1995) Daily estimates of soil ingestion in children. Environ Health Perspect 103(3):276–285

Steiling W, Buttgereit P, Hall B, O'Keeffe L, Safford B, Tozer S, Coroama M (2012) Skin exposure to deodorants/antiperspirants in aerosol form. Food Chem Toxicol 50(6):2206–2215

Steiling W, Bascompta M, Carthew P, Catalano G, Corea N, D'Haese A, Jacksong P, Kromidas L, Meurice P, Rothe H, Singal M (2014) Principle considerations for the risk assessment of sprayed consumer products. Toxicol Lett 227(1):41–49

Stoffenmanager (2016) Stoffenmanager web tool http://nano.stoffenmanager.nl/. Accessed 06.05.16

Stoffenmanager Nano (2016) Stoffenmanager web tool – Nano modul Version 1.0. http://nano.stoffenmanager.nl/. Accessed 06.05.16

Subramaniam RP, Asgharian B, Freijer JI, Miller FJ, Anjilvel S (2003) Analysis of lobar differences in particle deposition in the human lung. Inhal Toxicol 15(1):1–21

te Biesebeck JD, Nijkamp MM, Bokkers BGH, Wijnhoven SWP (2014) General fact sheet. General default parameters for estimating consumer exposure – Updated version 2014. RIVM Report 090013003. https://www.rivm.nl/bibliotheek/rapporten/090013003.pdf. Accessed 27 Oct 2019

ter Burg W, Bremmer HJ, Van Engelen JGM (2007) Do-it-yourself products fact sheet. To assess the risks for the consumer. RIVM report 320104007. http://www.rivm.nl/bibliotheek/rapporten/320104007.pdf. Accessed 27 Oct 2019

Tulve NS, Suggs JC, McCurdy T, Cohen Hubal EA, Moya J (2002) Frequency of mouthing behavior in young children. J Expo Anal Environ Epidemiol 12(4):259–264. https://doi.org/10.1038/sj.jea.7500225

Turner WL Jr, Millette JR, Boltin WR, Hopen TJ (2005) A standard approach to the characterization of common indoor dust constituents. Microscope 53(4):169–177

US-EPA (1990) Arthur D. Little Inc. Migration estimation model prepared for US EPA. Description in: https://archive.epa.gov/scipoly/sap/meetings/web/pdf/chapd-3.pdf. Accessed 24 Oct 2019

US-EPA (2008) Child-specific exposure factors handbook. National Center for Environmental Assessment, Washington, DC; EPA/600/R-06/096F. Available from the National Technical Information Service, Springfield, VA. https://ofmpub.epa.gov/eims/eimscomm.getfile?p_download_id=484738. Accessed 27 Oct 2019

US-EPA (2009a) Exposure factors handbook 2009 update: external review draft. US Environmental Protection Agency, Washington DC, USA

US-EPA (2009b) Reregistration eligibility decision (RED) for permethrin. RED Case No 2510. EPA 738-R-09-306

US-EPA (2011) Exposure factors handbook: 2011 edition. National Center for Environmental Assessment, Washington, DC; EPA/600/R-09/052F. Available from the National Technical Information Service, Springfield, VA. https://cfpub.epa.gov/si/si_public_record_report.cfm?Lab=NCEA&dirEntryId=236252. Accessed 27 Oct 2019

US-EPA (2016) E-FAST Internetpage. https://cfpub.epa.gov/si/si_public_record_report.cfm?Lab=NCEA&dirEntryId=236252

US-EPA (2017a) Expobox. https://www.epa.gov/expobox/exposure-assessment-tools-chemical-classes-nanomaterials. Accessed 31 Jan 2018

US-EPA (2017b) MCCEM (Multi Chamber Consumer Exposure Model). https://www.epa.gov/tsca-screening-tools/forms/mccem-multi-chamber-concentration-and-exposure-model-download-and-install. Accessed 29 Jan 2018

US-EPA (2017c) WPEM (wall paint emission model) https://cfpub.epa.gov/si/si_public_record_report.cfm?Lab=NCEA&dirEntryId=236252

US-EPA (2017d) Update for Chapter 5 of the exposure factors handbook: soil and dust ingestion. US-EPA Office of Research and Development, Washington, DC, EPA/600/R-17/384F. https://cfpub.epa.gov/si/si_public_record_report.cfm?Lab=NCEA&dirEntryId=236252

US-EPA (2019) Consumer Exposure Model (CEM). https://www.epa.gov/tsca-screening-tools/consumer-exposure-model-cem-version-20-users-guide. Accessed 8 Feb 2018

Valentine R, Kennedy GL (2008) Inhalation toxicology. In: Hayes AW (ed) Principles and methods of toxicology, 5th edn. Informa Healthcare, New York, pp 1407–1464

Van Engelen JG, Heinemeyer G, Rodriguez C (2007) Consumer exposure scenarios: development, challenges and possible solutions. J Expo Sci Environ Epidemiol 17(Suppl 1):S26–S33

Valks R, Conde-Salazar L, Malfeito J, Ledo S (2005) Contact dermatitis in hairdressers, 10 years later: patch-test results in 300 hairdressers (1994 to 2003) and comparison with previous study. Dermatitis 16:28–31

Van Duuren-Stuurmann B, Pelzer J, Moehlmann C, Berges M, Bard D, Wake D, Mark D, Jankowska E, Brouwer D (2010) A structured observational method to assess dermal exposure to manufactured nanoparticles. Int J Occup Environ Health 16(4):399–405. https://doi.org/10.1179/107735210799159978

Van Duuren-Stuurmann B, Vink SR, Verbist KJ, Heussen HG, Brouwer DH, Kroese DE, Van Niftrik MF, Tielemans E, Fransman W (2012) Stoffenmanager Nano version 1.0: a web-based tool for risk prioritization of airborne manufactured nano objects. Ann Occup Hyg 56 (5):525–541. https://doi.org/10.1093/annhyg/mer113

van Holderbeke M, Cornelis C, Bierkens JG, Torfs R (2008) Review of the soil ingestion pathway in human exposure assessment – final report. Study in support of the BeNeKempen project, subproject on harmonisation of the human health risk assessment methodology. OVAM, Openbare Vlaamse Afvalstoffenmaatschappij, Mechelen, BE

Van-Wendel-de-Joode B, Brouwer DH, Vermeulen R, Van Hemmen JJ, Heederik D, Kromhout H (2003) Ann Occup Hyg 47(1):71–87

Vernez D, Bruzzi R, Kupferschmidt H, De-Batz A, Droz P, Lazo R (2006) Acute respiratory syndrome after inhalation of waterproofing sprays: a posteriori exposure-response assessment in 102 cases. J Occup Environ Hyg 3:250–261

Waidyanatha S, Yeowell-O'Connell K, Rappaport SM (1998) A new assay for albumin and hemogobin adducts of 1,2- and 1,4-benzoquinones. Chem Biol Interact 115:117–139

Walters KA, Abdalghafor HM, Lane ME (2012) The human nail—barrier: characterisation and permeation enhancement. Int J Pharm 435:10–21

Webster TF, Harrad SJ, Millette JR, Holbrook RD, Davis JM, Stapleton HM, Allen JG, McClean MD, Ibarra C, Abdallah MA-E, Covaci A (2009) Identifying transfer mechanisms and sources of decabromodiphenyl ether (BDE 209) in indoor environments using environmental forensic microscopy. Environ Sci Technol 43(9):3067–3072. https://doi.org/10.1021/es803139w

Weerdesteijn MCH, Bremmer HJ, Zeilmaker MJ, van Veen MP (1999) Hygienic cleaning products used in the kitchen. Exposure and risks. Bilthoven, The Netherlands. RIVM Report 612810008

Weschler CJ, Nazaroff WW (2010) SVOC partitioning between the gas phase and settled dust indoors. Atmos Environ 44(30):3609–3620. https://doi.org/10.1016/j.atmosenv.2010.06.029

Weschler CJ, Nazaroff WW (2012) SVOC exposure indoors: fresh look at dermal pathways. Indoor Air 22:356–377

Weschler CJ, Nazaroff WW (2014) Dermal uptake of organic vapors commonly found in indoor air. Environ Sci Technol 48:1230–1237

Weschler CJ, Beko G, Koch HM, Salthammer T, Schripp T, Toftum J, Clausen G (2015) Transdermal uptake of diethyl phthalate and di(n-butyl) phthalate directly from air: experimental verification. Environ Health Perspect 123:928–934

WHO (1997) Vector control methods for use by individuals and communities. Geneva: World Health Organization. http://apps.who.int/iris/bitstream/10665/41968/1/9241544945_eng.pdf

Wijnhoven SWP, Dekkers S, Hagens WI, de Jong WH (2009) Exposure to nanomaterials in consumer products. RIVM Letter Report 340370001/2009. http://www.rivm.nl/bibliotheek/rapporten/340370001.pdf. Accessed 27 Oct 2019

Wilson R, Jones-Otazo H, Petrovic S, Mitchell I, Bonvalot Y, Williams D, Richardson GM (2013) Revisiting dust and soil ingestion rates based on hand-to-mouth transfer. Hum Ecol Risk Assess 19(1):158–188. https://doi.org/10.1080/10807039.2012.685807

Wolfram LJ, Maibach HI (1985) Percutaneous penetration of hair dyes. Arch Dermatol Res 277:235–241

Wormuth M, Scheringer M, Vollenweider M, Hungerbühler K (2006) What are the sources of exposure to eight frequently used phthalic acid esters in Europeans? Risk Anal 26(3):803–824. https://doi.org/10.1111/j.1539-6924.2006.00770.x

Wyke S, Duarte-Davidson R, Tempowski J, Heinemeyer G, Heiland A, Rüdiger T, Lindtner O, Desel H, Wagner R, Bauer K, Mathieu-Nolf M, Jumaeu F, Peucelle D, Cezard C, Nisse P, Linke JC, Pelclova P, Mrazova K (2009) Description of the Nature of the Accidental Misuse of Chemicals and Chemical products (DeNaMiC). Final internal Project Report (unpublished)

Xue J-P, Zartarian VG, Moya J, Freeman N, Beamer PI, Black K, Tulve N, Shalat S (2007) A meta-analysis of children's hand-to-mouth frequency data for estimating nondietary ingestion exposure. Risk Anal 27(2):411–420. https://onlinelibrary.wiley.com/doi/abs/10.1111/j.1539-6924.2007.00893.x#references-section

Yourick JJ, Bronaugh RL (2000) Percutaneous penetration and metabolism of 2-nitro-p-phenylenediamine in human and fuzzy rat skin. Toxicol Appl Pharmacol 166:13–23

Chapter 6
Data Availability and Data Generation Concepts

Gianfranco Brambilla, Astrid Heiland, Gerhard Heinemeyer, and Christian Sieke

6.1 Introduction

Gerhard Heinemeyer

This chapter gives an overview of the approaches and data that are needed for exposure assessment, with the aim to point out the information gaps and missing links (complete lack of data, lack of adequate data, lack of data particularly generated for exposure assessment). A common language and understanding of terms in exposure assessment is a central key to harmonise exposure assessment approaches.

This chapter covers a wide field of the challenges in the use of the existing data and the generation of new data. Most of the data used in exposure assessment are generated for other purposes than exposure. For example, food consumption data are primarily intended to evaluate the relationship between nutrition and illnesses, e.g. by over- or malnutrition. The intake of nutrients (see Sect. 4.20), fat, proteins and carbohydrates is an additional issue. Measurements of substances in food and consumer products help to avoid the exceedance of limit values or other levels of

G. Brambilla
Department Food Safety, Nutrition and Veterinary Public Health, National Institute of Health (ISS), Rome, Italy
e-mail: gianfranco.brambilla@iss.it

A. Heiland
Department of Exposure, German Federal Institute for Risk Assessment, Berlin, Germany
e-mail: astrid.heiland@bfr.bund.de

G. Heinemeyer (✉)
Department of Exposure, German Federal Institute for Risk Assessment (Retired), Berlin, Germany
e-mail: gerhardheinemeyer@o2mail.de

C. Sieke
Department of Pesticide Safety, German Federal Institute for Risk Assessment, Berlin, Germany
e-mail: christian.sieke@bfr.bund.de

G. Heinemeyer et al. (eds.), *The Practice of Consumer Exposure Assessment*,
https://doi.org/10.1007/978-3-319-96148-4_6

concern. Limited information about substances in mixtures and preparations is available from product databases (Heinemeyer and Hahn 2005). Only a limited number of many types of measurements of substances exist, e.g. migration data for articles. While it is a big challenge to discuss how these types of data can be used (in all special fields of exposure assessment), the data that fulfil the requirements of exposure assessment need to be created before they can be used.

The objective of this chapter is to describe a general concept for data generation, with the purpose of their use in exposure assessment. For example, the concepts associated with questionnaire studies developed to study food consumption may serve as a basis for transferring the approaches to study food consumption to those for studying consumer habits and exposures. The improvement of the overall knowledge and sets of data needed to perform exposure analyses is the main focus.

At the moment (November 2019), most of the data used for exposure assessments is not generated and collected for that purpose. Therefore, it should be discussed how existing data and methods can be improved for exposure calculations and adjusted to the needs of the assessment.

For exposure evaluations two basic principles must be separated which have diverging consequences for organising an assessment and the planning of an exposure evaluation.

First, there is the regulatory purpose for safeguarding the population. Here, the concepts applied are based on conservative scenarios and high-end exposure assessments, with the aim of developing estimates that, under usual and foreseeable conditions, would indicate or not indicate risks for the general population. This is the principal approach for the REACH regulation, and it stands behind the precautionary principle. This approach aims at "zero-risk". The purpose is to execute the respective legislation. It does not explicitly consider exposures from various other sources and from pathways not covered by the respective regulation.

In addition, the role of measures that reduce the risk (risk management measures, RMMs) need to be included. RMMs are used as variables in exposure models to indicate the option to reduce the exposure. Increasing ventilation is, for example, a risk-reducing measure. The corresponding change of the room ventilation rate (q) is a model variable (see Table 5.10). Risk management measures should be kept in mind due to their qualitative nature. It should also be considered that, in contrast to workers' exposure (which is characterised primarily by the implementation of risk management measures), risk management measures for consumer exposure evaluations can often not be applied in a similar way.

Secondly, it might be important to get an overview of the entire (also called the real or actual or total) exposure to compare and aggregate exposures from the different sources that are under different legislations. For consumers, the estimation of the real exposure is therefore an important issue as a point of departure for risk communication and management. Consumer exposure has many and different sources and pathways that need to be considered. If the particular regulation does not consider the entire set of exposures, risk management might fail. Therefore, seeking to assess the total exposure should be the central focus, leading to insights about how to approach the goal of achieving consumer safety via risk avoidance and reduction.

In an internal paper, RIVM scientists (Bolte et al. 2016) analysed the current state of exposure assessment. Based on a thorough evaluation of the different issues and elements of exposure assessment, this paper developed a concept to improve the exposure assessment in upcoming years. This concept can be expanded for the general situation in Europe (and worldwide).

A primary aim of the RIVM paper was to develop an approach for the improvement of the scientific basis of exposure assessment together with developing a concept for collaboration and integration. In the context of this book and among other important points mentioned in the RIVM report, the following three issues are of special importance as primary objectives:

- Improvement of the scientific expertise
- Sharing of the knowledge
- Improvement of data availability

First, the improvement of scientific expertise in exposure assessment is the main objective of the editors and authors of this book. It is the first time that almost all issues of exposure assessment have been assembled and discussed under one common roof. Regardless of the different regulations and respective special aspects of exposure assessment, it is important to note that every exposure assessment follows a common concept. Exposure assessment involves several scientific disciplines, from medicine, toxicology, chemistry and physics, nutrition, statistics, mathematics, epidemiology, public opinion research and psychology. Sharing the knowledge from each discipline and integrating it with the others are of central importance and usefulness.

6.2 How to Use Existing Data and Improve These for Exposure Assessment

Gerhard Heinemeyer

Basically, a quantitative consumer exposure estimate is the result of an assessment that includes the content of a substance in a product and the amount of that product associated with the exposure of interest. Taking into account the frequency and length (e.g. seconds, minutes or hours a day, for one day or for many days or weeks) of exposure decides whether exposure is short-term or long-term. This kind of assessment can be made for different kinds of products, as shown in Fig. 6.1.

Consumers may be exposed to the same substance from food, consumer products, textiles and cosmetics and during hobby and leisure, with challenges encountered in the case of pesticides and biocides. To establish a common basis for the exposure assessment, data are needed that fulfil a sufficient degree of comparability. A comparison of exposure results from different studies or even an aggregation from a set of studies is simply impossible if, for example, food exposure is estimated on the basis of questionnaire and survey data, while the exposure estimates from the use of consumer products are based on default assumptions. This conclusion must be

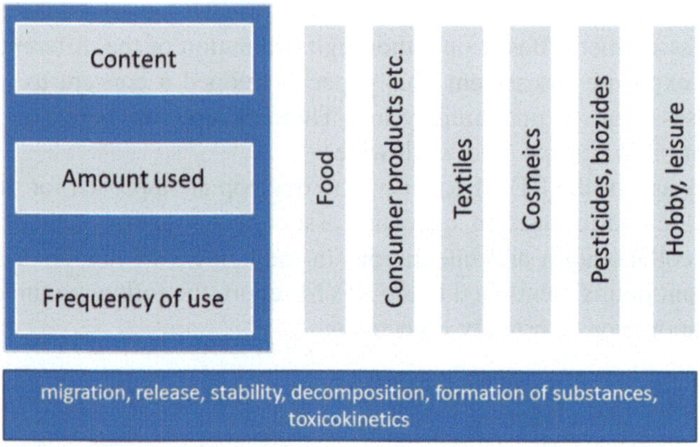

Fig. 6.1 The entire frame of data needs for consumer exposure assessment

made for any type of product, ranging from food contaminants to pesticide residues to consumer products, including toys, cosmetics, textiles and other sources that may be relevant as consumer products (Table 6.1).

In addition, the circumstances of the use/consumption of products must be described using data that characterise the pathway of an agent from the origin (e.g. using a product) to the contact with humans and the internal targets. This includes the changes of concentrations of the substances in the products during transport, storage and use/preparation, release and migration, as well as the internal processes (absorption, distribution, elimination). A common approach is needed for all of these processes to create and store data, to make them available and to apply them in exposure assessments.

The concept of the "exposome" encompasses the *totality* of human environmental (i.e. non-genetic) exposures[1] from conception onwards (Wild 2005; see also section "References"). Aligned with this concept, it can be said that everything with regard to consumer exposure and which is schematically shown in Fig. 6.8 covers a concept that could be called "Consumosome".

To compare the different approaches and regulations, a compilation and quality assessment of the available information is needed. Table 6.2 gives an overview of the available information in several regulations. With the exception of data used for exposure estimation via the consumption of food, more or fewer considerable data gaps are obvious. One concept to balance the lack of knowledge with the interest to invest time and other resources is to use the tiered approach. For example, a lower level of initial effort is needed for an assessment using the tiered approach, with the lowest tier based on expert judgements and conservative default (also called worst-case) assumptions. In some cases, migration data and data characterising the concentrations of substances in the products are available. For articles, toys in particular,

[1]https://en.wikipedia.org/wiki/Exposure_assessment

Table 6.1 Information base about different product types

	Concentrations in products	Product use data
Food	Regulatory limit values	Food consumption surveys with representative selected persons on different national levels and with different methodology
	Market surveillance data	
	Food monitoring data	
	Total diet study	
Mixtures for consumer use	Product ingredient information (product databases)	Assumptions based on expert judgement Product use recommendations Results from consumer questionnaires
	Measurements of substances in product	Data from studies measuring substances in products, e.g. the Danish EPA
	Release data	Data from studies that measured the release from products e.g. DEHP from baby suckers
Articles intended for consumer use (including toys)	Product ingredient information from general literature Limit values Measurements of substances in products	Assumptions based on expert judgement General available literature
Cosmetic products	Product databases Regulatory limit values Lists of ingredients Guidance documents for risk assessment	Assumptions based on expert judgement Product use recommendations
Chemicals	Default values Product database Broad scenario	Assumptions based on expert judgement Product use recommendations
Pesticides	Limit values due to approval	Food consumption survey Assumptions from expert judgement and conventions
Human biomonitoring	Concentrations of substances in fluids of the human body (urine, blood) The concentration of substances in urine collected over 24 h	

release/migration rates are important, and data from baby suckers and pacifiers are examples of what has been reported due to special requirements. However, systematic data (data for many types of articles) are lacking. In addition, the measurement methods vary. Product databases may help to identify critical substances in products (Heinemeyer and Hahn 2005). For pesticide approvals, strict protocols are prescribed that support systematic data collection. The information base is poor for other regulations.

In the BfR guidance document for uncertainty analysis in exposure assessment (BfR 2015), the data used for exposure assessment have been characterised according to their level of knowledge. On the one side are data characterised according to conventions and expert judgement, with the aim to support conservative assessments as the first step in the tiered approach. On the other side are empirical

Table 6.2 Data availability in the EU for exposure assessment

	Concentrations in products	Purpose of data collection
Concentrations of substances in food	Regulatory limit values	Derived values based on risk considerations
	Market surveillance data	Risk-oriented measurements to control the limit values
	Pesticide residue monitoring	Systematic surveillance
	Food monitoring data	Systematic surveillance
	Total diet study	Special approach to measure mean concentrations in prepared food in some countries
Consumption of food	Food consumption surveys	Systematic questionnaire approaches in all countries
	Food questionnaire studies	Risk-oriented questionnaire approach
Concentrations of substances in mixtures for consumer use	Product ingredient information (product databases)	Lists of ingredients of substances in mixtures (national and supranational registers)
	Product information according to § 16e chemical law (Germany)	Poison information centres
Concentrations and release data of substances in consumer articles	Rapid alert system for dangerous non-food products (RAPEX, RASFF)	European reporting system for dangerous substances in the frame of the products (Product Safety Directive, GPSD 2001/95EC)
	Market surveillance data	Systematic approach to measure substances in articles for market control
	Concentrations of substances in products	Danish EPA Laboratory data
Use of consumer products	Product use information	No systematic data available Use instructions Expert judgement
Guidance documents in accordance to directives	REACH Pesticides Cosmetics	Limit values, health-based guidance values, maximum concentration

data based on prospective or retrospective investigations or surveys. To date, most information used for exposure assessment is "floating" between these two extremes (Fig. 6.2) with varying extent and quality. Basically it can be stated that, going from the left to right, the appropriateness of the data increases and the uncertainty decreases. The data collected to characterise exposure are qualified by their own variability and uncertainty. One of the primary tasks for the future should to harmonise the data and the methodology to create them.

The data needs for exposure assessment will be discussed next. Due to the challenges of improving the data and because of conflicting requirements, a discussion about how to balance the scientific needs, the costs and the political demands needs to be included.

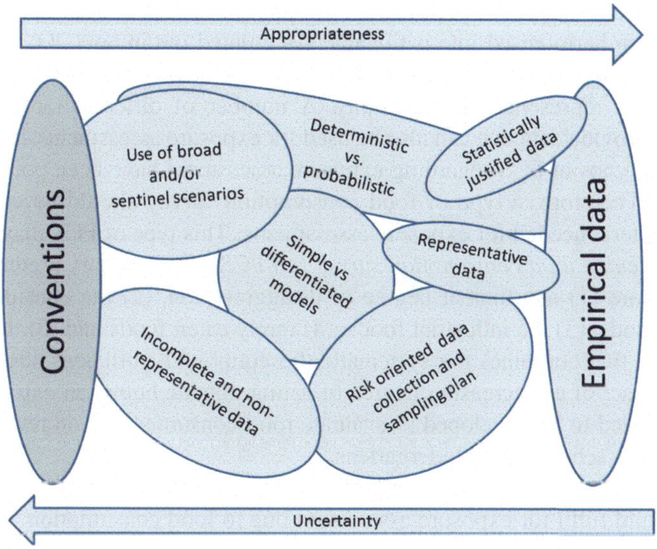

Fig. 6.2 The frame of exposure data

Another point that should be considered is how we can couple the different types of national and international surveys (health surveys, nutrition surveys, environmental surveys). In all of these projects, valuable data are created which can support exposure assessment; however, the different objectives, designs, conducts, data analyses and reporting performance of these studies and missed co-ordination significantly restrict the efficient use of these data for exposure analysis.

6.3 Food

Gerhard Heinemeyer

To characterise appropriately food consumption and food contamination data, reliable criteria for food consumption evaluations and for concentration measurements that should meet the prerequisites for exposure assessment need to be discussed and established.

6.3.1 Activity Pattern: Food Consumption Data

The improvement of the food consumption information needs to address the comparability and availability of the data. Food exposure is the field in exposure assessment where the databases have advanced the most. The reason is that nutrition surveys have not only been performed for many years and in many countries and have also been harmonised internationally. As pointed out in Sect. 4.6, many food consumption surveys have been performed at high quality standards and, consequently, under representative conditions. A number of different methodological approaches provide data that can also be used for exposure assessments. The reasons for particular gaps in FCS regarding exposure assessment have been pointed out in Sect. 4.6.6. Therefore, a type of food consumption survey should be created that fulfils the criteria needed for exposure assessments. This type of FCS may be called *exposure-oriented food consumption survey* (eo-FCS). The primary prerequisites of the eo-FCS are (1) a sufficient degree of disaggregation, (2) the consideration of composite foods, (3) the industrial foods, (4) rarely eaten foods and (5) the use of a classification that combines the systematic thesaurus with sufficient flexibility. To meet the impact of an increasing degree of eating outside home, an extra approach might be needed to be developed to evaluate food consumed, e.g. in restaurants, in cantinas and in schools and kindergartens.

Table 6.3 lists a number of important requirements that food consumption surveys should fulfil for exposure assessment due to food consumption.

An appropriate characterisation of food categories is needed to cover the latter mentioned foods. The LanguaL approach using a thesaurus and a system of descriptors (facets) seems promising. The system has sufficient flexibility to consider particular characteristics of the foods to be evaluated. It should have at least three to four subclasses and respective facets. For example, to evaluate cadmium exposure, wheat as raw commodity must first be identified by the category system as the original source and the type of cereal, e.g. bread (made from wheat flour), as the immediate source of the exposure. In the FoodEx2 classification system of EFSA (2011), this type of food can be found under the code A002F (grains and grain-based products - cereals and similar - cereals and cereal-like milling products and derivatives - wheat milling products and derivatives), as well as under A004V (grain and grain-based products - bread and similar products - wheat bread and rolls). As a first step, this category organisation can be taken as a standard.

A challenge for eo-FCS is to identify the foods that are not covered in the actual food groups in the FCS. In the modern world, more and more foods are industrially prepared, thus increasing the range of different food brands. These foods often, if not always, include various additives as ingredients. Official committees formulated the needs for special food consumption data years ago.

Food consumption data should allow the assessment of short-term and long-term exposures. For rare or individual short-term exposures, representative data might not be essential. However, a critical number of consumers should be interviewed to characterise, on a sound statistical basis, the incidences of these events, e.g. the 95th percentile of the portion size.

Table 6.3 lists the basic requirements to collect and describe the data in a sufficient degree. This applies for the characterisation of the different populations, e.g. the general population as a basis for comparison, the population not consuming the particular food (nonconsumers) vs. the consuming population and persons with particular eating habits (vegetarians and vegans), as well as for seasonal and

Table 6.3 Requirements for evaluation of dietary exposure assessments

Needs for dietary exposure surveys	Availability from systematic FCS
Characterisation of foods	
Adequate classification of food	Detail of food groups often inadequate or not available FoodEx2 classification scheme: at least four levels
High percentiles of portion size for short-term exposures	24 h recalls Dietary (weighing) records
Identification of regional foods	Regional representativity should be expressively stated
Identification of seasonal consumed food	In any food survey, if seasonal representativity is considered (it should be stated)
Identification of traditional foods	Not convincingly available, seldom eaten foods are often not covered by the survey
Identification of prepared and ready-to-eat food	Industrial prepared foods Additives and flavourings
Identification of food eaten outside home	Catering in schools, in kindergartens, at work Restaurants Cantinas
Incidences	
Numbers of individuals eating particular foods, with regard to gender and age	In systematic FCS not always available
Identification of seldom eaten food	Not available in systematic FCS
Populations	
Consumers versus nonconsumers	Only available if individual (primary) data are presented In statistical data often not available
Identification of particular populations	Basically possible but in general not available
Differentiation between children and adults	Special surveys for children and adults are available; interview methodology is differing
Statistics	
Central tendencies of dietary intake	Available in every FCS
High-end estimates	Available in every FCS
Comparability of surveys	
Survey questionnaire methodology	Hardly comparable between different national surveys
Food classification system	Limited comparability, the relevance of international proposed systems is increasing

traditional foods. A common classification system to characterise the food in the ways discussed above on an international basis is crucial.

In conclusion, data from food consumption surveys can be used for exposure assessment. However, they must be analysed in order of their applicability for that purpose. Providing data from the same survey obtained by different questionnaire methodologies is very helpful, e.g. 24 h recall and diet history/diary or at least FFQ or FFP.

The national bodies responsible for food consumption reporting have supported EFSA's comprehensive food consumption database. This database is described in Sect. 7.1.1 and classifies four subcategories (Sect. 4.4.1). The German data that can be found in EFSA's comprehensive database are much more detailed than in the basic MRI report (MRI 2008, 2015). Thus, using the EFSA data set instead of published national data should be taken into consideration. EFSA also extents food consumption survey data from national to pan European level (EFSA 2009b).

A basic question that should be addressed here is whether the data from EFSA's database are appropriate enough to estimate exposure? There are some examples where the data from the EFSA database will not meet the requirements of the assessment (see the cadmium example Sect. 4.4.2), although most of the evaluations of exposure should be adequate.

Special Approaches to Create Food Consumption Data for Exposure Assessment

Special types of questionnaire studies to evaluate particular special questions to clarify the non-regular occurrence of substances in food have been developed and applied. In these studies, only one particular issue was in the focus of the study, e.g. coumarin as an ingredient in cinnamon used for ginger bread and other baked goods that is typically consumed during Christmas time (Abraham et al. 2010), and the formation of 5-HMF and of acrolein in food (Abraham et al. 2011a, b). As a special example, energy drinks can be discussed which are rarely consumed within the general population but are regularly consumed at specific events and in high quantities among a special subpopulation (Lindtner et al. 2014).

Designing Food Consumption Studies on National and International Basis for Exposure Assessments

The main challenge in planning an exposure driven FCS is to ensure that the following criteria are met:

- Compatibility with existent (classical) food consumption studies
- Appropriate flexibility regarding to food list characterisation
- Coverage of industrial foods
- Compromise between scientifically justified needs and practical and financial limitations

When discussing compatibility questions, one should differentiate between the nutrition-oriented food consumption survey (no-FCS) and the exposure-oriented food consumption survey (eo-FCS). At the moment, only no-FCSs are available on national and international levels. An eo-FCS would extend the design that leads to downwards compatibility from the eo-FCS to the no-FCS. The major difference between no-FCS and eo-FCS is that the latter is more extended and detailed, which requires a more extended sample size. In earlier chapters the minimum needs for numerical estimations were presented (Sect. 4.6.5). Therefore, compatibility is not a problem to eo-FCS, if the same classification system is used. The LanguaL approach offers a flexible design that should be practicable for both purposes. In addition, it is helpful to apply several different methods for the questionnaire to differentiate between long-term and short-term exposures and to identify irregular and rarely eaten foods at an acceptable statistical level.

The extension of food categories and the identification of rarely eaten foods seem to represent a major challenge for the eo-FCS. New approaches and questionnaires should be developed. Also, a modular organisation of the survey might be discussed to ensure flexibility of the survey and to save costs. For example, the target population should be stratified, and the groups of particular interest should be interviewed separately from the general study population. The approach that has been practised by BfR to perform small questionnaire studies on demand seems a suitable option to answer questions that cannot be included in a big and systematic survey. Separating the FCS into modules might create the flexibility that is needed, yet while not increasing the costs. The impact of coupling FCS with human biomonitoring or environmental monitoring surveys should be considered.

Taking compatibility and flexibility into account, an eo-FCS is only limited by the financial resources. In Germany, BfR has performed about 10–15 additional and small questionnaire surveys with different purposes, partly with a very specific focus (energy drinks; see Sect. 4.6.6) but with limited costs. This type of approach seems promising for extended and flexible exposure-oriented food consumption surveys.

Also, food additives have an increasing importance due to an increase of consumption of industrial foods. This is not adequately covered in no-FCS, which should have consequences for future eo-FCS or in alternative food consumption evaluations.

In addition, food eaten outside home is also an issue of increasing interest. The majority of children, for example, spend the most time of the day in kindergartens and eat catered food. Therefore, this type of food can for some children represent a or the major part of nutrition in childhood. New and alternative approaches for evaluating food consumption than FCS in kindergartens and schools are needed to clarify nutrition behaviour.

6.3.2 Database of Concentrations of Substances in Food

One example of a set of concentration data is the one used for the cadmium opinion prepared by the EFSA (2009a) panel for contaminants (CONTAM). It is based on 137,202 samples analysed for cadmium concentrations, with about 43,000 samples from Germany, 40,000 from Slovakia and 18,000 from France. The rest were distributed over a wide range of countries. This figure clearly shows that the database, although consisting of a huge number of samples, is rather poor from a statistical view. The national data sources are from market control, which is a risk-oriented approach (see Sect. 4.8). These data do not fulfil requirements for representativeness, although the exposure has an intended tendency towards a conservative estimate due to risk orientation.

It should be also stated that the database shown for the cadmium example is not exposure-oriented. It is not even obvious that the mean and median estimates represent a central tendency of the concentrations of cadmium in food. The result therefore does not provide an indication of the real cadmium concentrations in food in Europe.

One exception of analytical data that should be mentioned are those that are systematic and not risk-oriented. For example, a small subsample of the German concentration data is collected in the German "food monitoring programme" and is characterised as a systematic approach to measure prospectively individual samples. The number of the samples is however low compared to the market control. These samples are appropriate to describe statistically sound central and high tendencies.

Example
The mean dietary cadmium exposure estimated on the above-mentioned data revealed 2.3 μg/kg of BW per week and the high exposure estimate 3.9 μg/kg of BW per week. For comparison, in Schwarz et al. (2014), data from the German food monitoring programme were used to estimate the cadmium exposure. An estimate based on these data revealed an exposure of 1.46 and 2.35 μg/kg of BW per week for average and high-end consumers, respectively, i.e. much lower than the EFSA (2009a) estimate.

Findings similar to those for cadmium can be obtained for other contaminants. EFSA has reacted and is trying to improve the database. However, the principal problem remains that a systematic and prospective approach is missing. Representativeness is the most important criterion among all quality criteria; however, representativeness is not a focus of any market control and surveillance approach.

Total diet studies can overcome these problems. As pointed out in Sect. 4.8.4, TDI delivers data that describe central tendencies of occurrence data of substances in food. For chronic risks, associated with long-term exposure, total diet studies seem to represent a practical alternative. It must be stated that the TDI is not appropriate for short-term risks. Data from individual measurements from market controls can therefore not be exchanged completely by TDS data.

The value of TDS data is limited by the systematic organisation of the study. Basically it follows the same principle as food classification. Even the TDS needs a thorough organisation of the deepness of the sampling. The approach of TDS is increasingly applied on the national levels but also on international levels. It seems crucial to apply harmonised food classification on the different levels. Otherwise results from TDS's will hardly be comparable.

In the future, the concept of TDS should be improved to analyse frequent occurring substances in foods. For rarely occurring substances, special approaches must be considered, too. In addition, every sample collection should be prospectively organised.

The concept of TDS can also ideally be applied for nutrients. Due to daily required intake and the clear systematics and classification, this category of food components seems ideally be suitably for the TDS approach.

New Emerging Issues

The change of eating behaviour represents a challenge of food consumption evaluation with the objective of exposure analysis. Appropriate data that characterise food consumption outside home are missing. This applies in particular for data characterising food consumption of children in kindergartens at early ages (>1 year). But also fast-food consumption and consumption of industrial manufactured food are increasing issues that should be considered in food consumption evaluations.

For relevant botanical contaminants, EU member states and professional stakeholder organisations are recommended to monitor and report on occurrence data. This applies also for process contaminants. On this basis exposure assessments can be updated and allow for further decisions in risk management. These substances should be measured by means of surveillance programmes as well as in TDS approaches.

A further challenge is to improve the knowledge about the occurrence of additives in industrial foods. Some examples of databases exist that support this objective, but the use of additives in a huge number of products on the market needs a particular approach. The BfR approach to build up a database of additives in resp. products (Diouf et al. 2014) should be extended. An appropriate system for aggregated classification of such food is also needed to reduce the workload to a justifiable degree.

6.4 Consumer Products

Gerhard Heinemeyer

Adequate data to perform sound exposure assessments for consumer products are rare. First, during and after use mixtures/preparations may decompose, and consequently substances are released and become available for exposure. In articles, substances must migrate to the outer surface to be released. Secondly, depending on the vapour pressure, substances may evaporate and deposit on, e.g. house dust. Thirdly, time activity of the people coming into contact with the substances must be described. There is missing information on every level of the exposure assessment: representative measurements of substances in products, migration data and time-activity data of people in their homes (Holzwarth et al. 2017).

6.4.1 Time-Activity Data: Targeting the Use of Consumer Products

Studies on use of consumer products that are comparable in size and focus to food consumption studies do not exist. Starting with the knowledge about methods and instruments used for food consumption studies, it should be discussed whether it might be possible to apply the experiences and the knowledge from FCS for studies evaluating consumer behaviour. From the view of the exposure assessor, the question might be asked what from the exposure point of view is the difference between eating some food, cleaning a bathroom or wearing a cloth? In any case exposure is a consequence of the amount released from the source, the frequency of use and the pathway from source to contact.

The time-activity questionnaire should address all these issues for all products and articles of interest.

In Fig. 6.3 the results of a questionnaire survey regarding the use of cleaning products in the household show that the consumers seldom follow use instructions (Heinemeyer et al. 2006). The figure also shows the different behaviour of males and females. Instructions for use, e.g. have obvious limited value in exposure assessment, due to strongly differing habits. As a consequence, separate consumer surveys are urgently needed to evaluate consumer behaviour by means of adequate questionnaire studies. BfR started a pilot study (Schneider et al. 2019) to clarify the conditions for a study (see Sect. 5.2.2).

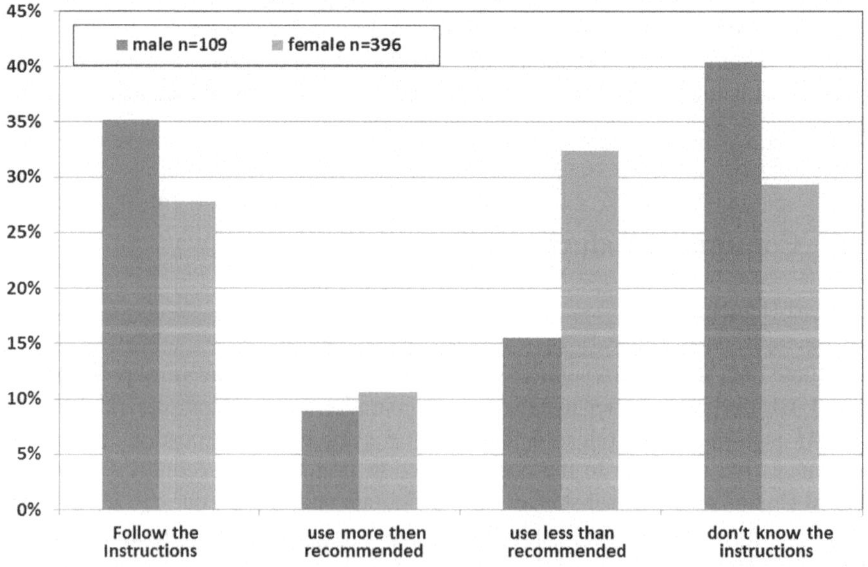

Fig. 6.3 Following the instructions for use of cleaning products in German households (© Bundesinstitut für Risikobewertung (BfR) 2018. All Rights Reserved)

6.4.2 Substances In and Releases from Consumer Products

As in food, the concentrations of substances in consumer products and articles are needed for consumer exposure assessment. Heinemeyer and Hahn (2005) pointed out the impact of product databases for consumer exposure assessment and its availability. The REACH regulation unfortunately reduced the availability of information about substances in consumer products, due to the transfer of responsibility for product safety to the manufacturers (see Sect. 5.6.1). The availability of data is restricted to a few databases, e.g. the Nordic countries product database. In Germany, in particular cases of substances in the REACH evaluation, the use of data in the BfR database of product information for poison centres can only be claimed via a strict and complicated procedure. Regarding the availability of data for consumer products safety, the REACH regulation is, therefore, a step backwards.

For articles, data from measurement programmes are available (e.g. Danish EPA 2001–2016), but these data focus on particular situations and do not provide a representative picture of substances in consumer products.

The most important information for exposure assessment is the release of substances from the articles, and this should be given increased attention with special focus on food packaging materials due to the increasing proportions of industrial foods. If, for example, a manufacturer prepares an article that contains a hazardous substance but he declares that the product is safe due to, e.g. coating of the surface that prevents the release, he has no responsibility to inform about the substance in the product. In practice, exposure from these products should be zero, and, of course, the manufacturer has responsibility for his product and must guarantee its safety.

6.4.3 Challenges for the Future

Gianfranco Brambilla

Apart from the progressive reduction of the uncertainties related to the food consumption databases and the statistical descriptors of the hazard occurrence in food, regulatory levels are ineffective when the health-related end-points rely, for instance, on the onset of antimicrobial resistance pathogens (ECDC, EFSA and EMA 2015) and on the onset of allergies and dismicrobisms (Thiemann et al. 2016): antimicrobial residues that could occur not occasionally at doses well below the MRLs and the analytical dynamic range of official checks, as consequence of the systematic antimicrobial use in conventional and intensive animal farming systems. Other emerging risks derive from combined exposure to different chemicals, in an aggregated way (inhalation, dermal absorption, intake from food and non-food matrices, such as dust) (EFSA 2016), whose relative contribution to the overall exposure may vary with the age (Winkens et al. 2017).

Within one health approach, the risk management should take an advantage of the so-called big data by merging of substance, exposure and health-related information from different open sources. Examples include:

- Drug consumption data (pharmaco-surveillance) from and pharmacovigilance data from EMA, Animal Health and Food Safety
- Human health surveillance data from EFSA and European Centre for Disease Prevention and Control (ECDC) (prevalence, incidence)
- Frequency of the occurrence/co-occurrence of residues above the analytical limit of detection in food commodities (national food safety authorities and EFSA)
- The use of chemicals in consumers products (European Chemicals Agency, ECHA)
- The presence of pollutants in air, soil and water (European Environment Agency, EEA)

Within an exposome framework (Rappaport 2011), the need for risk managers is to consider integrated information. Such aspects are especially valid for those substances, i.e. pesticides/biocides, that people are exposed to via different pathways and routes of entry (inhalation, dermal, dust intake, food intake), and for sensitive groups as bystanders (i.e. people living in agriculture areas, toddlers exposed for their mouthing behaviour to contaminants and biocides associated to house dust). These considerations imply that regulatory limits in food MRL should account for the relative contributions of the different food sources and the aggregate exposures and should also introduce risk communication tools based on open data access to endorse stewardship programmes and promote consumer empowerment. This would dramatically improve the risk reduction/risk acceptance policies and the accountability of those institutions involved in the risk assessment/risk management tasks (EFSA 2015).

6.5 Uncertainty Analysis

Gerhard Heinemeyer

Several guidance documents and reports have been published (WHO, EFSA, US-EPA, BfR; see Sect. 3.8). There is general consensus that uncertainty analysis is an indispensable part of the exposure assessment (WHO 2008). Uncertainty analysis should be regularly embedded as, e.g. in EFSA opinions, and some publications exist that provide practical examples. As shown by Heinemeyer et al. (2012), uncertainty analyses may take a considerable scale of work for the assessor, and practical experience in performing uncertainty analyses is needed. This regards to the analysis per se and its steps. The BfR guidance document proposes a distinction between qualitative and quantitative UA. The last report from EFSA (2018) requires both qualitative and quantitative in parallel. This discussion is driven by two different ideas that might be based on a more regulatory versus an academic view on UA. Separation of qualitative and quantitative UA is driven by the idea of a tiered approach and to go on step by step. In many cases, a qualitative UA identifies the sources and paths that primarily contribute to uncertainty, which might often be sufficient to propose risk management options. Therefore it may be sufficient to perform qualitative uncertainty considerations to derive proposals for risk communication and management before doing a quantitative UA.

The inclusion of the qualitative and quantitative UA in the same assessment is driven by a mathematical and statistical view as an objective of the overall uncertainty of the entire evaluation. This view provides an excellent characterisation of the statistics and separates variability from uncertainty of the overall picture but may not really indicate the sources of the uncertainty.

To conclude, there is an urgent need for practical examples that work with the either concepts of uncertainty analysis and compare them and their impact for risk management.

6.6 Harmonisation of Exposure Assessment Methodology

Gerhard Heinemeyer

Harmonisation is a central issue in exposure assessment. First of all, harmonisation means communication between exposure assessors. Recognising that some regulations are using different approaches to solve the same problem and using different terminology clearly indicates non-communication. It is simply a fact that most of the regulations are driven by the aim of setting limits. However substances are not aware of regulations, and they distribute according to their chemical and physical properties from their sources and over several pathways. The view of exposure from only one regulatory side is therefore too much restricted to the particular issue and does not convincingly consider reality.

Harmonisation of exposure should therefore have a focus on a common picture in which the basic information, scenario, model and parameters follow a common approach. In the *IPCS Harmonization Programme*, an expert group has prepared guidance documents for some important issues, with particular focus (and the first) on harmonisation of terminology (WHO 2004). It should also be noted that in many EFSA opinions, sources of exposures other than food have been considered.

6.6.1 Terminology of Exposure Parameters in Different Regulations

Although the parameters used in exposure modelling (as mentioned in Sect. 3.3) have common validity in exposure assessment, a different terminology is used due to the enormous diversity of regulations, on the national and international levels. The use of a consistent terminology is one of the key elements to ensure transparency in exposure assessment (WHO 2008). Exposure assessors may have difficulties in comparing model approaches and the respective results performed on the basis of different guidance and regulations. Harmonised or standardised terminology is not consistently followed, although international organisations have required this (see Sect. 7.4). To demonstrate the heterogeneity of denominating exposure parameters, three examples given in this book in the later chapters have been chosen for comparison. For example, dermal exposure is the primary pathway for the cosmetic products because of their application on skin.

Table 6.4 compiles some terms and respective abbreviations used in different fields and (in parallel) different international legislations. The reader gets the impression that there is something like a Babylonian confusion in the terms used. Many of the terms are very close together, others not.

Table 6.4 illustrates that the terminology used in the different regulations is not consistent. This regards to terms of the model parameters, its abbreviations and their descriptions. It should be noted that every regulation uses a particular nomenclature and terminology of terms and abbreviations that are difficult to be brought into a common context.

Dermal exposure is an important exposure path for household chemicals, e.g. cleaning products, or for articles. For example, the contacted skin area is a very important determinant for dermal exposure. The three selected regulations, however, are using the terms "skin surface area" (SSA), "surface area of the exposed skin" (A_{der}) and "contact area" ($A_{contact}$). The reason is that every regulation is worked out independent from each other. Due to the fact that in exposure models the abbreviations play a predominant role, the formulation of exposure models in a harmonised form is very difficult and frustrating. It should be mentioned that the term "contact area" might also apply for any other areas, e.g. the contact of food with container areas.

Table 6.5 lists selected terms as proposed in the IPCS guidance and those used in this book as well as some proposed in guidance documents regarded to particular regulations, to address the emerging need for further harmonisation. First of all, we can find terms that are consistently used, such as *exposure* and *exposure scenario*. Other terms proposed by IPCS are also used here, but there are deviations to terms mentioned in some regulations. For example, the term *frequency of use* is used instead of *exposure frequency*. Someone can argue that exposure frequency must not convincingly equal exposure frequency and shall be explained by the respective exposure scenario (see example below).

Table 6.4 Different descriptions of common parameters of exposure models for dermal exposure assessment in three separated regulations

Common abbrev	Description in resp. guidance docs	Dedicated abbr		Regulation	References
A_{der}	(Amount of) dermal absorption/ per area)	DAa	$\mu g/cm^2$	Cosmetics	SCCS
	Amount of substance on skin area per event	A_{der}	mg/cm^2	Chemicals	ECHA Guidance R 15
	Amount of substance loaded onto skin per unit skin area	L_{derm}	$\mu g\ cm^{-2}$	Articles	Product Safety
AREA	Skin surface area	SSA	cm^2	Cosmetics	SCCS
	Surface area of the exposed skin	A_{der}	cm^2	Chemicals	ECHA Guidance R.15
	Contact area	$A_{contact}$	cm^2	Articles	Product Safety
BW	Body weight	BW	kg	Cosmetics	SCCS
	Body weight	BW	kg	Chemicals	ECHA Guidance R.15
	Body weight	BW	kg	Articles	Product Safety
C_{der}	Concentration of the substance on the application site	C	%	Cosmetics	SCCS
	Dermal concentration of substance on skin	C_{der}	mg/cm^3	Chemicals	ECHA Guidance R.15
TH	Contact layer thickness	$\Delta\chi$	cm	Articles	Product Safety
	Thickness of product layer on skin (default 0.01 cm)	TH_{der}	cm	Chemicals	ECHA Guidance R.15
$Expo_{derm}$	External dermal exposure	$Exp_{ext.derm}$	µg/kg bw/dayy	Articles	Product Safety
	Amount of substance (external dose) that can potentially be taken up (account later for actual dermal absorption) per body weight	D_{der}	mg/kg bw/day	Chemicals	ECHA Guidance R.15
	Estimated daily exposure	A	mg/kg bw/day	Cosmetics	SCCS
	Systemic exposure dose	SED	mg/kg bw/day	Cosmetics	SCCS

The particular descriptions are pointed out for articles (Sect. 5.4), cosmetics (Sect. 5.5) and chemicals (Sect. 5.6)

Table 6.5 Selection of terms that are not consistently used in the IPCS guidance document, in this book and in different regulations

IPCS list of terms	This book	Other uses or failures
Exposure	Exposure	Exposure
Exposure scenario	Exposure scenario	Exposure scenario
Exposure frequency	Exposure frequency	Frequency, frequency of use
Exposure duration	Exposure duration; contact time	Contact time
Exposure surface	Exposure surface	Contact area
Exposure route	Exposure route	
Exposure pathway	Exposure pathways	
Exposure load		Dermal load
Acute exposure	Short-term exposure	Acute exposure
Chronic exposure	Long-term exposure	Chronic exposure
Bioavailability	Systemic availability	
	Dermal exposure	Dermal exposure
	Oral exposure	Oral exposure
	Inhalation exposure	Inhalation exposure
Exposure concentration		Dermal concentration Residue concentration

Example
A person is using a hobby product containing a volatile substance once a week. The duration of use is for 5 h in a particular room. After finishing work the person leaves the room, but comes back every next day. The person is exposed to the residual amounts of the volatile compound, but not using the product. This example shows that the exposure frequency may exceed the use frequency. The same also applies for use duration, exposure duration and contact time. The latter is the sum of all exposure durations after multiple uses, but the use duration may be smaller.

The comparison of terms proposed in the IPCS guidance document (WHO 2004) and terms used in particular regulations and in this book as well show that (1) identical terms are used or (2) terms proposed by IPCS are not used. In addition, the majority of terms are used as synonyms. In addition, a number of terms are used in this book, but not in the IPCS list. These two latter points certainly indicate the need for more and additional harmonisation work.

Other terms may be used synonymously, such as exposure area vs. contact area. It should be mentioned however that this area is not always human skin but also, e.g. the contact area of food in a food container (Table 6.5).

As pointed out earlier in this book, the terms *acute exposure* and *long-term exposure* should not be used. Instead the terms *short-term exposure* and *long-term exposure* are recommended.

To conclude, further work on the international basis is needed to harmonise the exposure terminology. This terminology should be proposed as a standard for all issues and fields of exposure assessment.

6.6.2 *Classification of Consumer Products*

Astrid Heiland and Gerhard Heinemeyer

Clear and systematic classification of products in terms of terminology is a central issue of harmonisation. Many different systems to characterise consumer products exist, but a transparent system that classifies products with the aim to support exposure analysis is lacking. Products classification should be developed that meets the requirements for scenario building. As mentioned above, the LanguaL system has been developed for food classification. The combination of a thesaurus consisting of a number of fixed levels may systematically characterise the product classes. The latter can be complemented by the facets that may characterise the individual aspects of the products. For example, room textiles are made by different kinds of fibres, which might by characterised by a facet "fibre" which is connected to the hierarchical and systematic thesaurus (Table 6.6).

Hierarchical category systems (Fig. 6.4) are decision trees, where the product characteristics, all or partly, are structured per definition on different levels. They are suitable for simple as well as for complex objects, but mostly with a loss of product information. Depending on the available data, the classification can stop at any subcategory.

Very recently, a proposal to systematically characterise "consumer products" was suggested (EU 2017; see also Sect. 5.2.1). A refined classification system was also presented by OECD in 2017 (OECD 2017). It should be mentioned that the systematic structure of these classification lists does not fully agree with the definition of "product", "article" and "mixture/preparation". There is some mixing of the terms and is not clear whether the term "product" covers the category "article" and "mixture". This might produce some confusion in the systematic view of consumer products.

Table 6.6 Example of classification for room textiles

Thesaurus			Facets		
Level			Facet 1 (material)	Facet 2 (coulour)	Facet 3 (fibre length)
Room textile	Curtain		Cotton	Red	
Room textile	Carpet	Carpeted floor	Wool	Grey	

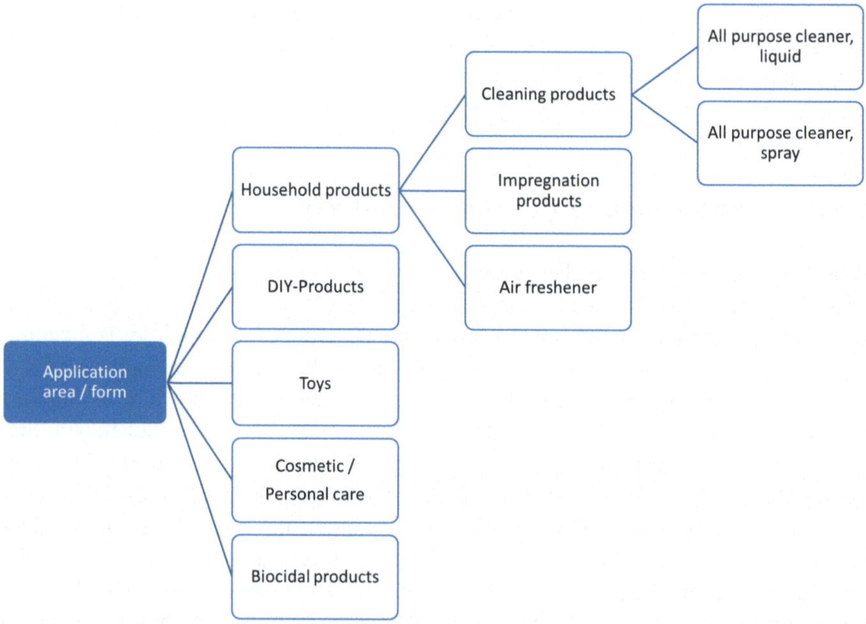

Fig. 6.4 Example for a hierarchical thesaurus for consumer products (© Bundesinstitut für Risikobewertung (BfR) 2018. All Rights Reserved)

Relational category systems (Fig. 6.5) allow a characterisation of an object with several descriptors (index or facets). These are related to the different levels of the hierarchical categories and can contain different descriptors. Such systems are useful, if many product properties should be considered. Most of the relational systems arranged in a hierarchical system are widely used in food classification (see Sect. 4.4) such as LanguaL and are also implemented in the use descriptor system under REACH.

Finally, the structure and the content of a category system are determined by the scope and the purpose of the product's use.

The "conflict of interest" existing between the scientific objectives of exposure assessment versus regulatory requirements might have consequences for classification issues. Regulation is often interested in broader (aggregated) ranges of characterisation of products, while the real exposure estimate depends on the details of the detailed characterisation (disaggregated) of the products (see Sect. 4.4). To achieve a clear differentiation, the classification system is important to define categories and levels stringently and precisely. This is particularly important where essential product use information is missing leading to erroneous results of exposure estimations. The above-mentioned lack of differentiation between articles and mixtures/preparations is a good example for inducing inadequacy by neglecting important subcategories for which the "toy" in Fig. 6.6 is a good example.

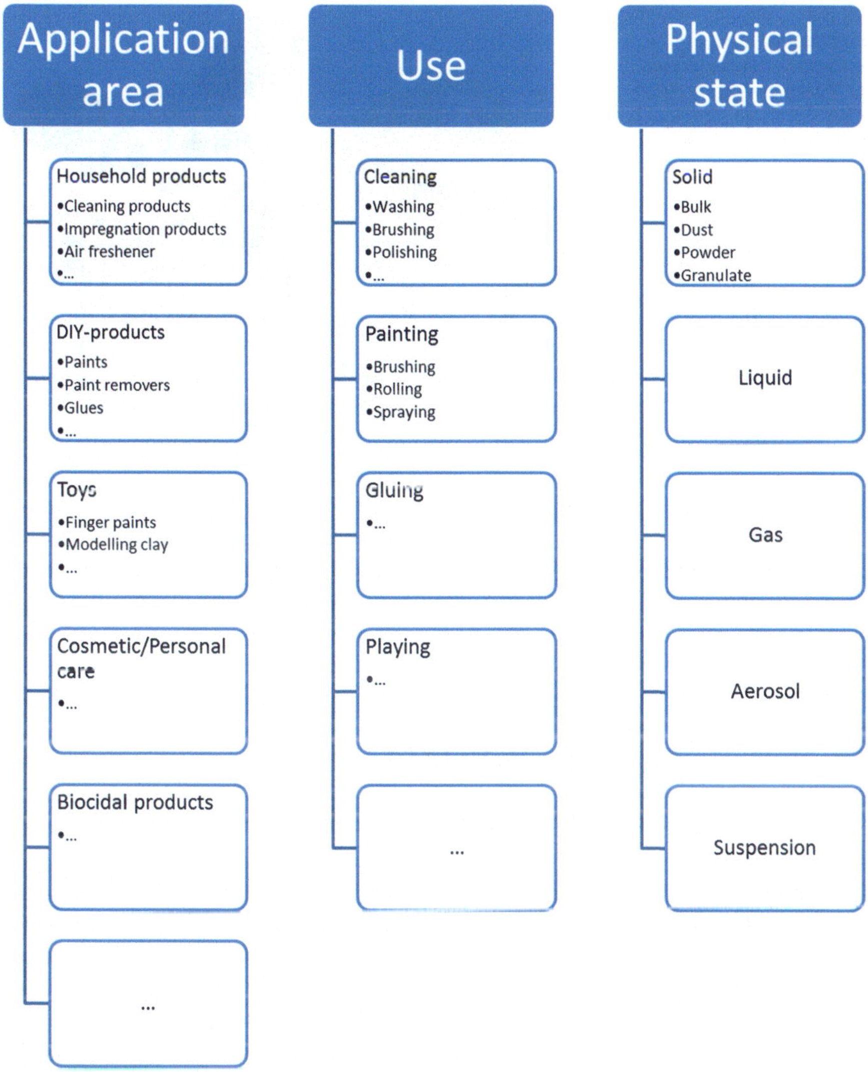

Fig. 6.5 Example for a multilevel relational descriptor system with an additional explaining facet approach (© Bundesinstitut für Risikobewertung (BfR) 2018. All Rights Reserved)

This product may be classified as a washcloth specially designed for children (Fig. 6.6). It is made from soft, fluffy terry cloth consists of 100% cotton. The colours are blue, yellow, black and white. It is sewn in form of glove with an animal head. Article category systems can be material-based (e.g. rubber, plastic, textile, wood, etc.) and/or structured by product functionalities (e.g. clothing, toys, stationery, furniture, etc.). Even if the category system is only based on the product functionality, more than one option of classification is thinkable, here personal care item or toy. The number of possibilities increases with the complexity of the

Fig. 6.6 Toys represent an important subcategory of articles (© Bundesinstitut für Risikobewertung (BfR) 2018. All Rights Reserved)

system, especially if several product characteristics are concerned. To settle various characteristics on the same level, e.g. "toy" (functionality) and "textile" (material), the uncertainties of a unique classification increase and can result in a mismatch without strict rules.

6.7 Exposure Scenarios and Models

Gerhard Heinemeyer and Christian Sieke

Regulatory exposure assessment has the focus on individual substances to be managed. However, in many cases, health effects may be associated to a number of chemicals acting together. Therefore, the assessment of cumulative exposures should be given particular attention. A first international accepted and applied concept to handle cumulative risks has been introduced for polychlorinated and polyfluorinated biphenyls by introducing the "toxicological equivalents" into the discussion. Cumulative (combined) exposure assessment evaluates the combined exposure of several single substances or groups of substances (Meek et al. 2011). A concept was presented to combine the assessment of cumulated risks by means of tiered approaches for exposure and hazard assessment, The most simple example is that several substances having similar or same hazard properties, but the considerations were also extended to substances with differing hazardous actions. Cumulative approaches always work with ranges of quantitative and qualitative characteristics of substances. Probabilistic methods may provide valuable instruments to handle.

Therefore, current approaches on assessing the cumulative exposure against multiple active substances rely on probabilistic methodologies. The major advantage compared to deterministic methodology is the consideration of all data available, including the full range of values and variability for each parameter. While the IEDI/NEDI and IESTI approaches select only a single representative value both for the consumption and for the residue concentration, probabilistic methods are capable of considering the complete daily consumption of all foods for each individual in combination with large occurrence data sets obtained from annual food monitoring programmes. Especially for the estimation of the cumulative dietary exposure, the consideration of all available data allows correlation of occurrence frequencies for multiple compounds already on sample level, making them capable of assessing both the long- and the short-term dietary exposure with the same tool.

Probabilistic models for pesticide residues are available in most regulatory systems. In the US, probabilistic methods and guidance to assess the cumulative dietary risk were established in 2002 (Office of Pesticide Programs U.S. Environmental Protection Agency (US-EPA) 2002). In Europe, the Dutch National Institute for Public Health and the Environment (RIVM) develops the Monte-Carlo Risk Assessment tool (MCRA; see also Sect. 4.10) (Dutch National Institute for Public Health and the Environment (van der Voet et al. 2015)), which is compliant to EFSA's Guidance on Probabilistic Modelling for Estimating the Dietary Exposure to Pesticide Residues (EFSA 2012).

Joint research between EFSA and RIVM initiated by the European Commission is still ongoing to adapt MCRA for the assessment of cumulative risks of pesticide residues, both by dietary exposure and by aggregate exposure considering all exposure routes including inhalation and dermal exposure. However, a commonly harmonised methodology for the assessment of mixtures has not been finalised yet and still demands further research both for the hazard characterisation and for the exposure estimation.

In addition cumulative considerations might be applicable when assessing other groups of chemicals, e.g. phthalates.

6.8 Communication and Co-operation Among Exposure Assessors

Gerhard Heinemeyer

As pointed out earlier and in Sect. 7.6, exposure assessment is an issue including a broad range of scientific disciplines. It is therefore important to ensure communication to ensure the exchange of minds between the different players.

6.8.1 Communication and Co-operation on National Levels

The organisation of tasks and co-operations between German agencies may serve as appropriate example to demonstrate how parties can communicate and co-operate among each other and manage related problems and challenges (Fig. 6.7). The Robert Koch Institute, under auspices of the ministry of health, is performing the German national health survey. This survey has a modular organisation, consisting on the central survey, a module with the objective of food consumption in children, a module focused on environmental influences on health and others. The children's FCS (KIESEL) is performed by the BfR, and the environmental survey is performed by the UBA. Although there is a close co-operation between the three agencies involved in the health survey and the modules, there is an urgent need to strengthen the adjustments of the modules with each other. This should be applied for a common population, as well as for the selection of substances that are of particular interest to relate the environmental occurrences with food consumption.

Data from further studies and surveys can be brought together in a big data pool, which is in the moment not existing. BfR performs the first German TDS. The TDS data are complemented by data created in the German food monitoring system and from market surveillance programmes (see Sect. 4.8.3). Unfortunately, a common data management is not foreseen.

This example shows that for Germany an infrastructure exists which however needs an impulse and the political willingness plus a financial frame to improve the co-operation.

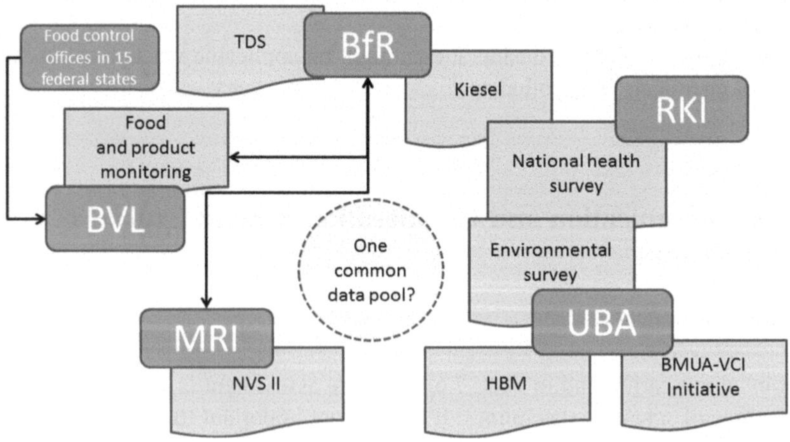

Fig. 6.7 Co-operation of national bodies (German example)

6.9 The Exposome

Gerhard Heinemeyer

It has been proposed by Lioy and Rappaport (2011) and Rappaport (2011) to discuss the exposure as a scenario of all different external influences on humans with subsequent internal effects. They called this concept the "exposome", which characterises the total issue of exposure science. As pointed out in Sect. 3.2.7, exposure characterises to entire way an agent takes to arrive from different sources via the exposure pathways in the human body. Consumer exposure compiles an important part of the exposure; it includes some environmental sources and a greater number of sources that are human made. Internal exposure can be seen as the final target. Expressed as a process that starts externally and proceeds step by step from the exposure source via the exposure pathways, a substance will enter the human body via the oral, dermal and inhalation routes and by passing the absorption barriers. It distributes in the human body and is eliminated via metabolism and urinary excretion. The latter serves as an important access for studying the exposure.

Figure 6.8 displays as a scheme the different direct and indirect pathway of consumer exposure. The environmental source is like a circle around the total consumer exposure scenario.

The exposure scenario as shown in Fig. 6.8 can be characterised by a mathematical mechanistic model:

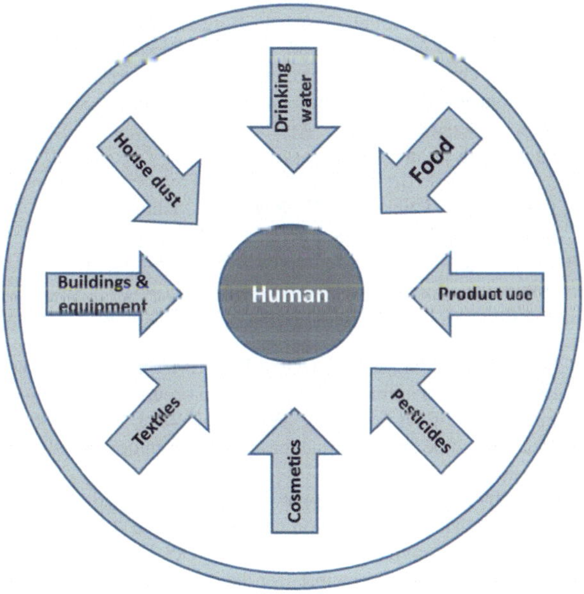

Fig. 6.8 Exposome, adjusted to consumer exposure

$$\text{Expo}_{\text{tot}} = \sum \text{Expo}_{\text{Food}} + \text{Expo}_{\text{ConsProd}} + \text{Expo}_{\text{Cosm}} + \text{Expo}_{\text{Text}} + \text{Expo}_{\text{Env}} + \text{Expo}_{\text{unknown}}$$

The individual parts of the exposure scheme and model can be characterised differently. This may apply for their source and pathways, but in particular by their degree of quality, uncertainty and harmonisation. It is an important task for exposure science to improve all that three elements of the exposure assessment with the aim of an exposome of common database and comparable result.

It should be taken into account that particular regulations, e.g. REACH, are not compatible with the exposome concept. In the REACH exposure assessment, the real exposure is seldom estimated, due to the concept of the safe product. Exposure assessment results revealed under REACH can therefore not be accounted for in exposome considerations. However, regarding the hierarchy of regulations for chemicals, REACH represents a roof under which the other regulations (partly) can be aggregated.

References

Abraham K, Wöhrlin F, Lindtner O, Heinemeyer G (2010) Toxicology and risk assessment of coumarin: focus on human data. Mol Nutr Food Res 54(2):170–306. Special issue: Special: novel approaches for risk assessment of phytochemicals in food

Abraham K, Gürtler R, Berg K, Heinemeyer G, Lampen A, Appel KE (2011a) Toxicology and risk assessment of 5-Hydroxymethylfurfural in food. Mol Nutr Food Res 55:667–678

Abraham K, Andres S, Palavinskas R, Berg K, Appel KE, Lampen A (2011b) Toxicology and risk assessment of acrolein in food. Mol Nutr Food Res 55:1277–1290

BfR (2015) Guidelines on uncertainty analysis in exposure assessments. Recommendation of the Committee for Exposure Assessment and Exposure Standardisation of the Federal Institute for Risk Assessment (BfR). Available at: http://www.bfr.bund.de/cm/350/guidelines-on-uncertainty-analysis-in-exposure-assessments.pdf. Accessed 29 Oct 2019

Bolte J, Bruinen de Bruin Y, van Engelen J, Evers E, Herremans J, Hogerwerf L, van der Linden T, Ossendorp B, Peijnenburg W, Piersma A, Schols E, Sijm D, Tielemans E, Traas T, Vermeire T, van der Vliet D, Dissel S, Lebret E (2016) RIVM exposure assessment strategy 2016–2020

Diouf F, Berg K, Ptok S, Lindtner O, Heinemeyer G, Heseker H (2014) German database on the occurrence of food additives: application for intake estimation of five food colours for toddlers and children. Food Additives & Contaminants: Part A 31(2):197–206. https://doi.org/10.1080/19440049.2013.865146

ECDC, EFSA, EMA (2015) ECDC/EFSA/EMA first joint report on the integrated analysis of the consumption of antimicrobial agents and occurrence of antimicrobial resistance in bacteria from humans and food-producing animals. EFSA J 13(1):4006, 114 pp. https://doi.org/10.2903/j.efsa.2015.4006. http://onlinelibrary.wiley.com/doi/10.2903/j.efsa.2015.4006/pdf. Accessed 29 Oct 2019

EFSA (2009a) Scientific opinion of the panel on contaminants in the food chain on a request from the European Commission on cadmium in food. EFSA J 980, 1–139. http://www.efsa.europa.eu/en/efsajournal/pub/980. Accessed 29 Oct 2019

EFSA (2009b) General principles for the collection of national food consumption data in the view of a pan-European dietary survey. EFSA J 7(12) 1–51. https://efsa.onlinelibrary.wiley.com/doi/abs/10.2903/j.efsa.2009.1435. Accessed 29 Oct 2019

EFSA (2011) Technical report. The food classification and description system FoodEx 2 (draft-revision 1). Supporting Publications 2011: 215, 438 pp. http://onlinelibrary.wiley.com/doi/10.2903/sp.efsa.2011.EN-215/abstract. Accessed 29 Oct 2019

EFSA (2012) Guidance on the use of probabilistic methodology for modelling dietary exposure to pesticide residues. EFSA J 10(10):2839, 95 pp. http://onlinelibrary.wiley.com/doi/10.2903/j.efsa.2012.2839/full. Accessed 29 Oct 2019

EFSA (2015) Shaping the future of food safety, together. Proceedings of the 2nd EFSA scientific conference. Milan, Italy, 14–16 October 2015. EFSA J 13(10):s1310, 137 pp. https://www.efsa.europa.eu/de/events/event/151014. Accessed 29 Oct 2019

EFSA (2016) Public consultation on the terms of reference of the Scientific Committee Working Group on "Harmonisation of risk assessment methodologies for human health and ecological risk assessment of combined exposure to multiple chemicals". Approved: 19 Oct 2016. https://www.efsa.europa.eu/en/consultations/call/161024a. Accessed 29 Oct 2019

EFSA (2018) Guidance on uncertainty analysis in scientific assessments. EFSA J 16(1):5123, 39 pp. https://doi.org/10.2903/j.efsa.2018.5123. http://onlinelibrary.wiley.com/doi/10.2903/j.efsa.2018.5122/abstract. Accessed 29 Oct 2019

EU (2017) Study on a Product Categorisation System for information to be submitted to Poison Centres, according to Article 45 (4) of EC Regulation No 1272/2008 (CLP Regulation). ISBN:978-92-79-63777-3. Final report https://doi.org/10.2873/87997. https://publications.europa.eu/en/publication-detail/-/publication/269f51ae-e6d4-11e7-9749-01aa75ed71a1/language-en. Accessed 29 Oct 2019

Heinemeyer G, Hahn A (2005) Use of product databases for risk assessment purposes. Toxicol Appl Pharmacol 207:S636–S644

Heinemeyer G, Lindtner O, Scholz, R, Rüdiger T (2006) Evaluation of behaviour of consumers for exposure assessment by questionnaire studies. ISEE/ISEA international conference. 2–6 Sept 2006, Paris

Heinemeyer G, Sommerfeld C, Springer A, Heiland A, Lindtner O, Greiner M, Heuer T, Krems C, Conrad A (2012) Estimation of dietary intake of bis(2-ethylhexyl) phthalate (DEHP) by consumption of food in the German population. Int J Hyg Environ Health 216:472–480

Holzwarth A, Rüdiger T, Sommerfeld C, Lindtner O, Heinemeyer G (2017) Verwendung von Luftbehandlungsprodukten in Privathaushalten – eine Verbraucherbefragung. Gefahrstoffe Reinhalt Luft 77:31–38

Lindtner O, Ehlscheid N, Berg K, Blume K, Dusemund B, Ehlers A, Niemann B, Rüdiger T, Heinemeyer G, Greiner M, Hallerbach B, Thömmes O, Thier S (2014) Event-related survey of high consumers of energy drinks. Federal Institute for Risk Assessment, BfR Wissenschaft, 5/2014

Lioy PJ, Rappaport SM (2011) Exposure science and the exposome: an opportunity for coherence in the environmental health sciences. Environ Health Perspect 119:A466–A467

Meek ME, Boobis AR, Crofton KE, Heinemeyer G, Van Raaij M, Vickers C (2011) Risk assessment of combined exposure to multiple chemicals: a WHO/IPCS framework. Regul Toxicol Pharm 60(2011):S1–S14. https://doi.org/10.1016/j.yrtph.2011.03.010. https://www.sciencedirect.com/science/article/pii/S0273230011000638?via%3Dihub. Accessed 29 Oct 2019

MRI (2008) Nationale Verzehrs Studie II, Ergebnisbericht, Teil 2. Die bundesweite Befragung zur Ernahrung von Jugendlichen und Erwachsenen. https://www.mri.bund.de/fileadmin/MRI/Institute/EV/NVSII_Abschlussbericht_Teil_2.pdf. Accessed 9 Mar 2018

MRI (2015) Nationale Verzehrsstudie II. Ergänzungsband zum Ergebnisbericht, Teil 1. Ausgewählte Ergebnisse nach Schichtindex. Karlsruhe, 2008. Internet: https://www.mri.bund.de/fileadmin/MRI/Institute/EV/NVS_II_Abschlussbericht_Teil_1_mit_Ergaenzungsbericht.pdf. Accessed 29 Oct 2019

OECD (2017) Internationally harmonised functional, product and article use categories. Series on testing & assessment no. 262. OECD Environment Directorate, Paris, FR. http://www.oecd.org/

officialdocuments/publicdisplaydocumentpdf/?cote=env/jm/mono%282017%2914&doclanguage=en. Accessed 29 Oct 2019

Rappaport SM (2011) Implications of the exposome for exposure science. J Exp Sci Environ Epidemiol 21:5–9

Schneider K, Recke S, Kaiser E, Götte S, Berkefeld H, Lässig J, Rüdiger T, Lindtner O, Oltmanns J (2019) Consumer behaviour survey for assessing exposure from consumer products: a feasibility study. J Expo Sci Environ Epidemiol 29(1):83–94. https://doi.org/10.1038/s41370-018-0040-2

Schwarz MA, Lindtner O, Blume K, Heinemeyer G, Schneider K (2014) Cadmium exposure from food: the German LExUKon project. Food Addit Contam Part A Chem Anal Control Expo Risk Assess 31(6):1038–1051. https://doi.org/10.1080/19440049.2014.905711

Thiemann S, Smit N, Strowig T (2016) Antibiotics and the intestinal microbiome: individual responses, resilience of the ecosystem, and the susceptibility to Infections. Curr Top Microbiol Immunol 398:123–146. https://doi.org/10.1007/82_2016_504

US-EPA (2002) Guidance on cumulative risk assessment of pesticide chemicals that have a common mechanism of toxicity. U.S. Environmental Protection Agency (US-EPA), Washington, DC. https://www.epa.gov/sites/production/files/2015-%2007/documents/guidance_on_common_mechanism.pdf. https://www.epa.gov/sites/production/files/2015-07/documents/guidance_on_common_mechanism.pdf. Accessed 8 March 2018

van der Voet H, de Boer WJ, Kruisselbrink JW, Goedhart PW, van der Heijden GWAM, Kennedy MC, Boon PE, van Klaveren JD (2015) The MCRA model for probabilistic single-compound and cumulative risk assessment of pesticides. Food Chem Toxicol 79:5–12. https://doi.org/10.1016/j.fct.2014.10.014

WHO (2004) IPCS risk assessment terminology. IPCS harmonization project document no. 1. WHO, Geneva. http://www.who.int/ipcs/methods/harmonization/areas/terminology/en/. Accessed 29 Oct 2019

WHO (2008) Uncertainty and data quality in exposure assessment. Part 1: guidance document on characterizing and communicating uncertainty in exposure assessment. IPCS harmonization project document no. 6. WHO, Geneva. http://www.who.int/ipcs/publications/methods/harmonization/en/. Accessed 29 Oct 2019

Wild CP (2005) Complementing the genome with an "exposome": the outstanding challenge of environmental exposure measurement in molecular epidemiology. Cancer Epidemiol Biomark Prev 14(8):1847–1850

Winkens K, Vestergren R, Berger U, Cousins IT (2017) Early life exposure to per- and polyfluoroalkyl substances (PFASs): a critical review. Emerg Contam 3:1–14. https://doi.org/10.1016/j.emcon.2017.05.001

Chapter 7
Activities Encountered by European and Other International Authorities

Andreas Ahrens, Jos Bessems, Yuri Bruinen de Bruin, Alison Connolly, Peter Fantke, Mary Gilsenan, Gerhard Heinemeyer, Matti Jantunen, Majlinda Lahaniatis, Demosthenes Papameletiou, Artur Radomyski, Vittorio Reina, Urs Schlüter, Yasmin Sommer, Anne Theobald, Natalie von Goetz, and Alexandre Zenié

A. Ahrens
European Chemicals Agency (ECHA), Helsinki, Finland
e-mail: Andreas.Ahrens@echa.europa.eu

J. Bessems
VITO Health, Mol, Belgium

ISES-Europe Board, Freiburg, Germany
e-mail: Jos.Bessems@ISES-Europe.org

Y. Bruinen de Bruin
European Commission Knowledge Management Service, European Commission Joint Research Centre, Ispra, Italy
e-mail: Yuri.Bruinen-de-bruin@ISES-Europe.org

A. Connolly
ISES-Europe Board, Freiburg, Germany

National University of Ireland, Galway (NUIG), Ireland
e-mail: Alison.Connolly@ISES-Europe.org

P. Fantke
ISES-Europe Board, Freiburg, Germany

Technical University of Denmark, Lyngby, Denmark
e-mail: Peter.Fanke@ISES-Europe.org

M. Gilsenan
Evidence Management Unit, European Food Safety Authority (EFSA), Parma, Italy
e-mail: Mary.GILSENAN@efsa.europa.eu

G. Heinemeyer (✉)
Department of Exposure, German Federal Institute for Risk Assessment (Retired), Berlin, Germany
e-mail: gerhardheinemeyer@o2mail.de

G. Heinemeyer et al. (eds.), *The Practice of Consumer Exposure Assessment*,
https://doi.org/10.1007/978-3-319-96148-4_7

The impact of this chapter is to inform the reader of this book about current activities of international bodies to improve the methods used in exposure assessment and the knowledge and information database. In addition the reader shall be given insight into the regulatory work and decision procedures.

7.1 EFSA

Mary Gilsenan, Majlinda Lahaniatis, and Anne Theobald

The European Food Safety Authority (EFSA) Parma, Italy is a separate legal entity independent from the European Union (EU) institutions and an integral part of the EU's food safety system (EU 2002). Founded in 2002, EFSA contributes to the safety of the EU food and feed chain and to a high level of protection of human health, mainly by (EFSA 2016a):

- Providing EU risk managers with independent, up-to-date and fit-for-purpose scientific advice on questions related to food and feed safety, animal health and welfare, plant health, nutrition, and environmental issues specific to the above
- Communicating to the public on its scientific outputs and the information on which they are based

M. Jantunen
Department of Environmental Health, National Institute for Health and Welfare (THL) (Retired), Kuopio, Finland
e-mail: matti.jantunen@janding.fi

M. Lahaniatis
International Affairs, German Federal Institute for Risk Assessment, Berlin, Germany
e-mail: Majlinda.Lahaniatis@bfr.bund.de

D. Papameletiou · A. Radomyski · V. Reina · A. Zenié
European Commission, Joint Research Centre (JRC), Ispra, Italy
e-mail: Alexandre.ZENIE@ec.europa.eu

U. Schlüter
ISES-Europe Board, Freiburg, Germany

BAUA, Dortmund, Germany
e-mail: Urs.Schlueter@ISES-Europe.org

Y. Sommer
Department of Exposure, German Federal Institute for Risk Assessment, Berlin, Germany
e-mail: yasmin.sommer@bfr.bund.de

A. Theobald
European Food Safety Authority (EFSA), Parma, Italy
e-mail: Anne.Theobald@efsa.europa.eu

N. von Goetz
ISES-Europe Board, Freiburg, Germany

ETH Zurich, Zurich, Switzerland
e-mail: Natalie.von-Goetz@ISES-Europe.org

- Cooperating with Member States, institutional partners and other interested parties/stakeholders in the EU to promote coherent advice and increase trust in the EU food safety system
- Developing standardised and harmonised methodologies, and collecting and analysing data to allow the identification, characterisation and monitoring of emerging risks that have a direct or indirect impact on food and feed safety.

Only the European Commission (EC), the European Parliament (EP) and the competent authorities in EU Member States (EU MS) are entitled to ask scientific advice from EFSA. Once a request for scientific advice from the above-mentioned actors is accepted by EFSA, it becomes a Mandate, which is assigned to one or more scientific panels or the Scientific Committee depending on the scientific area and the expertise needed.

EFSA's scientific advice is given in form of scientific opinions by one or more of EFSA's 10 scientific panels or its Scientific Committee. The members of the Scientific Committee and the panels are independent experts appointed through an open selection procedure mainly from European universities, public research organisations, national authorities and food safety agencies. EFSA staff may also produce scientific outputs on behalf of the agency, such as peer reviews of the risk assessment of active substances in pesticides, dietary exposure assessments of food-borne substances or responses to urgent requests for scientific advice. EFSA staff also monitors and analyses information and data on biological hazards, contaminants, veterinary drug residues, and emerging risks.

EFSA's panels and Scientific Committee also carry out scientific work on their own initiative. This is known as 'self-tasking' and mainly applied to examine emerging issues and new hazards, and to update its risk assessment methods and approaches.

The EFSA's scientific process[1] is composed of three steps: receipt of request, risk assessment and adoption of the opinion. In the risk assessment phase, the panel or the Scientific Committee sets up a Working Group of experts, which undertakes detailed scientific work based on scientific evidence (data and literature) and expertise. The Working Groups generally develop draft opinions and submit them to the panel for discussion and possible adoption. EFSA often holds public consultations on its draft opinions and guidance documents. The Scientific Committee or panel reviews the feedback received to draft the final opinion, which is adopted by a majority of their members with a quorum of at least two thirds of members present. The EFSA opinions are published in the EFSA Journal, which is an open-access, online scientific journal. The Authority is legally bound by European Union legislation on issues such as public access to documents (EU 2001).

EFSA adheres to a number of principles and practices aimed at ensuring the excellence of its scientific work. Transparency (access to documents) and openness (engagement) have been key values for EFSA since its creation. Adherence to these

[1]EFSA's scientific process. A step-by-step guide. Available at: http://www.efsa.europa.eu/interactive_pages/scientificprocess/ScientificProcess

values helps to legitimise EFSA's work and ensure accountability to society. In EFSA's context, openness and transparency allow civil society to be informed about EFSA's internal processes, have access to and understand its outputs, and engage in EFSA's scientific decision-making process in the appropriate consultative fora. The Open EFSA initiative,[2] launched in 2014, and the subsequent TERA project (Transparency and Engagement in Risk Assessment) aim at exploring and implementing several measures for enhancing transparency and engagement in EFSA's risk assessment workflow. With the entry into force of the so-called 'Transparency Regulation' (EU 2019), EFSA has a legal mandate for several tasks related to data management, e.g. submission of data in structured format which might have also an implication on future ways of engaging with the public and performing exposure and risk assessment.

Additional activities are undertaken by EFSA to further develop its risk assessments and align its methodologies with the latest developments in science. The collection, appraisal and analysis of evidence - data, literature and expert judgement - are key steps in preparing EFSA's scientific risk assessments. In the frame of the PROMETHEUS project (PROmoting METHods for Evidence Use in Scientific Risk assessments), EFSA developed new methods and approaches[3] for impartiality, methodological rigour, transparency and responsiveness when using scientific evidence (i.e. selecting, appraising and integrating evidence in scientific risk assessments). The new methodological framework for the use of evidence can also be applied by risk assessors across Europe and beyond. It involves a four-step process to achieve fit-for-purpose, evidence-based scientific advice for decision-makers:

- Plan (i.e. planning a strategy for the risk assessment upfront).
- Conduct (i.e. carrying out the assessment as far as possible in line with the strategy). In this context, data are selected, appraised and integrated according to predefined criteria, and methods are tailored to the objectives and context of the assessment.
- Verify (i.e. corroborating the assessment process with the strategy).
- Report (i.e. documenting and explaining the process, results and conclusions and ensuring accessibility of methods and data).

EFSA seeks to develop further guidance, recommendations, and processes to keep pace with international scientific developments and regulatory science needs.[4] For example, EFSA's Scientific Committee has developed draft guidance on how to assess uncertainty in a structured and systematic way with the aim to offer a toolbox of methodologies, both quantitative and qualitative, and related training (EFSA

---

[2]EFSA 2014. Discussion Paper Transformation to an 'Open EFSA' http://www.efsa.europa.eu/en/corporate/doc/openefsadiscussionpaper14.pdf

[3]Methods for evidence use in scientific assessments. Furthering impartiality, methodological rigour, transparency and responsiveness when using evidence. Available at: https://www.efsa.europa.eu/interactive_pages/prometheus/prometheus

[4]EFSA Guidance and assessment approaches (https://www.efsa.europa.eu/en/methodology/guidance)

2018). In 2016, EFSA's Scientific panels started to trial the Revised draft Guidance on Uncertainty in Scientific Risk assessment in at least one of their scientific risk assessments. Other guidance documents relate to biological relevance and on weight of evidence pertaining to its scientific assessments (EFSA 2017a).

EFSA has also developed Guidance on expert knowledge elicitation (EFSA 2014f) to be used in cases of limited empirical evidence for a risk assessment and where expert judgment in the areas covered by EFSA's remit is needed. The Guidance provides detailed protocols and establishes a phased process with assigned responsibilities at each step: defining the risk assessment problem, preparing for elicitation, and conducting an expert elicitation, culminating in complete documentation.

In order to continually monitor and improve the quality of its scientific work, EFSA has developed a robust quality management system (QMS) that includes self-review and customer feedback systems. It ensures that scientific processes are developed consistently and continuously improved across EFSA's panels and by staff.[5] In November 2016, EFSA obtained the ISO9001: 2015 certification attesting that its QMS is considered robust, well established and in compliance with international recognised standards.

EFSA estimates dietary exposure in two main areas: regulated products and contaminants.

- *Regulated products*: Includes substances intentionally added to and/or used on food and feed such as additives, enzymes, flavourings, and nutrient sources and other substances/products such as food contact materials, pesticides, genetically modified organisms, novel foods, food-related processes and processing aids.
- *Contaminants in food and feed*: EFSA also carries out risk assessments on a wide range of contaminants that can be present in food and feed due to food production, distribution, packaging or consumption, as well as those that might be present in the environment naturally or as a result of man-made activity.

The provision of European data and exposure assessments is important as many food products are regulated at the European level, hence calling for a European approach in exposure assessment. The challenge is to provide risk managers with assessments of exposure and risk that enable to establish maximum limits, indicative values for contaminants, or to establish limits for other food borne hazards. These values should ideally be conservative enough to protect the population with the highest expected exposure taking into account consumption patterns and occurrence data for the respective hazard.

In general dietary exposure is calculated for each European country using consumption data from the EFSA comprehensive database (see Sect. 6.3.1) combined with levels of e.g. food-borne hazards or nutrients. In order to be protective, usually a high percentile of exposure (e.g. the 95th percentile) is used in risk assessments. Dietary exposure for subpopulations such as infants is calculated separately for specific requests.

[5]How we work. Available at: https://www.efsa.europa.eu/en/about/howwework

7.1.1 EFSA Databases and Data Collection

In the late 1990s–early 2000s, European legislators identified the lack of an effective system of collection and analysis of data on the food supply chain at the European Community level as a major shortcoming. Therefore, it proposed in EFSA's Founding Regulation (EU 2002) that one of EFSA's tasks is to search for, collect, collate, analyse and summarise relevant scientific and technical data within its remit. This involves, in particular, the collection of data relating to: (a) food consumption and the exposure of individuals to hazards related to the consumption of food; (b) incidence and prevalence of biological risk; (c) contaminants in food and feed; (d) residues.

For the purpose of collecting data on occurrence of food-borne chemical hazards, as well as food consumption, EFSA relies on competent authorities in EU Member States, the food industry, academia, and research organisations. To reinforce this task, EFSA manages data Networks (with data provider representatives from European countries), e.g. on Chemical Occurrence Data, Pesticides Residues and on Food Consumption Data, to discuss data needs and common standards to transmit the data to EFSA, to provide training on data transmission requirements, and to share experiences and best practice. The minutes of the data network meetings are published on EFSA's website.[6] EFSA also engages with industry and other stakeholders to collect food additive and contaminant data for its dietary exposure assessments.[7] An extensive description of methods to collect and generate reliable data on concentrations of substances in food is given in Sect. 4.8. These data collections are mainly conducted at the national level.

EFSA created its *Comprehensive European Food Consumption Database* in 2010 from existing national information on food consumption at individual level. Competent organisations in European countries provided EFSA with data from the most recent national dietary surveys in their country. Survey results for children were mainly obtained through the EFSA Article 36 project 'Individual food consumption data and exposure assessment studies for children' (EXPOCHI) through the consortium (EFSA 2011). EFSA uses its food classification system 'FoodEx' (see Sect. 4.4.1) to categorise all foods and beverages included in the Comprehensive Database. Since 2012, all food consumption data submissions to EFSA have been reported in accordance with FoodEx2 food classification and description system that enables more precise reporting of consumption patterns. The database includes individual dietary records from some 100,000 subjects (infants to elderly) out of more than 510 million inhabitants in 23 out of 28 EU Member States (status: April 2017). The summary statistics are published in EFSA's scientific data warehouse.

[6]EFSA's Data collection and analysis networks https://www.efsa.europa.eu/en/data/networks

[7]EFSA stakeholder discussion group on food chemical occurrence data https://www.efsa.europa.eu/en/engage/stakeholders

EFSA and the national data providers of the comprehensive database have agreed on the use of the data. EFSA has the right to make use of the raw individual food consumption data for carrying out risk assessments and other scientific analyses within the activities related to EFSA's mandate. A formal authorisation from the data provider must be requested for any other use of the data in order to make the data also useful for the wider public including applicants (EFSA 2011).

One of the limitations of the comprehensive database is that due to the different methodologies used to collect data at national level, country-to-country comparisons of exposure estimates should be interpreted with caution. In 2011, EFSA launched its EU Menu project that aims to provide more standardised and higher quality data on food consumption using FoodEx2 food classification and description across the EU, according to the EU Menu methodology (EFSA 2014b). EFSA provides guidance as well as financial support to participating countries. Some 30 surveys covering different population groups (e.g. children and adults) are ongoing in the EU Menu project.

EFSA aims to gradually update its Comprehensive database with national surveys conducted according to the EU Menu project methodology.

On an international level, EFSA collaborates with the WHO (World Health Organisation) and the FAO (Food and Agricultural Organisation of the United Nations) to support efforts to harmonize food consumption data in the Codex region. In particular, EFSA provides support to extend the use of FoodEx2 food classification to cover food consumption preferences outside of Europe, in the context of the FAO/WHO Global Individual Food Consumption Data Tool (FAO/WHO GIFT).[8] Summary statistics from the EFSA comprehensive database have also been included in the FAO/WHO Chronic Individual Food Consumption Database - summary statistics (CIFOCOss)[9] hosted by the WHO.

Chemical Occurrence Databases

EFSA collects some one million analytical records on contaminants and food additives annually from data providers (public and private) in European countries through calls for data,[10] as well as some 20 million analytical results on pesticide residues in food. The majority of the data are from monitoring programmes. In order to standardise data on contaminants and pesticide residues transmitted to EFSA and to improve the data quality, EFSA developed the Standard Sample Description (SSD) (EFSA 2010a, 2013) in collaboration with Member States, which is the EFSA standard for transmission of analytical occurrence data to EFSA.

---

[8]FAO/WHO Global Individual Food consumption data Tool (FAO/WHO GIFT) http://www.fao.org/nutrition/assessment/food-consumption-database/en/

[9]The FAO/WHO Chronic individual food consumption database - Summary statistics (CIFOCOss) http://www.who.int/foodsafety/databases/en/

[10]EFSA's calls for data https://www.efsa.europa.eu/en/calls/data

The SSD data model contains approximately 80 standardised data elements (fields) that describe the characteristics of an analytical sample and result (e.g. laboratory sample code, analytical method, limit of detection of the analytical method, country of origin, etc.), of which approximately 20 are mandatory. In addition, it contains controlled terminologies (e.g. standard lists of analytical methods, names of chemicals, etc.) and business (validation) rules (e.g. whether data pertaining to data elements have been submitted in the required format) to ensure baseline data quality. The IT protocol to transmit data to EFSA using the SSD standard data model is described in a complementary guidance document, Guidance on Data Exchange (EFSA 2014c).

Data providers transmit data through the EFSA Data Collection Framework (DCF) web-based interface that provides functionalities of automatic validation of the incoming data. The system automatically checks for the correct completion of mandatory fields and compliance with business rules, after which data providers receive automatic feedback. After an ETL (Extract, Transform, Load) procedure, data are stored in EFSA's scientific data warehouse and later extracted for use in dietary exposure assessments. Data stored in EFSA's scientific data warehouse are publically accessible via EFSA's website at different levels of aggregation to third parties according to EFSA's data warehouse access rules (EFSA 2015a). An overview of data providers is also provided.[11]

Since 2007, EFSA awarded some 4 million euros to Member States organisations involved in data collection to support data standardisation at the European level as well as electronic transmission of data to EFSA across the following domains: zoonoses, contaminants, food additives, pesticides residues and veterinary drug residues.

The added value for Member State data providers is that occurrence data from several food safety domains can be reported to EFSA in a standardised format. Consequently, the SSD is becoming the accepted European standard for describing and reporting monitoring results for food-borne hazards. An additional benefit at the national level is that Member States accumulate a large volume of data from their national control and monitoring activities in a harmonised format that can also be used to support risk assessment activities at national level.

To avoid duplication of effort by European data providers, EFSA transmits contaminant occurrence data to the WHO GEMS/Food (Global Environment Monitoring System) for use in risk assessments of the joint FAO/WHO Expert Committee on Food Additives (JECFA).

As EFSA's SSD data model and the WHO GEMS/Food model to receive data for risk assessment purposes are not entirely aligned, some adaptation of EFSA data from SSD to GEMS/Food was required to transmit raw data from EFSA to WHO. In 2014, EFSA compared the two models and provided technical information regarding the transmission of contaminant occurrence data from the EFSA database into the electronic data submission system of WHO (EFSA 2015b).

[11]EFSA's data collection, standardisation and analysis https://www.efsa.europa.eu/en/science/data

7.1.2 *EFSA's Approaches to Dietary Exposure Assessment of Regulated Products and Contaminants*

This chapter briefly describes the principles of dietary exposure assessment applied by EFSA for regulated products and contaminants and gives some examples of exposure assessment approaches. In general, EFSA provides guidance to applicants/industry on the kind of scientific data and information needed to perform a risk assessment. These guidance documents include advice on how exposure shall be calculated.[12] It should however be borne in mind that science evolves and better/more precise data and methodologies may become available. In this case the EFSA guidance documents are updated and published.

The exposure estimates are performed by the working groups of the respective scientific panels and/or by EFSA staff. The panels are responsible for and officially adopt the final risk assessments, including the underlying exposures assessments.

Food Additives

EFSA has developed a screening tool for estimating long-term exposure to food additives. This is the Food Additives Intake Model (FAIM) (EFSA 2014a, see also Sect. 4.17). It allows applicants and EFSA experts to estimate mean and high exposure to food additives for different population groups throughout several European countries using the summary statistics from EFSA's Comprehensive Database.

In the case of an application for a new food additive or an extension of use for an already approved food additive, the applicant is requested to provide information on the proposed uses and use levels in foods in the data package submitted for the safety evaluation to EFSA (2012b). This information can be inserted in FAIM to estimate dietary exposure.

In the context of the re-evaluation programme for food additives permitted for use before 2009 and currently on the EU market, a stepwise approach to estimating exposure is followed. Specific maximum permitted levels (MPLs) have been established for most food additives in the respective EU legislation. Some of them can, however, be used according to the quantum satis (QS) principle - at the lowest level necessary to achieve the desired technological effect. In a first step exposure is estimated by combining these MPLs with the food consumption data. This may result in a very conservative exposure scenario because it assumes that an individual is eating all the foodstuffs that can contain a given food additive at its MPL. Such overestimation often leads to exceedance of the ADI (Acceptable Daily Intake) for the respective food additive. In order to obtain more realistic use data EFSA regularly publishes calls for data[13] for each food additive under re-evaluation to

[12]Submitting an application to EFSA: http://www.efsa.europa.eu/en/applications/about/submitting

[13]EFSA's calls for data https://www.efsa.europa.eu/en/calls/data

get information from interested business operators (e.g. food industry) to support the exposure assessments. Typical use levels from industry together with analytical data provided by the Member State competent authorities are combined with the food consumption data from the EFSA comprehensive database to estimate a more refined exposure scenario.

Flavourings

Chemically defined flavouring substances were assessed for previous risk assessments via flavouring group evaluations (FGEs) within the so-called 'Procedure'. Dietary exposure was based on: (1) poundage data (production volumes of a flavouring) reported by industry using the Maximised Survey-Derived Daily Intake (MSDI) method and (2) a combination of standard portion sizes with rough usage levels reported by industry using the modified Theoretical Added Maximum Daily Intake (mTAMDI) method. Both methods suffer from limitations such as the year-to-year variability in production volumes, the underreporting correction factor and the assumed percentage of consumers. Whereas the MSDI model tends to underestimate the intake of a flavouring substance for certain groups of consumers, the mTAMDI is regarded as a conservative estimate because it is based on the assumption that the consumer regularly eats and drinks several food products containing the same flavouring substance. However, with the modification to consider the normal instead of the upper reported use levels, it is less conservative than the TAMDI. As a consequence, there have been differences of several orders of magnitude in results between mTAMDI and MSDI for some of the FGEs (Arcella and Leclercq 2005). The mTAMDI value was only used as a tool to screen and prioritise the flavouring substances according to the need for refined intake data and was not considered in the 'Procedure'.

The CEF Panel has subsequently established a new exposure approach called Added Portions Exposure Technique (APET) method in its guidance document from 2010 (EFSA 2010b) which is applicable to all new flavourings, both chemically defined and other flavourings for which an evaluation and authorisation after 2010 has been sought. This methodology accommodates the need for harmonised exposure estimation with existing exposure assessment procedures for flavourings such as the Single Portion Exposure Technique (SPET) used by JECFA. The APET method is an adaptation of the TAMDI method with the main difference being use of standard portions established for each food category. Short-term exposure was in the past and is also now addressed. Novelties of the method are the estimation of the potential cumulative dietary exposure by taking into account the intake of structurally and metabolically related substances, the consideration of natural occurrence in processed and non-processed foods, as well as exposure from non-food sources. Specific scenarios are calculated for children in order to attribute their higher food consumption with respect to body weight.

Smoke Flavourings

Smoke flavourings are both legally and from a scientific point of view considered different from flavourings. They are mixtures of many chemical compounds and their anticipated uses as well as the use levels differ from those proposed for chemically defined flavouring substances. Therefore, as no specific methodology for the estimation of dietary exposure to Smoke Flavourings existed, the CEF Panel decided to test existing methodologies and develop new methodologies for this special type of flavourings. The details are described in the CEF opinion on Dietary exposure assessment methods for smoke flavouring Primary Products (EFSA 2009a).

The Smoke Flavouring Theoretical Added Maximum Daily Intake (SMK-TAMDI) assumes that the hypothetical consumer will consume a fixed amount of flavoured solid foods and liquids daily. The unique approach here is that a single group 'Beverages' is used for liquids whereas solid foods are divided in 'traditionally smoked solid foods' and 'other solid foods not traditionally smoked'.

The Smoke Flavouring EPIC (SMK-EPIC) model makes use of the information on the consumption of smoked foods available from the European Prospective Investigation into Cancer and Nutrition (EPIC) study. This study is one of the few cases in which the consumption levels of 'smoked meat' were assessed for different European countries. It is based on assumptions that a hypothetical high-level consumer of smoked meat is also an average consumer of the other traditionally smoked foods and an occasional consumer of smoked foods or beverages from each of the other categories.

To date the CEF Panel has evaluated 11 smoke flavourings and based its exposure assessments on Upper Use Levels as provided by the applicants, using the SMK-TAMDI and SMK-EPIC methods. The highest value among these exposure estimates was used when carrying out risk assessments to these products.

Food Enzymes

The CEF Panel had suggested in its guidance (EFSA 2009b) that a conservative method such as the 'Budget method' should be used to estimate potential human exposure to food enzymes and any other constituents or by-products of concern. The panel has taken note when applying the methodology that enzymes are added during processing of food and food ingredients.

Based on the initial experience the CEF Panel has proposed a tiered approach in line with the EFSA advice to use 'a stepwise or tiered approach in which the initial steps rely on conservative screening methods'. Instead the panel recommends the use of actual food consumption data reported in the EFSA comprehensive food consumption database, and recommends the development of a food process-based tool, based on summary statistics (EFSA 2016b).

Food Contact Materials

In contrast to chemicals that are directly added to and are present (intentionally or not) in food, starting materials or additives of plastic materials can unintentionally end up in food by migration. The magnitude of the potential so-called specific migration depends on a multitude of factors such as the nature of the packaging material, the nature of the food (e.g. dry, acid, fatty, or liquid), and the temperature and time of contact of the plastic material with food. Actual occurrence data of chemicals derived from food contact materials are very scarce as it is very cumbersome and resource intensive to measure these chemicals in a multitude of different complex food types. Therefore, several approaches have been taken to provide data on migration as a basis for occurrence data. It is assumed that the food contact material releases a substance into a food at the highest concentration level measured either in food simulants (for aqueous, fatty, alcoholic or acid food types), or calculated using worst-case scenarios (i.e. total mass transfer) or mathematical migration modelling. Dietary exposure to food contact materials is estimated on the basis of a model diet, which assumes that a person weighing 60 kg may consume 1 kg daily of food (including beverages) that is in contact with 6 dm^2 (surface of a 10 cm^3) of food contact material containing the chemical under assessment. For fatty foods, which are important for the migration of highly polar organic substances, it is assumed that a person consumes only up to 200 g of fat daily (Fat reduction factor).

Such an approach is necessary as in the past knowledge of the type of packaging for the different foodstuffs was neither systematically available nor collected. The EFSA comprehensive database, however, contains some national surveys for which more general information on the type of packaging is available. EFSA has also updated its FoodEx2 format in order to incorporate a facet for food packaging that will allow a better estimation of exposure from certain food contact materials. The issue is not trivial if one imagines that the final food (and its ingredients) may have been in contact during manufacture, storage, transport, etc., several times and with several different packaging materials. Also refined models will not be able to deal with such complex scenarios.

The CEF Panel has started to revise its guidance on food contact materials and identified, as one major area to revisit, the estimation of consumer exposure. The standard value of 6 dm^2/kg is appropriate to represent the surface to mass ratio of packaged foodstuffs based on consideration of several scientific publications where surface to volume of packed food was analysed. The Panel had different suggestions to replace the standard value with more refined scenarios. This proposal will provide the European Commission (EC) with the scientific basis for a discussion among risk managers on possible implications for risk management. It is intended that the European Commission will provide feedback to EFSA to prepare updated guidelines for data requirements (including exposure assessment) for the safety assessment of a substance to be used in food contact materials (EFSA 2016c).

Feed Additives

Feed additives may end up in the human diet indirectly through the consumption of tissue from mammals, birds or farmed fish, and via products, such as milk and eggs. The estimation of dietary exposure considers the additives (expressed as total relevant residues) in combination with a model diet for humans using default values for daily food consumption based on worst-case scenarios, e.g. daily consumption of 100 g of liver from mammals and/or birds (EFSA 2012a).

EFSA has recently published the Guidance for public consultation on the assessment of the safety of feed additives (EFSA 2017b). It proposes use of consumption data of edible tissues and products as derived from the EFSA Comprehensive European Food Consumption Database to assess exposure to residues from the use of feed additives in different EU countries, age classes and special population groups. A web-based tool will be made available by EFSA, supporting assessors in the calculation of long- and short-term exposure estimates according to the above methodology.

Genetically Modified (GMs) Foods

In the European Union (EU), foods derived from genetically modified organisms (GMOs) are subjected to a risk assessment before authorisation. Dietary exposure is one essential element of the risk assessment of GM foods.

Commission Implementing Regulation (EU) No. 503/2013 (EU 2013) requests that the anticipated dietary intake of genetically modified (GM) food is estimated by an applicant on the basis of representative consumption data, and that it should consider also particular consumer groups. Data on import and production quantities may provide additional information for the intake assessment.

A crude estimate of dietary exposure is used during hazard identification and characterisation to support the choice of dose regimes. If a hazard is identified, then a more refined dietary exposure is necessary to determine the associated risk. A refined exposure assessment is also needed to assess the consequences of consumption of GM foods with altered nutritional profiles. The exposure of a hazardous substance is based on estimates of the observed consumption of all foods that may contain that substance, preferably in representative samples of the population. This is followed by multiplication of the amounts of food consumed by the known or predicted concentrations of the substance in each food, to arrive at the total daily exposure to that substance by members of a defined population group. These data are then compared with toxicological data to assess risk. The outcome of the compositional analysis decides whether a detailed nutritional assessment is required for specific nutrient(s). The nutritional assessment of GM foods is intended to demonstrate that the GM food is not nutritionally disadvantageous to humans. This can be as a consequence of new constituents, or constituents that occur naturally in the concerned crop but at altered levels as a consequence of the genetic modification.

Novel Foods

Long-term dietary exposure is typically assessed for novel foods. Food consumption data from the literature (e.g. figures published from national dietary surveys) are multiplied by maximum intended use levels of the novel food and food ingredient to estimate dietary exposure. In practice, food consumption data from one Member State (e.g. UK National Diet and Nutrition Survey) is mainly used as it is open data. When necessary, EFSA validates applicants' exposure assessments submitted in the frame of an application by using data from the comprehensive database. The FAIM tool may be used by applicants for the intake assessment of novel foods used as ingredients where the food categories to which the ingredient is intended to be added match reasonably well the food categories included in the FAIM tool. It may be sufficient for the intake assessment if the resulting high percentile intakes of the concerned population groups are below the levels that may pose safety concern. In cases where more refined estimates are needed, the applicant may consider more detailed assessments (EFSA 2016d).

Nutrients Sources Added to Food

Applicants estimate the dietary exposure to nutrients added to food by summing the highest reported nutrient use level (per day) with average and high (based on the 95th or 97.5th percentile) dietary intakes derived from the scientific literature. Estimates are calculated for adults and children.

Other Substances Added to Food (e.g. Botanicals)

Dietary exposure to other substances added to food (e.g. botanicals and isoflavones) is estimated on a case-by-case basis. EFSA supports the ANS Panel working group on botanicals in food regarding ad hoc data collection on food supplement composition and use patterns. This information is usually complemented by a summary of the relevant literature and used for the exposure assessment.

Pesticides

While risk assessments and authorisation for active substances were harmonised in the early 1990s at the European level with the now repealed Council Directive 91/414/EEC (EU 1991), safety assessments and approvals of plant protection products (also referred to as 'pesticides') containing active substance(s) are still within the remit of the European Member States, albeit respecting the harmonised assessment principles for plant protection products established in Regulation (EC) No. 1107/2009 (EU 2009).

Within the risk assessments for humans, exposure to active substances is assessed not only via diet, but also for operators, workers, residents and bystanders exposed via other routes, e.g. dermal and inhalative (EFSA 2014d). It is a major objective of the editors of this book to address the impact of assessing the total exposure accumulated from all sources combined. This issue is particularly addressed in various chapters (Sects. 3.2, 3.5 and 3.6 (consumer exposure exposome)). This kind of exposure assessment has been addressed in Sect. 7.8 and here the focus will be on dietary exposure assessment.

In the context of risk assessment, prospective assessments for short- and long-term exposure are performed for many kinds of applications (new products, amendment of uses and renewal of authorisation) of active substances. For this purpose, EFSA developed a calculation model, PRIMo (Pesticide Residue Intake Model) that is based on national food consumption figures and unit weights provided by Member States and implements internationally agreed risk assessment methodologies. Revision 2 of the model allows users to perform the risk assessment in accordance with the internationally agreed methodologies, with some specific European deviations regarding the variability factors for acute risk assessment.[14] PRIMo revision 2 can also be used as a first 'screening tool' based on conservative assumptions using MRL values (EFSA 2007). If no consumer risk is identified with the screening methodology, it is not necessary to use more sophisticated calculation tools. For 'refined calculations', the MRLs can be replaced by other values such as Supervised Trials Median Residue (STMRs). Further refinements for the specific active substance include Highest Residue (HR-values) obtained from field trials or monitoring programmes and the consideration of processing factors.

As requested in Regulation (EC) No. 396/2005 (EU 2005), EFSA also prepares annual reports[15] on pesticide residues from the official control activities for food and feed carried out by the European Member States, Iceland and Norway. In addition to the analysis of MRL compliance, EFSA carries out short- and long-term dietary exposure assessments based on the actual monitoring data using the deterministic screening tool PRIMo. Through this screening EFSA estimates the exposure of European consumers to pesticide residues. Taking into account the recommendations presented in the report on pesticide residues, the European Commission, in its risk manager role, takes appropriate risk management actions, complementing the risk management actions implemented at the approval of a pesticide.

According to Regulation (EC) No. 1107/2009 (EU 2009), the known cumulative and synergistic effects shall be taken into account in the risk assessment. EFSA has proposed methodologies to address the potential effects from combined exposure to pesticides that share similar toxicological properties. The PPR Panel has adopted four scientific opinions and one guidance document on the use of a probabilistic

[14]PRIMo - Pesticide Residue Intake Model. http://www.efsa.europa.eu/applications/pesticides/tools

[15]European Union reports on pesticide residues. http://www.efsa.europa.eu/en/publications/advanced-search/?sub_subject=70065

methodology for modelling cumulative dietary exposure to pesticide residues in order to develop and apply a science-based methodology in this area. In its 2010 annual report on pesticide residues in food, EFSA has for the first time estimated the combined exposure for two groups of pesticides on the basis of monitoring data provided by all Member States (EFSA 2010c). Since the work on the establishment of cumulative assessment groups (CAGs) and the assessment methodology had not yet been finalised, this exercise was aimed to testing of the suitability of the monitoring data for this type of assessment. The PPR Panel also tested the cumulative risk assessment using selected pesticides from the Triazole Group. The scientific opinion describes the progressive steps of refinement in cumulative risk assessment. The refinement of the cumulative exposure assessment making use of deterministic and probabilistic methodologies in successive tiers is one part. The most important step in applying this methodology will however be the development of the cumulative assessment groups, which is still in progress. The development of tools to perform cumulative exposure calculations according to the methodology proposed by the PPR Panel was also covered by a project funded by the European Commission (ACROPOLIS project) using MCRA (Monte Carlo Risk Assessment) software.

Contaminants

Daily dietary intakes (average and high levels) of contaminants in food for the total population are estimated by combining individual food consumption data from the EFSA Comprehensive Food Consumption Database with contaminant occurrence data mainly from national monitoring and surveillance programmes. Data on contaminant occurrence in food are matched with the related consumption data using the FoodEx food classification. In the absence of monitoring data from Member States, literature data are typically used when available. Long-term dietary exposure is typically assessed and, when a toxicological evaluation indicates a need, acute dietary exposure is estimated by means of deterministic or probabilistic methods. Since 2010, all dietary exposure estimates in this area are carried out using the Comprehensive Database and Contaminant Occurrence Database.

Specific Examples

Aggregate exposure [exposure to a single chemical via all exposure routes (e.g. dermal, oral, and inhalation) and from different sources] has not been widely applied in EFSA Panel opinions to date. Two recent examples include the CEF Panel opinion on BPA (Bisphenol A) (EFSA 2015b) and the Scientific Committee opinion on Carvone (EFSA 2014e). The former included an estimate of combined exposure from the diet, drinking water, inhalation, as well as dermal contact to cosmetics and thermal paper; the latter included an estimate of combined exposure from pesticides, flavourings, feed as well as personal care products. In both cases, the exposure assessments were carried out using individual food consumption data from the EFSA's Comprehensive Database.

Some of EFSA's guidance documents also suggest providing information on known oral exposure from non-food sources (cosmetics, medicines, toys, etc.). However this information is only qualitatively included in scientific opinions, e.g. flavourings and food additives, and not taken into account in an overall exposure assessment from all source and all routes. Although this might be desirable from a scientific point of view, such assessments are very resource demanding and often are not necessary when it can be anticipated that exposure sources other than diet are not contributing much to the overall burden.

A summary of EFSA's approaches to exposure assessment is presented in Table 7.1.

7.2 ECHA

7.2.1 Introduction

Andreas Ahrens

The European Chemicals Agency (ECHA) in Helsinki was set up in 2007 in order to manage the implementation of the new European Chemicals legislation on *R*egistration, *E*valuation, *A*uthorisation and restriction of *CH*emicals (REACH). As of January 2018 about 560 employees from 25 EU Member States are working at the Agency.

REACH has introduced a number of new paradigms for regulating chemicals. Implementation of the paradigms is expected to significantly increase the safety of products and processes regarding risk from chemicals. The emphasis is on generation, dissemination and use of knowledge under the responsibility of industry, based on a harmonised level playing field across Europe. Where generation of more knowledge and risk management under responsibility of industry cannot work or does not sufficiently work in practice, REACH foresees complementary mechanisms for regulatory risk management.

It is ECHA's task to technically support the new system and its core processes. This *secretariat* function also includes the power to make decisions together with the Member States. To help with better understanding of ECHA's role in connection to consumer exposure assessment under REACH, this chapter will reflect on the main regulatory innovations, explain the key REACH processes, and provide a short overview of the different committees working with ECHA.

Table 7.1 EFSA's approaches to exposure assessment

Type of substance	Type of evaluation, if regulated product	Type of exposure	Exposure method/ model	Food consumption source	Occurrence data/use levels source	Exposure assessed for different age groups (Yes/No)	High consumers considered (Yes/No)
Contaminants	Not applicable	Acute and Chronic	Individual based model	Food consumption data from EFSA comprehensive database	Summary statistics from national monitoring and surveillance data	Yes	Yes
Enzymes	New applications	Chronic (typically)	FEIM	Summary statistics from the comprehensive database	Recommended, typical or maximum food enzyme use levels expressed as total organic solids per kg of raw material corrected by conversion factors	Yes	Yes
Feed additives	Re-evaluation	Chronic	Model diet	Theoretical diet	Summary statistics from national monitoring and surveillance data	Yes	Yes
Flavourings	Re-evaluation	Chronic	MSDI	None	poundage data	No	No
			mTAMDI	Model diet	Normal use levels	No	No
	New applications	Chronic and acute	APET	Model diet based on standard portions	Proposed use levels (added) + normal occurrence levels (if occurring as natural constituent)	Yes	No
Smoke flavourings	New applications	Chronic	SMK-TAMDI	Model diet	Normal use levels	No	Yes
	New applications	Chronic	SMK-EPIC	EPIC study	Normal use levels	No	Yes
Food additives	New applications	Chronic	FAIM	Summary statistics from the comprehensive database	Intended use levels	Yes	Yes
	Re-evaluation	Chronic	Individual based model	Food consumption data from EFSA comprehensive database	Summary statistics from national monitoring and surveillance data and/or	Yes	Yes

					reported use levels from industry		
Food contact materials	New applications	Chronic	Model diet	Theoretical diet (1 kg of food in contact with 6 dm^2) **or cm^2**	Maximum permitted migration level or data from targeted analytical projects	No	Yes
Genetically modified organisms (GMOs)	New compounds (e.g. newly expressed proteins) or altered endogenous compounds in GM food and feed	Acute and chronic	Case by case	Summary statistics from the literature (e.g. GEMS/food, FAO food balance sheet, EFSA comprehensive database	Provided by the applicant	Yes	Yes
Novel foods and ingredients	New applications	Chronic (typically)	Case by case	Summary statistics from the EFSA comprehensive database, or FAIM template; individual data from national food consumption surveys	Proposed use levels provided by the applicant	Yes	Yes
Nutrient sources	New applications	Chronic	Poundage method	Summary statistics from the literature	Highest supply use	Yes	Yes
Pesticides	New applications	Chronic (Tier 1)	TMDI	Summary statistics from national dietary surveys (PRIMo)	Maximum Residue Levels	Yes	No
		Chronic (Tier 2)	IEDI	Summary statistics from national dietary surveys (PRIMo)	Summary statistics from supervised trials or national monitoring and surveillance data	Yes	No
		Acute	IESTI	Summary statistics from national dietary surveys (PRIMo)	Summary statistics from supervised trials or national monitoring and surveillance data	Yes	Yes

7.2.2 What Is New?

REACH introduced a number of new principles and paradigms in regulating chemicals:

- The new rules apply to all substances manufactured or imported in quantities of one tonne or more, independent of when they had been placed on the market for the first time.
- Manufacturers and importers are obliged to register their substances with an obligatory set of data to characterise the intrinsic properties and the related hazards.
- REACH requires cooperation among registrants. Each registrant has to characterise the identity of his substance (including its composition), and all registrants of the same substance shall exchange their information on the substance, and submit a joint data set. Testing is required when information is lacking. For vertebrate testing REACH foresees that authorities evaluate industries' testing proposals beforehand.
- Exposure and risk assessment are required when hazards are identified.
- The responsibility to identify the existing uses of a substance and to demonstrate how the substance can be used safely is placed on industry (as the registrant's Chemical Safety Assessment). The safety assessment shall cover the whole life cycle of the substance from manufacture, the formulation into mixtures, and the use of mixtures down to the use of the articles into which the substance has been incorporated. The assessment is to be carried out for each use. In addition, the registrant should assess the cumulative exposure from the different uses of a substance.
- Suppliers are obliged to include use-specific safety advice into their safety data sheets (exposure scenarios), and the recipients are obliged to respond, i.e. follow the advice or carry out their own safety assessment. The risk management advice may address technical measures and personal protection equipment for workers, measures to control environmental releases from industrial sites, and advice for the design of consumer products (e.g. concentration of substance and the size and form of the package).
- The non-confidential information from the registration dossiers, as submitted by the registrants, is automatically disseminated via ECHA's website, in order to provide public access to information about the intrinsic hazards of the substances placed on the EU market, and their use patterns.

To make the registration mechanisms and the subsequent communication in the supply chain work under the responsibility of industry, REACH foresees a number of processes run by the authorities such as checking the completeness and compliance of registration dossiers (by ECHA).

As a second pillar of REACH, ECHA and Member States together set up a rolling action programme for evaluating substances (by Member States) that may require regulatory risk management at the community level (CoRAP). The selection of substances for CoRAP is risk-based, i.e. hazard and exposure concerns play a role.

The pre-existing regulatory risk management by harmonised classification and labelling and restrictions has been maintained under REACH.

One of the main novelties under REACH is a particular regime for substances of very high concern (SVHC). These are substances causing serious long-term effects, and in particular those where no safe level of exposure can be established, or where such level is very uncertain. Typical examples are genotoxic substances and substances toxic to reproduction and/or development, or substances persisting and bioaccumulating in the environment. In a first step, based on a recommendation by ECHA, substances with confirmed properties of very high concern are placed on the candidate list of substances for authorisation. In a second step, the substances from this list are selected for authorisation, which means that from a certain point in time (sunset date), only authorised uses are permitted, and this is only for a limited time span.

If such substances are contained in articles, such as textiles, electronic goods, toys, and furniture, special rules apply for notification to authorities and for communication to consumers. The rules are particularly relevant to substances in imported articles, as these are not covered by the REACH registration regime.

7.2.3 REACH Processes, Consumer Exposure Assessment, ECHA's Role

Implementation of these paradigms takes place through a number of REACH processes managed by ECHA and its Committees. The following section provides an overview on a few key processes.

Registration

For substances having production amounts greater than 10 tonnes per year, registrants submit a technical dossier and attach a Chemicals Safety Report (CSR) documenting the registrant's safety assessment. This includes classification and derivation of no-effect levels for the various toxicological endpoints, and determination of chemical–physical properties that drive the behaviour of the substance when used, when treated in waste water/air, or after release into the environment. Behaviour in the environment is of particular relevance for predicting indirect exposure of the population through the food chain. Where hazards have been identified, the CSR includes a description of the conditions of use, corresponding exposure estimates and a comparison with the derived no-effect levels. Registrants are requested to demonstrate how the substance should be used in products and processes, so that the expected exposure remains below the effect level. Based on this assessment, registrants communicate the conditions of safe use in the form of exposure scenarios to their customers, for example the producers of mixtures or articles for consumers.

ECHA provides guidance and tools to registrants to collect, assess, document and submit this information. The guidance and tools have been developed in consultation with industry and Member States. The key tools are REACH IT (system to run the registration business), IUCLID (tool for the technical dossier, i.e. the substance data set) and CHESAR (a tool to carry out and report the chemicals safety assessment based on the substance data set stored in IUCLID). In the 2016 update of IUCLID, much emphasis was put into increasing the transparency of the relationship between substance compositions/form, the tested material, and the outcome in terms of toxicological characterisation. Thus, the tools for data submission provided by ECHA impact the transparency and consistency of assessments.

ECHA checks the completeness of the incoming dossiers in an automated or semi-automated way. Incomplete dossiers are stopped and the registrant gets the opportunity to complete the dossier.

In order to avoid too many dossiers getting blocked and to support a minimum quality of the registration dossier in terms of consistency and plausibility, IUCLID is equipped with a validation assistant to help the registrant avoid obvious and unintended shortcomings of their dossier.

Registrants need to decide whether they want to support the use of their substance in mixtures and/or articles for consumers. If not, they do not include uses by consumers in the corresponding IUCLID section of their registration, and might explicitly advise against consumer uses in the safety data sheet. For supported consumer uses, the registrants must determine the safe concentration of the substance in the different consumer products and a safe amount per use event. They may also advise controlling exposure via package design, such as dissolvable packaging for portions of dishwashers detergents or very small amounts per package for hazardous adhesives.

Dissemination

Most of the information in the Technical Dossier is automatically disseminated via ECHA's website so that there is online access for the public and others to the information on (1) the properties of the substance, including classification, labelling, derived no-effect levels (for human health) and predicted no-effect levels (for the environment), and (2) its use pattern(s). Only information that is by default confidential (e.g. details on substance composition and on manufacturing and use processes), or that has explicitly been claimed confidential (including a justification) is not disseminated.

Also, the conditions of safe use, determined in the Chemical Safety Assessment, are accessible via ECHA's website, but only to the extent registrants have transferred this information voluntarily from their CSR into the IUCLID dossier. ECHA's safety assessment tool for registrants, CHESAR, is particularly equipped to transfer the results of the exposure assessment (the conditions of safe use, exposure estimates per activity, and corresponding risk characterisation) into the IUCLID dossier.

As a result of dissemination, it is for example possible to identify all substances that are reported to be used in mixtures or articles for consumers, and to identify the types of mixtures and articles in each case. The granularity and accuracy of the information differs a lot from registrant to registrant as there are no definitive legal requirements about how the uses of a substance are to be described.

The information disseminated via the ECHA website is not checked by authorities beforehand, and the current quality is quite diverse across the registrants. It is expected that registrants will improve the information depending on how much industry itself, NGOs, academia/ media, and the general public reviews and works with the information.

Dossier Compliance Check

ECHA is obliged to evaluate the compliance of the registration dossiers. This REACH process aims to ensure that the registrant community as a whole fulfils their duties, rather than checking correctness and accuracy of each single dossier and each single information requirement. In a compliance check, ECHA particularly concentrates on the substance identity (including the substance composition) and the information endpoints related to the identification of the presence or absence of major hazards. Examples include genetic toxicity, toxicity to reproduction or devel opment, repeated dose toxicity, aquatic toxicity, bioaccumulation, and environmental degradability.

The checks on the substance identity profile are particularly important for ensuring that the composition of the registered substance is clearly defined (including its impurities), and that the available test data and the corresponding hazard characterisation match this composition.

Uses, conditions of use, and exposure play a particular role in compliance checks when the registrant has adapted information requirements based on exposure considerations (i.e. waives required tests), or when determining whether an additional dermal or inhalation study would be required.

Substance Evaluation

Based on the community rolling action program (CoRAP), Member States evaluate substances when there is a concern of controlled risks at a community level. For example, this could be the case where significantly hazardous substances are reported to be used in consumer products (information from REACH registration database or from other sources). As of October 21, 2019 the CoRAP list contains 375 substances.

Classification and Labelling

Member States may also propose a harmonised classification of a substance, based on the criteria laid down in the CLP regulation. This proposal is then examined by ECHA's Risk Assessment Committee (RAC). Any classification as a CMR has by default the consequence that the substance is banned for use in mixtures for consumers.

Determine Candidates for Authorisation

Based on the intrinsic properties of the registered substances, ECHA or Member States make proposals for identifying substances as being of very high concern (SVHC), for example because they are genotoxic carcinogens, mutagens, toxic to reproduction or development, or because they persist and accumulate in the food chains. The high concern relates to the seriousness of the effect and the low dose at which it occurs, or to the fact that a safe level cannot be determined. The proposals are subject to a public consultation. Substances fulfilling the criteria are then included in the candidate list for authorisation. By October 21, 2019 two hundred and one substances are listed as candidates.

As soon as a substance has entered into this candidate list, all (consumer) articles containing this substance need to be notified to ECHA (unless already registered for that use in the EU), and consumers may ask the companies they purchase the article from for advice on safe use. Producers or importers of articles being confronted with such duties usually try to find less hazardous alternatives for the substance.

Inclusion of Substances into the Authorisation List

Once a year, ECHA recommends a number of substances from the candidate list to be included into the list of substances for authorisations. Priority is given to substances in widely dispersed usage and/or placed on market in high volumes, and/or being PBTs/vPVBs. The proposal is again subject to public consultation. By October 21, 2019 forty-three substances have been placed on the authorisation list. Once a substance has been included in the authorisation list, companies that want to continue its use after a certain sunset date have to apply for an authorisation with ECHA. Based on a safety assessment, applicants will explain how exposure is minimised on use and which risk(s) remain. For many substances (those without safe level), the applicant also needs to demonstrate that no safer alternative exists at present.

Restriction

Restrictions are a tool to protect human health and the environment from unacceptable risks posed by chemicals. Restrictions may limit or ban the manufacture, placing on the market, or use(s) of a substance. A restriction applies to any substance on its own, in a mixture or in an article, including those that do not require registration. It can also apply to imports.

A Member State, or ECHA on request of the European Commission, can propose restrictions if they find that the risks need to be addressed on a European Union-wide basis. ECHA can also propose a restriction on articles containing substances that are on the Authorisation list (Annex XIV). Anyone can comment on a proposal to restrict a substance.

Restrictions play a particular role for consumer exposure via articles. Where articles are manufactured outside the community, restrictions may prevent certain articles entering the community when they contain certain hazardous substances and using the articles may cause adverse health or environmental effects.

7.2.4 Committees

The above processes rely on the work of a number of committees associated to ECHA.

Committee for Risk Assessment

The Committee for Risk Assessment (RAC) prepares the opinions of ECHA related to the risks of substances to human health and the environment in the following REACH and CLP processes. The final decisions are taken by the European Commission.

RAC examines the proposals for harmonised classification and labelling and gives an opinion on the proposed harmonised classification of substances as carcinogenic, mutagenic, toxic for reproduction, or as a respiratory sensitiser, as well as other effects on a case-by-case basis.

The Committee also evaluates whether a proposed restriction on manufacture, placing on the market, or use of a substance is appropriate in reducing the risk to human health and the environment. This includes the assessment of comments submitted by third parties.

Finally, the RAC assesses the risk of a substance arising from the uses of a substance when an application for authorisation is submitted. This includes an assessment of the appropriateness and effectiveness of the risk management measures as described in the authorisation application, and if relevant, the risks of possible alternatives. Third party contributions linked to the application will also be assessed.

Member State Committee

The Member State Committee (MSC) participates in several REACH processes such as evaluation and authorisation. The MSC is responsible for resolving divergences of opinions among Member States and on proposals for the identification of SVHCs. The Committee provides opinions on ECHA's draft recommendation for the authorisation list (Annex XIV) and the draft Community Rolling Action Plan (CoRAP) for the substance evaluation process. If an agreement is not reached within the MSC, the matter is referred to the European Commission for decision-making.

When amendments related to ECHA draft decisions on testing proposals or dossier compliance checks are proposed by Member States, the MSC seeks unanimous agreement.

The MSC also seeks unanimous agreement on Member State draft decisions on substance evaluation when amendments are proposed to them by other Member States or ECHA. The MSC takes into account the comments of the registrants on the proposed amendments to the draft decisions based on dossier and substance evaluations. Once agreed by the MSC, ECHA finalises the decision and provides it to the registrant.

The MSC has an essential role in seeking agreement on the identification of Substances of Very High Concern (SVHCs). It also provides opinions on ECHA's draft recommendations on priority substances for inclusion in the authorisation list (Annex XIV).

ECHA takes into account the MSC opinion when finalising its recommendation to the Commission on priority substances for inclusion in the authorisation list.

Committee for Socio-economic Analysis

The Committee for Socio-economic Analysis (SEAC) prepares the opinions of ECHA related to the socio-economic impact of possible legislative actions on chemicals in the following REACH processes. The final decisions are taken by the European Commission.

It evaluates the socio-economic impact of the proposed restriction on manufacture, placing on the market or use of a substance. This includes the assessment of comments and socio-economic analyses submitted by third parties.

SEAC also assesses the socio-economic factors and the availability, suitability and technical feasibility of the alternatives associated with the uses of a substance when an application for authorisation is submitted. Third party contributions linked to the application will also be assessed.

The Forum

The Forum for Exchange of Information on Enforcement (Forum) is a network of the national authorities responsible for the enforcement of the REACH, CLP and PIC regulations.

The Forum sets its work programme based on the list of tasks specified in the REACH, CLP and PIC regulations. The practical work of the Forum is divided between ten working groups that provide solutions to specific areas of enforcement. The Forum holds three plenary meetings every year.

7.3 Activities of the European Commission: Joint Research Centre

7.3.1 The EIS-ChemRisks Toolbox[16]

Alexandre Zenié, Vittorio Reina, and Demosthenes Papameletiou

In 2003, DG JUST requested the JRC to develop a European information system on chemical risks from consumer products and articles (EIS-ChemRisks). This project aimed to collect information and provide tools and reference data to enable the design of harmonised exposure assessment procedures for consumer products and articles throughout the EU. The EIS-ChemRisks project was carried until 2009.

The work carried out led to an EU exposure assessment toolbox prototype, called EIS-ChemRisks that supports stakeholders (e.g. authorities, industries, research organisations, academia and NGOs) in the EU to carry out assessment of consumer

[16]Acknowledgement: Without the participation and the dedication of the following team members the EIS-ChemRisks toolbox would not have been possible: Pertti (Bert) Hakkinen, Dieter Schwela, Yuri Bruinen de Bruin, Carlos del Pozo, Irene Biavetti, Athanasios Arvanitis, Giorgios Giannopoulos, Majlinda Lahaniatis, Stylianos Kephalopoulos and Dimitris Kotzias. Particular thanks are expressed to Takis Daskaleros (DG SANCO).

Many thanks to the JRC colleagues from the European Chemicals Bureau (ECB), namely: Bert-Ove Lund, Jens Tørsløv and Frans M. Christensen.

Special thanks for the continuously involved experts, namely: Jacqueline van Engelen (RIVM), Gerhard Heinemeyer (BfR), Anne Catherine Viso (InVS), Lea Frimann Hansen (DEPA), Leif Bengtsson (KEMI), Matti Jantunen (KTL) and Carlos Rodriguez (P&G).

The following members of the IRAG sub-group have dedicated efforts on case studies carried out using the EIS-ChemRisks toolbox: Dirk van Aken [Food and Consumer Product Safety Authority, the Netherlands], Peter Hammer Sørensen [Environmental Protection Agency (DEPA), Denmark], Matthias Honnacker [Federal Institute for Occupational Safety and Health (BAuA), Germany], Janez Novak [Market Inspectorate, Slovenia], Corinne Mandin [Institut National de l'Environnement Industriel et des Risques (INERIS), France], Konstantinos Zisis [Ministry of Development, Greece] and Chris Van der Cruyssen [Federal Public Service Economy, Belgium] with the coordination of Jürgen Vogelgesang (DG SANCO). We are grateful to all of them.

exposure from the use of consumer products in a harmonised way. The EIS-ChemRisks toolbox prototype was used and tested by selected experts.

Concept

The main concept behind the European exposure assessment toolbox relies on scenario-based consumer exposure assessment over several rounds of stakeholders' dialogue. This aims to continuously improve existing consumer exposure scenarios as soon as new determinants, new algorithms or methods become available (Heinemeyer et al. 2006; van Engelen et al. 2007).

The adopted consumers' exposure assessment is conducted into six consecutive steps before storing the resulted scenario within the ExpoScenarios library:

1. Defining the general information including the source document, the disclaimer and the issue at stake
2. Setting the problem definition including the source, the chemical, the target population, the exposure event, the considered microenvironment, the risk management measures, the chemical release, the exposure pathways, the exposure routes and the building procedure on the basis of the guidance documents
3. Selecting the adequate mathematical algorithms evaluating the exposures in harmony with the concept
4. Selecting the reference data adequately by using chemical-specific determinants and exposure factors to be used within the selected algorithms
5. Evaluating the exposure results on the basis of previous steps
6. Highlighting commonalities with other exposure scenarios from the library in order to interpret the results.

Architecture

At the contrary of the linkages compendium approach adopted later on by the US-EPA's exposure assessment toolbox (EPA-Expo-Box) (Phillips et al. 2015), the EIS-ChemRisks toolbox integrated several databases and tools within the following six modules to act as a reference harmonised documenting system:

- ExpoData collecting data and information on chemical-specific exposure data (e.g. substance content in specific products and articles, related emission or migration data and related substance concentrations in environmental media or body parts or fluids)
- EU-ExpoFactors collecting data and information on *exposure factor* [e.g. anthropometric data (human body weight, breathing rates for various types of consumers), time use data and product use frequencies]
- Chem-Test inventorying exposure testing methods (e.g. measurement methods to quantify emission of chemicals from consumer products and articles)
- ExpoModels inventorying exposure models, relative algorithms (e.g. algorithms for assessing dermal exposure to chemicals in products used for a specific purpose such as cleaning) and adequate concepts

- ExpoScenarios collecting libraries of available consumers' exposures assessments compiled from EU and international risk assessment resources integrated by a scenario generator facility aiming to develop new exposure assessments using a standardised procedure and presentation template
- ExpoHealthData, cataloguing human health data associated to specific consumer exposure scenarios (e.g. physician reports in publications of human skin irritation and sensitisation associated with particular chemicals used in consumer products and articles).

Figure 7.1 displays the EIS-ChemRisks toolbox workflow, based on the concept of integration of several databases and tools in six modules.

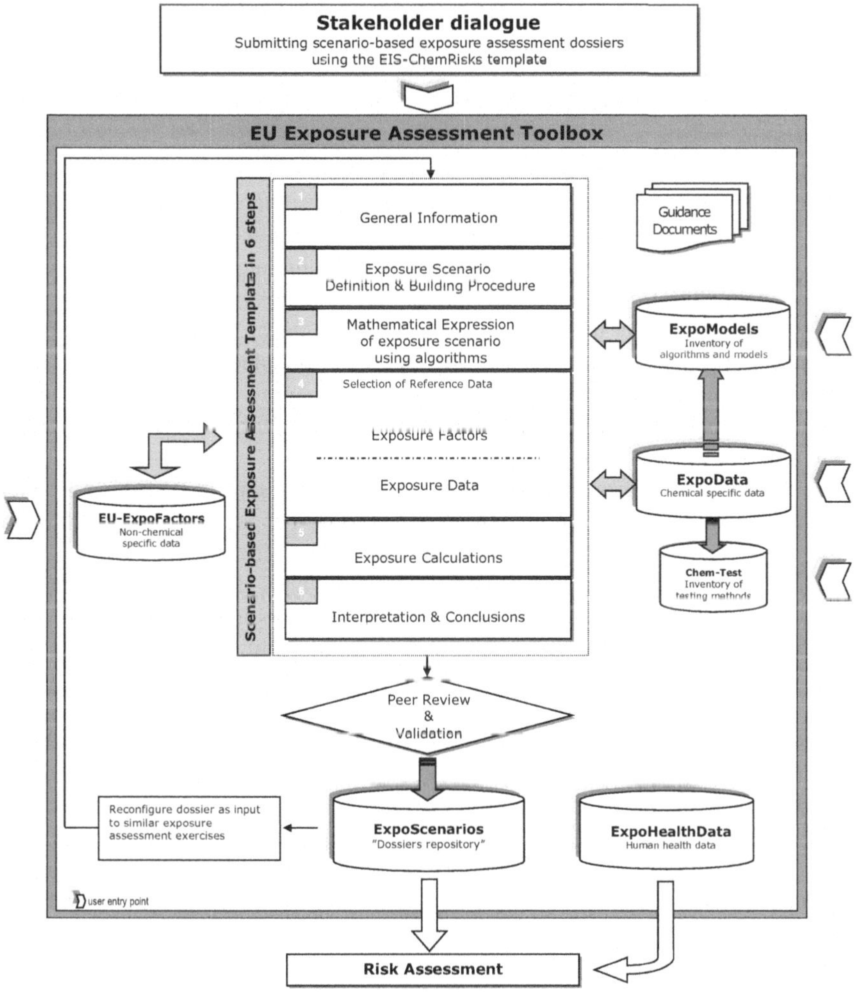

Fig. 7.1 The EIS-ChemRisks toolbox concept workflow

The toolbox approach was specifically designed with an open architecture which is modular and easily adaptable to specific needs arising from the implementation of particular management policies.

Implementation

The EIS-ChemRisks project maintained a home page in order to continuously collect user feedback as well as to disseminate general information and specific user guidance documents. Furthermore, the EU Exposure Assessment Network ensured that the needs of different stakeholders and of different policy areas were adequately catered for within the toolbox.

The toolbox represented a unique reference system for consumer exposure scenarios that were applied in exposure and risk assessments in the EU and globally. It was publicly accessible from the site http://www.jrc.ec.europa.eu/eis-chemrisks and several access levels were allowed in order to maintain the sensitivity of the information.

The toolbox allowed users to access information and make queries from 5771 primary source documents and 477 harmonised consumer exposure scenarios, in order to find exposure scenarios details, exposure determinants including exposure factors, testing methods and exposure models including algorithms (van Leeuwen and Vermeire 2007).

The EIS-ChemRisks toolbox enhanced the mapping of chemicals occurring in consumer products with a view to listing all chemicals which are reported in each consumer product, according to the available taxonomy. The mapping can be confined to a specific information source, e.g. RAPEX notification or risk assessment report.

The exposure scenarios library can also be filtered by chemical, consumer product and/or risk management measure (RMM) for consumer products defined within the exposure scenarios (Bruinen de Bruin et al. 2007). Furthermore, this library can be restrained to a geographic entity such as region, country, EU, worldwide, specific microenvironment such as indoor and a specific target population such as neonate, preterm neonate, full-term neonate, infant, toddler, child, adult, adolescent and elderly.

A standardized format for storing exposure scenarios, as implemented in the EIS-ChemRisks toolbox, allows the use of the EIS-ChemRisks toolbox as a documentation system serving as a library of the different types of consumer exposure scenarios. An example is given in Appendix 2.

In addition to enabling scenario-based exposure assessment in a standardized way, the EIS-ChemRisks project offered multidisciplinary training opportunities for exposure assessors within the EU. Training modules were developed, tested and applied by several experts.

Finally, the EIS-ChemRisks concept, as applied in the toolbox for consumer exposure to chemicals, provided a model for carrying out assessments related to physical hazards. Such formalisation of a consumer exposure assessment process was implemented in the Commission Decision 2010/15/EU of 16 December 2009, laying down guidelines for the management of the RAPEX notifications system of the GPSD directive (EU 2002, 2010).

7.3.2 *EXPO-Facts*

Alexandre Zenié, Vittorio Reina, and Artur Radomyski

Introduction

The EXPO Facts project was originally funded by the Long-Range Research Initiative (LRI) of the European Chemical Industry Council (CEFIC) (contract no. NMALRI-C1.5KTL-0207 and LRI-C1.5.1-KTLE-0512) and contracted to the Finish National Public Health Institute (KTL, Finland). It complemented the ECETOC sourcebook that focused on the UK (Zaleski and Gephart 2001). The project was undertaken in the period from 2002 to 2006 (Vuori et al. 2006). Given its EU-wide perspective, the EXPO-Facts database was then handed over to the European Commission's Joint Research Centre (JRC), and became available online on 15 January 2007 via the EU Science Hub.

Non-chemical exposure factors are considered as the actions, the media and the microenvironments that bring humans in contacts with agents (e.g. housing conditions, time use in different microenvironment and product use). They complement the chemical specific exposure data and serve as input for exposure models in order to quantify the human exposure to chemicals (Zaleski et al. 2016).

The EXPO-Facts database contains non-chemical exposure factors, data and information which are relevant to exposure and risk assessment of chemicals, performed in the context of several sectorial EU regulations on environment and public health (e.g. REACH, BPR, GPSD, PPPR, CPR, food safety and nutrition and food contact materials). It also contains data and information on population data covering all EU Member States plus Iceland, Norway and Switzerland (31 countries in total) with extensive links and references. As a common gateway, EXPO-Facts serves to gather and disseminate European-based reference exposure factors data and information that were previously scattered in various sources (e.g. databases, information systems of national institutes and international organisations such as WHO and UNECE, scientific articles in peer-reviewed journals, reports and surveys from national studies). The EU EXPO-Facts related data and information have been compiled into an online database from over 120 sources.

The EXPO-Facts database is mainly a tool for European exposure assessors and risk managers involved in public health and environmental health issues, particularly in the areas of indoor air quality, dietary exposures and safety of consumer products

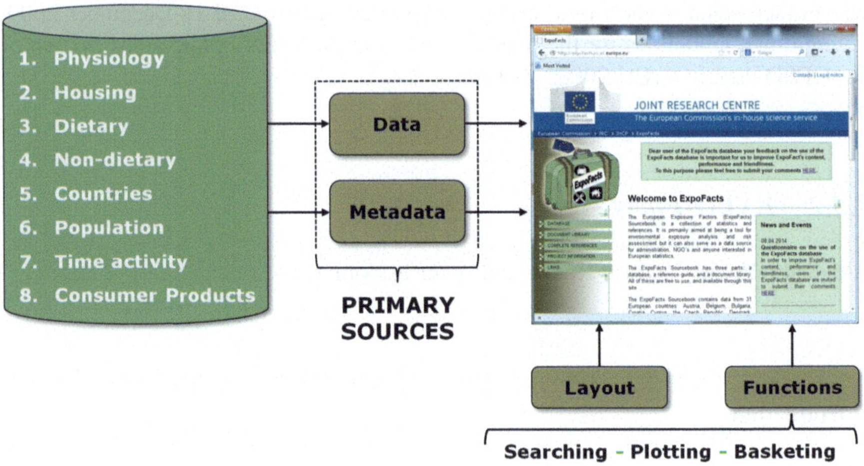

Fig. 7.2 Main components of the EXPO-Facts database

and articles. Data are mostly on diet, time-activity patterns, physiology, housing and demographic factors including non-dietary and consumer products (Fig. 7.2).

Components of the System

The database is built in straightforward manner. It offers several functionalities including categorising, querying, sorting, retrieving and visualising. Searching can be structural or through free text option that may be useful for survey-based data (Fig. 7.3). New functions are being developed to allow for the aggregation of data.

The EU EXPO-Facts is a web-based database that is freely accessible at: https://ec.europa.eu/jrc/en/expofacts.

Issues in Interpreting European Exposure Determinants' Values of the EXPO-Facts Database

Data in EXPO-Facts are not processed but presented as originally found in the sources, for that reason the use of statistical measures is very heterogeneous and source-dependent.

EXPO-Facts has the peculiarity to present data of many countries with some tables being country-specific (e.g. Ventilation rates measured in homes in the Netherlands and Food consumption in Germany, Fig. 7.4). It should be noted that the data from various sources were collected using different survey methodologies and may represent different degrees of representativeness (Table 7.2).

Fig. 7.3 Searching within the EXPO-Facts database

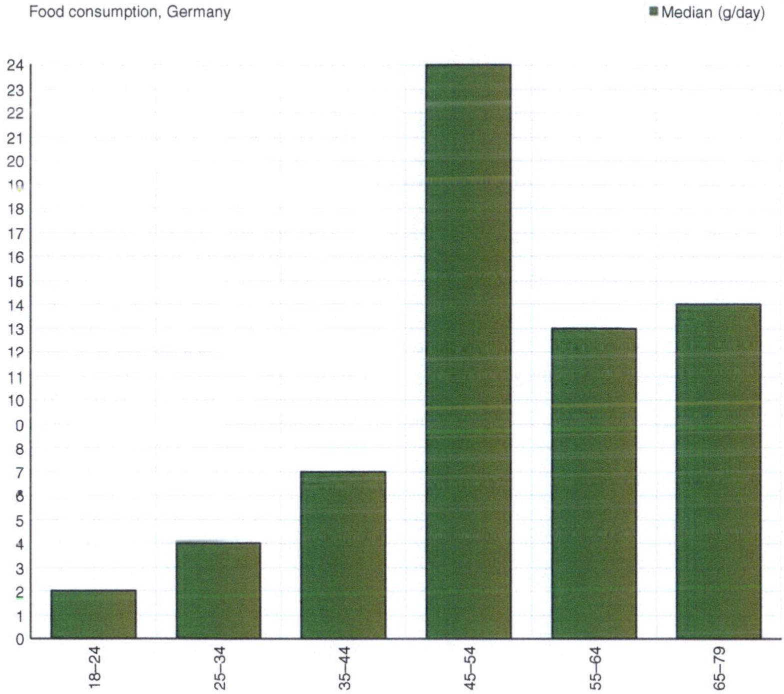

Fig. 7.4 Food consumption for different categories of age in Germany

Table 7.2 Example of inconsistencies between Italy and Denmark for water consumption exposure factor

Data year	Country	Age (years)	Gender (m/f/mf)	Product group	Product	Mean (g/d)	Median (g/d)	s.d.	Consumer (%)	Consumer only			# of sample	References
										Mean (g/d)	Median (g/d)	s.d.		
1994–1996	Italy	18–64	mf	Beverages	Mineral water	311.8	230	355.2	65.1	479.2	450	337.1	1978	Turrini et al. (2001)
2000–2001	Denmark	15–75	m	Beverages	Water (with or without gas)	466							393	Fagt et al. (2002)

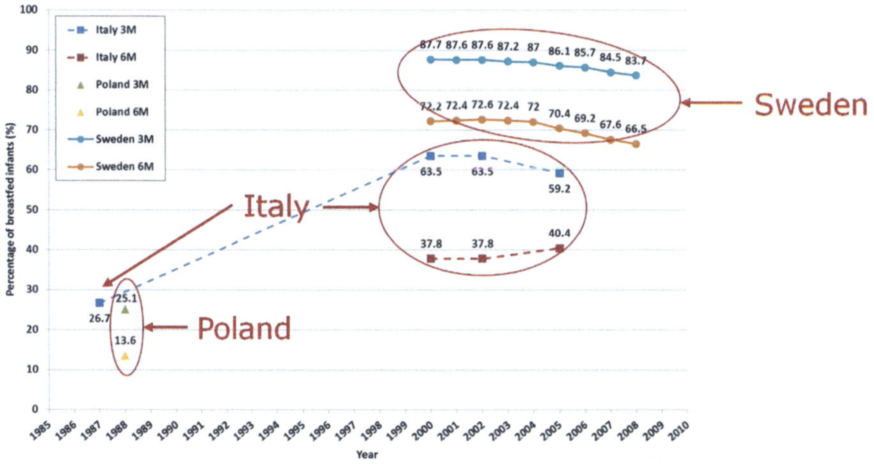

Fig. 7.5 Infants breastfed (%) at the age of 3 and 6 months in Italy, Poland and Sweden

Consequently, in some cases, direct comparisons by age group or population groups between different countries can be inconsistent because of the ways in which they provide details and how the information is harmonised (Fig. 7.5):

- Selecting data over the lifetime of a population within a specific country can be misleading if several different primary sources were used for this coverage.
- Comparing across EU Member States for a specific exposure factor is not meaningful if several different primary sources were used in this analysis.
- Averaging over time for analysing the trend of a specific exposure factor can also be misleading if several different primary sources were involved.

Recommendations for Systematic Approach to the Interpretation of the EXPO-Facts Exposure Determinants' Values

The following three recommendations should be systematically considered for a meaningful interpretation of the EXPO-Facts exposure determinants' values.

- *Checking the primary sources* in order to get full insight of the source values
- *Analysing the interval dimension* of the data and the sample size before averaging the data
- *Adopting advanced techniques* such as interval analysis for using exposure determinants' values.

The interval dimension of the data can be computationally preserved by using interval analysis in low or intermediate tiers of exposure assessment.

Conclusions

Although specific to the European population, the EXPO-Facts database includes explicit links to international exposure factors systems (i.e. Australia, Canada, China, Germany, Japan, Korea and the USA), with the aim of seeking harmonisation, interoperability and integration.

The interoperability between these exposure factors systems, where EXPO-Facts is playing a leading role due to its straightforward availability on the Internet, promotes on the one hand the development of standardised protocols for collecting data and on the other hand the correct interpretation and use of the exposure determinant values in harmonised manner.

A future development of the EXPO-Facts database could consider the inclusion of other factors, ranging from external dose to internal dose and target tissues, including biomarkers and factors needed for pharmacokinetic modelling to allow for dose reconstruction from human biomonitoring data.

7.4 Activities by the WHO

Gerhard Heinemeyer and Matti Jantunen

7.4.1 Exposure Assessment Harmonisation in the Frame of the International Programme on Chemical Safety

Based on the IFCS Bahia Declaration and the World Summit on Sustainable Development that called for common approaches to chemical risk assessment, the WHO initiated the international programme on chemical safety (IPCS). The aim of this project is to harmonize assessment approaches of risk from exposure to chemicals. Within this programme, the documents produced are generally in the form of guidance to be adopted for use by regulatory schemes. The work is not specific to a particular category of chemicals, but is applicable to the assessment of industrial chemicals, pesticides, food additives and contaminants, industrial by-products, etc., in general.

The project is directed to:

- The general public to enhance confidence and trust in assessment and regulation. An increased access to information would allow greater participation in risk management decisions.

- Governments for increased access to risk assessment in global setting, cost savings through burden sharing, better risk assessment promoting safer chemical use.
- Industry in order to reduce testing costs, increase common data sets and consistency in interpretation of data, and improve risk characterisation leading to safe chemical use, and the
- Scientific community to promote sound science as the basis for decisions, and advances in innovative methodologies driven by identification of critical needs for data and new research.

An Exposure Assessment Planning Working Group was initiated to make proposals for emerging issues regarding exposure assessment. At the first meeting (1999 in Athens, Greece), four primary issues to advance harmonisation in exposure assessment were identified.

Working groups were formed to work out the following guidance documents:

1. Harmonisation of Chemical Exposure Assessment Terminology: The chairs of this project were Michael Callahan and Karen Hammerstrom (US-EPA).
2. Principles of Characterizing and Applying Exposure Models with Matti Jantunen as chair (Institute for Health and Consumer Protection at the Joint Research Centre, European Commission, Italy).
3. Uncertainty Analysis in Exposure Assessment chaired by Gerhard Heinemeyer (Federal Institute for Risk Assessment, Germany).
4. Data Quality Assurance in Exposure Assessment, which was performed in parallel with the uncertainty project as recommended by the Steering Committee, chaired by William C. Griffith (Institute for Risk Analysis and Risk Communication, University of Washington, Seattle, USA).

Project No. 1: Harmonisation of Exposure Assessment Terminology: (WHO 2004)
The Scope of this project was to develop a consistent framework and definitions for terminology in the field of chemical exposure assessment, with an objective to improve communication by harmonising definitions.

The group took exposure assessment-related terms from a collection of 57 glossaries for exposure and related fields, and developed a logically consistent framework of definitions for 39 key terms. Toxicological terms, risk-related terms, chemical names, procedural terms, and risk management terms were generally excluded. The Glossary was adopted by the International Society of Exposure Analysis (now the International Society of Exposure Science) in September 2004.

Project No. 2: Harmonisation of Exposure Modelling (WHO 2005)
First, the working group developed a concept for a conceptual or mathematical representation of the exposure process by means of the IPCS definitions.

Models are mathematical representations of a natural system intended to mimic the behaviour of the real system, allowing for description of empirical data and predictions about untested states of the system.

Models can be characterised and applied based on the following principles:

- Terminology and definition
- 'Models' as tools for exposure assessment
- Mathematical/numerical representations of exposure scenarios
- Hierarchical aggregations of lower level (e.g. emission) models into a higher level (e.g. exposure) model.

A set of characteristics of a model that should be provided by the model developer and considered by the model user to be considered was developed:

1. Description of the purpose of the model and its components
2. Individual- or population-level analysis (level of aggregation)
3. Modelled time resolution
4. Applicability to diverse exposure scenarios
5. Description of data inputs
6. Model processes
7. Model code and platform
8. Model performance and evaluation summaries
9. Description of model outputs
10. Model sensitivity and uncertainty.

The guidance document 'Principles of characterising and applying exposure models' was issued by WHO in 2005.

Project No. 3: Uncertainties in Exposure Assessment (WHO 2008, Part I)
With this project and the corresponding preparation of a guidance document, the IPCS started to consider uncertainty as a central issue in exposure assessment. Uncertainty represents the lack of knowledge about factors affecting exposure or risk and can lead to inaccurate or biased estimates of exposure. Uncertainty must be separated from variability. The document is targeted for practical-level exposure assessors with limited experience in uncertainty analysis.

Focusing on the following issues, the guidance document in the form of a monograph gives a broad overview on uncertainty analysis:

- Scenario uncertainty
- Parameter uncertainty
- Model (construct) uncertainty
- Principal methods of uncertainty analysis
- Discuss harmonisation aspects
- Tiered approach for exposure assessment.

Several guidance documents that deal with uncertainty analysis reference the IPCS document, as seen in Sect. 3.8.

Project No. 4: Data Quality and Exposure Assessment (WHO 2008, Part II)
This document addresses:

- Collection and use of exposure data
- Reasons for quality consideration needs
- Description of the principles of data quality
- Brief illustrative examples that highlight the importance of the key principles.

Quality assessment and uncertainty analysis are addressed together in one guidance document. A particular section of this book is also relevant (Sect. 3.7).

As stated in the introduction of this book, two of the editors participated in the work of the exposure harmonisation group. The consideration of the principles laid down in the IPCS guidance documents are therefore of the greatest concern. As pointed out in several chapters, the objectives of harmonisation have not yet been reached and a lot of commitment and work is still needed.

In addition, the IPCS activity on '*Risk Assessment of Combined Exposures to Multiple Chemicals*' should be mentioned. This project developed a concept for a framework of risks assessment approaches for combined exposures. The project addressed a tiered approach starting with a broad scenario of combinations of various substances up be tiered for a defined scenario of well-known combinations of substances. The results of the project were presented at an international workshop organised by ILSI/HESI and hosted by OECD in 2011 in Paris, France (WHO OECD ILSI/HESI 2011), as well as a comprehensive report (Meek et al. 2011) (see also Sect. 7.6.4).

7.5 Activities by OECD

Yasmin Sommer

The functions of the Organisation for Economic Co-operation and Development (OECD) have shifted over the years of its existence. Today it is a globe-spanning intergovernmental organisation which provides the setting to discuss issues affecting the economic and social well-being including chemical safety. Here governments can exchange expertise, coordinate and harmonise policies, and set standards.

OECD's work is mostly carried out in specialised committees and their subsidiary bodies. They are composed of delegates from different countries, the European Commission, Industry/Trade Associations, and Non-Governmental Organisations (NGOs), and served by the OECD Secretariat, which is located in the Headquarters in Paris, France. More information about the mission and the way OECD works is available from the OECD website (http://www.oecd.org/about/).

The 'Chemicals Management Programme' is one part of OECD's chemical activities. It includes nanomaterials, pesticides, and biocides, and is overseen by the Chemical Committee. Consumer Exposure Assessment is addressed by several of the subsidiary bodies contributing to this work programme, e.g. the Working

Party on Nanomaterials and the Working Party on Exposure Assessment (formerly known as Task Force on Exposure Assessment).

The Working Party on Nanomaterials was established in 2006. It develops methods and strategies for the identification and management of the potential risk of nanomaterials to the environment and human health. Among its activities is a project compiling the available tools and models used for assessing consumer exposure to manufactured nanomaterials. This project also aims to evaluate their applicability in exposure assessments characterising their performance. The Working Party on Exposure Assessment's scope was extended to include human health in 2008. Its mandate is to facilitate sharing, develop, disseminate, compare and, if possible, harmonise exposure-related information and exposure mitigation-related information, and methodologies and tools for assessment of the impact of the release of chemicals during their life cycle on the environment and on various human populations. In current practice 'exposure-related information' are the OECD Emission Scenario Documents (further explained below) and a work item on Internationally Harmonized Use Codes (Sect. 5.1.2). The term 'methodology and tools' comprises databases, guidance documents, harmonized templates, and exposure models.

Some examples of work items with relevance to consumer exposure are given below in the sections highlighting OECD work types. The Working Party on Exposure Assessment has currently taken up work items on Children's Exposure to chemicals and releases of chemicals from products. In cooperation with the Working Party on Hazard Assessment, it has also taken up work on combined exposures to multiple chemicals. Further information on the outcome of these efforts will become available over the next years via the respective sections of OECD's website on chemical safety. As a member of the United Nations Inter-Organisation Programme for the Sound Management of Chemicals (IOMC), OECD makes findings and workshop reports within the 'Chemicals Management Programme' publicly available. They are published free of charge on the OECD website.

The OECD website can be consulted for further information on the programme (http://www.oecd.org/chemicalsafety/). The publications on exposure issues are accessible from dedicated website subsections on the exposure assessment programme (http://www.oecd.org/env/ehs/risk-assessment/exposureassessment.htm) or website subsections on the OECD Series on Testing and Assessment (http://www.oecd.org/env/ehs/testing/seriesontestingandassessmentadoptedguidanceandreviewdocuments.htm).

Information regarding OECD activities and findings on the safety of manufactured nanomaterials can be obtained via a different section of the OECD website (http://www.oecd.org/science/nanosafety/) and a subsequent link to the publications within the OECD series on the Safety of Manufactured Nanomaterials (http://www.oecd.org/env/ehs/nanosafety/publications-series-safety-manufactured-nanomaterials.htm).

7.5.1 OECD Emission Scenario Document (ESD)

The documents in the OECD Emission Scenario Document (ESD) series cover release estimations of chemicals during their life cycle including production, processing, use, service-life, recycling, treatment, and disposal phase. Primarily developed for environmental exposure assessment benefits, they also provide data for consumer exposure assessments, e.g. when information on typical product compositions is given as in the ESD No. 3 on plastic additives (OECD 2009). A list of all ESDs and access to them is available (http://www.oecd.org/env/ehs/risk-assessment/emissionscenariodocuments.htm).

7.5.2 OECD Harmonised Templates (OHTs)

The OECD Harmonised Templates (OHTs) are standard data formats. They prescribe which information can be entered into and maintained in a database (e.g. IUCLID). Their principal users are developers of database systems. The benefit of the OHTs for assessors and regulators is that they facilitate a standardised electronic exchange of information between different stakeholders and databases. Most OHTs cover test study summary information on the properties or effects on human health and the environment of a chemical. The first set of OHTs concerning information relevant for exposure assessment became available in 2016. It covers 'use and exposure information' (OECD 2016). However, further work is required to address additional types of data in exposure assessment. All available OHTs are listed on the OECD website and accessible via it (https://www.oecd.org/ehs/templates/).

7.5.3 OECD Guidance Documents

OECD Guidance Documents provide counsel for performing the specified tasks and help to standardise procedures. One example is the 'OECD Guidance Document on Reporting Summary Information on Environmental, Occupational and Consumer Exposure' (OECD 2003). It contains three formats for the reporting of summary exposure information that allow sufficient flexibility to be used in various chemical assessment programmes. The formats enable accommodation of both quantitative (e.g. release rate and resulting exposure) and qualitative information (e.g. descriptions of the source of release and control measures). The user needs to determine the purpose, scope, and level of detail of the summary and complete the formats accordingly. Possible entries in the fields are explained in the guidance. Format A covers General Information and Overview of Exposure. This always needs to be completed and contains basic information about the chemical/category and an overall perspective on potential exposures. Available monitoring or modeling data can be summarised in the other two formats, Format B: Monitoring Evaluations and/or Format C: Modeling Evaluations.

More recently, the 'OECD Guidance Document for Exposure Assessment Based on Environmental Monitoring' (OECD 2013a) was published. It addresses both environmental and human exposure, and describes the basic methodology used to conduct an exposure assessment based on environmental monitoring data. The document focuses on common approaches despite the existence of fundamental differences between human and environmental exposure assessment methodologies. It aims to summarise the main facts and issues to be considered, e.g. the ways of using monitoring data in exposure assessments for differing purposes, data collection, sensitivity, and the quality of monitoring activities. It also includes a list of monitoring programmes in some OECD members states.

7.5.4 OECD Surveys and Workshops

Surveys are conducted to collect information and sometimes ratings on a specified subject from the participants. Subsequently, responses are compiled and also regularly evaluated. The outcome of a survey is always a source of information, e.g. 'Descriptions of Existing Models and Tools Used for Exposure Assessment' (OECD 2013b). In addition, survey results can serve OECD as a starting point to determine recommendations for further activities such as workshops or projects.

OECD workshops can be organised in cooperation with other organisations. One example relevant for consumer exposure is the 2011 'WHO OECD ILSI/HESI International Workshop on Risk Assessment of Combined Exposures to Multiple Chemicals' (OECD 2011a, b, c, d). It was organised by WHO and OECD, with input from the International Life Sciences Institute (ILSI). The workshop's purpose was to inform participants about the corresponding WHO framework and explore the application of the framework by discussing a number of case studies. In addition, information about current issues in combined exposure assessment was shared and the need for further work to be undertaken was identified. Meanwhile a formal programme to work on combined exposures to multiple chemicals was launched. In collaboration of the Working Party on Hazard Assessment and Working Party on Exposure Assessment, a first result, which includes considerations for the exposure characterisation informing assessments of combined exposures, was published in 2018 (OECD 2018).

Information on the results of surveys and workshops can be obtained on the OECD website (http://www.oecd.org/chemicalsafety/testing/seriesontestingandassessmentworkshopreportsmeetingreports.htm).

7.6 Anchoring Exposure Science in Europe

Yuri Bruinen de Bruin, Jos Bessems, Peter Fantke, Natalie von Goetz, Urs Schlüter, and Alison Connolly

7.6.1 Introduction

Exposure assessment of stressors, e.g. chemicals, viruses, microbes, noise, and radiation, aims to understand the extent to which humans, organisms or non-living items in the environment are exposed. It contributes to exposure science in understanding situations that pose a potential risk to human health and the environment. Equally important to exposure assessment is to understand the hazards of such stressors that may lead to unwanted health and environmental effects. Both, exposure and hazard assessment sciences, are indispensable parts of risk assessment and subsequent analyses such as impact assessment and socio-economic analysis (Vermeire et al. 2017).

Many disciplines (NRC, 2012) need or at least benefit from knowledge in the exposure science domain (Fig. 7.5a, b). Exposure assessment is implemented in many EU regulatory legislative domains in the area of chemicals. As such, it is an important issue for many stakeholders such as (chemical) industry, national research institutes, NGOs, health insurance companies, (consultancy) companies in the occupational and public health arena, academia, research institutes, national and international policy makers, national and international agencies and authorities (Fig. 7.6).

7.6.2 The International Society of Exposure Science (ISES): Corporate

The role of many of these stakeholders is in one or more ways, and to a larger or smaller levels of detail, to pay attention to exposure as part of predicting, preventing or limiting risks, and to consider it as an essential element in the management or mitigation of the risks of chemicals and other stressors. Demands on the exposure assessment in the context of the various stressor-related disciplines are very diverse and to a large extent determined by the various exposed populations such as consumers, workers and ecosystems. In order to exchange knowledge and information globally, to discuss, learn and network the International Society of Exposure Analysis (ISEA) [subsequently renamed the International Society of Exposure Science (ISES)] was set up in 1989. The Society's mission is to meet the humanity's needs for public health and environmental protection through a global community of exposure science professionals. ISES encourages the open exchange of information,

Fig. 7.6 (**a**) Related and overlapping disciplines that benefit from exposure science and (**b**) EU Regulatory domains for which exposure knowledge is crucial. Disciplines that benefit of Exposure Science

provides opportunities for career development, and acknowledges and promotes excellence in the practice of exposure assessments and research in the field of exposure science. This statement reflects ISES's dedication to promoting, protecting, and enhancing the role of exposure science in diverse fields and professional disciplines across academia, government, industry, and policy makers (https://www.intlexposurescience.org).

As part of this mission, ISES aims at the advancement of exposure science related to environmental contaminants, both for human populations and ecosystems. Its objective is to foster and advance exposure science related to environmental contamination in human populations and ecosystems; promote communication among exposure analysts, epidemiologists, toxicologists, and other scientists; recommend approaches to methodological concerns; and strengthen the impact of exposure assessment and analysis on environmental policy. ISES also aims to disseminate results for the general public, workers, consumers, etc. using proper science communication instruments.

ISES stands for a knowledge society having societal relevance. It is internationally embedded and via its members methodologically innovative. ISES mission is to provide state-of-the-art science and demand-driven policy support to protect human and environmental health. To reach this mission, core values are expertise, reliability, independence, self-consciousness, purposefulness, vigilance, international mindedness, innovativeness, flexibility, partnership and entrepreneurship. The Society's multidisciplinary expertise and international reach make it the premier professional society for practitioners associated with all aspects of exposure science (research, teaching, policy, communication, and outreach).

7.6.3 ISES Annual Meetings

Over the years, the annual meeting has become the cornerstone of the Society's direct interaction with its members and the place where those interested in exposure science can meet, exchange ideas, and develop ideas for the future. At the beginning, all meetings were held in collaboration with different societies (joint meetings) because it was believed that ISES was too small to have a meeting on its own. Additionally, this approach helped to increase the Society's visibility amongst other societies and disciplines. The year 1997 marked the first time ISES had its own annual meeting. Annual meetings are now held routinely and the goal is to have a joint meeting every third year with another society. The need for ISES 'only' meetings was essential for determining how well ISES was achieving its goals and mission. In future years the Society plans to host specialty conferences devoted to particular areas of interest and new areas of concern to members and others.

7.6.4 The International Society of Exposure Science (ISES): Regional Chapters

The Society encourages members in different geographic regions to form regional chapters to:

- Facilitate meetings among members who reside in close proximity
- Provide guidance on local scientific, regulatory and policy issues that require an understanding of exposure to environmental or occupational agents, and
- Provide a forum for promoting student participation in the field of exposure science.

North America

Tri-State Regional Chapter (New Jersey, New York, Pennsylvania).

This Chapter's aim is to foster and advance the science of exposure analysis related to environmental contaminants, both for human populations and ecosystems, and to promote communication among scientists, regulatory personnel, general public and other individuals or organisations (http://isestristate.org/).

California Chapter

The objectives of this Chapter are to foster and promote the goals, purposes, and objectives of the International Society of Exposure Science among scientists and at institutions within California. In particular, the objectives of the Chapter are to further understanding, awareness, and appropriate applications of exposure science, and to promote an exchange of ideas, communication and practical experiences among members of the academic, professional, industrial, and regulatory communities engaged in exposure science, e.g. exposure scientists, epidemiologists, toxicologists, and other scientists in academia, government, and industry. The Chapter aims to act as a resource for and to provide support to the Society (https://www.isescalifornia.org/).

Asia

The East Asia Chapter is for members within the East Asian countries, like Japan, China, Taiwan, Korea, etc. The objectives of the Chapter are to: (a) foster, promote and advance the science of exposure analysis related to environmental stressors in human populations and ecosystems, (b) promote communication among scientists, policy makers, the general public and other individuals or organisations interested (items listed in the first objective), (c) strengthen the impact of exposure science on environmental policy making, (d) serve as the focal point for interaction of members of the Society and other interested individuals from academia, government and industry in the East Asian countries, i.e. to provide communication and interaction opportunities for researchers, particularly young investigators and scientists associated with government whose opportunity to travel to international meetings is limited, (e) further improve the understanding, awareness, and appropriate applications of exposure science, and to promote an exchange of ideas, communication and practical experiences among members of the academic, professional, industrial and regulatory communities engaged in exposure science, e.g. exposure scientists, epidemiologists, toxicologists and other scientists in academia, government and industry, (f) disseminate exposure information and concepts to all interested individuals, (g) foster integration and interaction of the various disciplines involved in exposure science, (h) hold scientific and educational meetings and workshops, and (i) act as a resource for and provide support to the Society.

7.6.5 Europe

The European Regional Chapter (further referred to as ISES-Europe) is the youngest Regional Chapter of ISES. The increasing need to have a European programme to promote increased generation of exposure information followed by independent

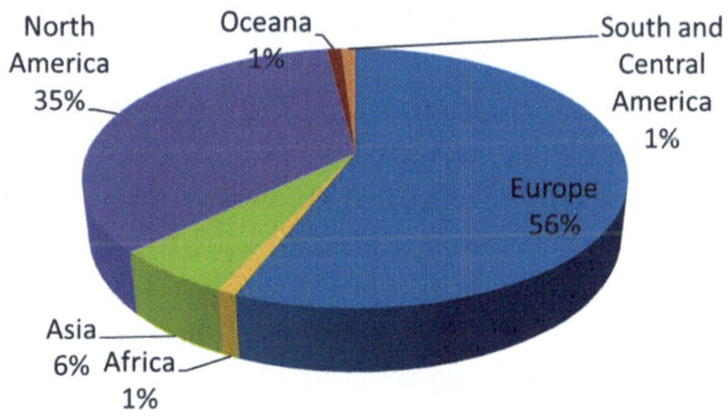

Fig. 7.7 Geographical distribution of meeting participants

assessment and use by risk managers and policy makers was the main driver to set up the European Regional Chapter. ISES-Europe is regarded to be crucial in starting off such a programme in Europe at the moment; no other independent organisation is likely to fulfil this job. In addition, many if not all objectives of such a European programme are fully in line with the general objective of the Society as a whole. In this respect, the establishment of ISES-Europe is the first step in the strategy towards the establishment of that European programme for exposure science (http://ises-europe.org/).

The 2016 meeting in Utrecht, the Netherlands was the biggest meeting since the start of the Society with participants from 50 different countries (see Fig. 7.7).

During the 2016 annual meeting in Utrecht, efforts were made to establish ISES-Europe as an answer to requests to explore this from many ISES members in Europe. Based on the input obtained, ISES-Europe was officially established in mid-2017, and elections were held to establish the ISES-Europe Board among the European members of ISES. Once established the Board endorsed the draft mission and draft bylaws as prepared before the Utrecht meeting. Subsequently and based on this, the ISES-Europe Board reflected on the definition of exposure science in Europe and came up with six key objectives.

Mission of ISES-Europe ISES-Europe's mission is to integrate exposure science into European regulations and industry practice, and to anchor it in academic research and education, in order to foster innovation and to create a safe and sustainable future for humans and the environment.

What Is Exposure Science? Exposure science in its broadest sense studies the contact between stressors (primarily chemical, biological, and physical agents) and receptors (e.g. molecules, cells, organs, humans, other living organisms, and non-living items like buildings), and the associated pathways and processes potentially leading to negative effects on human health and the natural and built environment.

ISES-Europe's Key Objectives

1. Build a European Programme to generate, assess, exchange, and communicate experimental and model-based exposure data in support of a European strategy for exposure science in the twenty-first century
2. Advance exposure science with focus on closing knowledge gaps, building science-based and operational data and methods, fostering integration and interaction of disciplines involved in exposure science, and anchoring exposure information in decision-making by all stakeholders
3. Provide guidance and recommendations for exposure assessment in science and policy
4. Foster the inclusion of realistic exposure information in risk assessment, alternatives assessment and chemical substitution, high-throughput risk screening, life cycle assessment, and other regulatory and non-regulatory frameworks and management tools
5. Strengthen the impact of exposure science on human health and environmental policies
6. Support trans-agency and trans-institutional coordination, education, and engaging a broader European stakeholder community including triggering and supporting a European infrastructure to facilitate sharing, generation and dissemination of exposure information.

7.6.6 An Increasing Need for International Exposure Information

REACH, the current EU Regulation on chemicals management, created a mandate to deliver adequate exposure information to foster the safe use and management of chemicals. Exposure information is also required in other European regulatory frameworks (Endsdaily, 25/05/2016 ec.europa.eu/jrc/en/research-topic/human-exposure, echa.europa.eu, www.efsa.europa.eu, osha.europa.eu/en) (including regulations on plant protection products (ec.europa.eu/jrc/en/research-topic/human-exposure, echa.europa.eu, www.efsa.europa.eu, osha.europa.eu/en) biocidal products (echa.europa.eu/regulations/biocidal-products-regulation/understanding-bpr), construction products (ec.europa.eu/growth/sectors/construction/product-regulation_en), general product safety (ec.europa.eu/consumers/consumers_safety/product_safety_legislation/general_product_safety_directive/index_en.htm), classification, labelling and packaging (ec.europa.eu/growth/sectors/chemicals/classification-labelling_en), control of air quality (ec.europa.eu/environment/air/quality/existing_leg.htm) and major accident hazards (ec.europa.eu/environment/seveso/), etc. Current EU chemicals frameworks such as the ones listed above are complex and often in need of appropriate exposure information. Therefore, the starting focus of ISES-Europe has been on chemicals. At a later stage, physical and biological stressors may be incorporated.

Outside the EU, several countries in or near Europe are following or have even implemented complete legislative instruments. One example of a multilateral agreement is the participation of Norway, Switzerland, Liechtenstein and Iceland in EFTA (European Free Trade Association). Also, there are bilateral agreements between a Member State and the EU. Further away, other countries have also modified their chemicals legislation. They include China, Korea, Malaysia, Russia, Taiwan and Turkey, with some of the legislation having elements similar to REACH. These developments also increase the need for adequate exposure information.

Within the European Union, notification and registration processes of chemicals by industry that include submission of confidential and non-confidential information on (foreseeable) exposures and hazards have released a huge amount of data. Many of the non-confidential data (albeit with varying degrees of uncertainty) can now be found via websites of agencies like ECHA (Search for Chemicals option on https://echa.europa.eu/information-on-chemicals) and EFSA (e.g. OpenFoodTox, EFSA Comprehensive European Food Consumption Database https://www.efsa.europa.eu/en/data/chemical-hazards-data). As a result, these data have 'gone public' and have global value. For example, companies are now avoiding substances of very high concern (SVHCs).

Chemical safety has moved beyond the environment, health and safety departments of chemical companies. Being more and more of a real and recognized business issue, it is getting onto the radar of CEOs. It is critical that these leaders approach REACH and other chemicals legislative frameworks like a healthcare system: 'it never ends' (Chemical Watch 2016). Furthermore, EU strategies to move towards a non-toxic environment by 2050, to strive towards a circular and bio-based economy, and to promote green and sustainable chemistry add additional challenges requiring adequate exposure information (Decision No. 1386/2013/EU). The creation of a European Programme to promote increased generation of exposure information and fit-for-purpose exposure assessment tools followed by independent assessments and use by risk managers and policy makers is the main driver for the ISES-Europe Regional Chapter.

7.6.7 The Vision of Exposure Science in Europe

Experience has shown that regulatory anchoring and guidance is pivotal for a proper generation of any data, be it hazard data or exposure data. If not, industrial frontrunners might pay a price if there is no immediate return on investment: it is not always clear upfront for a specific chemical whether additional data generation on top of the legally required data pays off. Thus, clear regulatory anchoring which obliges all companies where the legislation applies (and indirectly far beyond) is crucial to create a level playing field for industry. In addition, it will automatically foster awareness of the usefulness of more and better exposure knowledge. Because of this, the regulatory aspects should not be neglected, but incorporated in the mission of ISES-Europe (Fig. 7.8).

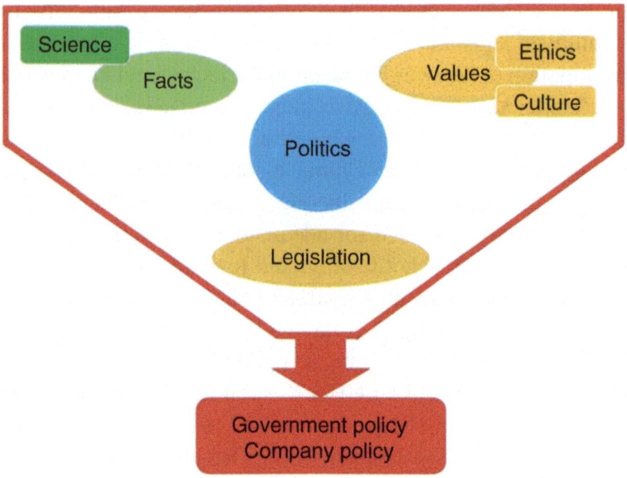

Fig. 7.8 Complex interaction between stakeholder groups that drives policy

ISES-Europe serves as an independent platform where stakeholders from society can exchange information and construct dialogues aiming at the identification of current and future scientific and policy needs and opportunities. ISES-Europe seeks to use the best dissemination strategies to ensure that needs and opportunities are communicated to relevant stakeholder groups such as academia, research funders, CEOs and competent authorities (Fig. 7.8).

ISES-Europe strives to become Europe's Reference Network on Exposure Science. One of the main tasks will be to develop a European strategy on exposure science describing ISES-Europe's vision and spelling out the exposure science activities and priorities to achieve its goals until 2030 by making a road map 2020–2025–2030. The primary focus will be to set up an overall strategy for ISES-Europe focusing on exposure-related strategic themes including a vision, mission and ambition on issues such as a European programme to: (1) assess exact and predictive exposure data; (2) enhance a better connection to Big Data and understanding its future potentials; (3) provision of a platform for more communication to explain the benefits of understanding and managing exposure to stakeholders such as academia, research institutes, national and supranational (e.g. EU) authorities and agencies, NGOs, insurance companies, industry and workers representatives, making exposure science a business case that stays within the attention of CEOs; (4) fitness check exposure science to deliver information for current policies; (5) incorporation of exposure in business strategies (not only RIE, health and safety); and (6) dealing with the socio-economic and sustainability implications of regulatory action, and more.

The ISES 2016 meeting revealed that participants representing various European stakeholders support the setting up of education and training opportunities, the development of a formal certification in Exposure Science, the creation of an exposure information exchange platform, having access to scientific networks and continental/regional meetings on exposure science, influencing the research and regulatory agenda and facilitating interdisciplinary collaboration and exchange. ISES-Europe is in the process of setting up its structure to build a European programme on exposure science including working groups, a European reference network on exposure science and is also in the outreach phase for engagement of stakeholders, the organisation of workshops and meetings, the setting up of a European Strategy on Exposure Sciences in 2030 with a roadmap 2020–2025–2030 and enlarging membership. For more information please visit: http://ises-europe.org/ and https://www.facebook.com/ISESEUROPE

References

Arcella D, Leclercq C (2005) Assessment of dietary intake of flavouring substances within the procedure for their safety evaluation: advantages and limitations of estimates obtained by means of a per capita method. Food Chem Toxicol 43:105–116

Bruinen de Bruin Y, Hakkinen P, Lahaniatis M, Papameletiou D, Del Pozo C, Reina V, van Engelen J, Heinemeyer G, Viso AC, Rodriguez C, Jantunen M (2007) Risk management measures for chemicals in consumer products: documentation, assessment, and communication across the supply chain. J Expo Sci Environ Epidemiol 17:S55-S66. http://www.nature.com/jes/journal/v17/n1s/pdf/7500587a.pdf. Accessed 27 Oct 2019

Chemical Watch (2016) REACH—a never-ending story. https://chemicalwatch.com/46844/reach-a-never-ending-story. Accessed 27 Oct 2019

EFSA (2007) Reasoned opinion on the potential chronic and acute risk to consumers health arising from proposed temporary EU MRLs. EFSA J 5(3):RN-32, 1141 pp. https://doi.org/10.2903/j.efsa.2007.32r. http://onlinelibrary.wiley.com/doi/10.2903/j.efsa.2007.32r/full. Accessed 27 Oct 2019

EFSA (2009a) Scientific opinion of the panel on food contact materials, enzymes, flavourings and processing aids (CEF) on dietary exposure assessment methods for smoke flavouring primary products. EFSA J RN-284:1–30. https://doi.org/10.2903/j.efsa.2009.248r. http://onlinelibrary.wiley.com/doi/10.2903/j.efsa.2009.248r/abstract. Accessed 27 Oct 2019

EFSA (2009b) Guidance of EFSA prepared by the scientific panel of food contact material, enzymes, flavourings and processing aids on the submission of a dossier on food enzymes. EFSA J 1305: 1–26. https://doi.org/10.2903/j.efsa.2009.1305. http://onlinelibrary.wiley.com/doi/10.2903/j.efsa.2009.1305/full. Accessed 27 Oct 2019

EFSA (2010a) Standard sample description for food and feed. EFSA J 8(1):1457, 54 pp. https://www.efsa.europa.eu/de/efsajournal/pub/1457. Accessed 27 Oct 2019

EFSA (2010b) Guidance on the data required for the risk assessment of flavourings to be used in or on foods. EFSA J 8(6):1623, 38 pp. https://doi.org/10.2093/j.efsa.2010.1623. http://onlinelibrary.wiley.com/doi/10.2903/j.efsa.2010.1623/pdf. Accessed 27 Oct 2019

EFSA (2010c) The 2010 European Union Report on Pesticide residues in food. EFSA J 11(3):3130, 808 pp. https://doi.org/10.2903/j.efsa.2013.3130. http://www.efsa.europa.eu/en/efsajournal/pub/3130. Accessed 27 Oct 2019

EFSA (2011) Use of the EFSA comprehensive European food consumption database in exposure assessment. EFSA J 9(3):2097. https://www.efsa.europa.eu/en/efsajournal/pub/2097. Accessed 27 Oct 2019

EFSA (2012a) Guidance of the FEEDAP panel for establishing the safety of additives for the consumer. EFSA J 10(1):2537, 12 pp. https://doi.org/10.2903/j.efsa.2012.2537. http://onlinelibrary.wiley.com/doi/10.2903/j.efsa.2012.2537/abstract. Accessed 27 Oct 2019

EFSA (2012b) Guidance for submission for food additive evaluations. EFSA J 10(7):2760. http://onlinelibrary.wiley.com/doi/10.2903/j.efsa.2012.2760/epdf. Accessed 27 Oct 2019

EFSA (2013) Standard sample description ver. 2.0. EFSA J 11(10):3424, 114 pp. https://doi.org/10.2903/j.efsa.2013.3424. Accessed 27 Oct 2019

EFSA (2014a) Food additives intake model (FAIM): comments received from stakeholders and EFSA's views. EFSA Support Publ 2014:EN-566, 25 pp. https://doi.org/10.2903/sp.efsa.2014.EN-566. http://onlinelibrary.wiley.com:doi/10.2903/sp.efsa.2014.EN-566/full. Accessed 27 Oct 2019

EFSA (2014b) Guidance on the EU Menu methodology. EFSA J 12(12):3944, 77 pp. https://doi.org/10.2903/j.efsa.2014.3944. http://onlinelibrary.wiley.com/doi/10.2903/j.efsa.2014.3944/abstract. Accessed 27 Oct 2019

EFSA (2014c) Guidance on data exchange version 2.0. EFSA J 12(12):3945, 173 pp. https://doi.org/10.2903/j.efsa.2014.3945. http://onlinelibrary.wiley.com/doi/10.2903/j.efsa.2014.3945/abstract. Accessed 27 Oct 2019

EFSA (2014d) Guidance on the assessment of exposure of operators, workers, residents and bystanders in risk assessment for plant protection products. EFSA J 2014;12(10):3874, 55 pp. https://doi.org/10.2903/j.efsa.2014.3874. http://onlinelibrary.wiley.com/doi/10.2903/j.efsa.2014.3874/abstract. Accessed 27 Oct 2019

EFSA (2014e) Scientific Opinion on the safety assessment of carvone, considering all sources of exposure. EFSA J 12(7):3806, 74 pp. https://doi.org/10.2903/j.efsa.2014.3806. http://onlinelibrary.wiley.com/doi/10.2903/j.efsa.2014.3806/abstract. Accessed 27 Oct 2019

EFSA (2014f) Guidance on expert knowledge elicitation in food and feed safety risk assessment. EFSA J 12(6):3734, 278 pp. https://doi.org/10.2903/j.efsa.2014.3734. http://onlinelibrary.wiley.com/doi/10.2903/j.efsa.2014.3734/abstract. Accessed 27 Oct 2019

EFSA (2015a) The EFSA Data Warehouse access rules. EFSA Support Publ 12(2):EN-768, 18 pp. https://doi.org/10.2903/sp.efsa.2015.EN-768. http://onlinelibrary.wiley.com/doi/10.2903/sp.efsa.2015.EN-768/abstract. Accessed 27 Oct 2019

EFSA (2015b) Scientific opinion of the CEF panel on the risks to public health related to the presence of bisphenol A (BPA) in foodstuffs: executive summary. EFSA J 13(1):3978, 23 pp. https://doi.org/10.2903/j.efsa.2015.3978. http://onlinelibrary.wiley.com/doi/10.2903/j.efsa.2015.3978/abstract. Accessed 27 Oct 2019

EFSA (2016a) Corporate publication. EFSA strategy 2020. Trusted science for safe food. http://www.efsa.europa.eu/sites/default/files/corporate_publications/files/strategy2020.pdf. Accessed 27 Oct 2019

EFSA (2016b) Exposure assessment of food enzymes. Statement of the EFSA panel on food contact materials, enzymes, flavourings and processing aids. EFSA J 14(11):4581, 9 pp. https://doi.org/10.2903/j.efsa.2016.4581. http://onlinelibrary.wiley.com/doi/10.2903/j.efsa.2016.4581/abstract. Accessed 27 Oct 2019

EFSA (2016c) Scientific opinion of the CEF Panel on recent developments in the risk assessment of chemicals in food and their potential impact on the safety assessment of substances used in food contact materials. EFSA J 14(1):4357, 28 pp. https://doi.org/10.2903/j.efsa.2016.4357. http://onlinelibrary.wiley.com/doi/10.2903/j.efsa.2016.4357/abstract. Accessed 27 Oct 2019

EFSA (2016d) Guidance of the NDA panel on the preparation and presentation of an application for authorisation of a novel food in the context of Regulation (EU) 2015/2283. EFSA J 14(11):4594, 24 pp. https://doi.org/10.2903/j.efsa.2016.4594. http://onlinelibrary.wiley.com/doi/10.2903/j.efsa.2016.4594/full. Accessed 27 Oct 2019

EFSA (2017a) Guidance on the use of the weight of evidence approach in scientific assessments. EFSA J 15(8):4971, 69 pp. https://doi.org/10.2903/j.efsa.2017.4971. https://www.efsa.europa.eu/en/efsajournal/pub/4971. Accessed 27 Oct 2019

EFSA (2017b) Guidance on the assessment of the safety of feed additives for the consumer. EFSA J 15(10):5022, 17 pp. https://doi.org/10.2903/j.efsa.2017.5022. http://onlinelibrary.wiley.com/doi/10.2903/j.efsa.2017.5022/full. Accessed 27 Oct 2019

EFSA (2018) Guidance on uncertainty analysis in scientific assessments. EFSA J 16(1):5123, 39 pp. https://doi.org/10.2903/j.efsa.2018.5123. http://onlinelibrary.wiley.com/doi/10.2903/j.efsa.2018.5122/abstract. Accessed 27 Oct 2019

ENDS Europe DAILY (2016) ECHA calls for stricter chemicals registration rules. (27/05/2016) https://echa.europa.eu/news-and-events/news-ale. Accessed 27 Oct 2019

EU (1991) Council Directive 91/414/EEC of 15 July 1991 concerning the placing of plant protection products on the market. http://eur-lex.europa.eu/legal-content/EN/TXT/?uri=celex%3A31991L0414. Accessed 27 Oct 2019

EU (2001) Regulation (EC) No 1049/2001 of the European Parliament and of the Council of 30 May 2001 regarding public access to European Parliament, Council and Commission documents. http://eur-lex.europa.eu/legal-content/EN/TXT/HTML/?uri=LEGISSUM:l14546&from=EN. Accessed 27 Oct 2019

EU (2002) Regulation (EC) No 178/2002 of the European Parliament and of the Council of 28 January 2002 laying down the general principles and requirements of food law, establishing the European Food Safety Authority and laying down procedures in matters of food safety. OJ (L31):1–24. 1.2.2002. http://eur-lex.europa.eu/legal-content/EN/TXT/HTML/?uri=CELEX:32002R0178&from=EN. Accessed 27 Oct 2019

EU (2005) Regulation (EC) No 396/2005 of the European Parliament and of the Council of 23 February 2005 on maximum residue levels of pesticides in or on food and feed of plant and animal origin and amending Council Directive 91/414/EE. http://eur-lex.europa.eu/legal-content/EN/TXT/HTML/?uri=CELEX:32005R0396&from=en. Accessed 27 Oct 2019

EU (2009) Regulation (EC) No 1107/2009 of the European Parliament and of the Council of 21 October 2009 concerning the placing of plant protection products on the market and repealing Council Directives 79/117/EEC and 91/414/EEC http://eur-lex.europa.eu/legal-content/EN/TXT/?uri=CELEX:32009R1107. Accessed 27 Oct 2019

EU (2010) Commission Decision of 16 December 2009 laying down guidelines for the management of the Community Rapid Information System 'RAPEX' established under Article 12 and of the notification procedure established under Article 11 of Directive 2001/95/EC (the General Product Safety Directive). Off J Eur Union 53(L22):1–64, 26 January 2010. Accessed 27 Oct 2019

EU (2013) Commission Implementing Regulation (EU) No 503/2013 of 3 April 2013 on applications for authorisation of genetically modified food and feed in accordance with Regulation (EC) No 1829/2003 of the European Parliament and of the Council and amending Commission Regulations (EC) No 641/2004 and (EC) No 1981/2006 (1)

EU (2019) Regulation (EU) No 2019/1381 of the European Parliament and of the Council of 20 June 2019 on the transparency and sustainability of the EU risk assessment in the food chain and amending Regulations (EC) No 178/2002, (EC) No 1829/2003, (EC) No 1831/2003, (EC) No 2065/2003, (EC) No 1935/2004, (EC) No 1331/2008, (EC) No 1107/2009, (EU) 2015/2283 and Directive 2001/18/EC. https://eur-lex.europa.eu/legal-content/EN/TXT/?uri=uriserv:OJ.L_.2019.231.01.0001.01.ENG&toc=OJ:L:2019:231:TOC. Accessed 28 Oct 2019

Fagt S, Matthiessen J, Trolle E, Lyhne N, Christensen T, Hinsch HJ, Hartkopp H, Biltoft-Jensen A, Møller A, Daae AS (2002) [Dietary habits in Denmark 2000–2001: developments in dietary consumption, purchases and habits in Denmark] Danskernes kostvaner 2000–2001: Udviklingen i danskernes kost-forbrug, indkøb og vaner, Danish food directorate, Department of nutrition, Food Report 2002, 10, June 2002, ISBN 87–91189–37-3. http://www.food.dtu.dk/-/media/Institutter/Foedevareinstituttet/Publikationer/Pub-2002/Danskernes-kostvaner-2000-2001.ashx?la=da

Heinemeyer G, van Engelen J, Papameletiou D (2006) Consumer exposure scenarios under reach. Epidemiology 17(6):S63. Abstract. http://journals.lww.com/epidem/Fulltext/2006/11001/Consumer_Exposure_Scenarios_Under_Reach.135.aspx?trendmd-shared=0. Accessed 27 Oct 2019

Meek ME, Boobis AR, Crofton KE, Heinemeyer G, Van Raaij M, Vickers C (2011) Risk assessment of combined exposure to multiple chemicals: a WHO/IPCS framework. Regul

Toxicol Pharmacol 60(2011):S1–S14. https://doi.org/10.1016/j.yrtph.2011.03.010. https://www.ncbi.nlm.nih.gov/pubmed/21466831

NRC (2012) Exposure science in the 21st century. A vision and a strategy. National Academies Press, Washington, DC. isbn:978-0-309-26468-6

OECD (2003) Guidance document on reporting summary information on environmental, occupational and consumer exposure. Environment, Health and Safety Publications, Series on testing and assessment no. 42. ENV/JM/MONO(2003)16. http://www.oecd.org/officialdocuments/publicdisplaydocumentpdf/?cote=env/jm/mono(2003)16&doclanguage=en. Accessed 27 Oct 2019

OECD (2009) Emission scenario document on plastic additives. Environment, Health and Safety Publications, Series on Emission Scenario Documents No. 3. ENV/JM/MONO(2004)8/REV1. http://www.oecd.org/officialdocuments/publicdisplaydocumentpdf/?cote=env/jm/mono(2004)8/rev1&doclanguage=en. Accessed 27 Oct 2019

OECD (2011a) WHO OECD ILSI/HESI international workshop on risk assessment of combined exposures to multiple chemicals: workshop report. Environment, Health and Safety Publications, Series on Testing and Assessment No. 140. ENV/JM/MONO(2011)103. http://www.oecd.org/officialdocuments/publicdisplaydocumentpdf/?cote=env/jm/mono(2011)10&doclanguage=en. Accessed 27 Oct 2019

OECD (2011b) WHO OECD ILSI/HESI international workshop on risk assessment of combined exposures to multiple chemicals: annex 2. Environment, Health and Safety Publications, Series on testing and assessment no. 140. ENV/JM/MONO(2011)10/ANN2. http://www.oecd.org/officialdocuments/publicdisplaydocumentpdf/?cote=env/jm/mono(2011)10/ann2&doclanguage=en. Accessed 27 Oct 2019

OECD (2011c) WHO OECD ILSI/HESI international workshop on risk assessment of combined exposures to multiple chemicals: annex 3. Environment, Health and Safety Publications, Series on testing and assessment no. 140. ENV/JM/MONO(2011)10/ANN3. http://www.oecd.org/officialdocuments/publicdisplaydocumentpdf/?cote=env/jm/mono(2011)10/ann3&doclanguage=en. Accessed 27 Oct 2019

OECD (2011d) WHO OECD ILSI/HESI international workshop on risk assessment of combined exposures to multiple chemicals: annex 7. Environment, Health and Safety Publications, Series on testing and assessment no. 140. ENV/JM/MONO(2011)10/ANN7. http://www.oecd.org/officialdocuments/publicdisplaydocumentpdf/?cote=env/jm/mono(2011)10/ann7&doclanguage=en. Accessed 27 Oct 2019

OECD (2013a) Guidance document for exposure assessment based on environmental monitoring. Environment, Health and Safety Publications, Series on testing and assessment no. 185. ENV/JM/MONO(2013)7. http://www.oecd.org/officialdocuments/publicdisplaydocumentpdf/?cote=env/jm/mono(2013)7&doclanguage=en. Accessed 27 Oct 2019

OECD (2013b) Descriptions of existing models and tools used for exposure assessment—results of OECD survey. Environment, Health and Safety Publications, Series on testing and assessment no. 182. ENV/JM/MONO(2012)37. http://www.oecd.org/officialdocuments/publicdisplaydocumentpdf/?cote=env/jm/mono(2012)37&doclanguage=en. Accessed 27 Oct 2019

OECD (2016) Harmonised templates 301 to 306: use and exposure information. http://www.oecd.org/ehs/templates/harmonised-templates-use-exposure-information.htm. Accessed 27 Oct 2019

OECD (2018) Considerations for assessing the risks of combined exposure to multiple chemicals. Series on testing and assessment no. 296, Environment, Health and Safety Division, Environment Directorate. ENV/JM/MONO(2018)37. http://www.oecd.org/chemicalsafety/risk-assessment/considerations-for-assessing-the-risks-of-combined-exposure-to-multiple-chemicals.pdf. Accessed 27 Oct 2019

Phillips L, Johnson M, Deener K, Bonanni C (2015) EPA's exposure assessment toolbox (EPA-Expo-Box). J Environ Inf 25(2):81–84. https://doi.org/10.3808/jei.201400269

Turrini A, Saba A, Perrone D, Cialfa E, D'Amicis A (2001) Food consumption patterns in Italy: the INN-CA Study 1994–1996. Eur J Clin Nutr 55(7):571–588 van Engelen JGM, Heinemeyer G, Rodriguez C (2007) Consumer exposure scenarios: development, challenges and possible solutions. J Expo Sci Environ Epidemiol 17:S26–S33. http://www.nature.com/jes/journal/v17/n1s/pdf/7500577a.pdf

van Leeuwen CJ, Vermeire TG (2007) Risk assessment of chemicals: an introduction, 2nd edn. Springer, Cham. isbn:978-1-4020-6101-1

Vermeire T, Bruinen de Bruin Y, Traas T, Bolte J, van Engelen J, Evers E, Herremans J, Hogerwerf L, van der Linden T, Ossendorp B, Peijnenburg W, Piersma A, Schols E, Sijm D, Tielemans E, Traas T, van der Vliet D, Dissel S, Lebret E (2017) RIVM exposure assessment strategy 2016–2020 (internal paper)

Vuori V, Zaleski RT, Jantunen MJ (2006) ExpoFacts—an overview of European exposure factors data. Risk Anal 26(3):831–843. https://doi.org/10.1111/j.1539-6924.2006.00772.x

WHO (2004) IPCS risk assessment terminology. IPCS harmonization project document no. 1. WHO, Geneva. isbn:92 4 156267 6. http://www.who.int/ipcs/methods/harmonization/areas/terminology/en/. Accessed 27 Oct 2019

WHO (2005) Principles of characterizing and applying human exposure models. IPCS harmonization project document no. 3, WHO, Geneva. isbn:92 4 156311.7. http://www.who.int/ipcs/publications/methods/harmonization/en/. Accessed 27 Oct 2019

WHO (2008) Uncertainty and data quality in exposure assessment. Part 1: guidance document on characterizing and communicating uncertainty in exposure assessment. IPCS harmonization project document no. 6. WHO, Geneva. isbn:978 92 4 156376 5 http://www.who.int/ipcs/publications/methods/harmonization/en/. Accessed 27 Oct 2019

WHO OECD ILSI/HESI (2011) Report of the international workshop on risk assessment of combined exposures to multiple chemicals: annex 2. Environment, Health and Safety Publications, Series on testing and assessment no. 140, ENV/JM/MONO(2011)10/ANNEX 2, 15–16 Feb 2011, Paris. http://www.oecd.org/officialdocuments/publicdisplaydocumentpdf/?cote=ENV/JM/MONO(2011)10/ANN2&docLanguage=En. Accessed 27 Oct 2019

Zaleski RT, Gephart L (2001) Exposure factors sourcebook for European populations (with focus on UK data). Technical report no. 79. European Centre for Ecotoxicology and Toxicology of Chemicals (ECETOC)

Zaleski RT, Egeghy PP, Hakkinen PJ (2016) Exploring global exposure factors resources for use in consumer exposure assessments. Int J Environ Res Public Health 13(7):744. https://doi.org/10.3390/ijerph13070744

Correction to: General Aspects of Exposure Evaluation

Christiaan Delmaar, Gerhard Heinemeyer, Matti Jantunen, Klaus Schneider, and Michael Schümann

Correction to:
Chapter 3 in: G. Heinemeyer et al. (eds.), *The Practice of Consumer Exposure Assessment*, https://doi.org/10.1007/978-3-319-96148-4_3

The chapter was inadvertently published with an incorrect city "Berlin" in the affiliation of the author Klaus Schneider. This has been updated to reflect the correct city as "Freiburg".

The updated version of this chapter can be found at
https://doi.org/10.1007/978-3-319-96148-4_3

G. Heinemeyer et al. (eds.), *The Practice of Consumer Exposure Assessment*,
https://doi.org/10.1007/978-3-319-96148-4_8

Appendices

Appendix A: Web Resources

1. The NCI Measurement Error Webinar Series reviews concepts estimating usual dietary intakes from diet and supplements, and the use of multiple dietary assessment instruments, and provides a comprehensive discussion on measurement error.
   - Available at: https://epi.grants.cancer.gov/events/measurement-error/
2. The Advanced Dietary Analyses part of NHANES Dietary Web Tutorial provides concepts for estimating usual intake of dietary and supplement intake.
   - Available at: https://www.cdc.gov/nchs/tutorials/dietary/advanced/index.htm
3. The Dietary Assessment Primer provides a comprehensive overview and guidance for major self-report dietary assessment instruments (Thompson et al. 2015).
   - Available at: https://dietassessmentprimer.cancer.gov/
4. The Nutritools provides guidance for researchers for applying dietary assessment instruments to measure food and nutrient intake.
   - Available at: https://www.nutritools.org/
5. The Diet, Anthropometry and Physical Activity Measurement Toolkit aims to assist researchers to select and apply suitable dietary and physical activity assessment instruments for their respective purpose.
   - Available at: http://www.dapa-toolkit.mrc.ac.uk/index.php

G. Heinemeyer et al. (eds.), *The Practice of Consumer Exposure Assessment*,
https://doi.org/10.1007/978-3-319-96148-4

Appendix B: Example of the Documentation of Exposure Information as Extracted from the EIS-ChemRisks Toolbox

Children's dermal, inhalation, oral, and aggregate exposure to PERMETHRIN from WOOL CARPETS

1. General information		
Source document	Berger-Preiss et al. (2002)	
Disclaimer	This EIS-ChemRisks exposure scenario refers exclusively to information from the source document related to the chemical, the particular product and the highlighted exposure event.	
Purpose	☒ As a technical resource for use in proactive risk assessment	☐ Biocides
	☐ Council Regulation (EEC) 793/93	☐ Danish EPA studies
	☐ GPSD	☐ HERA
	☐ Health Canada	☐ VCCEP
	☐ REACH	☐ Other
	To investigate human exposure to permethrin from treated woollen textile floor coverings and possible adverse health effects.	
Type of assessment	Proactive risk assessment	
Type of consumer use	Reasonably foreseeable use	
Stakeholder/rapporteur of the source documents	• Berger-Preiss, E., Fraunhofer Institute of Toxicology and Aerosol Research • Levsen, K., Fraunhofer Institute of Toxicology and Aerosol Research • Leng, G., Heinrich-Heine University Düsseldorf, Institute of Hygiene, Bayer AG. Leverkusen • Idel, H., Heinrich-Heine University Düsseldorf, Institute of Hygiene • Sugiri, D., Medical Institute of Environmental Hygiene • Ranft, U., Medical Institute of Environmental Hygiene	

2. Issue at stake	
General statement	Pyrethroids, including permethrin, are among the most frequently used insecticides. In Germany, indoor exposure to pyrethroids originates mainly from two sources: treatment of home textiles, where woollen textile floor coverings play the most important role and pest control operations. Today, the pyrethroid permethrin is mainly used for the treatment of woven wool carpets (rugs) and wall-to-wall textile floor coverings with wool yarn (in the following referred to as "wool carpets"). Permethrin is known to have a low mammalian toxicity. At the same time, it has been suspected that wool carpets in private homes which were treated with permethrin may lead to adverse health effects in the inhabitants.

(continued)

2. Issue at stake	
	This study was designed to: • Determine the concentration of permethrin in house dust and in the air (suspended particles) in homes with wool carpets. • Estimate the extent of permethrin bioavailable from wool carpets. • Determine if there is any correlation between the concentration of permethrin in suspended particles and the concentration of its metabolites in the inhabitants' biofluids (e.g., urine). • Determine if there is any correlation between the concentration of permethrin and health effects (symptoms) reported by the inhabitants.
Review of the source document	Peer-reviewed journal: *International Journal of Hygiene and Environmental Health*

3. Issue at stake	
Source	
Product as named in the source document: Wool carpet	EIS-ChemRisks source taxonomy: 1. Building materials 2. Carpets, wall-to-wall 3. Carpet natural fibers
Agent/Chemical	
Permethrin	CAS Number: 52645-53-1 - EINECS Number: 258-067-9
Target	
Children	Human subject addressed: Child category (between 5 and 14 years old): Age: <6 years old Gender: Male and female
Exposure Event	
Task/activity	Living at home Children living at home with wool carpets for 24 hours
Contact duration	24 hours exposure of children living at home, exposed by: – Dermal contact with permethrin in house dust, while playing and crawling on wool carpets. – Inhalation of permethrin in fine particles suspended in indoor air. – Ingestion of permethrin in house dust, while playing and crawling on wool carpets, and practising hand-to-mouth activity.
MicroEnvironment	
Indoor.	Residence
Geographic Applicability:	
Specific country(ies):	Germany
Seasonal applicability:	• Autumn • Spring • Summer • Winter
Risk Management Measures	
Manufacture of wool yarn leads to incorporation of permethrin into that yarn, limiting ability to be released from carpet except by abrasion.	

(continued)

3. Issue at stake		
Chemical Release		
Emission/release from source/ product matrix	Qualitative characterization: The insecticide is incorporated in the fiber during its manufacture. Permethrin in the house dust is mainly due to carpet fiber abrasion. Permethrin found in suspended particles may result from a transfer of the agent from the wool fiber into the gas phase, recondensation onto suspended particles and from the resuspension of house dust as fine particles. Resuspension of house dust as fine particles is assumed to be the main permethrin release to the air because statistical analysis demonstrates that the indoor air concentration of permethrin does not depend on the indoor particle concentration. Permethrin in house dust being mainly due to carpet fiber abrasion, this mechanism is assumed to be the cause of the permethrin air concentrations. Quantitative information: Significant correlation between the permethrin concentration in wool fibers and the fine and coarse dust, respectively was found, e.g., for the fine dust $r_{\mathrm{Spearman}} = 0.86$ and $p = 0.0001$ As a result of the very low vapor pressure of permethrin (45 μPa at 25 °C, it is unlikely that detectable amounts of permethrin are transferred from the carpet into the gas phase.	
Exposure Pathways		
Description	Permethrin is released from the carpet and found in house dust (sediment dust) by wool fiber's abrasion. As a consequence, children are exposed to permethrin by dermal (crawling activity) and oral (hand-to-mouth activity) routes. Permethrin in fine particles of house dust is suspended in indoor air. As a consequence, children and adults are exposed to permethrin by inhalation route.	
Dermal pathway	Inhalation pathway	Oral pathway
Exposure route		
Dermal exposure	Inhalation exposure	Oral exposure
Step(s)		
STEP 1: Permethrin is released from the carpet and found in house dust, mainly due to the wool fibers abrasion.	STEP 1: Permethrin is released from the carpet and found in house dust (sediment dust), mainly due to the wool fibers abrasion. Also, present in fine particles of house dust suspended in the indoor air.	STEP 1: Permethrin is released from the carpet and found in house dust, mainly due to the wool fibers abrasion.
STEP 2: Permethrin in house dust can enter in direct contact with the skin of children playing on a wool carpet (children crawling).	STEP 2: Permethrin in fine particles in the indoor air will be inhaled.	STEP 2: Permethrin in house dust can be ingested by children playing on a wool carpet (hand-to-mouth activities).
Exposure routes		
• Dermal exposure • Inhalation exposure • Oral exposure		

4. Exposure modelling	
Exposure model	
Exposure assessment concept:	The used model includes simple calculations based on a worst-case approach to estimate potential daily uptakes of permethrin by inhalation, ingestion, and dermal absorption. It uses some exposure factors, some exposure data, and the permethrin concentrations measured in indoor air and house dust: • Children's daily dermal uptake of permethrin from wool carpets is calculated on the basis of permethrin concentration measured in house dust (fine dust <2 mm), dust concentration in carpets, daily playing area, dermal bioavailability, and body weight. • Children's daily inhalation uptake of permethrin from wool carpets is calculated on the basis of permethrin concentration measured in the indoor air (suspended particles), breath volume, inhalation bioavailability, and body weight. • Children's daily oral uptake of permethrin from wool carpets is calculated on the basis of permethrin concentration measured in house dust (fine dust <2 mm), daily dust uptake, oral bioavailability, and body weight. • Children's daily aggregate uptake of permethrin from wool carpets is calculated cumulating daily inhalation, oral, and dermal uptakes.
Mathematical Expression/ Algorithms:	Children's daily dermal uptake of permethrin from wool carpets
Daily dermal uptake ($\mu g/kg_{sw}/day$) $= \frac{\text{permethrin concentration in house dust } (mg/kg_{dust}) \times \text{dust concentration in carpets } (mg_{dust}/m^2) \times \text{daily playing area } (m^2/day) \times \text{dermal bioavailability } (\%) \times 1000\ (\mu g/mg)}{\text{body weight } (kg_{sw}) \times 1,000,000\ (mg_{dust}/kg_{dust})}$	
	Children's daily inhalation uptake of permethrin from wool carpets
Daily inhalation uptake ($\mu g/kg_{BW}/day$) $= \frac{\text{permethrin concentration in the air } (ng/m^3) \times \text{breath volume } (m^3/day) \times \text{inhalation bioavailability } (\%)}{\text{body weight } (kg_{BW}) \times 1000\ (ng/\mu g)}$	
	Children's daily oral uptake of permethrin from wool carpets
Daily oral uptake ($\mu g/kg_{BW}/day$) $= \frac{\text{permethrin concentration in house dust } (mg/kg_{dust}) \times \text{daily dust uptake } (mg_{dust}/day) \times \text{oral bioavailability } (\%) \times 1000\ (\mu g/mg)}{\text{body weight } (kg_{BW}) \times 1,000,000\ (mg_{dust}/kg_{dust})}$	
	Children's daily aggregate uptake of permethrin from wool carpets
Daily aggregate uptake ($\mu g/kg_{BW}/day$) $=$ daily dermal uptake ($\mu g/kg_{sw}/day$) $\times$ dailly inhalation uptake ($\mu g/kg_{BW}/day$) $+$ daily oral uptake ($\mu g/kg_{BW}/day$)	

5. Exposure Determinants Selection		
Nonchemical-specific Exposure Data	Value	Source of Information
(Anthropometric/ physiological parameters) body weight of a child	15 kg/b.w. (single point default)	Source document
(Respiratory rates/ volume) Respiratory/ breathing rates of a child	6 m^3/day (single point default)	Source document
(Soil/particles/dust contact Factors) House dust ingestion rates of a child	100 mg/m^2 (single point default)	Source document
(Residential/living environment factors) Other: Daily playing area assumed to be a wool carpet	10 m^2/day (single point default)	Source document Note from EIS-ChemRisks: In the source document, the calculation is incorrect because it is based on "playing area" determinant expressed in m^2. Thus, it is assumed that the correct unit is m^2/day.
(Residential/living environment factors) Other: Dust surface mass on carpet	500 mg/m^2 (single point default)	Source document
(Time–activity patterns (nonproduct specific)) Duration of exposure activity: Daily residence time of children (<6 years old)	24 h (single point default)	Source document Note from the source document: In a conservative scenario, it is assumed that the daily residence time of the inhabitants in the home with wool carpets is 24 h. Note from EIS-ChemRisks: Although not explicitly used in the calculation, but this value is assumed. However, a more specific value is required for a more realistic daily residence time.
Chemical-specific Exposure Data		
Inhalation exposure Concentration of an AGENT in a MEDIUM: Concentration of permethrin in indoor air (suspended particles)	5.8 ng/m^3 (90th percentile) Method applied: Determination of the concentration of permethrin in suspended particulate matter, house dust (and in wool carpets) [ChemTest]	Source document Note from the source document: The suspended particles were collected on Pallflex filters by sampling approximately 10 m^3 at a rate of 2.6–3 m^3/h. Sampling was performed at a height of 1 m, where the sampler was placed in the center of the room. Note from EIS-ChemRisks: In each of the 80 homes, the indoor air was sampled twice for suspended particles. This value refers to the distribution of permethrin air concentrations measured during the first sample (excluding one outlier). Permethrin air concentrations were higher during the first sample than during the second one.

(continued)

5. Exposure Determinants Selection		
Dermal exposure AGENT UPTAKE: Bioavailability after uptake	2% (percentage)	Source document
Inhalation exposure AGENT UPTAKE: Bioavailability after uptake	100% (percentage)	Source document
Oral exposure AGENT UPTAKE: Bioavailability after uptake	100% (percentage)	Source document
Dermal exposure and oral exposure Concentration of an AGENT in a MEDIUM: Permethrin concentration in house dust	129.1 mg/kg (90th percentile) Method applied: Determination of permethrin concentration in the house dust of wool carpets [ChemTest]	Source document Note from the source document: House dust was collected using a vacuum cleaner (Philips HR 8733) with a dust bag. The study participants collected the house dust (2 ± 3 weeks) only from the wool carpets. The dust amounted to 10 ± 770 g (median: 84 g and 95th percentile: 305 g). In 25% of the fine dust (<2 mm) and 30% of the coarse dust (>2 mm) samples, the permethrin content was below the LOQ of 1 mg/kg. In the fine dust, an arithmetic mean of 53.7 mg/kg a median of 9.7 mg/kg and a 90th percentile of 129.1 mg/kg were found. On average, the permethrin concentration in the fine dust is 1.3 times higher than in coarse dust. In Germany, there is a general background level of permethrin in house dust. As wool is used only in 8 ± 10% of the textile floor coverings, the background concentration of permethrin in house dust may mainly result from wool carpets while pest control operations in the home are another important source.

6. Exposure Assessment and Comparison
Exposure Assessment Calculations
Children's daily dermal uptake of permethrin from wool carpets $\text{Daily dermal uptake} = \frac{129.1\ (\text{mg}/\text{kg}_{\text{dust}})\times 500\ \left(\text{mg}_{\text{dust}}/\text{m}^2\right)\times 10\ \left(\text{m}^2/\text{day}\right)\times 2\%\times 1000\ (\mu\text{g}/\text{mg})}{15\ (\text{kg}_{\text{BW}})\times 1,000,000\ (\text{mg}_{\text{dust}}/\text{kg}_{\text{dust}})} = 0.86\ \ (\mu\text{g}/\text{kg}_{\text{BW}}/\text{day})$
Children's daily inhalation uptake of permethrin from wool carpets $\text{Daily inhalation uptake} = \frac{5.8\ \left(\text{ng}/\text{m}^3\right)\times 6\ \left(\text{m}^3/\text{day}\right)\times 100\%}{15\ (\text{kg}_{\text{BW}})\times 1000\ (\text{ng}/\mu\text{g})} = 0.0023\ \ (\mu\text{g}/\text{kg}_{\text{BW}}/\text{day})$
Children's daily oral uptake of permethrin from wool carpets $\text{Daily oral uptake} = \frac{129.1\ (\text{mg}/\text{kg}_{\text{dust}})\times 100\ (\text{mg}_{\text{dust}}/\text{day})\times 100\%\times 1000\ (\mu\text{g}/\text{mg})}{15\ (\text{kg}_{\text{BW}})\times 1,000,000\ (\text{mg}_{\text{dust}}/\text{kg}_{\text{dust}})} = 0.86\ \ (\mu\text{g}/\text{kg}_{\text{BW}}/\text{day})$
Children's daily aggregate uptake of permethrin from wool carpets $\text{Daily aggregate uptake} = 0.86\ (\mu\text{g}/\text{kg}_{\text{BW}}/\text{day}) + 0.0023\ (\mu\text{g}/\text{kg}_{\text{BW}}/\text{day}) + 0.86\ (\mu\text{g}/\text{kg}_{\text{BW}}/\text{day}) = 1.72(\mu\text{g}/\text{kg}_{\text{BW}}/\text{day})$

Exposure Assessment Results	
Children's daily dermal uptake of permethrin from wool carpets	0.86 μg/kg b.w. per day
Children's daily inhalation uptake of permethrin from wool carpets	0.0023 μg/kg b.w. per day
Children's daily oral uptake of permethrin from wool carpets	0.86 μg/kg b.w. per day
Children's daily aggregate uptake of permethrin from wool carpets	1.72 μg/kg b.w. per day

Comparison with other Assessments		
Children's daily aggregate uptake of permethrin from wool carpets	1.2 μg/kg BW/day	Source document
Children's daily oral uptake of permethrin from wool carpets	0.9 μg/kg BW/day	Source document

References

Berger-Preiss E, Levsen K, Leng G, Idel H, Sugiri D, Ranft U (2002) Indoor pyrethroid exposure in homes with woollen textile floor coverings. Int J Hyg Environ Health 205(6):459–472

Thompson FE, Kirkpatrick SI, Subar AF, Reedy J, Schap TE, Wilson MM, Krebs-Smith SM (2015) The National Cancer Institute's dietary assessment primer: a resource for diet research. J Acad Nutr Diet 115(12):1986–1995

Index

This index should be understood as examples of important terms relevant in exposure evaluation and for selected special questions.
Very frequent occurring terms (ECHA, EFSA, REACH etc.) are not included.

G. Heinemeyer et al. (eds.), *The Practice of Consumer Exposure Assessment*, https://doi.org/10.1007/978-3-319-96148-4

The manufacturer's authorised representative in the EU is Springer Nature Customer Service Centre GmbH, Europaplatz 3, 69115 Heidelberg, Germany. If you have any concerns regarding our products, please contact ProductSafety@springernature.com

Printed and bound by CPI Group (UK) Ltd, Croydon, CR0 4YY
05/10/2026
02228339-0007